HANDBOOK OF HYDRAULIC ENGINEERING

ELLIS HORWOOD SERIES IN CIVIL ENGINEERING

"HANDBOOK OF HYDRAULIC ENGINEERING"

ARMANDO LENCASTRE, Ph.D.
Professor of Hydraulics
New University of Lisbon, Portugal

Translation Editor:
PATRICK HOLMES, Ph.D., C.Eng.
Professor of Hydraulics
Imperial College of Science and Technology
University of London

ELLIS HORWOOD LIMITED
Publishers · Chichester

Halsted Press: a division of
JOHN WILEY & SONS
New York · Chichester · Brisbane · Toronto

This English Language Edition first published in 1987 by
ELLIS HORWOOD LIMITED
Market Cross House, Cooper Street,
Chichester, West Sussex, PO19 1EB, England
The publisher's colophon is reproduced from James Gillison's drawing of the ancient Market Cross, Chichester.

Distributors:

Australia and New Zealand:
JACARANDA WILEY LIMITED
GPO Box 859, Brisbane, Queensland 4001, Australia

Canada:
JOHN WILEY & SONS CANADA LIMITED
22 Worcester Road, Rexdale, Ontario, Canada

Europe and Africa:
JOHN WILEY & SONS LIMITED
Baffins Lane, Chichester, West Sussex, England

North and South America and the rest of the world:
Halsted Press: a division of
JOHN WILEY & SONS
605 Third Avenue, New York, NY 10158, USA

© **1987 English Language Edition Ellis Horwood Limited**
First published in Portuguese under the title *Hidráulica Geral,* by Hidroprojecto, Av. Marechal Craveiro Lopes, Lisbon, Portugal

British Library Cataloguing in Publication Data
Lencastre, Armando
Handbook of hydraulic engineering. —
(Ellis Horwood series in civil engineering)
1. Hydraulic engineering
I. Title   Hidráulica Geral. *English*
627   TC145
Library of Congress Card No. 87–3179

ISBN 0–7458–0095–5 (Ellis Horwood Limited)
ISBN 0–470–20828–7 (Halsted Press)

Phototypeset in Times by Ellis Horwood Limited
Printed in Great Britain by R. J. Acford, Chichester

Table of Contents

8 Flow Measurements. Orifices and Weirs

Tables and graphs

List of symbols

Subject index

Preface

The stated aims of the first edition of *Manual de Hidráulica Geral* (in Portuguese, 1957) were as follows:

"To be brief without omission, to be concise without being obscure — this has been the principal difficulty in preparing this Manual. The vast range of subject matter has made it essential to select, condense and arrange; the need for clarity and facility in applications has imposed detailed treatment and sometimes repetition.

To achieve clarity while being concise, to provide abundant facts while being brief, has thus been the author's main objective.

In the achievement of this aim alone lies any merit that it may have".

Twenty-six years later it became necessary for it to be revised, brought up to date and enlarged, with the same aims in mind, and this resulted in a new book with the title *Hidráulica Geral,* (1983) which substituted the previous one.

When the first edition came out, I was Senior Assistant Lecturer in Hydraulics at the Instituto Superior Técnico (Lisbon, 1950–1961) and as such was responsible for teaching the subject in design classes. I therefore prepared the material that I finally published in the form of a Manual of Hydraulics. Interest in the book far exceeded my expectations in Portugal, and also attracted attention abroad, which I had not anticipated: a French edition, included in the collection of the *Direction des Études et Recherches d'Electricité de France,* published by Eyrolles (1961) with ten successive printings, the last being in 1986; a Spanish edition by Dossat (1962); a second Portuguese edition (1969); a Turkish edition published by Istanbul University (1970); a Brazilian edition, sponsored by São Paulo State University (1972) and, more recently, a Greek edition (1978).

In the meantime, although I had passed all the board examinations required at the time for continuing a teaching career, I left the Instituto Superior Técnico in 1961 on obtaining the degree of Senior Research Officer at the Laboratório Nacional de Engenharia Civil in Lisbon, where I had been engaged in applied research (1950 to 1968). After a period in which I devoted myself exclusively to project design and consultancy (Hidroprojecto, Consulting Engineers), I accepted the invitation to return to teaching as Guest Professor at Lisbon New University.

However, I never lost sight of what I had written in the preface to the first edition of the *Manual* (1957):

> "It is to be hoped that interest shown in this edition may encourage the author and publishers to produce a larger work, with the completion of certain chapters and inclusion of new subject matter".

Further effort led to a new and thoroughly remodelled book, even though sales of the previous one has been considered extraordinarily satisfactory by the various publishers.

Throughout the period when the *Manual* was in circulation, about 26 years, it was found that apart from being used in practical applications and in teaching of design at universities and colleges (the latter use being its original aim), it was in general use by pupils as a theoretical textbook and as such was adopted in some schools. I therefore considered that the new book, *Hidráulica Geral,* should, without forgetting the aims of the first book, deal at greater length with the theoretical part.

However, when considering the publication of an edition in English, the existence of a number of excellent textbooks on fluid mechanics and hydraulics was recognized. Therefore, this edition contains somewhat less of the theoretical aspects whilst retaining the large range of information more directly related to practical problems of design.

In preparing this book I received various kinds of help from Hidroprojecto's colleagues, and I should like in particular to express my thanks to the engineers João Fonseca, João Salsinha, José Carvalho, Melo Franco, Oliveira Lemos and last but not least Matias Ramos, who was responsible for revising the Portuguese edition and English translation. I also wish to thank the drawing, typing and copying teams of Hidroprojecto.

Additionally, in preparing the English edition, the help and collaboration of Professor Patrick Holmes has been much appreciated and important.

Hidroprojecto, Lisbon   Armando Lencastre
September, 1986

Foreword to the English edition

An understanding of hydraulic principles combined with the knowledge and experience of practising engineers is essential to good hydraulic engineering design.

This Handbook, based on previous Portuguese editions by Professor Lencastre, provides that combination together with a wealth of data and information to assist the designer. It has been both a pleasure and an education to work with the author in the preparation of an English edition.

September, 1986   Patrick Holmes
                                                    Imperial College, London

Please note that, only in the Tables and Graphs section at the end of the book, the decimal comma rather than the decimal point has been used.

1

Physical properties of fluids

A. PHYSICAL CONSTANTS

1.1 Definitions

Fluids are bodies without their own shape that can flow, i.e. they can undergo great variations of shape under the action of forces; the weaker the force, the slower the variation.

Both *liquids* and *gases* are fluids. Their equilibrium and their movements, known as flow, are studied in the mechanics of fluids.

Liquids occupy a determined volume, cannot be subjected to *traction†* *and are relatively incompressible*.

Gases always occupy the maximum volume available to them and are highly compressible. When, however, the velocity of flow is small, in comparison with the velocity of sound, that flow can be studied by considering gases as being incompressible.

There are many materials such as powdery media, asphalt, plastics, etc., which have properties that are intermediate between the properties of solids and those of fluids. They are studied in the mechanics of soils, rheology, etc., [1, 2, 3].

1.2 Weight and mass

In current language, the notions of weight and mass are sometimes confused; however, from the physical point of view, they represent two different things. The *mass* of a body is a characteristic of the quantity of

---

† This property, however, is not absolute. In laboratory experiments it has been possible to subject liquids to considerable traction. In the case of water, it is mainly the presence of numerous undissolved gaseous particles that prevents it from resisting traction.

matter which that body contains, i.e. of the inertia which the body presents to movement; the *weight* of the body represents the action (force) that gravity exerts on it.

Between G weight and m mass of a body there is the fundamental vectorial relationship

$$\mathbf{G} = m\mathbf{g} \tag{1.1}$$

which corresponds to the scalar equation

$$G = Mg \tag{1.1a}$$

in which g is the gravitational acceleration.

1.3 The *Système International d'Unités* — SI

This system of units is an internationally agreed version of the metric system. There are six basic units and not only their names but also their symbols have been internationally agreed (see Table 1.1).

Table 1.1

Quantity	Basic units	Symbol
Length	metre	m
Mass	kilogram	kg
Time	second	s
Electric current	ampere	A
Temperature	kelvin	K[†]
Luminous intensity	candela	cd

† Formerly °K. Temperature is normally expressed in degrees Celsius, formerly known as degrees Centigrade (°C).
  The relationship between temperature Celsius, °C, and Kelvins, K, is: K = °C + 273.15.
  The relationship between temperature Celsius, °C, and Fahrenheit, °F, is: °C × 9/5 + 32 = °F.

From these basic units, all others are derived. For example: *area — square metres* (m²); *velocity — metres per second* (m/s); *density — kilogram per cubic metre* (kg/m³).

There are units with a special name, the most important being shown in Table 1.2.

1.4 Density

Density, ρ, is the mass contained in a unit volume. It has the dimensions ML^{-3}. In the SI system, it is expressed in kg/m³.[†]

  † In the traditional system it is expressed in a metric unit of mass per cubic metre (m.u.m./m³). For $\mathbf{g} = 9.81$ m/s², the density of water at 4°C is $\rho = 102$ m.u.m./m³.

Table 1.2

Quantity	SI units	Symbol	Expression in terms of other SI units	Expresssion in terms of other basic units
Force	Newton	N		$m.kg.s^{-2}$
Pressure	Pascal	Pa	N/m^2	$m^{-1}kg\,s^{-2}$
Energy, work, quantity of heat	Joule	J	$N.m$	$m^2.kg.s^{-2}$
Power	Watt	W	$N.m/s$	$m^2.kg\,s^{-3}$
Frequency	Hertz	Hz	cycle/s	s^{-1}
Dynamic viscosity	Poiseuille	Pl	$N.s/m^2$	$m^{-1}kg\,s^{-1}$

The density of water at 4°C is $\rho = 1000\,kg/m^3$; at 20°C it is $\rho \approx 998.2\,kg/m^3 \approx 1000\,kg/m^3$.

Table 8 gives the density of water at various temperatures. Table 9 gives the density of salt water at various temperatures and for various salinities.

1.5 Specific weight

Specific weight, γ, is the weight, that is, the gravitational attractive force acting on the matter contained in a unit volume.

Between specific weight and density there is the fundamental relationship: $\gamma = \rho g$.

In the SI system, specific weight is expressed in newtons per cubic metre: N/m^3.

The specific weight of water at 4°C is $\gamma = \rho g = 1000 \times 9.81\,N/m^3 \approx 10\,000\,N/m^3$.

The specific weight of water at different temperatures is given in Table 8. In natural watercourses, the specific weight may be greater, owing to the presence of solid material in suspension. For fairly clear water it will be approximately $\gamma = 9900\,N/m^3$; in cases of very turbid water it may be as high as $\gamma = 11\,800\,N/m^3$.

1.6 Relative density

Relative density, δ, is the ratio between the mass (or weight) of a given volume of the substance considered and the mass (or weight) of an equal volume of water at a temperature of 4°C.† As indicated by its definition δ is non-dimensional.

---

† The maximum density of water occurs at a temperature of 4°C. In many applications, it is of little importance whether the reference temperature is specified, since the density of water only varies by 0.003 when the temperature varies between 4°C and 25°C.

1.7 Coefficient of dynamic viscosity

The coefficient of dynamic viscosity, μ, is the parameter that represents the existence of tangential forces in liquids in movement. Supposing two plates of area, A, moving at a distance apart of Δn, and at a relative velocity of V, the force required for the displacement is:†

$$\Delta F = \mu A \frac{\Delta V}{\Delta n} \tag{1.2}$$

or, in terms of shear stress,

$$\tau = \frac{\Delta F}{A} = \mu \frac{\Delta V}{\Delta n} \tag{1.2a}$$

μ is known as the coefficient of dynamic viscosity. The dimensions of μ are $L^{-1}MT^{-1}$. In the SI system, μ is expressed in *poiseuilles* (Pl). $1\,\text{Pl} = 1\,\text{N.s/m}^2$. In the c.g.s. system, the unit is the *poise* (dyne.s/cm^2). The unit generally used is the *centipoise*, equivalent to one hundredth part of the *poise*. A *poise* is equal to $0.1\,\text{N.s/m}^2$. For water at 20°C, $\mu \approx 10^{-3}\,\text{N.s/m}^2$.

Table 8 gives the value of μ for fresh water at various temperatures.

Fluids that obey the law $\tau = \mu dV/dn$ are called Newtonian fluids.

For non-Newtonian fluids different laws are used, such as:

$$\tau^n = \mu \frac{dV}{dn} \tag{1.2b}$$

applicable to paints, varnishes, milk, blood, etc., or

$$\tau = \tau_o + \mu \frac{dV}{dn} \tag{1.2c}$$

applicable, usually, to pastes and plastic materials.

The preceding ratios are valid when the viscosity is independent of the state of agitation.

There are liquids which have a high viscosity when at rest; the viscosity diminishes when the liquid is subject to violent agitation at a constant temperature and such liquids are termed *thixotropic*. Examples of this are bitumen, cellulose compounds, glues, fats, molasses, soaps, tar, etc.

A liquid is said to be *expansile* if its viscosity increases with agitation, at a constant temperature, as is the case with clay pastes, jams and other similar fluids.

The viscosity of many thixotropic and expansile fluids returns to its original value when agitation ceases. The degree of recovery varies with the nature of the fluid.

In pumping systems it is very important to know up to what point the viscosity of the liquid may be affected by agitation, in order to be able correctly to calculate the head losses.

A perfect, or ideal liquid, with zero viscosity, does not exist in nature. A liquid at rest or in movement in which there are no relative displacements of the elements that constitute it behaves like a perfect liquid.

---

† Only in laminar flow. See section 2.2.

1.8 Coefficient of kinematic viscosity

The coefficient of kinematic viscosity, v, is the ratio of the coefficient of dynamic viscosity to the fluid density: $v = \mu/\rho$. The dimensions of v are $L^2.T^{-1}$. In the SI system it is expressed in m²/s. In the c.g.s. system the unit is the *stoke* — St(cm²/s). As a rule the unit used is the *centistoke* — cSt, equivalent to one-hundredth part of the stoke. One stoke is equal to 10^{-4} m²/s.

For water at 20°C, $v = 10^{-6}$ m²/s.

Table 8 gives the value of v, at various temperatures, for fresh water. Table 9 gives the value of v for salt water.

The kinematic viscosity of liquids varies considerably with the temperature. The influence of pressure is negligible.

1.9 Surface tension. Capillarity

A molecule inside a liquid is subject to molecular forces exerted on it by surrounding molecules. These forces vary, owing to molecular agitation; their average value, however, over a finite time interval is zero.

A molecule on the free liquid surface in contact with the atmosphere will no longer be subject to the action of symmetrical forces, since it will no longer be symmetrically surrounded by other molecules; the resultant of those forces is thus different from zero and gives rise to surface tension; its direction is normal to the surface of the liquid.

Any molecule on the surface, or at the interface between two fluids, has an energy corresponding to the work done for the molecule to reach the surface. The separation surface behaves as if it were an elastic membrane under tension. This *surface tension*, σ, is defined as the force in the liquid surface normal to a line of unit length drawn in the surface.

The dimensions of surface tension, σ, are MT^{-2}. In the SI system it is expressed in N/m. Surface tension for water is given in Table 8.

The surface tension phenomenon of main interest in hydraulics is the capillarity that occurs on the free surface of tubes of small bore (say less than 3 mm). In these tubes the free surface is raised, with the formation of a *concave* meniscus when the liquid wets the wall, and lowered with the formation of a *convex* meniscus if the liquid does not wet the wall. This raising or lowering, measured in relation to the point on the meniscus with a horizontal tangent, is given by:

$$h = \frac{2\sigma}{\gamma r}\cos\theta \tag{1.3}$$

in which r is the radius of the tube and θ is the angle of contact between the liquid and the wall of the tube. This angle is practically zero if the liquid completely wets the wall of the tube (as happens in the case of distilled water, with the glass of the wall absolutely clean). The preceding expression may be written, for a given liquid, in the form $h = k/d$ (Jurin's ratio). For water k $\approx$ 30 mm^2, with slight variation for temperature, and with h and d in mm, d being the tube diameter. For mercury, $k = -14$ mm^2 and is practically independent of temperature.

The ratio $\omega = \sigma/\rho$ is called *kinematic capillarity*. It has the dimensions L^3T^{-2}. In the SI system it is expressed in m^3/s^2.

1.10   Pressure

In a fluid at rest, the resultant force exerted on a particle is zero. The tension of the surface element of the fluid is normal to that element and, at a given point, identical in all directions. This tension is known as pressure.

Pressure has the dimensions of a force per unit area, dimensions $ML^{-1}T^{-2}$. In the SI system it is expressed in N/m^2.

Pressure, p, measured in relation to atmospheric pressure, is called *gauge pressure*.† *Absolute pressure*, p_a is the sum of gauge pressure, p and atmospheric pressure, p_0.

In hydraulics it is sometimes convenient to express pressure in terms of height of a liquid column.

Consider a straight prism of liquid at rest, with vertical generatrices, height h and base area A. The force that the liquid exerts on the base of the prism is equal to the weight of the liquid, i.e. γAh; the pressure (force per unit area) will thus be: $p = \gamma h$. We see, therefore, that the pressure, p, is associated with a height of liquid, $h = p/\gamma$.

Example: an oil of specific weight $\gamma = 8000$ N/m^3 is subjected to a pressure of 40 N/cm^2. Express this pressure in liquid column.

Solution:

$$p = 40 \text{ N/cm}^2 = 400\,000 \text{ N/m}^2 \,, \qquad \gamma = 8000 \text{ N/m}^3$$

$$h = \frac{p}{\gamma} = \frac{400\,000}{8000} = 50 \text{ m oil column}$$

Table 10 shows the value of atmospheric pressure for various altitudes and in various systems of units.

The value of the atmospheric pressure, p_0, under normal conditions, at sea level, in different units is as follows:

$$p_0 = 1 \text{ atmosphere} = 10.134 \text{ N/cm}^2$$
$$= 760 \text{ mm mercury column} = 10.33 \text{ m water column.}$$

In practical use the value assumed for p_0 is 10 N/cm^2 or 10 m water column.

1.11   Modulus of elasticity

This modulus, also termed as *bulk modulus*, is defined as the change in pressure intensity divided by the corresponding change in volume per unit volume.

† Or simply *pressure*.

$$\varepsilon = \frac{\Delta p}{\Delta \rho / \rho} \tag{1.4}$$

It has the dimensions of a pressure and is expressed in the same units.

Given two states of a liquid, defined by (ρ_1, p_1) and (ρ_2, p_2), there will be the following ratio:

$$p_2 - p_1 = \varepsilon \log \frac{\rho_2}{\rho_1} \tag{1.4a}$$

When a volume, e, of liquid is subjected to a variation in pressure, Δp, its volume changes by Δe.

We then have the following ratio:

$$\frac{\Delta e}{e} = -\frac{\Delta p}{\varepsilon} \tag{1.4b}$$

The quotient $\eta = \varepsilon / \rho$ is known as *kinematic elasticity*. Its dimensions are $L^2 T^{-2}$ and it is expressed in the SI system in m^2/s^2.

The approximate value of the modulus of elasticity of water, at various temperatures and at atmospheric pressure, is given in Table 8. For water at 20°C, $\varepsilon = 2.1 \times 10^9 \, N/m^2$.

The value of ε increases by about 2% for each increase in pressure of about $700 \, N/cm^2$. The modulus of elasticity of seawater is on average about 9% higher.

For other liquids, the modulus of elasticity can be obtained, approximately; by multiplying the values given for water by the following values: salt water 1.1; glycerine 2.1; mercury 12.6; oil 0.6 to 0.9.†

It will be seen that the modulus of elasticity of water, and of liquids in general, is very high. In order to increase the density of water by 1%, an increase in pressure of $2 \times 10^7 \, N/m^2$ is required, i.e. an increase of about 200 atmospheres.

For gases it is necessary to define two moduli of elasticity; one corresponding to the *isothermal* transformations (at constant temperature); the other corresponding to the *adiabatic* transformations (without any heat exchange with the exterior). In an isothermal transformation, the modulus of elasticity, expressed in the same system of units, is equal to the absolute pressure: $\varepsilon = p_a$; in an adiabatic transformation $\varepsilon = K p_a$, in which K is the adiabatic constant (ratio between the specific heats at constant pressure and at constant volume).‡

The modulus of elasticity of gases thus varies considerably with pressure, unlike liquids.

1.12 Velocity of elastic waves

The velocity of elastic waves, c, is the velocity at which a variation in pressure (e.g. an acoustic wave) is propagated within a liquid.

$$c = \sqrt{\frac{\varepsilon}{\rho}} = \sqrt{\eta} \tag{1.5}$$

† See [11].
‡ See [11].

In water at 10°C, c has the value of 1425 m/s.

Only in an absolutely incompressible liquid will the velocity be infinite.

1.13   Solubility of gases in water

At ordinary pressure and temperature water is capable of keeping air in solution up to about 2% of its volume.

For a given gas the ratio between the maximum volume of dissolved gas and the volume of the solvent liquid is known as the coefficient of solubility.

According to *Henry's law*, at constant temperature the coefficient of solubility is also constant.

In the case of mixtures, according to *Dalton's law*, each gas behaves as if it alone were in contact with the liquid. The air dissolved in water is thus more oxygenated that atmospheric air since the ratio between the coefficients of solubility of oxygen and of nitrogen is two to one.

Whenever for any reason there is a lowering of pressure the released mixture of gases will be very rich in oxygen.

1.14   Vapour pressure

All liquids tend to evaporate (or vaporize). This is because there is at the free surface a continual movement of molecules out of the liquid. Molecules leaving the liquid give rise to the *vapour pressure*, the magnitude of which corresponds to the rate at which molecules escape from the surface.† The space is said to be *saturated* when it can hold no more vapour. The value of vapour pressure for which this is so is the *saturation pressure*.

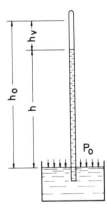

Fig. 1.1

The saturation vapour pressure of a substance depends only on its temperature, as shown in Table 8, and becomes equal to atmospheric

---

† The pressure exerted by a gas or vapour which, together with others, occupies a given space, is equal to the pressure that it would exert if alone it occupied that space. The pressure of the mixture is the sum of the pressure of the various components.

pressure at boiling point. The maximum water column height, h, which at a given temperature can be borne by atmospheric pressure, p_0, is the height corresponding to atmospheric pressure, h_0, less the height, h_v, corresponding to the water vapour pressure at that temperature† (Fig. 1.1).

B.   DIMENSIONAL ANALYSIS

1.15   General considerations. Geometric parameters

In order to have a clear grasp of how different parameters affect the flow of fluids, it is important to know the methods of dimensional analysis. Starting from the fact that physical laws, if they exist, must be independent of the units used to determine the numerical values of the various parameters, the theory of dimensional analysis makes it possible to determine the simplest form of those laws.

We have already considered the physical constants of fluids. The geometric parameters most commonly used in Hydraulics are given below.

— *Wetted area*, A: Cross-sectional area occupied by the flow.
— *Wetted perimeter*, P: Perimeter of the wet section in contact with the flowing liquid. In free-surface flows the part in contact with the air is not included in the wetted perimeter.
— *Hydraulic radius*, $R = A/P$: Quotient of the wetted area and the wetted perimeter.
— *Hydraulic diameter*, $D = 4R$: in circular pipes the hydraulic diameter is the diameter of the pipe.

For free-surface flows the following geometric elements are also defined:

— *Top width*, L: Width of the cross section, at the water surface.
— *Depth of flow*, h: Distance between the bottom of the channel and the free surface.
— *Mean depth*, $h_m = A/L$: Quotient of the wetted area and the top width.
— *Depth of the centre of gravity*, y: Depth to the centre of gravity of the cross-section of the stream.

1.16   Principle of homogeneity

The principle of homogeneity states that the two members of any relationship of a physical kind must have the same dimensions, so that the theoretical relationship that describes the physical phenomenon will remain independent of the system of units. A relationship of this type is called *dimensionally homogeneous*.

---

† This consideration is very important for studying pumping and the phenomena of cavitation.

Example: Consider a flow through a rectangular spillway. The discharge will be proportional to the length, l, of the spillway and will also be a function of the hydraulic head, H, over the spillway and of the gravitational acceleration, g.

C being a coefficient without dimensions, this may therefore be expressed as follows:

$$Q = Clf(H, g) \tag{1.6}$$

If it is assumed that

$$f(H, g) = H^\alpha g^\beta \tag{1.7}$$

we shall have:

$$Q = ClH^\alpha g^\beta \tag{1.7a}$$

The principle of homogeneity makes it possible to determine the exponents α and β. We have:

$$L^3T^{-1} = LL^\alpha(LT^{-2})^\beta = L^{1+\alpha+\beta}T^{-2\beta} \tag{1.7b}$$

hence

$$3 = 1 + \alpha + \beta ; \qquad -1 = -2\beta \tag{1.7c}$$

and thus

$$\beta = \frac{1}{2} ; \qquad \alpha = \frac{3}{2} \tag{1.7d}$$

which gives

$$Q = Clg^{1/2}H^{3/2} \tag{1.7e}$$

The value of C determined experimentally for a weir, would serve for all geometrically similar weirs, within the hypotheses assumed. As a rule there are other parameters that influence the value of C, namely the head, H.

Therefore C will not be constant (see Chapter 8) and it will be necessary to carry out not one but several experimental determinations, in which those parameters will be varied.

However, from this principle of homogeneity the general law governing the phenomenon is known and this greatly facilitates our study of it.

1.17   Π theorem or Vaschy–Buckingham theorem†

Consider a physical variable, G_1, a function of a certain number of other variables, $G_2, G_3, \ldots G_i \ldots G_n$. The dimensions of each of the variables, G_i, in relation to a unit system, LMT for example, will be $L^{\alpha i}M^{\beta i}T^{\gamma i}$.

The term *matrix of the dimensions* of the different variables, in relation to the unit system considered, is the term given to the matrix of the exponents α_i, β_i, γ_i whose number of columns is equal to the number n of variables considered (see Section 1.18).

The Π theorem is stated as follows:

† See [9].

If the variables characterizing a physical phenomenon are n and if the order of the matrix† of dimensions is r, the simplest form of the relationship existing between these n variables is a ratio between $n-r$ products without dimensions that can be formed with the n variables considered. A commonly used rule, which corresponds to the statement of the Π or Vaschy–Buckingham theorem and is simpler to use than the preceding one — though it sometimes leads to errors — is the following:

If there are n variables which characterize the phenomenon and these variables contain r fundamental dimensions (for example M, L, T) the equation relating the variables will contain $n-r$ non-dimensional groups.

As a consequence of this theorem, it may be said that n variables are fundamental if their determinant is non-zero.

1.18   Non-dimensional parameters commonly used in hydraulics‡

The various variables that occur in hydraulics or fluid mechanics are limited in number: the non-dimensional groups into which they can be formed are, in most cases, simple products that will be referred to later. Some of these products, which often occur in the fluid mechanics, have been given the name of the author who introduced them for the first time.

In a hydraulic phenomenon, the variables that may occur are: a certain number of geometric variables, a, b, c ...; the kinematic and dynamic characteristics, such as velocity, V, and the pressure variations, Δp; the gravitational acceleration, g; the physical properties of the fluid, such as density, ρ, viscosity, μ, surface tension, σ, and the modulus of elasticity, ε. The dimensional matrix concerned will be:

	a	b	c	V	ρ	g	Δp	μ	σ	ε
L	1	1	1	1	-3	1	-1	-1	0	-1
M	0	0	0	0	1	0	1	1	1	1
T	0	0	0	-1	0	-2	-2	-1	-2	-2

The order of this matrix is 3 and as fundamental variables we can take any three of those variables, subject only to the condition that they shall be dimensionally independent, i.e. that the corresponding determinant shall be non-zero.

Let us choose as fundamentals, for example, the variables a, V and ρ, whose dimensional determinant is non-zero.

$$\begin{vmatrix} 1 & 1 & -3 \\ 0 & 0 & 1 \\ 0 & -1 & 0 \end{vmatrix} = 1 \neq 0$$

---

† As is known from the matrix calculus, the order of a matrix is equal to the order of the non-zero determinant that can be extracted from the matrix.
‡ Closely followed by [9].

The non-dimensional parameters will thus be:

(a) relating to the *geometric characteristics* b and c, it may easily be seen that the following non-dimensional parameters can be obtained:

$$\Pi_1 = \frac{b}{a} ; \quad \Pi_2 = \frac{c}{a}$$

(b) relating to the *variation in pressure* it will be:

$\Pi_3 = a^x V^y \rho^z \Delta p$ or dimensionally, bearing in mind that the dimensions of Δp are those of a force (MLT^{-2}) per unit area (L^2).

$$(\Pi_3) = L^x (LT^{-1})^y (ML^{-3})^z . (MLT^{-2} . L^{-2}) \tag{1.8}$$

giving

$$\begin{array}{lll} x + y - 3z - 1 = 0 & & y = -2 \\ z + 1 = 0 & \text{hence} & z = -1 \\ -y - 2 = 0 & & x = 0 \end{array} \tag{1.8a}$$

or

$$\Pi_3 = \mathbf{E_u} = \frac{\Delta p}{\rho V^2} \tag{1.9}$$

This non-dimensional parameter is usually known as the **Euler number**.[†]

(c) relating to the *physical characteristics*, the following are obtained in an analogous manner:

— for *gravity*, **Froude number**[‡] (the ratio of inertia force to gravity forces);

$$\Pi_4 = \mathbf{F_r} = \frac{V}{\sqrt{ga}} \tag{1.10}$$

— for *viscosity*, **Reynolds number**[§] (the ratio of inertia forces to viscous forces);

$$\Pi_5 = \mathbf{R_e} = \frac{\rho \, Va}{\mu} = \frac{Va}{\nu} \tag{1.11}$$

[†] Euler, L. (1707–1783).
[‡] Froude, W. (1810–1879).
[§] Reynolds, O. (1842–1912).

— for *surface tension*, **Weber number**† (the ratio of inertia forces to surface tension forces);

$$\Pi_6 = \mathbf{W_e} = \rho \frac{V^2 a}{\sigma} = \frac{V^2 a}{\omega} \qquad (1.12)$$

— for *elasticity*, **Cauchy number**‡ (the ratio of inertia forces to elastic forces).

$$\Pi_7 = \mathbf{C_a} = \gamma \frac{V^2}{\varepsilon} = \frac{V^2}{\eta} \qquad (1.13)$$

Accordingly, the general equation of the phenomenon can be written as follows:

$$F\left(\frac{b}{a}, \frac{c}{a}, \mathbf{E_u}, \mathbf{F_r}, \mathbf{R_e}, \mathbf{W_e}, \mathbf{C_a}\right) = 0 \qquad (1.14)$$

The possibilities of dimensional analysis are confined to obtaining the above mentioned ratios. It is now necessary, in the light of a specific problem and for physical understanding of the phenomenon, to determine experimentally (or theoretically, if possible) the function F.

Apart from these non-dimensional parameters it is possible to define others, dependent on the phenomena to be studied, such as the following:

— **Dean number**, for studying head losses at curves, in laminar flow;

$$\mathbf{D_e} = \frac{Ud}{v} \sqrt{\frac{d}{2r}} \qquad (1.15)$$

in which d is the internal diameter of the curve and r the radious of curvature.
— **Leroux number**, for studying cavitation

$$\mathbf{L_e} = \frac{p_a - p_v}{\rho U^2} \qquad (1.16)$$

in which p_a is the absolute pressure and p_v the water vapour pressure at the temperature considered.
— **Mach number**§, for studying compressible fluids

$$\mathbf{M_a} = \frac{U}{c} \qquad (1.17)$$

---

† Weber, W. (1804–1891).
‡ Cauchy, A. (1789–1857).
§ Mach, E. (1838–1916).

in which c is the acoustic velocity in the fluid considered.
— **Strouhal number**, for studying vortex shedding from an immersed body.

$$S_t = \frac{f_t \cdot a}{U} \qquad (1.18)$$

in which f_t is the frequency of the vortices.

The Prandtl, Grashof and Eckert numbers are important in the study of heat exchange.

1.19 Example

Consider a steady flow in a long straight pipe of constant circular cross-section. It will be assumed that the flow is fully developed so that its characteristics are the same at all cross-sections.

The pressure loss, Δp, in a length, L, of the pipe is sought.

The variables concerned are:

Δp — pressure loss
L — length
D — hydraulic diameter equal to the geometric diameter d
ε — roughness characteristic of the internal pipe surface
U — mean velocity
ρ — density
v — kinematic viscosity

It is assumed that the other characteristics of the fluid (surface tension, for example) are not involved in the phenomenon.

Gravity, g is not considered, since the flow is under pressure (in a free surface flow it would be necessary to consider it).

The relevant dimensional matrix will be as follows:

	Δp	L	D	ε	U	ρ	v
L	-1	1	1	1	1	-3	2
M	1	0	0	0	0	1	0
T	-2	0	0	0	-1	0	-1

The physical relationship characterizing the pressure loss is thus a ratio between $7 - 3 = 4$ independent dimensionless products involving the seven variables:

$$\frac{\Delta p}{\rho U^2} = \Phi\left(\frac{L}{D}, \frac{UD}{v}, \frac{\varepsilon}{D}\right) \qquad (1.19)$$

Since Δp is proportional to L, as the flow is identical in all sections, the preceding equation may be written as follows:

$$\frac{\Delta p}{\rho U^2} = \frac{L}{D} f\left(\frac{UD}{v}, \frac{\varepsilon}{D}\right) \qquad (1.19a)$$

or

$$\Delta p = f \frac{L}{D} \rho U^2 \qquad (1.19b)$$

in which f is a function of UD/ν and of ε/D.

If we measure the pressure loss in head of fluid, i.e. $\Delta p = \rho g \Delta H$, the head loss, ΔH, can be written as follows:

$$\Delta H = f\left(\mathbf{R}_e, \frac{\varepsilon}{D}\right) \times \frac{L}{D} \frac{U^2}{2g} \tag{1.19c}$$

Remarks:

(1) the introduction of $2g$ in the last formula shown is an artifice of calculation, regardless of the physical law;

(2) when the roughness of the pipe was referred to (dimensions of the roughness) it was implicitly presumed that this was of a well defined type. In effect, f is determined for various types of roughness, and in order to be able to apply these values to different types of roughness, it is necessary to determine experimentally the relationship between the latter and the former;

(3) whenever a new variable occurs in the phenomenon, the number of dimensionless products is increased by one. For example, in the case of a pipe twisted into a spiral, with diameter d_s and pitch a, the law that gives the head loss will be:

$$\Delta H = f\left(\frac{UD}{\nu}, \frac{\varepsilon}{D}, \frac{d_s}{D}, \frac{a}{D}\right) \frac{L}{D} \frac{U^2}{2g} \tag{1.20}$$

The complete series of dimensionless products can be formed in various ways, but physical understanding of the phenomena must help the choice of the best method of establishing it.

It is logical, for example, to refer Δp to dynamic pressure $\rho U^2/2$, just as it is reasonable to refer ε to D (preferably to L, since this length is arbitrary).

BIBLIOGRAPHY

[1] Reboux, P., *Phénomènes de Fluidisation* (Ass. Française de Fluidisation, Paris).

[2] Reiner, M., *Rhéologie Théorique*. Paris, 1955.

[3] Terzaghi, K., Peck, R. B., *Soil Mechanics in Engineering Practice*. John Wiley & Sons, Inc., New York, 1967.

[4] Perry's, John H., *Chemical Engineers Handbook*, McGraw-Hill Book Company, 4th edition, New York, 1969.

[5] Boll, M., *Tables Numériques Universelles*. Dunod, Paris, 1947.

[6] Chenais, S. and J., *Propriétés Physiques de l'Eau*. Allier, Grenoble, 1939.

[7] *Handbook of Chemistry and Physics*. Chemical Rubber Publishing Co. Cleveland, 37th edition.

[8] Valembois, J., *Memento d'Hydraulique Pratique*. Eyrolles, Paris, 1958.

[9] Langhaar, H. L., *Annalyse Dimensionelle et Théorie des Maquettes*. Dunod, Paris, 1965.

[10] Bridgman, P. W., *Dimensional Analysis*. Yale University Press, 1931.

[11] Rouse, H., *Engineering Hydraulics*. John Wiley and Sons, Inc., 1962.

2

Theoretical bases of hydraulics

A. KINEMATICS OF FLUID MOTION. TYPES OF FLOWS

2.1 Posing the problem

Hydraulics is a branch of physical sciences whose purpose is the study of liquids in motion.

If a liquid flows in contact with the atmosphere there is said to be a *open-channel flow*: this occurs in a river or channel, for example. If the flow takes place in a closed pipe, occupying the whole section of the pipe, and generally with pressures different from atmospheric pressure, there is said to be a *pressure flow*: this is the case with flow in conduits. If the liquid flows through a porous medium, it is strictly neither under pressure flow nor is it an *open channel flow* and is then known as *flow in a porous medium* or filtration flow; this occurs in aquifers. As variants of these types of flow there are flows through orifices, over spillways, etc.

Fluid mechanics constitutes the theoretical basis of hydraulics. Real flows are, however, very difficult to analyse theoretically. The science of hydraulics thus has a certain degree of empiricism and experimentation is a fundamental aspect of it. In any theoretical formulation it is therefore essential to bear in mind the basic hypotheses.

2.2 Laminar and turbulent flow

The distinction between two types of flows in real liquids immediately gives an idea of the difficulties in theoretical analysis of liquids in motion.

In fact, two distinct types of flow can occur: *laminar or viscous flow*, in which each particle describes a well defined trajectory, with a velocity only in the direction of the flow; *turbulent or hydraulic flow* (this being the most usual in hydraulic phenomena), in which each particle, apart from the velocity in the direction of the flow, is animated by fluctuating crosscurrent velocities. Thus Reynolds number, R_e, is the characteristic parameter: for lower values of R_e, the flow is laminar, for higher values the flow is turbulent.

In order to ensure better understanding of these definitions, Reynolds' classic experiment is briefly described below.

Clean water is passed through a tube of transparent glass, at the beginning of which a small, highly coloured filament is introduced, with a direction coincident with the axis of the tube (Fig. 2.1). If the velocity of the

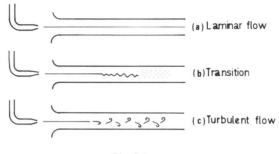

Fig. 2.1.

water in the tube is relatively low, the coloured filament remains straight and coincident with the axis of the tube; when the velocity of the water in the tube is increased, the coloured filament becomes mixed with the whole mass of water, to which it gives a uniform slight colouring.

Turbulent flow resembles a set of whirling movements, consisting of vortices of various dimensions and various frequencies which are super-imposed, in a random manner within the flow. Turbulence in a fluid is thus manifested by the irregular state of the flow, in which the various quantities show a variation that is random in space and in time, but in such a way that it is statistically possible to establish mean values; that is to say, in turbulent flow fluid particles do not remain in layers, but move in a heterogeneous fashion through the flow, sliding past other particles and colliding with some in a entirely haphazard manner but with a degree of regularity in time; and at a given moment, a certain form of flow is repeated with some regularity in space. It can be said that turbulence occurs as a *stationary random* process or as a *quasi-stationary* process which, for the sake of simplicity, is also called *steady*.

Thus the time-averaged parameters of the flow may be constant or slowly varying whilst the instantaneous parameters vary greatly.

2.3 Trajectory of a particle — Lagrangian variables†
One way of analysing the movement of a liquid is to follow the movement of an individual particle.

† Lagrange, J. L. (1736–1813).

Trajectory of the particle or *path line* is the term used for the locus of the successive positions occupied by the particle, during the course of time.

Let **P**, with coordinates x, y, z, be the position of an individual particle, at time t. Let $\mathbf{P}_0$ be the position occupied at time t_0.

If d**P** is the displacement in the interval of time dt, the differential equation of the movement of the particle is, vectorially:

$$\mathrm{d}\mathbf{P} = \mathbf{V}\,\mathrm{d}t \tag{2.1}$$

in which **V** is the velocity vector of components† (V_x, V_y, V_z).

The equation of the trajectory of the particle will be:

$$\mathbf{P} = \mathbf{P}(\mathbf{P}_0, t) \tag{2.1a}$$

which gives the position of the particle starting from an initial point $\mathbf{P}_0$.

The sets of variables thus defined $\mathbf{P}(x, y, z)$ are known as *Lagrangian variables*.

Owing to the complexity of the particle motion, this method of analysis is not the most convenient. It is only used when the initial position of the particle is important, for example, in the study of waves.

In the case of turbulent flows, integration of the equation of the trajectory is impossible.

2.4 Eulerian variables. Mean values of velocity, in time. Steady flow and unsteady flow

In practice, instead of following a particle, it is easier to define a velocity vector at each point and each moment of a flow.

$$\mathbf{V} = \mathbf{V}(\mathbf{P}, t) \tag{2.2}$$

The sets of variables thus defined $\mathbf{V}(v_x, v_y, v_z)$ are known as *Eulerian variables*: it is they that are normally used in hydraulics. If the velocity at each point is independent of time, there is a *steady flow*, otherwise the flow is *unsteady*.

In turbulent flow, the *instantaneous velocity*, **V**, has a random variation with time; it is, however, possible to determine a mean *velocity* $\bar{\mathbf{V}}$ so that at each moment, **V** is the sum of its mean value and a value of the *velocity fluctuation*, **V'**.

$$\mathbf{V} = \bar{\mathbf{V}} + \mathbf{V'} \tag{2.3}$$

The *mean velocity* at point P is thus the mean value, over a given time interval, of the instantaneous velocities at that same point.

In other words, in order to determine the mean value of the velocity at a point in a flow, it is necessary to use a certain measuring time, and this is sometimes very long, as occurs in the very ·

† In cylindrical coordinates (r, θ, z), in which $x = r\cos\theta$; $y = r\sin\theta$; $z = z$; the velocity will have the following components: radial: $V_r = \mathrm{d}r/\mathrm{d}t$; tangential $V_t = r\mathrm{d}\theta/\mathrm{d}t$; axial $V_z = \mathrm{d}z/\mathrm{d}t$.

irregular flows of natural watercourses in which as much as 10 to 15 minutes observation may be needed for measuring the mean value.

A turbulent flow is known, for the sake of simplicity, as a steady flow where the mean velocities are invariant with time; otherwise it is called unsteady.

2.5 Streamlines

Streamlines is the name given to the lines that are tangents, at any point and any moment, to the fluid velocity vector.

In turbulent flow, it is only of interest to study the streamlines corresponding to the mean fields of velocity.

For steady flows the path lines and streamlines coincide.
For a better understanding of the difference between path line and streamline, Fig. 2.2 is given.

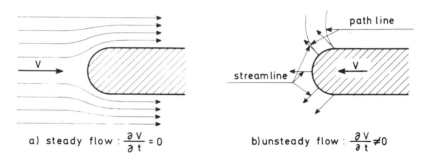

Fig. 2.2.

In Fig. 2.2a an obstacle is placed in a current with velocity **V**, in steady flow; the path lines and the streamlines coincide.
In Fig. 2.2b a boat moves in a liquid at rest, with velocity **V**. The movement generated in the liquid is not steady, since in each section the state of movement of the particles depends on the time of passage of the boat; the streamlines do not coincide with the trajectories.

2.6 Flowrate or discharge. Mean velocity in a section. Uniform and nonuniform flow

In a field of velocities **V**, consider an area A; let **n** be the unit vector normal to each element dA (Fig. 2.3(a)).

The *flowrate* or *discharge* over area A is:†

$$Q = \iint_A \mathbf{V} \cdot \mathbf{n} \, dA \qquad (2.4)$$

† **V.n** represents the *internal product*, i.e. a scalar, product of the modulus of one vector by projection of the other on it. Since **n** is unitary, **V.n** represents the projection of the modulus of V on the normal **n**, and will be represented by V_n. This product is zero if the vectors are perpendicular.

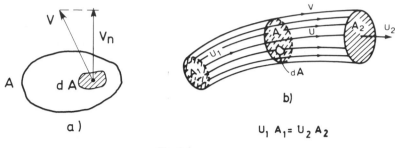

Fig. 2.3.

In the case of turbulent flow, it is only worth considering the value of the discharge corresponding to the mean velocity, $\bar{V}$, since the net flux corresponding to the velocity fluctuations is zero over a time interval, T.

$$Q = \iint_A (\bar{V} + V') \cdot n \, dA = \iint_A \bar{V} \cdot n \, dA \qquad (2.5)$$

Discharge is thus the volume of liquid passing a given cross-section area in unit time. It is usually expressed in cubic meters per second (m³/s), but also in litres per second (l/s), litres per minute (l/min) or cubic metres per hour (m³/hr).

Also the *mass discharge*, ρQ, can be defined as the mass fluid passing a given cross section in the unit time. It is usually expressed in kg/s.

Lines joining points of equal mean velocity in time, $\bar{V}$, are known as *isotachs*. The mean value of mean velocity $\bar{V}$ at the different points of a cross-section is known as *mean velocity*, U, in that *section*, whose modulus will be U.

We thus have (Fig. 2.3(a)):

$$U = \frac{1}{A} \iint_A V_n \, dA \qquad \text{or} \qquad Q = U \cdot A \qquad (2.6)$$

Streamtube is the term used for the set of streamlines which make up of a closed contour and across which there can therefore be no flow (Fig. 2.3(b)).

The area, A, intersecting a streamtube perpendicular to the streamlines, constitutes a *straight cross-section* of the flow. If this cross-section is infinitesimal, it will give rise to a *stream filament*.

If the mean velocity does not vary from section to section, there is a *uniform flow*. If it varies a *non-uniform flow* exists which may be *accelerated* or *retarded*.

2.7 Equation of continuity

The equation of continuity represents the following evident physical fact, that of the conservation of mass: the variation in the mass of a fluid, contained in a certain control volume e and limited by a control surface — the boundary of the control volume — during a certain time dt is equal to the flux of the mass of fluid over the area, A, during the same time, that is to say, it is equal to the mass of fluid entering it, less the mass of fluid leaving, during the time dt.

The differential equation will be as follows:

$$\frac{\partial \rho}{\partial t} = \operatorname{div}(\rho \mathbf{V}) \tag{2.7}$$

In the case of incompressible flows at constant temperature, $\rho = $ constant, so that we have:†

$$\operatorname{div} \mathbf{V} = \frac{\partial V_x}{\partial x} + \frac{\partial V_y}{\partial y} + \frac{\partial V_z}{\partial_z} = 0 \tag{2.8}$$

The equation of continuity applied to a streamtube, limited by straight cross-sections, A_1 and A_2, normal to the mean velocities U_1 and U_2 in those cross-sections, will be (Fig. 2.3b):

$$U_1 . A_1 = U_2 . A_2 \tag{2.9}$$

That is to say, the discharge $Q = U_1 A_1$ that enters is equal to the discharge $Q = U_2 A_2$ which leaves.

Bearing in mind Fig. 2.3, in which $A_2 \geqslant A_1$, the equation of continuity will give us $U_2 \leqslant U_1$, i.e. the flow is nonuniform (retarded).

Turbulence has no significance in the equation of continuity, as the mean value of the velocity fluctuations of turbulence is zero, i.e. the difference between the quantity of mass leaving and entering in a given control volume, due to the turbulence, is zero over a sufficient length of time.

2.8 Equation of state

The equation of state relates pressure p, temperature T, and density ρ, and is of the following type:

---

† In cylindrical coordinates (r, θ, z), the equation of continuity for steady motion of an incompressible fluid is written:

$$\frac{\partial}{\partial r}(r V_r) + \frac{\partial V_\theta}{\partial \theta} + r\frac{\partial V_z}{\partial z} = 0$$

If the motion has an axis of rotation, i.e. is independent of θ, we have $\partial V_\theta / \partial \theta = 0$. V_θ represents the tangential velocity V_t.

$$F(p, T, \rho) = 0 \qquad (2.10)$$

In hydraulic phenomena, considering the temperature to be constant, the equation of state can be written as follows:

$$\frac{\mathrm{d}\rho}{\mathrm{d}p} = \frac{\rho}{\varepsilon} \qquad (2.10a)$$

in which ε is the modulus of elasticity (see Section 1.11).

Considering incompressible liquids ($\varepsilon \to \infty$), which though not exact is valid for many problems, except for those connected with water hammer, the equation of state is reduced to:

$$\rho = \text{constant} \qquad (2.10b)$$

2.9 Rotational and irrotational flows

The possible variations in the form of an *elementary cube* of liquid, when it starts moving, are shown schematically in Fig. 2.4:

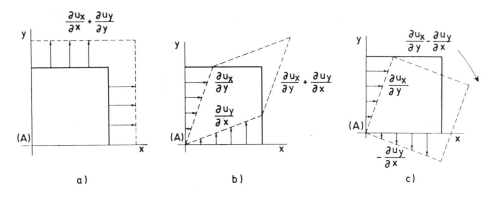

Fig. 2.4.

(a) *cubic expansion* — zero in incompressible fluids ($\rho = $ constant);
(b) *angular deformation* — represented mathematically by the tensor of the deformations;
(c) *rotation* — as a block, represented mathematically by the vector rot **V**.†

† The components of vector rot **V** are:
$$(\partial V_x/\partial z - \partial V_z/\partial x) \; ; \quad (\partial V_y/\partial x - \partial V_x/\partial y) \; ; \quad (\partial V_z/\partial y - \partial V_y/\partial z)$$

If rot **V** = 0, the flow is said to be *irrotational*; otherwise it is said to be *rotational*.

The most common flows in hydraulics are rotational, as will be seen later. In effect, rotation is a result of viscosity, i.e. the tangential forces exerted by particles on one another.

In some cases, which will be dealt with later, real flows may resemble irrotational flows.

Examples:

(a) In a rectilinear flow, if there is uniform distribution of velocities, a fluid element of rectangular, or more precisely, cubic shape remains rectangular (Fig. 2.5(a)). This is only possible if there is no friction next to the fixed wall, represented in the figure, or at the start of the flow when the effect of the wall is not yet felt.

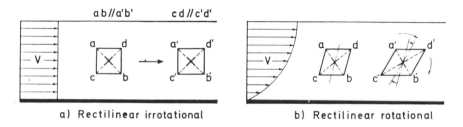

a) Rectilinear irrotational   b) Rectilinear rotational

Fig. 2.5.

If the distribution of velocities is not uniform, which is usually the case, the element undergoes a rotation as a whole, the diagonals do not remain orthogonal, and the flow is rotational (Fig. 2.5(b)).

(b) In a curvilinear flow, the element may be displaced as shown in Fig. 2.6(b), rotating on itself (non-parallelism of the diagonals) and the flow is then rotational.

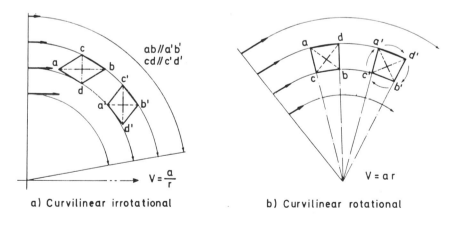

a) Curvilinear irrotational   b) Curvilinear rotational

Fig. 2.6.

The flow may, however, be irrotational provided that there is a suitable distribution of velocities, and the nearer it is to the centre of rotation (irrotational vortex — which will be studied in due course), the higher will be the velocity. In this case the diagonals of the element remain parallel and the element becomes deformed, without being subject to a rotation as a whole (Fig. 2.6(a)).

B.   DYNAMICS. GENERAL EQUATION OF MOTION

2.10   Posing the problem

Up to now an analysis has been made of the fluid motion without considering the forces that may give rise to the motion, i.e. velocities and accelerations. We now come to the fundamental equation of dynamics, $F = Ma$, which relates forces with acceleration.

Navier[†] and Stokes[‡] were responsible for deducing a general equation of fluid motion.

Acting on an elementary particle are: *body forces*, $\rho\mathbf{F}$, such as its weight, $\rho\mathbf{g}$; *inertial forces*, also proportional to the mass, $\rho d\mathbf{V}/dt$; *surface forces* that limit the fluid element, or pressure, $\mathbf{P}$. The body forces, $\rho\mathbf{F}$, depend on the nature of the problem and in most cases are reduced to weight. As regards inertial forces, it will be the flow conditions of the motion that will define them. Surface forces, resulting from the state of surface tension, have a normal component, *hydrostatic pressure*, and tangential components, due to viscosity, which are defined by the tensor of the *viscous stresses*. There are also turbulent *stresses*.

2.11   General Navier–Stokes equation

Equilibrium of the forces referred to above leads to the vectorial equation.

$$\rho\left(\mathbf{F} - \frac{d\mathbf{V}}{dt}\right) = \operatorname{grad} p - \mu\nabla^2\mathbf{V} - \frac{1}{3}\mu\operatorname{grad}\theta \qquad (2.11)$$

This is the general equation of viscous fluid motion known as the *Navier–Stokes equation*, integration of which is very difficult. It is possible to integrate it in particular cases of laminar flows but in most it is necessary to resort to experiment. Nevertheless, a knowledge of the significance of each of its terms can be of considerable help in trying to understand hydraulic phenomena and in conducting experiments. Accordingly:

— $\rho\mathbf{F}$ represents the body forces; in the case of a liquid flowing in the field of gravity, $\rho\mathbf{F}$ is the weight, so that $F_x = F_y = 0$; $F_z = g$, with g, the gravitational acceleration;
— $\rho d\mathbf{V}/dt$ represents the inertial forces;

† Navier, L. M. H. (1785–1836).
‡ Stokes, G. (1819–1905).

— grad p is the vector of components $\partial p / \partial x_i$. It corresponds to the derivative or inclination of the pressure in the direction of the flow;†
— the term $\mu \nabla^2 \mathbf{V}$ represents diffusion of the vector $\mathbf{V}$ within the flow: in other words, it represents the action of one particle on the others owing to the effect of viscosity.

For a better understanding of the signficance of $\nabla^2 \mathbf{V}$, it must be remembered that $\nabla^2 \mathbf{V} = \text{div (grad } \mathbf{V})$.

The term $1/3 (\mu \, \text{grad} \, \theta) = 1/3 [\text{grad (div } \mathbf{V})]$ represents the influence of the compressibility and in the case of incompressible liquids $\theta = \text{div } \mathbf{V} = 0$.

2.12   Flows in the field of gravity

Given the hypothesis of the external forces being derived from a potential, ξ, so that $\mathbf{F} = \text{grad } \xi$, then, in the case of incompressible liquids equation 2.11 becomes:

$$- \rho \, \text{grad} \, \xi + \text{grad} \, p \; = \; - \rho \frac{d\mathbf{V}}{dt} + \mu \nabla^2 \mathbf{V} \; . \tag{2.12}$$

If the potential is that of gravity, i.e. $\xi = - gz + \text{const.}$, then dividing throughout by $\gamma = \rho g$:

$$\text{grad} \left(z + \frac{p}{\gamma} \right) \; = \; - \frac{1}{g} \frac{d\mathbf{V}}{dt} + \frac{\mu}{\gamma} \nabla^2 \, \mathbf{V} \; . \tag{2.12a}$$

Let us see the physical significance of $(z + p/\gamma)$.

Consider a particle of volume e, density ρ and specific weight $\gamma = \rho g$. The mass of the particle will be $m = \rho e$ and its weight γe.

If the particle concerned is at an elevation z, above a horizontal datum, it will have a potential energy of position in relation to that plane which is given by $W_z = \gamma ez$. Considering unit weight, we thus have:

$$E_z \; = \; \frac{W_z}{\gamma e} \; = \; z \; . \tag{2.13}$$

The same particle subjected to a pressure, p, has a potential energy of pressure $W_p = pe$; in the same way, therefore:

---

† As is known from the vector algebra, the gradient of a scalar function is a vector, so that the derivative of that function in a given direction is a component of the gradient of the function in that direction. The gradient therefore corresponds to the maximum value of the variation of the function. It may be said that it corresponds to a line of greatest inclination in relation to a surface:

$$\text{grad} \, p \; = \; \frac{\partial p}{\partial x_1} \mathbf{i} + \frac{\partial p}{\partial x_2} \mathbf{j} + \frac{\partial p}{\partial x_3} \mathbf{k} \; .$$

$$E_p = \frac{W_P}{\gamma e} = \frac{p}{\gamma} \qquad (2.13a)$$

Accordingly, the sum $z + p/\gamma$ represents the potential energy per unit weight (energy of position and of pressure) whose variation, measured by the gradient, relates to the variations in velocity and to friction losses.

In the case of a perfect or ideal liquid, which does not exist, we should have, $\mu = 0$, so that:

$$\mathrm{grad}\left(z + \frac{p}{\gamma}\right) = -\frac{1}{g}\frac{d\mathbf{V}}{dt} \qquad (2.14)$$

In a liquid at rest $\mathbf{V} = 0$ and $\mathrm{grad}\,(z + p/\gamma) = 0$, so that:

$$z + \frac{p}{\gamma} = \mathrm{const.} \qquad (2.14a)$$

which represents the pressure variation in a liquid at rest, that is, a hydrostatic distribution of pressures (see Section 3.1).

2.13   Example. Poiseuille's† equation for flow in pipes, in steady laminar flow

The Navier–Stokes equations can help to resolve some problems, in the case of laminar flow.

Consider, for example, their application to a cylindrical pipe of constant circular section, with radius r_0, and a steady laminar flow.

Let the axis of the pipe coincide with the x − axis. We have y = r and z = r, as the pipe has a circular cross-section. In addition $U_x = V$; $U_y = 0$; $U_z = 0$.

Since V is a function of time and space:

$$dV/dt = \partial V/\partial t + \partial V/\partial x . \partial x/\partial t = \partial V/\partial t + \partial V/\partial x . V$$

In the case of a steady flow, $\partial V/\partial t = 0$; and since the section is constant and the motion uniform, $\partial V/\partial x = 0$, that is to say, $dV/dt = 0$, and equation 2.12 takes the following form:

$$\mathrm{grad}\left(z + \frac{p}{\gamma}\right) = \frac{\mu}{\gamma}\nabla^2 V \qquad (2.15)$$

or

† Poiseuille, J. L. (1799–1869).

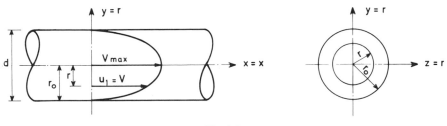

Fig. 2.7.

$$\frac{\partial}{\partial x}\left(z+\frac{p}{\gamma}\right) = \frac{\mu}{\gamma}\left(\frac{\partial^2 V}{\partial r^2}+\frac{\partial^2 V}{\partial r^2}\right) = 2\frac{\mu}{\gamma}\frac{\partial^2 V}{\partial r^2} \qquad (2.15a)$$

This equation was integrated directly by Poiseuille.
By representing the head variation $\partial(z + p/\gamma)/\partial x$ by i:

$$\frac{\partial^2 V}{\partial r^2} = \frac{\gamma i}{2\mu} \qquad (2.16)$$

On integrating, this gives:

$$\frac{\partial V}{\partial r} = \frac{\gamma i}{2\mu}r + c_1 \qquad (2.16a)$$

Since the velocity will be maximum in the centre, we shall have $r = 0$, $\partial V/\partial r = 0$ and $c_1 = 0$.
Integrating once more:

$$V = \frac{\gamma i}{4\mu}r^2 + c_1 \qquad (2.16b)$$

and since for $r = r_0$, $V = 0$.

$$V = \frac{\gamma i}{4\mu}(r_0^2 - r^2) \qquad (2.16c)$$

Equation 2.16(c) shows that the distribution of velocities is parabolic. The maximum value of the velocity, V_M, will be:

$$V_M = \frac{\gamma i}{4\mu}r_0^2 \qquad (2.16d)$$

and the mean value, U, will be:

$$U = \frac{\gamma}{8\mu}r_0^2i = \frac{g}{8\upsilon}r_0^2i = Ki \qquad\qquad (2.16e)$$

Equation 2.16(e) is Poiseuille's well known formula demonstrating that in laminar flow the mean velocity is proportional directly to the gradient of $(z + p/\gamma)$, i.e. proportional to the head loss. Conversely:

$$i = \frac{8\upsilon}{gr_0^2}U = \frac{32\upsilon}{gd^2}U \qquad\qquad (2.16f)$$

that is to say, the head loss is proportional to the velocity.

The distribution of velocities in terms of mean velocity is shown by simple algebraic operations:

$$V = 2U\left[1 - \left(\frac{r}{r_0}\right)^2\right] \qquad\qquad (2.16g)$$

Application:

Determine i in a pipe with $d = 0.1$ m, through which flows a thick oil whose kinematic viscosity is $\upsilon = 200$ centistokes $= 2 \times 10^{-4}$ m²/s and whose relative density is $\delta = 0.88$. The mean velocity of flow is $U = 1$ m/s.
The Reynolds number will be:

$$\mathbf{R_e} = Ud/\upsilon = (1 \times 0.1)/2 \times 10^{-4} = 500 \; ;$$

in this case the flow may be regarded as laminar (see Section 4.2).
From Equation 2.16f we shall thus have:

$$i = \frac{8\upsilon U}{gr_0^2} = \frac{8 \times 2 \times 10^{-4}}{9 \times 8 \times 0.05^{-2}} = 0.065$$

2.14 Turbulent flow. Mixing length

In turbulent flow it is considered that the instantaneous components of the velocity $u_i = \bar{u}_i + u_i'$ obey the Navier–Stokes equations.

Substituting in each element of those equations its mean value and the value of the fluctuations due to turbulence, there appear terms which represent new stresses that will be added to the stresses of viscous origin, thus giving a great increase in head losses.

Everything takes place as if, in turbulent flow, to the viscosity of the fluid is added a turbulent viscosity that is a property of the flow and not of the fluid. When the turbulence has fully developed, these turbulent stresses† are far greater than the viscous stresses and the flow is practically independent of viscosity, μ.

The complexity of the equations does not permit their integration, and it is usually necessary to resort to experiment, although new numerical models are being developed as an alternative procedure.

† Also called *Reynolds stresses*.

In turbulent flows, the particles undergo disordered transverse motion that tends to make the velocities uniform.

The distance at which two layers can be influenced by one another is known as the *mixing length* (Prandtl),[†] which is a measure of the length scale of the turbulence. The notion of coefficient of viscosity, μ, defined in 1.7, is replaced by a very complex factor called the *coefficient of turbulence*, or *eddy viscosity*.

2.15 Euler's equations along a path line
Consider the path line (Fig. 2.8).

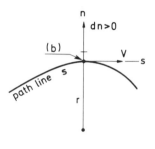

Fig. 2.8.

Construct a system of axes that at each point consists of the tangent to the path line (s), the normal (n) and the bi-normal (b), (intrinsic coordinates). Let r be the principal radius of curvature at the point considered.

The components of **V** will thus be: $V_s = V$; $V_n = 0$, $V_b = 0$.

Since $\mu = \rho v$ and $\gamma = \rho g$, equations 2.12 are written as follows:[‡]

$$\frac{\partial}{\partial s}\left(z + \frac{p}{\gamma}\right) = -\frac{1}{g}\frac{dV}{dt} + \frac{v}{g}\nabla^2 V \; ; \tag{2.18}$$

$$\frac{\partial}{\partial n}\left(z + \frac{p}{\gamma}\right) = \frac{1}{g}\frac{V^2}{r} \; ; \tag{2.18a}$$

$$\frac{\partial}{\partial b}\left(z + \frac{p}{\gamma}\right) = 0 \; ; \tag{2.18b}$$

In the case of the ideal or perfect liquid ($\mu = 0$), the term $\nabla^2 V$ is zero, and we have the equations deduced directly by Euler for perfect liquids.

Let us analyse in detail the physical significance of Euler's equations:

[†] Prandtl, L. (1875–1953).
[‡] It must be borne in mind that $\partial V/\partial n = V^2/r$ and $\partial V/\partial b = 0$.

(a) the first equation has an overall energy significance that will be brought
    out in the following paragraphs;
(b) the second equation shows that in a curve there exists a negative
    gradient of the potential energy, $(z + p/\gamma)$, that is to say, this value
    diminishes towards the centre of the curvature, and conversely. If the
    flow is rectilinear or has a negligible curvature, $(r \to \infty)$, then $z + p/\gamma =$
    constant, i.e. there will be a hydrostatic pressure distribution;
(c) the third equation shows that along the bi-normal the pressure distribu-
    tion is hydrostatic.

C.   ENERGY OF FLOWS. BERNOULLI'S† THEOREM

2.16   Types of energy

Energy or work, W, is defined in mechanics as the product of a force and a
displacement. Its dimensions are thus L^2MT^{-2} and the unit of energy in the
SI system is the joule, J.

In hydraulic problems energy is usually related, as mentioned in Section
2.12, to unit weight and is known simply as *head*, E, which consequently has
the dimensions of a length. In the **SI** system it is expressed in meters.

If the particle has a velocity, V, its kinetic energy is $W_c = mV^2/2$. The
kinetic energy per unit weight will then be as follows, taking e as the volume
of the particle.

$$E_c = \frac{W_c}{\gamma e} = \frac{1}{2} \cdot \frac{mV^2}{\gamma e} = \frac{1}{2}\frac{\rho}{\gamma}V^2 = \frac{V^2}{2g} \qquad (2.19)$$

We have already seen that the potential energy per unit weight is z, just
as the pressure energy per unit weight is p/γ. Thus a particle of liquid having
velocity, V, subjected to a pressure, p‡, and placed at an elevation, z, above
a horizontal datum, will have, per unit weight, the following types of energy
or head:

Type of head	Hydraulic designation	Representation
Due to position	Elevation above a datum or *elevation head*	$E_z = z$
Due to pressure	Pressure expressed in height of liquid or *pressure head*	$E_p = p/\gamma$
Kinetic	*Velocity head*	$E_c = V^2/2g$

---

† Bernoulli, D. (1700–1782).

‡ When nothing is said to the contrary, it is presumed that this is gauge pressure (see Section
1.10), i.e. measured in relation to atmospheric pressure.

The total energy per unit weight will thus be:

$$E = z + \frac{p}{\gamma} + \frac{V^2}{2g} \tag{2.20}$$

The pressure head represents the height of a column of liquid that through its weight can give rise to pressure p.

The velocity head represents the height, h, from which an element of fluid must fall freely, in vacuum, in order to reach velocity V.

2.17 Bernoulli's theorem: energy of a particle along its path line

Taking equation 2.18 again, and bearing in mind that $dV/dt = \partial V/\partial t + V\partial V/\partial s$ and also that $V\partial V/\partial s = \partial/\partial s(V^2/2)$, we have:

$$\frac{\partial}{\partial s}\left(z + \frac{p}{\gamma} + \frac{V^2}{2g}\right) = -\frac{1}{g}\frac{\partial V}{\partial t} + \frac{\upsilon}{g}\nabla^2 V \tag{2.21}$$

The first element of the equation has an essentially overall energy significance, as has already been mentioned. It represents the variation in the total energy discharged per unit weight or total head, E, of a particle along its trajectory.

Although a perfect liquid, i.e. without viscosity ($\mu = 0$) does not exist in nature, there are cases in which the liquid behaves as if it were perfect, e.g. a liquid at rest in which the viscosity is not felt. Furthermore, a flow starting from a state of rest will have an initial region in which the effects of viscosity are not yet significant, for example, the flow over a spillway or the flow from a reservoir to a pipe or channel. In this case the flow may resemble a perfect liquid.

In these circumstances the viscosity terms are removed from the equation and we have:

$$\frac{\partial}{\partial s}\left(z + \frac{p}{\gamma} + \frac{V^2}{2g}\right) = -\frac{1}{g}\frac{\partial V}{\partial t} \tag{2.21a}$$

In the case of steady flow, $\partial V/\partial t = 0$, so that:

$$E = z + \frac{p}{\gamma} + \frac{V^2}{2g} = \text{const.} \tag{2.22}$$

this being the expression that represents *Bernoulli's theorem*: in the case of an incompressible liquid in steady flow, in which the friction forces and, consequently, energy losses can be disregarded, the total head of a particle is maintained along its trajectory.

As can be seen from the statement itself, Bernoulli's theorem is the direct result of the principle of the conservation of energy: if there is no friction, the particle moves without loss of energy.

2.18 Energy line and piezometric line

Consider a streamline in a steady flow.† At each point of this streamline situated at elevation z above a datum, the different particles which successively occupy that point are subject to a pressure p and have a velocity, V, to which correspond the energy conditions defined previously.

It is thus possible to define, through Euler's variables, the energy conditions at each point of a streamline, independently of the particles occupying that point.

To sum up: in relation to each point of a streamline the following specific heads or energies are defined:

— *Piezometric head*: $E_e = z + p/\gamma$
— *Dynamic or velocity head*: $E_c = V^2/2g$
— *Total head or energy*: $E = z + p/\gamma + V^2/2g$

If along a streamline, on a vertical from the horizontal datum, lengths are marked to represent the static head, we shall obtain the *piezometric head line* corresponding to the streamline considered.

Likewise, if the total head is marked, we obtain the *total head line* or simply energy line.

The energy line is distant from the piezometric line by a length, measured on the vertical, equal to the *velocity head*.

2.19 Energy or head in section of the flow

Total head can be defined not only at a point on a streamline, but also in a straight section of a flow, provided that the streamlines have a very small curvature, such that they can be considered practically straight and parallel. The static head, $z + p/\gamma$, in this case has the same value for the whole straight section;‡ the velocity, V, however, may vary from one point to another of the straight section. Dynamic head is thus defined as the quotient, $W_c/\gamma Q$, of the power, W_c, that passes through the section in kinetic form, and the weight rate of flow, γQ.

By substituting the mean velocity, U, for the various velocities, V, of the particles, a correction factor of the kinetic energy, α, is introduced. This is known as *Coriolis*§ *coefficient*, defined as the ratio between the real kinetic

---

† As has been said, this term includes quasi-steady turbulent flow, in which case the mean value of the velocity at each point is considered, and thus is equivalent to disregarding the actual energy of the turbulence.

‡ See Section 2.15, equation 2.18(a). In this case there is said to be a hydrostatic distribution of pressure. If this hypothesis is not valid it is possible to establish a corrective term to account for the effects of the curvature of the streamlines.

§ Coriolis, G. (1792–1843).

energy of the flow and the kinetic energy of a fictitious flow, in which all the particles are assumed to move at the mean velocity U.

The total head in a straight section will thus be:

$$E = z + \frac{p}{\gamma} + \alpha\frac{U^2}{2g} \tag{2.23}$$

In a similar way to the above, the *piezometric head line* and *energy line* are defined as relative or absolute (Fig. 2.9) according to whether the

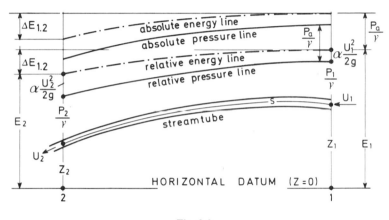

Fig. 2.9.

pressure is considered to be relative or absolute (see section 1.10).

If the distribution of the velocities is uniform, then $\alpha = 1$; if not, then $\alpha > 1$; in practical cases, in turbulent flow, α varies between 1.05 and 1.20, though it may reach very high values (see Table 12). Sometimes, in practice, this corrective coefficient can be dispensed with, making it equal to unity.

Since the real kinetic passing through the section is:

$$W_c = \iint_A \frac{1}{2}V^2\rho V\,\mathrm{d}A = \frac{1}{2}\rho\iint_A V^3\,\mathrm{d}A \tag{2.24}$$

and since the kinetic energy due to considering the average velocity, U, is:

$$W'_c = \iint_A \frac{1}{2}U^2\rho U\,\mathrm{d}A = \frac{1}{2}\rho U^3\iint_A \mathrm{d}A = \frac{\rho U^3 A}{2} \tag{2.24a}$$

we have the expression:

$$\alpha = \frac{W_c}{W'_c} = \frac{1}{AU^3}\iint_A V^3\, dA \qquad\qquad (2.25)$$

Determination of the kinetic coefficient α is based on its mathematical definition, given by equation 2.25 and can be found analytically given an expression for the distribution of velocities (see following example). Generally speaking, it is easier to make use of a graphical method, provided that we know the distribution of velocities (see Fig. 2.10) relating to a natural watercourse.

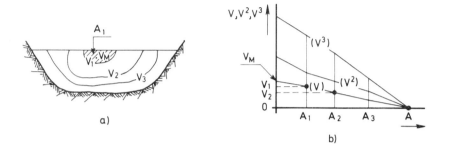

Fig. 2.10.

In effect, by marking the abscissae of the various areas $A_1, A_2 \dots$ in which the velocity is higher than $V_1, V_2 \dots$ and at ordinates the velocities $V_1, V_2 \dots$ as indicated in the figure, the curve (V) is obtained. In the same way, by marking at ordinates $V_1^2, V_2^2 \dots$, the curve (V^2) is obtained; and marking $V_1^3, V_2^3 \dots$ the curve (V^3) is obtained.

The mean velocity, U, is obtained by dividing the area between curve (V) and the axis OA by the value of A. The value of α is obtained by dividing the area between curve (V^3) and the axis OA by the value of U^3A.

Example:

Determine the velocity head of the flow in the example in Section 2.13.
The velocity distribution is (equation 2.16(g)).

$$V = 2U\left[1 - \left(\frac{r}{r_0}\right)^2\right]$$

We shall then have (equation 2.25), bearing in mind that $A = \pi r^2$ and $dA = 2\pi r\, dr$.

$$\alpha = \frac{1}{U^3A}\iint_A V^3\, dA = \frac{1}{U^3A}\int_0^{r_0}\left\{2U\left[1 - \left(\frac{r}{r_0}\right)^2\right]\right\}^3 2\pi r\, dr = 2$$

$$E_c = \frac{\alpha U^2}{2g} = 2\times\frac{1^2}{2\times 9.8} \approx 0.1\,\text{m}\ .$$

The value $\alpha = 2$ is thus valid for a circular pipe in laminar flow.

In turbulent flow, the value of α is close to unity, since turbulence tends to make the velocities more uniform.

2.20   Application of Bernoulli's equation to a streamtube

Bernoulli's equation applies to a streamtube, provided that the total head is considered in each section:

$$E = z + \frac{p}{\gamma} + \alpha\frac{U^2}{2g} \qquad (2.26)$$

Between two sections of the flow, in the case of perfect liquids or in the case of real liquids when it is possible to disregard the head losses, we shall have.

$$E_1 = E_2 . \qquad (2.26a)$$

The head loss between sections 1 and 2 will, in general be:

$$\Delta E_{12} = E_1 - E_2 = \left(z_1 + \frac{p_1}{\gamma} + \alpha\frac{U_1^2}{2}\right) - \left(z_2 + \frac{p_2}{\gamma} + \alpha\frac{U_2^2}{2g}\right) \quad (2.26b)$$

i is taken to represent the *head loss per unit weight and per unit length*.

We thus have:

$$\Delta E_{12} = \int_1^2 i\,ds \qquad (2.26c)$$

If the loss is constant throughout the length then $\Delta E_{12} = i\Delta s$, Δs being the distance between the sections 1 and 2.

This energy loss can in some cases of laminar flow be determined analytically (see Section 2.13). In turbulent flow the calculation becomes impossible, and i is then determined experimentally by measuring E_1 and E_2 at two sections of the flow, between which i can be considered constant.

Determination of head losses has been one of the major concerns of hydraulic experiments. It will be studied in later chapters.

The power loss between sections 1 and 2 is:

$$\Delta P_{12} = \gamma Q \Delta E_{12} \qquad (2.27)$$

This energy may be lost owing to the internal friction of the fluid and friction of the fluid on the walls: in these cases it is degraded into heat. The variation in energy, ΔE_{12}, may also be due to the fact of a hydraulic machine being inserted in the circuit and receiving energy, thus reducing the energy of the fluid (e.g. a turbine); or yielding energy, thus increasing the energy of the fluid (e.g. a pump).

Examples:

(1) A pipe gets wider between section 1 where the diameter is $d_1 = 480$ mm and section 2 situated 2.0 m higher than 1, where the diameter $d_2 = 945$ mm. The discharge is $Q = 180$ l/s. The pressure at point 1 is 30 N/cm$^2 = 300\,000$ Pa. The density of the liquid discharged is $\rho = 1000$ kg/m^3.

Calculate:

(a) the velocities at 1 and 2;
(b) the pressure at 2, assuming the energy loss to be negligible.

Solution:

(a) according to the equation of continuity, $Q = U_1 A_1 = U_2 A_2$, whence

$$U_1 = \frac{Q}{A_1} = \frac{0.180}{0.18} = 1.00 \text{ m/s} \; ; \qquad U_2 = \frac{Q}{A_2} = \frac{0.180}{0.70} = 0.26 \text{ m/s}$$

(b) by applying Bernoulli's theorem between sections 1 and 2, taking as datum the horizontal plane passing through 1, making $\alpha = 1$ and bearing in mind that $\gamma = 9800$ N/m^3 we have:

$$\frac{1.00^2}{2 \times 9.8} + \frac{300\,000}{9.8 \times 1000} + 0 = \frac{0.26^2}{2 \times 9.8} + \frac{p_2}{\gamma} + 2.0$$

so that: $p_2/\gamma = 0.051 + 30.612 - 0.003 - 2 = 28.66$ m (water column), and $p_2 = 28.66 \times 9800$ Pa $= 280.9$ kPa $= 28.09$ N/cm^2.

(2) A hydraulic turbine is fed with a discharge $Q = 424$ l/s by a horizontal penstock with internal diameter $d_1 = 0.30$ m. In a section 1 of this penstock, just upstream of the turbine, the manometric pressure is $p_1 = 6.89$ N/cm$^2 = 7.03$ m $= 68.9$ kPa.

The outlet from the turbine is through a truncated cone conduit. In a section 2 of this conduit, with diameter $d_2 = 0.45$ m situated 1.5 m below section 1, a negative pressure $p_2 = -4.16$ N/cm$^2 = -4.22$ m $= -41.6$ kPa has been measured. Calculate the energy that the turbine can supply, assuming that its efficiency is 0.85.

Solution:

$$U_1 = \frac{Q}{A_1} = 6 \text{ m/s} \; ; \qquad U_2 = \frac{Q}{A_2} = 2.7 \text{ m/s}$$

The energy loss per unit weight (energy converted into mechanical energy by means of the turbine and energy degraded into heat by friction), taking as datum the horizontal plane passing through 2, is as follows:

$$E_{12} = \left(\frac{U_1^2}{2g} + \frac{p_1}{\gamma} + z_1 \right) - \left(\frac{U_2^2}{2g} + \frac{p_2}{\gamma} + z_2 \right) =$$

$$= (1.84 + 7.03 + 1.5) - (0.37 - 4.22 + 0) = 14.22 \text{ m} \; .$$

The energy absorbed by the turbine:

$$P_{ab} = \gamma Q \Delta E_{12} = 9800 \times 0.424 \times 14.22 = 60.293 \text{ kW}$$

The energy supplied by the turbine is therefore:

$$P_u = 60.293 \times 0.85 = 51.249 \text{ kW} \; .$$

D. MOMENTUM. EULER'S THEOREM

2.21 Momentum

The *momentum* of an elemental fluid system or particle of mass, m, moving at a velocity, $\mathbf{V}$, is the product $m\mathbf{V}$. This product has the dimensions FT.

The momentum of a flow will be the sum of the momentum of its particles. In a given section a certain amount of fluid mass flows per unit time and this makes it possible to establish the concept of rate of change of momentum through that section whose dimensions are those of a force F.

The *flux of momentum* through the section of area A will thus be:

$$\mathbf{M} = \iint_A \rho \mathbf{V} V_n \, dA \tag{2.28}$$

In order to be able to take into account the mean velocity, $\mathbf{U}$, in substituting for the various velocities, $\mathbf{V}$, of the particles, a correction factor of the momentum β, is introduced. This is known as *Boussinesq's coefficient*,[†] defined as the ratio between the real momentum of the flow and the momentum of a fictitious flow in which all the particles would move at a mean velocity, $\mathbf{U}$.

The momentum per unit time, in a section in which there is a flow rate Q at a mean velocity $\mathbf{U}$, will thus be:

$$\mathbf{M} = \beta \rho Q \mathbf{U} \tag{2.29}$$

in which it must be assumed that $\mathbf{U}$ has the same direction and orientation as that resulting from the various velocities, $\mathbf{V}$.

Should there be a uniform distribution of velocities, then β is equal to 1; otherwise β is slightly higher than 1. Generally the value of β is closer to 1 than the value of the coefficient of velocity, α (see Table 12).

The mathematical expression which gives the value of β is, from its definition, as follows:

$$\beta = \frac{1}{AU^2} \iint_A V^2 \, dA \tag{2.30}$$

Bearing in mind Fig. 2.10b, the value of β is obtained by dividing the area between the curve (V^2) and axis OA by the value of $U^2 A$.

Applied to the same example of Section 2.13, we have:

$$\beta = \frac{1}{AU^2} \iint_A V^2 \, dA = \frac{1}{AU^2} \int_0^{r_0} \left[2U\left(1 - \frac{r^2}{r_0^2}\right) \right]^2 2\pi r \, dr = 1.33 \; ;$$

this being the value of β in the case of laminar flow in a circular pipe.

---

† Boussinesq, J. (1842–1929).

2.22 Euler's theorem

Consider a control volume of fluid (Fig. 2.11a). The forces acting on it are

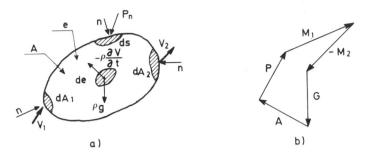

Fig. 2.11.

described below.

— The force resulting from the *body forces*: if the liquid is subject only to
the action of gravity, $\mathbf{F} = \rho\mathbf{g}$, and the first term is reduced to the weight,
$\mathbf{G}$, of the volume e.
— The force resulting from the *inertia forces* due to local accelerations,
$\partial V/\partial t$ will be represented by $\mathbf{A}$. In the case of steady flow, $\mathbf{A} = 0$.
— The *flux of momentum* through A and represented by $\mathbf{M}$. In terms of
mean velocity, as we have seen, $\mathbf{M} = \beta\rho Q\mathbf{U}$, which can be broken down
into two parts: $\mathbf{M}_1$ corresponding to the velocity, $\mathbf{U}_1$, of the liquid which
enters, and $\mathbf{M}_2$ corresponding to the velocity, $\mathbf{U}_2$, which leaves the
control volume.
— The *surface forces* that act at the control surface: these forces being
represented by $\mathbf{P}$ or $-\mathbf{P}$, considered to be positive or negative whether
they are directed towards or away from the control volume. These
external forces may be normal or tangential to the control surface A.

The equation of equilibrium is thus written vectorially (Fig. 2.11(b)):

$$\mathbf{G} + \mathbf{A} + \mathbf{P} + \mathbf{M}_1 - \mathbf{M}_2 = 0 \tag{2.31}$$

It is in this form that it is usually applied in hydraulics.
It must not be forgotten that this is a vectorial equation (Fig. 2.11(b)).
It can be used, in vectorial or component form, on one or more
judiciously chosen axes. It is necessary to bear in mind the directions of the
forces and not forget any of them.
The interest in this theorem is due to the fact that it is not necessary to
know the internal friction forces in the fluid; it is therefore applicable even

when there are energy losses inside the volume of fluid considered. On the other hand, it is not possible to forget, apart from the normal stresses, the tangential *shear stresses* exerted by the walls on the fluid, though in many cases they can be ignored (e.g. in the study of the hydraulic jump — see Chap. 6) or can be determined from the theorem itself.

In turbulent flow the preceding equations are valid for instantaneous velocity, and in practice also for mean velocity.

2.23 Application of Euler's theorem to a streamtube (steady flow)

Consider the volume limited by a streamtube and by two normal sections A_1 and A_2 (Fig. 2.12) with velocities U_1 and U_2, respectively, in steady flow and subjected to pressures p_1 and p_2.

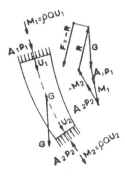

Fig. 2.12.

Let R represent the result of the pressures exerted by the walls of the tube on the liquid. The resultant pressure exerted by the surface limiting the volume will be:

$$\mathbf{P} = \mathbf{R} + A_1\mathbf{p}_1 + A_2\mathbf{p}_2$$

By applying the vectorial equation (2.31) we obtain the following:

$$\mathbf{G} + \mathbf{R} + A_1\mathbf{p}_1 + A_2\mathbf{p}_2 + \mathbf{M}_1 - \mathbf{M}_2 = 0 \qquad (2.32)$$

The force exerted by the liquid on the walls of the tube is:

$$\mathbf{F} = -\mathbf{R} = \mathbf{G} + A_1\mathbf{p}_1 + A_2\mathbf{p}_2 + \mathbf{M}_1 - \mathbf{M}_2 \qquad (2.32a)$$

2.24 Application of Euler's theorem to a non-rectilinear pipe

Consider a curved pipe on a horizontal axis subjected to pressure p, assumed to be constant, and divide $\mathbf{F}$ into two components; $\mathbf{F_V}$, and $\mathbf{F_H}$, the vertical and horizontal components. We have $p_1 = p_2 = p$, and then:

$$\mathbf{F_V} = \mathbf{G}$$

$$\mathbf{F_H} = A_1\mathbf{p_1} + A_2\mathbf{p_2} + \mathbf{M_1} - \mathbf{M_2} \tag{2.32b}$$

If the section of the pipe and the radius of the curve are constant, $\mathbf{F_H}$ will be directed along the bisector of the angle at the centre, θ, of the curve.

If $h = p/\gamma$ is the head, in metres, at the curve, we shall have:

$$\mathbf{F_H} = 2A(\rho U^2 + \gamma h)\sin\frac{\theta}{2} = 2A\gamma\left(2\frac{U^2}{2g} + h\right)\sin\frac{\theta}{2} \tag{2.33}$$

Assuming the pipe to be circular with diameter d, the previous expression becomes:

$$\mathbf{F_H} = 2\frac{\pi d^2}{4}\gamma K\sin\frac{\theta}{2} \tag{2.33a}$$

in which:

$$K = 2\frac{U_2}{2g} + h \tag{2.34}$$

2.25 Example

A metal pipe of $0.76\,\mathrm{m}$ internal diameter has at its end, subject to head $h = 190\,\mathrm{m}$, a bifurcation, as shown in Fig. 2.13, each branch being fitted with a valve. The branch of the bifurcation has an internal diameter of $0.50\,\mathrm{m}$. The pipe is horizontal and is anchored to a concrete block, the stability of which is to be determined.

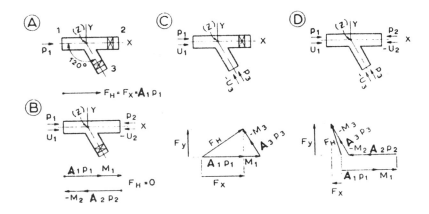

Fig. 2.13.

Calculate, in magnitude and position, the forces exerted on the anchorage block for the following conditions:

(a) valves 2 and 3 closed;
(b) valve 2 discharging 4.8 m³/s and valve 3 closed;
(c) valve 3 discharging 0.62 m³/s and valve 2 closed;
(d) both valves open and discharging the flow rate already mentioned.

Solution:

$$A_1 \ = \ A_2 \ = \ 0.454\,\text{m}^2 \ ; \qquad A_3 \ = \ 0.196\,\text{m}^2 \ ;$$

$$p_1 \ = \ p_2 \ = \ p_3 \ = \ p \ = \ \gamma h \ = \ 9.8 \times 1000 \times 190\,\text{N/m}^2 \ = \ 1862\,\text{kPa}$$

Consider the axis system, $x\,y\,z$, indicated in the figure, in which z is the vertical axis. By direct application of Euler's equation, for the different cases, taking F_x, F_y and F_z to represent the components of **F** according to the coordinate axes, we have moduli of those components:

(a) $\mathbf{F_X} \ = \ A_1 \cdot p_1 \ = \ 845\,\text{kN} \ ; \quad \mathbf{F_y} = 0 \ ; \quad \mathbf{F_H} = \mathbf{F_x} \ ; \quad \mathbf{F_z} = \mathbf{G}$

(weight of liquid in the control volume in the pipe)

(b) $A_1 \cdot \mathbf{p}_1 \ = \ -A_2 \cdot \mathbf{p}_2 \ ; \qquad \mathbf{U}_1 \ = \ \mathbf{U}_2$

thus

$$\mathbf{M}_1 \ = \ \mathbf{M}_2 \ .$$

Whence

$$\mathbf{F}_x \ = \ (A_1 \mathbf{p}_1 + \mathbf{M}_1) + (A_2 \mathbf{p}_2 - \mathbf{M}_2) \ = \ 0 \ ; \quad \mathbf{F}_z = \mathbf{G} \ .$$

(c) $A_1 p_1 \ = \ 845\,\text{kN} \ ; \qquad A_3 p_3 \ = \ 365\,\text{kN} \ ;$

$$\mathbf{M}_1 \ = \ \rho Q U_1 \ = \frac{Q^2}{A_1} = \ 1000 \times \frac{0.62^2}{0.454} \ = \ 0.85\,\text{kN} \ .$$

$$\mathbf{M}_3 \ = \ \rho \frac{Q_3^2}{A_3} \ = \ 2\,\text{kN} \ .$$

Thus:

$$\mathbf{F}_x \ = \ (A_1 \mathbf{p}_1 + \mathbf{M}_1) + (A_3 \mathbf{p}_3 - \mathbf{M}_3) \cos 120° \ = \ 663\,\text{kN} \ ;$$

$$\mathbf{F}_y \ = \ (A_3 \mathbf{p}_3 - \mathbf{M}_3) \sin 120° \ = \ 318\,\text{kN} \ ;$$

$$\mathbf{F}_H \ = \ \sqrt{(F_x^2 + F_y^2)} \ = \ 735\,\text{kN} \ ; \qquad \mathbf{F}_z \ = \ \mathbf{G} \ .$$

(d) $Q_2 \ = \ 4.8\,\text{m}^3/\text{s} \ ; \qquad Q_3 \ = \ 0.62\,\text{m}^3/\text{s} \ ; \qquad Q_1 \ = \ Q_2 + Q_3 \ = 5.42\,\text{m}^3/\text{s} \ ;$
$A_1 p_1 = A_2 p_2 = 846\ \text{kN}; \ A_3 p_3 = 365\ \text{kN};$

$$M_1 = \rho \frac{Q_1^2}{A_1} = 65\text{kN} \ ; \quad M_2 = \rho \frac{Q_2^2}{A_2} = 51\,\text{kN} \ ; \quad M_3 = \rho \frac{Q_3^2}{A_3} = 2\,\text{kN}$$

Thus:

$$\mathbf{F}_x \ = \ (A_1 \mathbf{p}_1 + \mathbf{M}_1) + (A_2 \mathbf{p}_2 - \mathbf{M}_2) + (A_3 \mathbf{p}_3) \cos 120° = \ - 170 \ \text{kN};$$
$$\mathbf{F}_y \ = (A_3 \mathbf{p}_3 - \mathbf{M}_3) \sin 120° \ = \ 318 \ \text{kN};$$
$$\mathbf{F}_H = \ \sqrt{F_x^2 + F_y^2} \ = \ 360 \ \text{kN};$$
$$\mathbf{F}_z \ = \ \mathbf{G}$$

E. FLOW ESTABLISHMENT

2.26 Boundary layers

In the flow of an ideal liquid (zero viscosity), there is no interaction between the liquid in motion and a solid wall. On the other hand, in the case of real liquids, the effect of viscosity is such that the velocity of the fluid near the wall approaches the velocity of the wall: that is to say, if the wall is at rest (e.g. as in the case of a pipe) there will be a small layer of fluid that remains practically stationary, while the velocity increases rapidly towards the centre. This means that a conventional velocity profile must show a velocity of zero at the boundary.

Fig. 2.14 shows the mechanism of *boundary layer* growth from a pipe entrance.

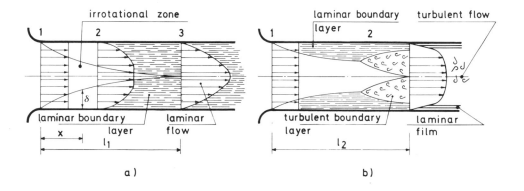

Fig. 2.14.

Qualitatively speaking, the phenomenon is similar in the transition from a reservoir to a channel or in flow over a spillway. In the initial section, section *1*, the liquid starts from rest; the flow may also be treated as *irrotational* and the velocities are very approximately equal in the whole section.

Downstream, the liquid in contact with the wall is retarded and the *laminar boundary layer* begins; the liquid particles in the central zone of the pipe, which have not felt the effect of friction (the core), undergo an acceleration in such a way that the discharge remains constant. If $R_e < 2000$, the boundary layer continues to be laminar, growing with distance downstream until it occupies the whole section of the pipe.

The thickness of the laminar boundary layer increases according to the formula:

$$\delta = \frac{Kx}{\sqrt{\mathbf{R}_{ex}}} \tag{2.35}$$

in which $K \approx 5$ and $\mathbf{R}_{ex} = Ux/\upsilon$.

The distance l_1 from the origin as far as the section in which the boundary layer occupies the whole pipe is:

$$l_1 = 0.01\frac{d^2U}{\upsilon} \tag{2.36}$$

In the case of a channel with water depth h it is:

$$l_1' = 0.04\frac{h^2U}{\upsilon} \tag{2.37}$$

If $\mathbf{R}_e > 2000$, the disturbances starting in the laminar boundary layer will increase, giving rise to a *turbulent boundary layer* (Fig. 2.14b). Next to the wall, however, particularly smooth walls, the transverse fluctuating component cannot be very great since transverse movements are countered by the presence of the wall. This means that when the boundary layer becomes turbulent there exists along the wall a *laminar film* or *viscous sublayer*. The separation zone between the laminar layer and the turbulent boundary layer constitutes a *transition zone*. The point marking the start of the transition zone, i.e. the point from which the flow becomes turbulent, is known as the *transition point*.

The distance, l_2, at which the turbulent boundary layer occupies the whole section of the pipe, is given by:

$$l_2 = 1.5d\sqrt[5]{\mathbf{R}_{ex}} \tag{2.38}$$

In the case of a channel it is:

$$l_2' = 3h\sqrt[5]{\mathbf{R}_{ex}} \tag{2.39}$$

h being the depth of water and $\mathbf{R}_{ex} = Ux/\upsilon$.

2.27 Separation of the flow

Consider a curved surface immersed in a parallel flow; case 2 of Fig. 2.16 can be used as an example. Its presence causes diversion and concentration of the streamlines and, consequently, an increase in the velocity, this being shown in detail in Fig. 2.15.

According to Bernoulli's theorem, if the total energy and elevation z remain constant, an increase in velocity will have a corresponding diminution in pressure. This reduction in pressure is favourable to the flow and this

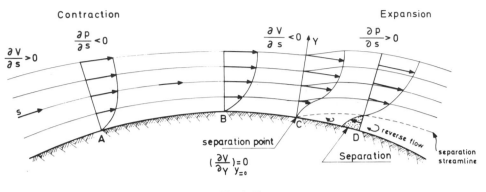

Fig. 2.15.

occurs between A and B, a zone in which owing to the *contraction* caused by the solid boundary, the velocity increases:

$$\frac{\partial V}{\partial s} < 0 \; ; \qquad \frac{\partial p}{\partial s} > 0$$

Downstream of B the flow undergoes an *expansion*, the velocity decreases and the pressure increases.

At a certain location, the fluid at the wall is brought to rest (point C); downstream of this point, the velocities near the wall may be negative and the flow separates from the wall, promoting instability, eddy formation and large energy dissipation. This phenomenon is called *separation*; point C is the *separation point*.

We thus find that separation occurs when there is an *adverse pressure gradient*; this happens whenever we have a decelerating motion.

Euler's number, $E_u = \Delta p/\rho V^2$, is the non-dimensional parameter that characterizes the pressure variations (see Section 1.18) and, consequently, the separation phenomena.

Fig. 2.16 gives some examples of flows with and without separation.

Separation always results in additional head losses which should be avoided. Use can therefore be made of the following means, which are developed in subsequent chapters: in submerged bodies, giving them aerodynamic shapes; at diverging sections, by not exceeding appropriate angles; in spillways, using profiles which are not too steep. It is also possible to draw off the boundary layer in the separation zone, by opening slots in the wall (Fig. 4.3a) or injecting fluid into the boundary layer (Fig. 4.3b).

2.28   Forces on immersed bodies. Stokes formula
(a)   *General formula* — A fluid flowing past an immersed body exerts two types of forces: *surface drag*, due to friction of the fluid against the body surface; *form drag or pressure drag* due to separation of the boundary layer

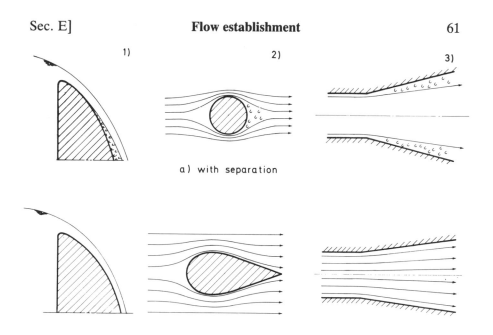

a) with separation

b) without separation

Fig. 2.16.

and establishment of a pressure defect in the wake. These two types of forces are usually grouped into one, the total force being given by:

$$F = C\gamma A \frac{V^2}{2g}$$ (2.40)

in which:

F — total drag force
C — drag coefficient depending on the shape, dimensions, and surface roughness of the body and the Reynolds number of the flow
A — maximum cross-section of the body, in a plane perpendicular to V
V — velocity of the flow.

Table 13 gives the values of C in some cases.

(b) *Fall of a sphere. Stokes formula* — In the case of a sphere that falls into a viscous fluid at rest, the Reynolds number characteristic of the motion is $\mathbf{R}_e = Ud/\upsilon$, in which d is the diameter of the sphere.

For the value $\mathbf{R}_e < 1.0$, the coefficient, C, of formula (2.40) is $C = 24/\mathbf{R}_e$. In this condition:

$$F = C\gamma A \frac{V^2}{2g} = 24\frac{\upsilon}{Vd}\gamma\frac{\pi d^2}{4}\frac{V^2}{2g} = 3\rho\upsilon\pi Vd \ .$$ (2.40a)

The weight, **W**, of the sphere is counteracted by Archimedes' buoyancy, **I** = **P** and the movement is counteracted by the resistance, **F**. The difference between weight and buoyancy is called apparent weight. When these forces are balanced, the sphere reaches a terminal velocity.

If this terminal velocity is reached in laminar flow, it is easy to deduce its value. Taking γ and γ_1 as the specific weights of the liquid and the sphere, we have:

$$\frac{\pi d^3}{6}(\gamma_1 - \gamma) = 3\rho\upsilon\pi \, Vd \tag{2.41}$$

whence, for $\mathbf{R_e} < 1.0$

$$V = \frac{d^2 g}{18\upsilon}\left(\frac{\gamma_1 - \gamma}{\gamma}\right) \tag{2.41a}$$

This formula can also be used to determine the viscosity of the fluid and is known as *Stokes formula*.

F.   IRROTATIONAL FLOWS

2.29   General considerations

In Section 2.9 it was explained that flow is irrotational when rot **V** = 0. Taking into account the components of rot **V**, we then have:

$$\frac{\partial V_z}{\partial y} = \frac{\partial V_y}{\partial z} \; ; \qquad \frac{\partial V_x}{\partial z} = \frac{\partial V_z}{\partial x} \; ; \qquad \frac{\partial V_y}{\partial x} = \frac{\partial V_x}{\partial y} \tag{2.42}$$

These ratios show that[†]

$$V_x = \frac{\partial \phi}{\partial x}; \; V_y = \frac{\partial \phi}{\partial y}; \; V_z = \frac{\partial \phi}{\partial z} \tag{2.43}$$

that is to say, that there is a function ϕ known as *velocity potential function* so that:

$$V = -\operatorname{grad} \phi \tag{2.44}$$

Bearing in mind the equation of continuity (2.8) we now have div grad $\phi = 0$ or $\nabla^2\phi = 0$; that is to say, the function ϕ is a zero Laplacian function or a harmonic function.

[†]   In effect, from 2.43 we have $\partial V_x/\partial y = \partial^2\phi/\partial x\partial y$ and $\partial V_y/\partial x = \partial^2\phi/\partial y\partial x$ whence $\partial V_x/\partial y = \partial V_y/\partial x$ which satisfies (2.42).

As has been explained (Section 2.9), a real (viscous) liquid can never flow in irrotational motion; viscosity and irrotationality are incompatible.

However, experience has shown that in certain cases the motion of real liquids is very similar to irrotational flow. Examples of this are: flow starting from rest (flow through an orifice in the wall of a reservoir, flow near to the crest of a spillway); rapidly accelerated motions (convergent flows) and, generally speaking all those in which the effects of viscosity can be disregarded and where, consequently, the energy loss is very small, namely flows in porous media.

Equipotential surface is the name given to surfaces which correspond to constant values of the velocity potential function, i.e. $\phi = constant$.

The velocity at a point is directed along to the normal to the equipotential surfaces. The streamlines, tangents to the velocity vector, are therefore orthogonal to the equipotential surfaces.

If a stream surface is replaced by a solid wall, the form of flow does not change. This property, also known as the *solidification principle*, allows the forms of flow with fixed contours to be studied, as will be seen in due course. This is, in fact, the most interesting field of irrotational flow, especially in the case of short sections, in which the effect of form is far more important than the effect of friction, which can therefore be disregarded.

2.30   Flows in space

(a) *Rectilinear flow* — Consider a uniform flow in space directed along the axis OX, with constant velocity equal to a (Fig. 2.17a). The velocity

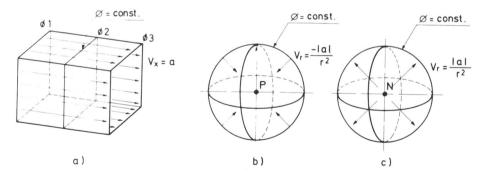

Fig. 2.17.

potential function is $\phi = ax + c$. In effect:

$$V_x = \frac{\partial\phi}{\partial x} = a \;; \qquad V_y = \frac{\partial\phi}{\partial y} = 0 \;; \qquad V_z = \frac{\partial\phi}{\partial z} = 0 \quad (2.45)$$

On the equipotential surfaces $\phi = $ const., that is to say, $ax + c = $ const., which represent planes normal to the direction of the flow.

(b) *Source and sink* — Now consider a velocity potential function, in spherical coordinates (θ, β, r) — (longitude, latitude, distance from the centre†) defined by $\phi = a/r + b$.

$$V_r = \frac{\partial \phi}{\partial r} = -\frac{a}{r^2} ; \qquad \frac{\partial \phi}{\partial \theta} = 0 ; \qquad \frac{\partial \phi}{\partial \beta} = 0 \qquad (2.46)$$

Thus the flow will only have radial velocities (spherical motion) which will be greater the nearer they are to the centre, i.e. there is a point in space from where the flow diverges or on which it converges; if $a < 0$, this will be a *sink* (Fig. 2.17b); if $a > 0$, it will be a *source* (Fig. 2.17c). For $r = 0$, the velocity would theoretically be infinite, but this does not occur since it is impossible to realise $r = 0$. The source and sink therefore represent points of singularity where the equation of continuity is not valid. The equipotentials are concentric spheres with radius a/r.

The flowrate Q that passes over each surface is:

$$Q = 4\pi r^2 \frac{|a|}{r^2} = 4\pi |a| \qquad (2.47)$$

whence:

$$|a| = \frac{Q}{4\pi} \qquad \text{and} \qquad \phi = -\frac{Q}{4\pi r} + b \qquad (2.48)$$

The flow of a source in space represents very well what happens when a small orifice is opened in the wall of a large reservoir. There will be hemispheres centred at the orifice, as equipotential surfaces; the streamlines are radial and are directed to the orifice.

(c) *Dipole or doublet* — Consider a source N and a sink P, at a distance b from one another. The coordinates are taken to be cylindrical,‡ with the line joining the centre of the source and that of the sink (Fig. 2.18a) as axis, OZ, and the middle of the segment NP as the origin of the

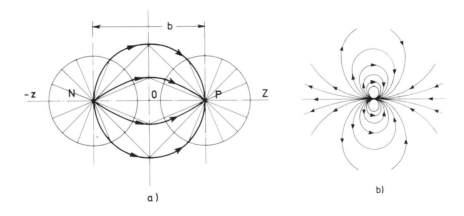

a)

b)

Fig. 2.18.

coordinates.

Should the flowrate of the source be equal to that of the sink, the equipotential surfaces, ϕ, are given by the sum of the equipotentials ϕ_1 of the source and ϕ_2 of the sink:

† The transform from spherical to Cartesian coordinates is:

$$x = r\cos\theta \sin\beta; \ y = r\sin\theta \sin\beta; \ z = r\cos\beta.$$

‡ $x = r\cos\theta; \ y = r\sin\theta; \ z = z.$

$$\phi = \phi_1 + \phi_2 = \frac{Q}{4\pi \sqrt{r^2 + \left(z - \frac{b}{2}\right)^2}} - \frac{Q}{4\pi \sqrt{r^2 + \left(z + \frac{b}{2}\right)^2}} \qquad (2.49)$$

The stream surfaces are represented in Fig. 2.18, which represents a cross-section through a plane passing through the axis OZ; they constitute surfaces of revolution around OZ.

If b tends to zero, increasing Q so that the product $Q.b$ is constant, a dipole is obtained in which $Q.b$ is the dipole moment (Fig. 2.18b).

The velocity potential function of the dipole is:

$$\phi = \frac{2bz}{(r^2 + z^2)^{3/2}} \qquad (2.50)$$

(d) *Flow round a semi-finite body* — The form of this flow is obtained by superimposing the flow of a source in the rectilinear flow which gives the stream surface shown in Fig. 2.19(a).

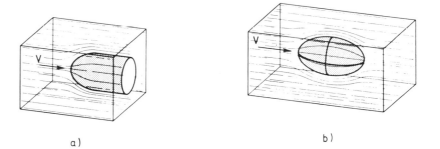

Fig. 2.19.

Analytically the procedure is to add ϕ_1 of the rectilinear flow to ϕ_2 of the flow of the source.

(e) *Flow round an ellipsoid* — Superposition of a source-sink together on a uniform flow makes it possible to obtain the flow round an ellipsoid (Fig. 2.19(b)).

(f) *Flow round a sphere* — Superposition of a dipole on a uniform flow gives the flow round a sphere, i.e. the ellipsoid of the previous example is transformed into a sphere.

2.31   Vortex motion

Vortex motion is a type of flow that requires special attention. Consider the velocity potential ϕ in cylindrical coordinates, so that:

$$\phi = \phi_1(r, z) + a\theta \qquad (2.51)$$

a being a constant and ϕ_1 a potential function independent of θ.

Accordingly, $\mathbf{V} = \text{grad}\,\phi$, so that, with V_t representing the tangential velocity, with V_r representing the velocity along a vector radius and V_z the velocity along the axis OZ:

$$V_t = \frac{1}{r}\frac{\partial\phi}{\partial\theta} = \frac{a}{r} \qquad \text{(independent of } \theta\text{)} \tag{2.52}$$

$$V_r = \frac{\partial\phi}{\partial r} = \frac{\partial\phi_1}{\partial r} \qquad \text{(independent of } \theta\text{)} \tag{2.52a}$$

$$V_z = \frac{\partial\phi}{\partial z} = \frac{\partial\phi_1}{\partial z} \qquad \text{(independent of } \theta\text{)} \tag{2.53}$$

that is to say, the components of the velocity in this flow are independent of θ, i.e. the flow has an axis of rotation that coincides with OZ; r represents the distance from the axis.

The value $\Gamma = 2\pi r V_t$ is known as the *circulation*, and is constant. In equation 2.51, $a = \Gamma/2\pi$.

The component $\mathbf{V}_m$ of the velocity is vectorially, along a meridian ($\theta = $ constant):

$$\mathbf{V}_m = \mathbf{V}_t + \mathbf{V}_z \tag{2.54}$$

The particles, which at a given moment are on the same parallel (with the same coordinates r and z, with any θ), during the whole movement remain on the same surface of revolution cross-section through, independent of a and V_t.

The meridian cross-section through this surface is the *stream function*: it is not a streamline since the streamlines of this flow have a spiral form round the axis OZ.

In Fig. 2.20, the point M that occupies a given position, M_0, will successively occupy the

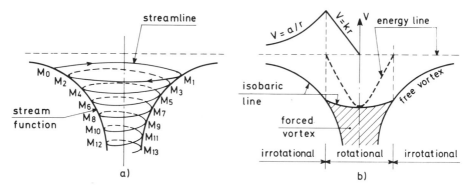

Fig. 2.20.

positions M_1, M_2 ... on a stream function.

The same will occur for another point placed on the same parallel.

Although the vortex is an irrotational flow, since the velocity field is derived from a potential, there is a single line, the axis, at which the velocity is theoretically infinite and at which the flow is rotational.

In all irrotational flows, the motion is conservative, i.e. the energy is constant at all points of the flow. The head corresponds to the undisturbed free surface.

Infinite velocity at the axis is impossible, physically speaking: in reality, the central part of the flow is occupied by a mass of water that rotates as a whole (rotational motion) with a distribution of velocities $V_t = kr$. In this flow the velocity is zero at the axis ($r = 0$) (Fig. 2.20(b)). This flow is known as a *forced vortex*, as distinct from the irrotational vortex, known as a *free vortex*.

In turbines and pumps, in order to prevent the formation of this forced (rotational) vortex, a solid of revolution is placed at the axis, the generatrices of the solid coinciding with a stream function of the irrotational flow; the aim is for the velocities thus obtained to be compatible with maintenance of the irrotational flow.

A vortex often appears in the outlet orifices of reservoirs (cf. what occurs in a wash basin or bathtub), in water intakes, pump sumps, etc. It may be asked why a vortex occurs in these conditions: the first fact shown by experiment is that the free vortex is formed with an anticlockwise rotational direction in the northern hemisphere; the opposite occurs in the southern hemisphere. The free vortex is, in fact, the result of the asymmetry of the friction forces due to Coriolis acceleration, which is due to the Earth's rotation.

Examples of movements in the atmosphere that may resemble vortices are whirlwinds, tornadoes, typhoons, etc.

2.32   Plane flows. Use of the complex variable functions

Plane flow, or two-dimensional flow, is the name given to the motion in which the phenomena are repeated on parallel planes, so that if it is known what is happening on one plane the entire flow is defined.

The simplifications that are the result of considering flows to be irrotational are particularly useful in plane flows:

— the equipotential surfaces are transformed into equipotential lines of equation $\phi = K_1$ (constant);
— it is possible to define a stream function, ψ, so that the equations of the streamlines are written $\psi = K_2$ (constant).

We then have:[†]

$$V_x = \frac{\partial \phi}{\partial x} = \frac{\partial \psi}{\partial y}$$

$$V_y = \frac{\partial \phi}{\partial y} = -\frac{\partial \psi}{\partial x} \qquad (2.55)$$

These expressions, of course, show that ϕ and ψ can resemble the real part and imaginary part of a function $w(z)$, of complex variable, $z = x + iy$, so that:

$$w(z) = \phi(x, y) + i\psi(x, y) \qquad (2.56)$$

Equations (2.55) are the analytic conditions[‡] of the function w, which is known as a complex potential function.

[†] Since the flow is irrotational, we have $\partial V_x / \partial y = \partial V_y / \partial x$; from the equation of continuity we have $\partial V_x / \partial x = -\partial V_y / \partial y$. The two conditions are met by (2.55).
[‡] It must be remembered that a function is termed analytic at a point when it has a single derivative at that point, i.e. when there is:

$$\lim_{\Delta z = 0} \frac{\Delta w}{\Delta z}$$

with a single given finite value, whatever the way in which z tends to zero.

A study of analytic functions is therefore very useful for determining the streamlines and equipotential lines.

In effect, by examining analytic functions of various types, different forms of plane flows can be described.

The derivate of w is related to the velocity V as follows:

$$\frac{dw}{dz} = \frac{\partial \phi}{\partial x} + i\frac{\partial \psi}{\partial x} = V_x - iV_y = Ve^{-i\theta} \tag{2.57}$$

in which V is the velocity and θ the angle that it makes with the axis OX.†

Examples:

(a) *Rectilinear flow field* — Consider the function

$$w = az \tag{2.58}$$

Thus:

$$w = \phi + i\psi = a(x + iy) \tag{2.58a}$$

$$\phi = ax \; ; \quad \psi = ay \tag{2.58b}$$

$$V_x = \frac{\partial \phi}{\partial x} = a \; ; \quad V_y = \frac{\partial \phi}{\partial y} = 0 \tag{2.58c}$$

This is a uniform flow, parallel to OX and with velocity $V = a$ (Fig. 2.21).

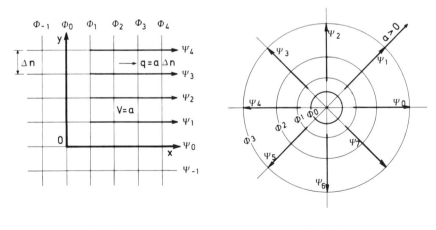

Fig. 2.21.   Fig. 2.22

† It must be remembered that $e^{i\theta} = \cos\theta + i\sin\theta$ whence $e^{-i\theta} = \cos\theta - i\sin\theta$. Between the Cartesian coordinates x, y, and the polar coordinates r and θ there are the following relationships: $r^2 = x^2 + y^2$; $\theta = \arctan y/x$; $x = r\cos\theta$; $y = r\sin\theta$. We also have $z = x + iy = re^{i\theta}$.

The equipotentials are rectilinear lines given by $\phi = ax$; the streamlines are also rectilinear lines defined by $\psi = ay$. These rectilinear lines are obtained by assigning discrete values to ϕ and ψ. The flowrate between two streamlines is $a \cdot \Delta n$, Δn being the distance between two adjacent streamlines, ψ_n and ψ_{n+1}. This motion coincides with the parallel flow in three dimensions.

(b) *Sink or source* — Consider the function

$$w = a \ln z \ . \tag{2.59}$$

Thus:

$$w = \phi + i\psi = a \ln r e^{i\theta} = a \ln r + ai\theta \tag{2.59a}$$

$$\phi = a \ln r \qquad \psi = a\theta \tag{2.59b}$$

Whence:

$$V_r = \frac{\partial \phi}{\partial r} = \frac{a}{r} \ ; \qquad V_t = \frac{1}{r}\frac{\partial \phi}{\partial \theta} = 0 \tag{2.59c}$$

The flow is radial and is represented graphically in Fig. 2.22 for positive a (source); if a is negative, the direction of the streamlines is contrary (sink). This flow, which is cylindrical, must not be confused with sinks or sources in space, which are spherical. The equipotential lines are such that $a \ln r = $ constant, i.e. they are circles; the streamlines are semirectilinear lines, passing through the origin.

The flowrate, per unit thickness of the layer, will be:

$$q = 2\pi r V_r = 2\pi a \ , \qquad \text{so that,} \qquad a = \frac{q}{2\pi} \tag{2.59d}$$

Between two consecutive streamlines, ψ_n and ψ_{n+1}, the flowrate will be q/n, n being the number of equal divisions of the circumference.

(c) *Source and sink of equal strength* — In the flowfield produced by a source and a sink of equal strength the total flowrate passes from one to the other and thus features by a family of streamlines originating at the source at ending in the sink.

The matter can be resolved analytically, but an example is given of its resolution by graphical means (Fig. 2.23).

Around the source and the sink, the circumference 2π is divided into n equal values; $\theta_1, \theta_2, \theta_3, \ldots, \theta_i$, are taken to be the n values corresponding to the streamlines of the source; $\theta'_1, \theta'_2, \theta'_3 \ldots \theta'_j \ldots$ are taken to be the n values corresponding to the streamlines of the sink, marked in the opposite direction to those of the source.

A streamline of the combined flow is obtained in the following way: the intersection of θ_i and θ'_j is found; this point is on a streamline whose index will be $k = i + j$. The set of points with the same k thus defines a streamline.

As with the three-dimensional flows, if the source coincides with the sink, a *plane dipole* is obtained.

(d) *Superposition of two sinks* — The graphical construction is made in Fig. 2.24 as indicated in the preceding example (the numbering of the streamlines, however, is done in the same direction). Drawing in the streamline corresponding to the symmetry axis, the solution represents a well next to an impermeable wall.

(e) *Superposition of a source on a rectilinear flow* — This is the case of the superposition of a radial flow and rectilinear flow, studied in examples (a) and (b). Since it is a question of superposing two potential flows, the equipotentials and streamlines are obtained by adding,

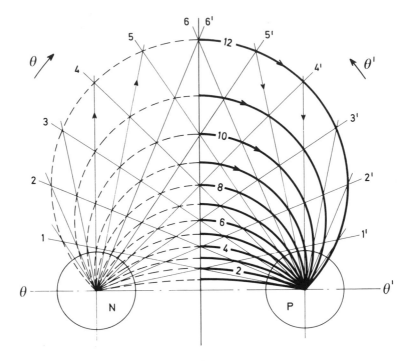

Fig. 2.23.

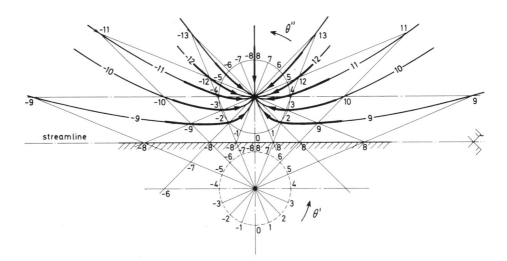

Fig. 2.24.

respectively, the equipotentials and streamlines of those two flows, which can be done graphically or analytically, giving the composition shown in Fig. 2.25.

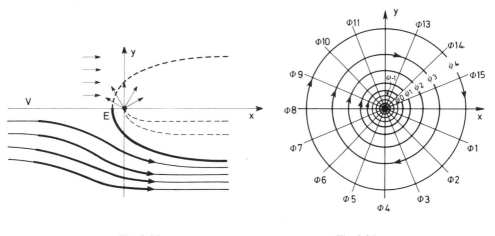

Fig. 2.25.   Fig. 2.26

Stagnation point, E, is the name given to the point at which the velocity of the source is equal and opposite to the velocity of the rectilinear flow, yielding a net velocity of zero for the combined flowfield. By solidifying the streamline that passes through this point, we obtain the flow around a semi-infinite cylindrical body. This is not to be confused with example of Fig. 2.19(a), in which the body has three dimensions, whereas here it is a cylinder (two dimensions).

(f) *Vortex* — Consider the function

$$w = ki \ln z \tag{2.60}$$

Thus:

$$w = \phi + i\psi = ki \ln re^{i\theta} = -k\theta + ik \ln r \tag{2.60a}$$

Whence:

$$\phi = -k\theta \qquad \psi = k \ln r \tag{2.60b}$$

and

$$V_r = \frac{\partial \phi}{\partial r} = 0 ; \qquad V_t = \frac{1}{r}\frac{\partial \phi}{\partial \theta} = -\frac{k}{r} \tag{2.60c}$$

that is to say, the equipotentials are radial and the streamlines are circumferential.

This is a vortex with constant circulation $\Gamma = 2\pi r V_t = -2\pi k$ (Fig. 2.26) and with centre at the origin of the axes. If k is negative, the direction of the circulation is contrary to that shown in the figure.

It can easily be seen that the expression $w = ik \ln (z - z_0)$ represents a vortex with intensity k, with centre at point z_0.

These plane vortices must not be confused with the three-dimensional vortex described in

Section 2.31. In fact, only the three-dimensional vortex exists in nature, and the plane vortex represents an approximation to it.

(g) *Rectilinear flow round a stationary cylinder* — Consider the function:

$$w = V(z + a^2/z) \tag{2.61}$$

Thus:

$$w = \phi + i\psi = V(x + iy) + \frac{Va^2(x - iy)}{x^2 + y^2} \tag{2.61a}$$

Whence:

$$\phi = Vx\left(1 + \frac{a^2}{x^2 + y^2}\right) \; ; \quad \psi = Vy\left(1 - \frac{a^2}{x^2 + y^2}\right) \tag{2.61b}$$

The graphic representation for $|z| > a$ (Fig. 2.27) indicates that we have the flow with

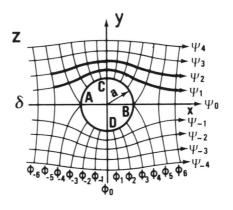

Fig. 2.27.

velocity V round a cylinder of radius a.

(h) *Rectilinear flow round a rotating cylinder* — Consider the equation:

$$w = V\left(z + \frac{a^2}{z}\right) + \frac{i\Gamma}{2} \ln \frac{z}{a} \tag{2.62}$$

This is the case of superposition of a vortex of circulation Γ (Fig. 2.28) on a flow of velocity V around a cylinder with radius a. In other words, it is the potential flow around a cylinder with a rotation velocity of $\omega = \Gamma/2\pi a^2$.

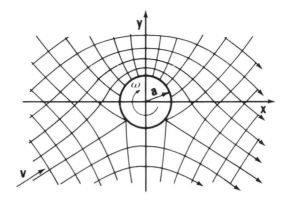

Fig. 2.28.

(j) *Changes of direction* — Consider the function

$$w = az^{\pi/\alpha} \tag{2.63}$$

Thus:

$$w = \phi + i\psi = ar^{\pi/\alpha}e^{i\pi\theta/\alpha} = ar^{\pi/\alpha}\left(\cos\frac{\pi\theta}{\alpha} + i\sin\frac{\pi\theta}{\alpha}\right) \tag{2.63a}$$

Whence:

$$\phi = ar^{\pi/\alpha}\cos\frac{\pi\theta}{\alpha} \; ; \qquad \psi = ar^{\pi/\alpha}\sin\frac{\pi\theta}{\alpha} \tag{2.63b}$$

The shape of the equipotentials, ϕ, and the streamlines, ψ, depends on the angle α. Fig. 2.29 shows the flowfields for several values of α.
If $\alpha = \pi/2$, we have

$$\phi = ar^2\cos 2\theta = a(x^2 - y^2) \; ; \qquad \psi = ar^2\sin 2\theta = 2axy \; . \tag{2.63c}$$

ϕ and ψ, constants, correspond to hyperbolas and parabolas that have the graphic representation in greater detail in Fig. 2.30. This is, therefore, the potential flow at a right-angled corner.
The components of the velocity are:

$$V_x = \frac{\partial\phi}{\partial x} = 2ax \; ; \qquad V_y = -\frac{\partial\psi}{\partial x} = -2ay \tag{2.63d}$$

and its modulus is:

$$V = \sqrt{V_x^2 + V_y^2} = 2a\sqrt{x^2 + y^2} = 2ar \tag{2.63e}$$

One of the streamlines can be replaced by a solid contour (wall), which makes it possible to determine a hydraulically correct form for the corner.

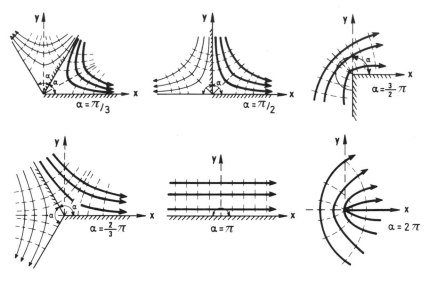

Fig. 2.29.

(j) *Flow entrance into a channel* — Consider the equation

$$z = w + e^w \qquad (2.64)$$

Thus:

$$
\begin{aligned}
z = x + iy &= \phi + i\psi + e^{(\phi + i\psi)} \\
&= \phi + i\psi + e^{\phi}(\cos\psi + i\sin\psi) \qquad (2.64a) \\
&= \phi + e^{\phi}\cos\psi + i(\psi + e^{\phi}\sin\psi)
\end{aligned}
$$

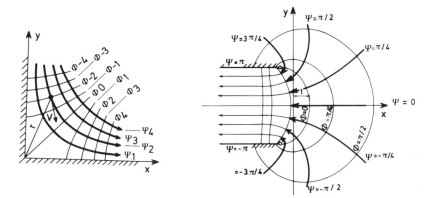

Fig. 2.30.   Fig. 2.31

Whence:

$$x = \phi + e^\phi \cos \psi \; ; \qquad y = \psi + e^\phi \sin \psi \qquad\qquad (2.64b)$$

Give constant values to ϕ and ψ; thus we have values of x and y that represent the coordinates of the points of intersection of ϕ and ψ (Fig. 2.31). This is then the potential flow in the entrance to a channel with parallel walls.

The following table exemplifies the way of obtaining, by points, the curves ϕ and ψ. For example, ϕ_1 and ψ_0 pass through point $x = 1$ and $y = 0$; ϕ_1 and ψ_1 pass through point $x = 0.71$ and $y = 1.49$.

$\psi_k \rightarrow$ $\phi_i \downarrow$	$\psi_0 = 0$		$\psi_1 = \pi/4$		$\psi_2 = \pi/2$		$\psi_3 = 3\pi/4$		$\psi_4 = \pi$	
	x	y	x	y	x	y	x	y	x	y
$\phi_1 = 0$	1	0	0.71	1.49	0	2.57	-0.71	3.06	1	3.14
$\phi_2 = \pi/4$	2.98	0	2.98	2.34	0.79	3.76	-0.77	3.91	-1.41	3.14
$\phi_3 = \pi/2$	6.38	0	4.97	4.19	1.57	6.38	-1.83	5.76	-3.24	3.14
$\phi_4 = 3\pi/4$	12.91	0	9.82	8.25	2.36	12.12	-5.10	9.82	-8.19	3.14

2.33   Plane flows. Use of conformal representation

It has been seen how an analytic function of a complex variable represents the equipotential lines and streamlines of a plane flow.

$$w(z) = f(x + iy) = \phi(x, y) + i\psi(x, y) \qquad\qquad (2.65)$$

in which the equipotentials, $\phi = $ const., and the streamlines, $\psi = $ const., are orthogonal to one another.

Now suppose that to each point, $z = x + iy$, on the plane xy, is given another corresponding point, $Z = X + iY$ on another plane XY, by means of functions, $Z = Z(z) = A + iB$, that is to say, $X = A(x, y)$ and $Y = B(x, y)$.

A family of curves of (x, y) will have a corresponding family of curves of (X, Y). The transformation will be conformal, if the angles of the curves are kept constant between one another. Thus, if on one plane (x, y) two curves cut one another orthogonally, their transforms on the plane (X, Y) will also cut one another orthogonally. One scheme of equipotential lines and streamlines will therefore have another corresponding scheme of equipotential lines and streamlines.

The condition for the transformation being conformal is that the function, $Z(z) = A + iB$, shall be an analytic function, i.e. that:

$$\frac{\partial A}{\partial x} = \frac{\partial B}{\partial y} \; ; \qquad \frac{\partial A}{\partial y} = -\frac{\partial B}{\partial x} \qquad\qquad (2.66)$$

2.34 Joukowsky's transformation

Joukowsky's transformation is one of the most useful conformal transformations for fluid flows. Its analytic expression is:

$$Z = b\left(z + \frac{a^2}{z}\right) = b\left(r e^{i\theta} + \frac{a^2}{r} \cdot e^{-i\theta}\right) \qquad\qquad (2.67)$$

Consider a point z on the plane xy (Fig. 2.32(a)), of modulus r and

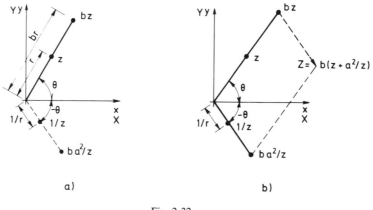

Fig. 2.32.

argument θ; by maintaining the argument θ and marking a modulus equal to br, the point bz is obtained.

The operation $1/z = zr^{-1}e^{-i\theta}$ gives a point whose modulus is $1/r$ and whose argument is $-\theta$. By multiplying modulus $1/z$ by ba^2, the point ba^2/z is obtained.

From the vectorial sum of bz and ba^2/z, we get $Z = b(z + a^2/z)$, (Fig. 2.32(b)).

Example:

A radial flow to a circular opening with radius a (part of the flowfield of the sink) is transformed into the flow to a rectilinear slot of length $2a$ (Fig. 2.33). In effect we then have:

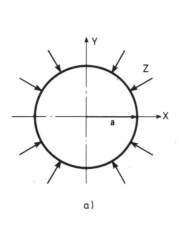

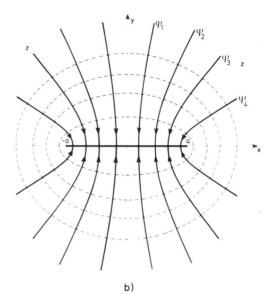

Fig. 2.33.

$$Z = 1/2(z + a^2/z) \tag{2.68}$$

which gives

$$z = Z + \sqrt{Z^2 - a^2} \tag{2.68a}$$

The function of the sink (Eq. 2.59 and 2.59d) then being transformed into:

$$W(Z) = \phi + i\psi = \frac{q}{2\pi}\ln z = \frac{q}{2\pi}\ln[Z + \sqrt{Z^2 - a^2}] =$$

$$= \frac{q}{2\pi}\left[\operatorname{arc ch}\left(\frac{Z}{a}\right) + \ln a\right] \tag{2.69}$$

By separating the real and imaginary parts, we find that the equipotentials are homofocal ellipses with foci $x = \pm a$ and the streamlines are hyperbolas of foci $x = \pm a$.

2.35   Plane flow. Graphical methods

Given below are some methods for plotting equipotentials and streamlines without resorting to complex variable functions, which is not always possible.

(a) *Small squares method* — Consider the functions ϕ and ψ, taking values in arithmetic progression, increasing by $\delta\phi = \delta\psi$. Since the lines cut one another orthogonally, provided that $\delta\phi = \delta\psi$ are small, their intersection will produce small squares or forms that are close to small squares.

For ease of plotting, a circle inscribed in these small squares must be a tangent to the middle of each side (Fig. 2.34a).

If $\delta\psi$ is the distance between two streamlines, the flowrate between them will be $\delta q = V\delta\psi$.

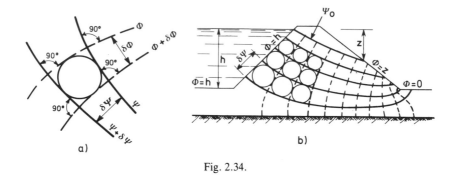

Fig. 2.34.

Fig. 2.34b shows how the method is applied to flow through a dam. Considering $\phi = h$ on the upstream face and $\phi = 0$ downstream, calculation is made of the line of saturation corresponding to a streamline, ψ_0, on which ϕ takes the values $\phi = z$, where z is the distance from the head plane. Tracing of the isometric network is begun by plotting the streamlines and equipotentials so that they comply with the condition of Fig. 2.34. Only by practice can sensitivity be acquired for correct tracing of the isometric network.

(b) *Prasil's method* — This method is applicable when a streamline ψ_0 = constant is known, as well as the values taken by ϕ on this line or reciprocally.

The procedure is the same: the known streamline, ψ_0, is graduated in intervals to make them correspond to the same variation, $\delta\phi$, thus obtaining a series of values in arithmetic progression: $\phi_1, \phi_2 \ldots$ in a proportion $\delta\phi$. M_1 and M_2 are taken as the point of ψ_0, corresponding to ϕ_1 and ϕ_2. From M_1 and M_2 lines are traced at 45° to the normal and their intersection gives the point P_1 that belongs to the new streamlines $\psi_1 = \psi_0 + \delta\psi$ (Fig. 2.35(a)).

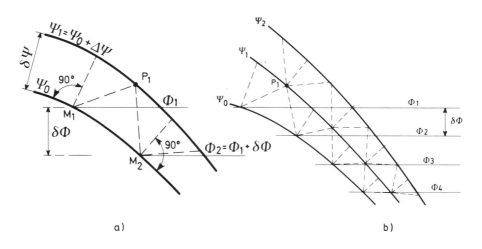

Fig. 2.35.

Fig. 2.35(b) shows how the method is applied to flow over a spillway. The known streamline, ψ_0, corresponds to the crest of the spillway; the equipotentials are horizontal and equal distances $\delta\phi$ apart.

BIBLIOGRAPHY

[1] Bakmeteff, B. A. *Mécanique de l'Ecoulement Turbulent des Fluides.* Dunod, Paris, 1941.

[2] De Marchi, G. *Idraulica.* Ulrico Hospli, Milano, 1974.

[3] Prandtl, L. *Guide à travers la Mechanique des Fluides.* Dunod, Paris, 1952.

[4] Rouse, H. *Elementary Mechanics of Fluids.* John Wiley & Sons, New York, 1946.

[5] Streeter, V. L. *Fluid Dynamics.* McGraw-Hill, New York, 1948.

[6] Schlichting, H. *Boundary Layer Theory.* McGraw-Hill, New York, 1960.

[7] Comolet, R. *Mécanique Expérimentale des Fluides.* Masson, Paris, 1963.

[8] Carlier, M. *Hydraulique Générale et Appliquée.* Eyrolles, Paris, 1972.

[9] Brand, L. *Vector and Tensor Analysis.* John Wiley & Sons, New York, 1947.

3

Hydrostatics

3.1 Fundamental equations. Fluid at rest subject only to the action of gravity

In a liquid at rest or in which there is no relative motion between the elements constituting it, no shear stress can exist, and thus it behaves like a perfect liquid.

Accordingly, equation 2.11 is written as follows:

$$\rho\left(\mathbf{F} - \frac{\mathrm{d}\mathbf{V}}{\mathrm{d}t}\right) = grad\ p \tag{3.1}$$

In the case of a fluid at rest subject only to the action of gravity, the external forces per unit mass, $\mathbf{F}$, are reduced to the weight of the liquid per unit mass, $\mathbf{g}$, and are therefore derived from a potential. It has been seen (Section 2.12) that in this case equation 3.1 was written:

$$grad\left(z + \frac{p}{\gamma}\right) = 0 \tag{3.1a}$$

whence:

$$z + \frac{p}{\gamma} = \text{const.} \tag{3.2}$$

This expression is known as the *fundamental equation of hydrostatics*.

3.2 Distribution of pressures

In the case of incompressible fluids, the characteristic equation is: $\rho =$ constant.

Also $\gamma = \rho g =$ constant, since for all practical differences in height the variation of g is negligible.

In this case it is easy to demonstrate that:

(1) The isobaric surfaces, i.e. the surfaces of equal pressure, are horizontal planes. In effect, for $p/\gamma =$ constant, we obtain (3.2):

$$z = \text{const.} \tag{3.3}$$

(2) The difference in pressure, $p_A - p_B$, between any two points in a liquid at rest depends only on the difference in elevations between the points and on the specific weight of the liquid (Fig. 3.1), so that:

$$p_A - p_B = \gamma(z_B - z_A) = \gamma\Delta z_{A,B} \tag{3.4}$$

(3) The separation surfaces of immiscible liquids are horizontal planes, since the isobaric surfaces are surfaces of equal relative density.

(4) In a liquid in equilibrium, variations in pressure are integrally transmitted to all points of the liquid mass. This statement constitutes *Pascal's pinciple*†. In effect, if at any point A the pressure undergoes a variation Δp_A, the corresponding variation at a point B will be such that:

$$(p_A + \Delta p_A) - (p_B + \Delta p_B) = \gamma(z_B - z_A) \tag{3.5}$$

A comparison with 3.4 shows that: $\Delta p_A = \Delta p_B$.
Taking the free liquid surface ($p = 0$) as horizontal datum‡, we get:

$$p = \gamma h \tag{3.6}$$

in which h is the distance from the free liquid surface to the point considered. The distribution of pressures is linear (Fig. 3.2). It can also be said that

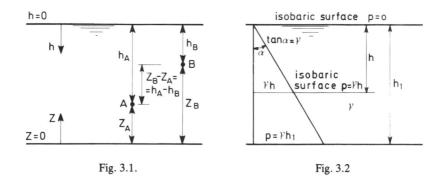

Fig. 3.1.   Fig. 3.2

the pressure, measured in height of liquid, is equal to h.
In the case of immiscible liquids that are superimposed, with different

† Pascal, B. (1623–1662).
‡ These are gauge pressures, i.e. without taking account of atmospheric pressure (see Section 1.10).

specific weights, the distribution of pressures along a vertical is shown graphically in Fig. 3.3. Pressure p at a point which is at depth h between h_2

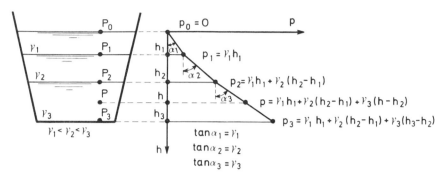

Fig. 3.3.

and h_3, for example, is:

$$p = \gamma_1 h_1 + \gamma_2(h_2 - h_1) + \gamma_3(h - h_2) \tag{3.6a}$$

3.3 Forces on submerged plane surfaces

In the case of any horizontal, vertical or sloping *plane surface* (Fig. 3.4),

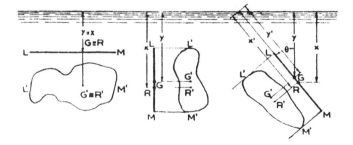

Fig. 3.4.

there is on each face a resultant of the elementary pressures that is known as *resultant force*, normal to the surface, whose value is:

$$I = \gamma y A \tag{3.7}$$

in which A is the area of the surface and y is the depth of its centroid below the free surface. This equation indicates that the magnitude of the resultant force on (one side of) any submerged plane may be calculated by multiplying the area, A, by the pressure at its centroid γy.

In fact, we have:

$$I = \iint_A p dA = \iint_A \gamma h dA = \gamma \iint_A h dA \qquad (3.7a)$$

The depth, y, of the centroid below the free surface is such that:

$$\iint_A h dA = yA \qquad (3.8)$$

This equation represents the first moment of the area, A, about the intersection of its plane with that of the free surface.

If the surface is submerged, the total force exerted on one of the faces is equal and opposite to that exerted on the other face, and the resultant force is zero. If the surface has one face in contact with the liquid and another in contact with the atmosphere, the result is given by (3.7).

In the case of a *surface on a sloping plane* it can also be written as follows:

$$I = \gamma y' A \cos \theta \qquad (3.9)$$

in which y' is the distance from the centroid to the intersection of the sloping plane with the free surface and θ is the angle of the plane with the vertical.

The distance, x, to the free surface, measured vertically, from point R of application of the resultant, I, also called *centre of pressure*, is[†]:

(a) *Horizontal plane*: the point of application, R, coincides with the centroid, G, so that $x = y$.

(b) *Vertical plane*:

$$x = y + \frac{K^2}{y} \qquad (3.10)$$

(c) *Sloping plane*: $x = y + K^2 \cos^2 \theta / y$ or $x' = y' + K^2/y'$

K is the radius of gyration of the surface considered in relation to a horizontal axis passing through the centroid.

† Determination of the point of application of parallel forces is a problem which is dealt with in statics, so that it has not been considered necessary to demonstrate it here.

The preceding expressions show that the centre of pressure either coincides with the centroid or is lower.

Table 14 gives the values of the area, A, of the distance, V, that defines the position of the centroid, G, and of the square of radius of gyration, K^2, of some simple geometrical figures.

3.4 Examples

(1) Determine the pressure, p, and the resultant force, I, at the bottom of a vessel whose area is $A = 0.5$ m^2 and height 2 m, in the following cases: (a) the vessel is completely full of water; (b) the vessel has water up to 2/3 of its height and olive oil ($\rho = 800$ kg/m^3) in the remaining third.

Solution:

(a) $p = \gamma h = \rho g h = 1000 \times 9.8 \times 2 = 19\ 600$ N/m$^2 = 19.6$ k Pa $= 2$m (water column)
        $I = \rho g h A = 1000 \times 9.8 \times 2 \times 0.5 = 0.5 = 9800$ N $= 9.8$ kN.

(b) $p = \gamma_1 h + \gamma_2 (h_2 - h_1) = 800 \times 9.8 \times 2 \times 1/3 + 1000 \times 9.8 \times 2 \times 2/3 =$
                $18\ 293$ N/m$^2 = 18.293$ kPa $= 1.83$ m (water column)
        $I = 18\ 293 \times 0.5 = 9147$ N $= 9.147$ kN.

(2) A square plane surface with a side of 0.20 m is placed on the wall of a reservoir, with a slope of 45°. The depth of the centroid is $y = 1.5$ m. Determine I and the depth of its centre of pressure.

Solution:

$$I = 1000 \times 9.8 \times 1.5 \times 0.04 = 600 \text{ N}$$

Table 14 shows that $K^2 = h^2/12$, whence

$$x = 1.5 + \frac{0.2^2}{12} \times \left(\frac{\sqrt{2}}{2}\right)^2 \times \frac{1}{1.5} = 1.501 \text{ m}.$$

3.5 Resultant force on rectangular surfaces with two horizontal sides

In the case of vertical rectangular surfaces with two horizontal sides, the distribution of pressures can be graphically represented, as is shown in Fig. 3.5. The resultant force per unit width, is:

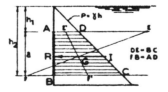

Fig. 3.5.

$$I = \gamma \frac{h_2^2 - h_1^2}{2} = \gamma a \frac{h_2 + h_1}{2} \tag{3.11}$$

which is equivalent to the area of the trapezium $A\,B\,C\,D$. When $h_1 = 0$ we have:

$$I = \gamma \frac{h_2^2}{2} \tag{3.11a}$$

The resultant force passes through the centroid, G, of the trapezium, which has been determined graphically in the same figure†. Its centre of pressure is at R.

If the rectangular plane surface has a slope of θ in relation to the vertical, the resultant force, I, continues to be represented by the area of the trapezium $ABCD$ and passes through its centroid (Fig. 3.6).

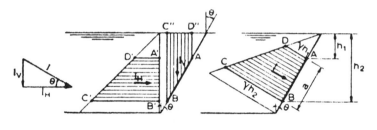

Fig. 3.6.

The horizontal component, I_H, is represented by the trapezium $A'\,B'\,C'\,D'$ and the vertical component, I_V, by the trapezium $ABC''D''$ (Fig. 3.6). We then have:

$$I = \gamma \frac{h_2^2 - h_1^2}{2} = \gamma a \frac{h_1 + h_2}{2} \tag{3.11b}$$

$$I_V = I \sin \theta; \quad I_H = I \cos \theta; \quad I = \sqrt{I_V^2 + I_H^2} \tag{3.12}$$

3.6 Buoyancy in immersed bodies: Archimedes' principle

If Euler's theorem (see Section 2.22) is applied to a volume, e, limited by a surface, A, inside a liquid at rest, the resultant, Π, of the pressures on that surface, or the buoyancy, I, will be given by:

---

† Determine the middles E' of AD and F' of BC; join E' to F'. Extend side AD by a length $DE = BC$; extend side CB by a length $BF = AD$; join E to F. The intersection of EF and $E'F'$ determines the centroid G.

$$\mathbf{I} = -\mathbf{G} \tag{3.13}$$

This equation represents *Archimedes' principle*[†], which consists of two parts, and may be stated as follows: *a body of volume, e, immersed in a fluid at rest, experiences a buoyancy force equal to the weight of fluid displaced by the body; consequently a floating body displaces a weight of fluid equal in magnitude to its own weight.*

3.7 Forces on submerged curved surfaces

In the case of any submerged curved surface enclosing a volume, e, the resultant of the pressure is determined, as we have seen, by direct application of Archimedes' principle.

Should the surface not be closed, i.e. not limit a volume, there will not in general be only a resultant force, but rather a *resultant force* and a *resultant moment*. From a practical point of view, the resultants are determined by computation of their vertical and horizontal components, which in general are not coplanar, and give rise, as has already been said, to a resultant force and a resultant moment.

Also in accordance with Euler's theorem, it may easily be concluded that the sum of the *vertically* directed components of the resultant on a surface A (see Fig. 3.7) is equal to the weight of the liquid contained in the volume limited by the free surface, by surface A and by the cylindrical surface with vertical generatrices that is supported on the contour c that limits surface A. If e is this volume, we then have:

$$I_V = \gamma e \tag{3.14}$$

The sum of the components in a *horizontal* direction H, of the forces on surface area A, are obtained as follows: consider the cylinder with horizontal generatrix of direction H that is supported on the contour, c, of A; consider a straight section, A_H, of that cylinder. If y is depth of the centroid, G, below to the free surface, we thus have:

$$I_H = \gamma y A_H \tag{3.15}$$

Consider, on the horizontal plane, two axes, x and y, taking z to be the vertical axis so that the three axes define a tri-rectangular trihedron. The components I_x, I_y and I_z of the force I, resulting from the pressures are determined as has previously been explained. Taking (y_1, z_1), (x_2, z_2) and (x_3, y_3) as the coordinates defining the lines of action I_x, I_y, I_z, respectively, the resultant moment, B, can be broken down into three components, B_x, B_y and B_z, with axes parallel to the coordinate axes and in such a way that

---

† Archimedes (287–212 B.C.).

$B_x = I_y\, z_2 - I_z\, y_3;\ B_y = I_z\, x_3 - I_x z_1\ ;\ B_z = I_x\, y_1 - I_y\, x_2$. We shall also have $I = \sqrt{I_x^2 + I_y^2 + I_z^2};\ B = \sqrt{B_x^2 + B_y^2 + B_z^2}$.

The resultant force, I, and the resultant moment, B between them form an angle, ψ, so that:

$$\cos\psi = \frac{I_x B_x + I_y B_y + I_z B_z}{IB} \tag{3.16}$$

In the case of spherical or cylindrical surfaces, there will be a resultant of the forces that passes through the centre of the sphere or through the axis of revolution of the cylinder. In the case of a cylinder of revolution with d diameter and unit length, subject to internal or external uniform pressures, p (Fig. 3.8), owing to its symmetry, the result of the pressures will be zero.

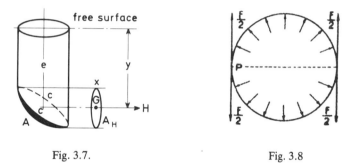

Fig. 3.7.   Fig. 3.8

The tensile or compressive force, F, that is exerted per unit length in a longitudinal cross section is $F = pd$.

3.8 Examples
(1) A hollow, watertight, cylindrical gate, with a length of 10 metres and a radius of 2 metres, based on a seat AB, forms a barrier in a channel with upstream and downstream conditions as shown in Fig. 3.9. The gate is supported at the ends of its axis on two piers and to these

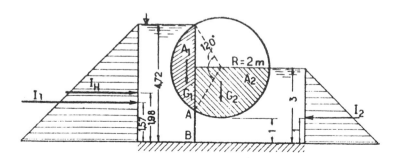

Fig. 3.9.

transmits a horizontal force for which it is intended to find the value. It is also wished to determine the minimium weight that the gate must have in order for it not to be lifted by the thrust of the water, under the conditions indicated, supposing that such a displacement were possible and disregarding the friction.

Solution:
The horizontal thrust on the piers is the horizontal component of the total force on the gate and this is, in turn, the difference between the two forces, I_1 and I_2.
Taking $\gamma = 10\ 000$ N/m³, we thus have:

$$I_1 = 10\ 000 \times \frac{4.72^2}{2} \times 10 = 1114 \text{ kN}; \quad I_2 = 10\ 000 \times \frac{3^2}{2} \times 10 = 450 \text{ kN}; \quad I_H =$$
$$= I_1 - I_2 = 664 \text{ kN};$$

The position of I_H is obtained considering moments about point B:

$$I_1 \times 1.57 - I_2 \times 1 = I_H d; \text{ thus } d = 1.96 \text{ m}.$$

In order for the gate not to float, it will have to have a minimum weight that is equal to the vertical component of the thrust. This component is equal to the weight of the liquid, $G = G_1 + G_2$, contained in the traced areas, A_1 and A_2, respectively:

$$G_1 = \gamma \frac{R^2}{2} L \left(\frac{2\pi}{3} - \sin \frac{\pi}{3} \right) \approx 241 \text{ kN}$$
$$G_2 = \frac{\gamma \pi R^2 I - G_1}{2} \approx 496 \text{ kN}; \quad G = G_1 + G_2 \approx 737 \text{ kN}$$

(2) Determine the vertical thrust on the hemisphere with radius r in Fig. 3.10.

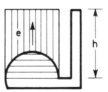

Fig. 3.10.

Solution:
The vertical thrust, directed upwards, is equivalent to the weight of the water contained in the volume e:

$$I_V \left(\pi r^2 h - \frac{1}{2} \times \frac{4}{3} \pi r^3 \right)$$

(3) Determine the resultant of the pressures acting on surface area A shown in Fig. 3.11.

Solution:
  In cases (a) and (b)

$$I_V = \gamma(e_1 - e_2); \quad I_H = 0; \quad I = I_V \tag{3.17}$$

The line of action of I passes through the centroid of the volume $e = l_1 - e_2$.

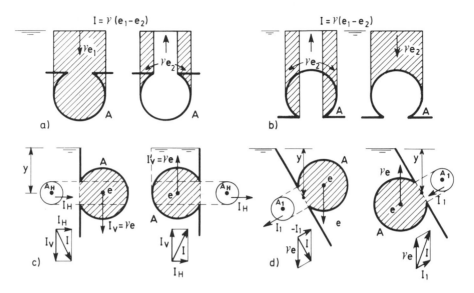

Fig. 3.11.

In case (c)

$$I_V = \gamma e; \quad I_H = \gamma y A_H; \quad I = \sqrt{I_H^2 + I_V^2} \tag{3.17a}$$

The line of action of I_H is at a distance x from the free surface, which is given by the formulae of Section 3.3.

The line of action of I_V passes through the centroid of volume e.

In general there will also be a resultant moment, which is easy to calculate as soon as the points of application of I_V and I_H are known.

In case (d)

It would be possible to use reasoning analogous to the foregoing, and determine I_V *and* I_H. It is also possible to reason as follows:

The total force on the whole surface $A + A_1$ in accordance with Archimedes' principle, is equal to γe, directed upwards if the fluid is outside the volume and downwards if it is not.

Vectorially, we shall also have:

$$\gamma e = \mathbf{I} + \mathbf{I}_1 \tag{3.17b}$$

in which I and I_1 are the forces on surfaces A and A_1, respectively.

Moreover, I_1 can be calculated in accordance with section 2.3.

$$I_1 = \gamma y A_1 \tag{3.17c}$$

Vectorially, we then have:

$$\mathbf{I} = \gamma e - \mathbf{I}_1 \tag{3.17d}$$

An analogous reasoning could also have been used in cases (a), (b) and (c).

3.9 Equilibrium in a liquid subject to fields of forces other than that of gravity

In these cases the fundamental equation 3.1 must be applied as illustrated below.

$$\rho \left(\mathbf{F} - \frac{d\mathbf{V}}{dt} \right) = \text{grad } p$$

Examples:
(1) Consider a container that moves with an acceleration $dV/dt = a$ along a sloping plane that makes an angle α with the horizontal , as shown in Fig. 3.12.

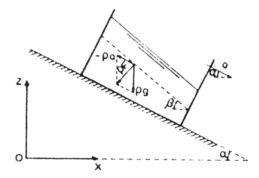

Fig. 3.12.

We then have:

$$\rho \mathbf{F} = -\rho \mathbf{a} + \rho \mathbf{g} \tag{3.18}$$

In cartesian coordinates the equilbrium will be:

$$-\rho a \cos \alpha = \frac{\partial p}{\partial x}$$
$$-\rho g + \rho a \sin \alpha = \frac{\partial p}{\partial z} \tag{3.18a}$$

Since:

$$dp = \frac{\partial p}{\partial x} dx + \frac{\partial p}{\partial z} dz \tag{3.19}$$

the isobaric surfaces $dp = 0$ will be written, in the differential form

$$-\rho a \cos\alpha \, dx - \rho(g - a \sin\alpha)dz = 0 \tag{3.19a}$$

and this when integrated gives

$$-\rho a \cos\alpha \ x - \rho(g - a \sin\alpha)z = \text{const.} \qquad (3.19\text{b})$$

Whence

$$z = -\frac{a \cos\alpha}{g - a \sin\alpha} x + \text{const.} \qquad (3.19\text{c})$$

Such surfaces make an angle β with the horizontal defined by:

$$\tan\beta = \frac{a \cos\alpha}{g - a \sin\alpha} \qquad (3.20)$$

If the motion is uniform, $a = 0$, $tan \ \beta = 0$ and $\beta = 0$, that is to say, the free surface is horizontal.

(2) Consider a cylindrical containder that moves with a uniform rotational movement, ω, around a vertical axis, z, as shown in Fig. 3.13.

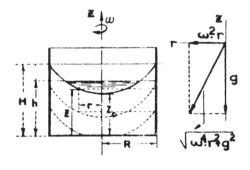

Fig. 3.13.

Each particle is subject to its own weight and to centrifugal force, so that we have:

$$\rho\mathbf{F} = \rho\mathbf{g} - \rho\omega^2\mathbf{r} \qquad (3.21)$$

Considering a system of axes, or and oz, connected to the container, and resolving the forces according to these axes, we have:

$$\begin{aligned} F_t &= -\rho\omega^2 r \\ F_z &= -\rho g \end{aligned} \qquad (3.21\text{a})$$

The equilibrium equations will be:

$$\rho\omega^2 r = \frac{\partial p}{\partial r}; \ -\rho g = \frac{\partial p}{\partial z} \qquad (3.22)$$

since

$$dp = \frac{\partial p}{\partial r}\, dr + \frac{\partial p}{\partial z}\, dz = \rho\omega^2 r dr - \rho g dz \tag{3.23}$$

the pressure at a point within the liquid will be

$$p = \frac{\rho\omega^2}{2}\, r^2 - \rho g z + \text{const.} \tag{3.23a}$$

and the isobaric surfaces

$$\frac{\rho\omega^2}{2}\, r^2 - xgz = \text{const.} \tag{3.24}$$

$$z = \frac{\omega^2}{2g}\, r^2 + \text{const.} \tag{3.24a}$$

which are paraboloids of revolution, having as an axis, the rotation axis.

BIBLIOGRAPHY

[1] Bouasse, H. *Hydrostatique*, Delagrave, 1923.
[2] Puppini, V. *Idraulica*, N. Zachelli, Bologna, 1947.
[3] Comolet, R. *Mécanique Experimentale des Fluides*, Vol. 1, Masson, Paris, 1961.
[4] Carlier, M. *Hydraulique Générale et Appliquée*, Eyrolles, Paris, 1972.
[5] Massey, B. S. *Mechanics of Fluids*, 2nd edition, Van Nostrand Reinhold, London, 1970.

4

Steady flow in pipes

A. FRICTION HEAD LOSSES

4.1 General equation

As derived by dimensional analysis (Section 1.19), head losses due to friction in a steady, fully developed flow in a rectilinear pipe are governed by an equation (1.19) of the type:

$$\Delta H = f \frac{L}{D} \frac{U^2}{2g} \tag{4.1}$$

in which:

— ΔH is the *head loss* or energy loss expressed in liquid column per unit weight discharged;
— f, which is known as the *friction factor*, is non-dimensional and is a function both of the Reynolds number, $\mathbf{R}_e$, and the *relative roughness*, ε/D, in which ε is the *absolute roughness* of the pipe (see Section 4.3);
— D, is the *hydraulic diameter* of the cross-section that in the case of circular pipes equals the geometrical diameter, d†;
— L is the length of the pipe.

The friction head loss in a pipe of unit length wil be represented by i, thus:

$$i = \frac{\Delta H}{L} = f \frac{1}{D} \frac{U^2}{2g} \tag{4.2}$$

Accordingly, i represents the *friction loss per unit weight discharged per unit length* and is non-dimensional.

---

† $D = 4R$, in which R is the hydraulic radius equal to A/P (see section 1.15).

The previous expression can also be put in the form:

$$i = aQ^2 \tag{4.3}$$

The dimensions of a are $L^{-6}T^2$. In the case of circular pipes, for D in metres and Q in m^3/s, we have:

$$a = f\frac{16}{2g\pi^2} \times \frac{1}{D^5} = 0.0826\, fD^{-5} \tag{4.4}$$

4.2 Reynolds number. Laminar and turbulent flow. Viscous sublayer

In order to determine the Reynolds number, it is necessary to know the viscosity of the fluid. For this purpose Table 8 can be consulted.

For $\mathbf{R}_e$ values of less than approximately 2000, the flow is laminar. For values above that, the flow generally ceases to be laminar (see Section 2.2)†.

The flow of water is seldom laminar. In practice, at 15°C the kinematic viscosity is $v = 1.15 \times 10^{-6}$ m^2/s; under these conditions, $\mathbf{R}_e$ will be greater than 2000 whenever:

$$U > 0.0023/D \tag{4.5}$$

For $D = 0.1$ m, we have $U > 0.023$ m/s; for $D = 1.0$ m, we have $U = 0.002$ m/s.

From laminar flow to turbulent flow a *transition region* occurs the characteristics of which vary according to the roughness of the wall. Points in this region may lie either on the laminar curve, or on one of the turbulent curves, or somewhere between.

In the case of *laminar flow* in pipes (see Section 2.13), the distribution of velocities follows a parabolic law and is zero at the walls, with a maximum in the centre where $V_M = 2\,U$. In *turbulent flow*, the transverse movements of the particles tend to make the velocities uniform. As order of magnitude, the maximum velocity, V_M, varies between 1.25 U and 1.10 U.

Nevertheless, as was explained in Section 2.26, even in turbulent flow there continues to exist by the wall a layer known as the *viscous sublayer*, in which the flow is laminar, with a parabolic distribution of velocities corresponding to laminar flow. The thickness of the viscous sublayer varies inversely with the Reynolds number: in hydraulic flows, as a rule this thickness is very small and can be expressed in tenths of a millimetre.

---

† In laboratory experiments in which every care is taken to avoid any disturbance, laminar flows have been obtained for $\mathbf{R}_e = 70\,000$, though in an unstable regime. In practical cases, for $\mathbf{R}_e$ between 2000 and 4000, the flow regime is very unstable.

4.3 Absolute and relative roughness

Absolute roughness, ε, is given by measurement of the roughnesses of the wall of the pipe. Relative roughness, ε/D, is the quotient of absolute roughness, ε, and the diameter D.

In practice, the absolute roughness is not uniform† and has to be defined by a mean value which, from the point of view of head loss, corresponds to a uniform roughness. Attempts are being made to find a method for determining these values directly. Within the scope of present knowledge, however, it is by observation of existing pipes and conduits that an equivalent uniform roughness is selected, this being related to a given type of material and finish. Determination of ε is important, especially for large concrete, steel and wooden pipes (see Table 17).

— In *concrete pipes*, the value of the absolute roughness depends mainly on the finish and on the frequency and alignment of the joints.
— In *welded steel pipes*, the value of ε mainly depends on the type of lining and manner of applying it.
— In *riveted steel pipes*, the lining is of secondary importance; the predominant factor is the manner of rivetting, the number of longitudinal and transverse rows and their distance apart.
— In *wooden pipes*, the alignment of the joints of principle importance.

4.4 Smooth pipes and rough pipes

The concept of viscous sublayer having been established, whenever the roughness of the wall is less than the thickness of this sublayer, such roughness does not have any influence in relation to turbulence and the flow is said to take place in a *smooth pipe*.

Otherwise the roughness of the wall extends into the turbulent zone of the flow, accentuating the turbulence and thus having an influence on the energy loss; it is then considered that the flow takes place in a *rough pipe*.

Turbulent flow may thus occur in smooth pipes — *smooth turbulent flow*; or in rough pipes — *rough turbulent flow*.

4.5 Friction factor, f. Moody diagram

(a)   *Laminar flow* — In the case of *laminar flow*, f is independent of the relative roughness and is only a function of the Reynolds number; it is given by the expression:

$$f = \frac{64}{\mathbf{R}_e} \tag{4.6}$$

In a logarithmic diagram, this expression is represented by a straight line known as *Poiseuille straight line* (Graph 18).

By substituting the values of f in (4.1), we obtain the Poiseuille formula valid for any fluid in *circular pipes* with laminar flow (see Section 2.13, equation 2.16):

---

† In the case of certain laboratory experiments (Nikuradse, for example), the absolute roughness was uniform and was obtained from grains of known diameter distributed uniformly on the walls of the pipes.

$$i = \frac{32}{g} \frac{\nu U}{D^2} = 3.26 \frac{\nu U}{D^2} = 4.15 \frac{\nu Q}{D^4} \tag{4.7}$$

In the case of flow through a pipe having a *non-circular cross-section*, this becomes:

$$i = \frac{\nu U}{agA} \text{ or } U = a \frac{gAi}{\nu} \tag{4.7a}$$

in which a is a coefficient depending on the shape of the section, see Table 19.

(b)   *Turbulent flow in smooth pipes* — If the flow is turbulent in smooth pipes, there are various expressions that represent the value of f.

→   *Karman–Prandtl† equation* — for smooth pipes:

$$\frac{1}{\sqrt{f}} = 2 \log_{10} \mathbf{R}_e \sqrt{f} - 0.8 \tag{4.8}$$

whose values are very close to those given by *Nikuradse's* formmula, which has been considered valid for $\mathbf{R}_e > 10^5$.

— *Nikuradse's equation* — for smooth pipes and $\mathbf{R}_e > 10^5$.

$$f = \frac{0.221}{\mathbf{R}_e^{0.237}} + 0.0032 \tag{4.9}$$

— *Karman–Nikuradse equation*

$$\frac{1}{\sqrt{f}} = 2 \log \frac{\mathbf{R}_e \sqrt{f}}{2.51} \tag{4.10}$$

— *Blasius's equation* — which has a far simpler structure:

$$f = \frac{0.3164}{\mathbf{R}_e^{0.25}} \tag{4.11}$$

For a long time this equation was considered valid only for $\mathbf{R}_e < 10^5$, but more recent experiments (1972) by Levin‡ on new materials of extraordinary smoothness, such as polyurethane, vinyl, araldite, etc., and the high $\mathbf{R}_e$ values ($\approx 10^8$), show that the experimental values are below the values of the formulae of Karman–Prandtl and Nikuradse and are far closer to the values given by Basius's formula.

(c)   *Turbulent flow in rough pipes* — For *rough pipes*, the *Karman–Prandtl* formula (1935) and many others are indicated, this formula being based mainly on artifical roughnesses created in the laboratory.

† Karman, T. (1881–1963).
‡ Levin, L., *Etude Hydraulique de Huit Revêtements Intérieurs de Conduites Forcées*. La Houille Blanche No. 4/1972.

$$\frac{1}{\sqrt{f}} = 2 \log_{10} \frac{D}{2\varepsilon} + 1.74 \tag{4.12}$$

Colebrook and *White* were responsible for the systematic study of industrial piping, on the basis of which they presented the formula:

$$\frac{1}{\sqrt{f}} = -2 \log_{10} \left(\frac{\varepsilon}{3.7 \, D} + \frac{2.51}{R_e \sqrt{f}} \right) \tag{4.13}$$

(d) *Moody's Universal Diagram* — Basing his studies on Nikuradse's experiments Prandt's and Karman's mathematical analysis, Colebrook's and White's results (1939) and on a great many experiments in industrial pipes, Moody established a logarithmic diagram in which f is given as a function of the Reynolds number and of the relative roughness, ε/D.

This diagram, known as the *Moody diagram*, is represented in Graph 18, in which there have already been introduced the results of Levin's experiments on ultra-smooth pipes.

The Moody diagram is applicable to any fluid and any type of flow. The difficulty in using it sometimes lies in selecting the value of the absolute roughness, ε.

In the case of non-circular pipes, the diameter, D, is subject to a corrective factor that may take the following values†; square section, 1.00 to 1.17; equilateral triangular section, 1.30; wide rectangular or narrow annular section, 0.84.

Examples:
(1) Determine the friction head loss in a welded steel pipe, coated with bitumen. The diameter of the pipe is $d = 1$ m, the length $L = 1000$ m. In the pipe there is a discharge $Q = 0.785$ m³/s at 20°C ($v = 1.01 \times 10^{-6}$ m²/s — Table 8).

Solution:

$$A = \frac{\pi d^2}{4} = 0.785 \text{ m}^2; \ U = \frac{Q}{A} = 1 \text{ m/s}; \ R_e = \frac{Ud}{v} = 0.99 \times 10^6;$$

Table 16 shows that the extreme values of absolute roughness are 0.0003 m and 0.0009 m.
From the Moody diagram, for $R_e = 0.99 \times 10^6$ and $\varepsilon/D = 0.0009$, $f = 0.0192$, and for $R_e = 0.99 \times 10^6$ and $\varepsilon/d = 0.0003$, $f = 0.0157$, and $U^2/2g = 0.051$ m.
Thus $\Delta H = f \times L/D \times U^2/2g = f \times 1000 \times 0.051 = 51 \ f$. The value of ΔH will be between $0.0157 \times 51 = 0.80$ m and $0.0192 \times 51 = 0.98$ m. Fuller knowledge of the state of conservation of the pipe and the nature of the problem to be studied would guide the project designer in choosing an appropriate value of f.
(2) Determine the friction head loss in a pipe with a diameter of 100 mm, made of new galvanized iron, in which flows lighting kerosene at 40°C, relative density $\delta = 0.813$, $v = 1.54 \times 10^{-6}$ m²/s, at a discharge of 20 l/s.

† Obtained by Marchi, E. (1967), quoted by [11].

Thus $U = Q/A = 2.55$ m/s and $\mathbf{R}_e = 1.6 \times 10^5$.
From Graph 17 for $\varepsilon/d = 0.0015 = 1.5 \times 10^{-3}$ and $\mathbf{R}_e = 1.6 \times 10^5$, $f = 0.023$.
For $U = 2.54$ m/s, $U^2/2g = 0.331$ m.

We thus have $i = 0.023 \times \dfrac{1}{0.1} \times 0.331 = 0.076$.

4.6 Empirical formulae

In determining friction losses, there are various empirical formulae for which a large number of experiments confer a certain value when they are applied within the field for which they were deduced.

They also have the advantage that they can more easily be expressed in the form of tables or graphs.

The main formulae of this type are as follows.

(a) *Manning–Strickler formula* — This formula, is known in the English-speaking world as *Manning's†* formula, although on the continent of Europe it is sometimes known also, as *Strickler's* formula. It has the advantage of being monomial, and can thus be calculated logarithmically; it is written:

$$U = K_s R^{2/3} i^{1/2} \qquad (4.14)$$

$$i = \frac{U^2}{K_s^2 R^{4/3}} \qquad (4.14a)$$

This formula is valid in metric units, i.e. the velocity U, is measured in metres per second and R, hydraulic radius (see Section 1.15), in metres‡.

K_s, is the coefficient that represents roughness. The smoother the pipe, the greater it will be, and it has the dimensions $m^{1/3}/s$.

It can also be written:

$$i = b \ U^2 = a Q^2 \qquad (4.15)$$

in which:

$$b = \frac{6.35}{K_s^2} D^{-1.333}; \qquad a = \frac{10.3}{K_s^2} D^{-5.333} \qquad (4.15a)$$

---

† Manning, R. (1816–1897).
‡ It is to be noted that in imperial units, the Manning–Strickler formula is usually written:

$$U = 1.486 \ K_S R^{2/3} i^{1/2}$$

with U in feet per second and R in feet.
    The numerical factor makes it possible to use the same numerical values for K_s as are used when the formula is in metric units.

For resolution of Manning's formula, see Table 22.

From expression (4.4) and (4.16) it is easy to relate Strickler's roughness coefficient K_s to the friction factor, f, of the Moody diagram.

$$f = \frac{124.6}{K_s^2} D^{-1/3} \tag{4.17}$$

(b) *Chézy equation* — Chézy's equation, which was originally deduced for flow in channels, was later generalized to steady, incompressible rough turbulent flows in pipes. It is written as follows:

$$U = C\sqrt{Ri} \text{ or } i = \frac{U^2}{C^2 R} = \frac{Q^2}{C^2 R A^2} \tag{4.18}$$

in which R is the hydraulic radius (see Section 1.16) and C is an experimental coefficient usually known as Chézy's coefficient whose dimensions are $L^{1/2}T^{-1}$. It can also be written $i = aQ^2$ in which:

$$a = \frac{1}{C^2 R A^2} \tag{4.18a}$$

There are the following relationships between the friction factor, f, Strickler's coefficient, K_s, and Chézy's coefficient, C:

$$f = \frac{8g}{C^2} \; ; \; C = K_s R^{1/6} \tag{4.19}$$

Several experimenters have arrived at expressions for C, functions of the hydraulic radius, R, and of the coefficient that characterizes the roughness of the walls of the conduit. It is to be noted that such expressions apply only to flows of water, in rough turbulent flow, and that the corresponding formulae must be used with the units for which those coefficients were determined.

As an example, the expressions proposed by *Bazin* and *Kutter* are given below:

$$\begin{array}{cc} \textit{Bazin} & \textit{Kutter} \\[2mm] C = \dfrac{87\sqrt{R}}{K_B + \sqrt{R}} & C = \dfrac{100\sqrt{R}}{K_k + \sqrt{R}} \end{array} \tag{4.20}$$

The constants K_B and K_k represent the roughness of the walls† and have the dimensions $L^{1/2}$. Their values are given in Tables 19 and 20, respectively, in metric units (R expressed in metres and U in metres per second).

(c) *Darcy's formula* — This formula, deduced for cast iron pipes, is written:

$$i = \frac{4}{D} bU^2 \tag{4.21}$$

in which, for cast iron pipes in service, we have:

$$b = a' + \frac{b'}{D} = 0.507 \times 10^{-3} + 12.94 \times 10^{-6} \times \frac{1}{D} \tag{4.21a}$$

the dimensions of b being $L^{-1}T^2$.
Also:

$$i = aQ^2 \tag{4.21b}$$

In which:

$$a = \frac{64b}{\pi^2 D^5} \tag{4.21c}$$

Table 23 gives the values of a, in metric units, obtained from Darcy's formula.

For new cast iron pipes, it is necessary to take half the values indicated. For tarred plate pipes, a third of the values shown must be taken.

Darcy's formula overestimates values for diameters of less than 0.10 m, and under-estimates values for diameters of over 0.80 m. The most exact values fall in the zone of 0.1 to 0.2 m.

In modern pipes of ductile cast iron ageing is not so noticeable but the influence of the internal lining must also be taken into account.

4.7 Compatibility between the empirical formulae and the friction factor, f (Moody diagram)

Formula 4.17 relates f to Strickler's coefficient, K_s. Formula 4.19 relates f to Chézy's coefficient, C, and with equation (4.20) to Bazin's coefficient, K_B.

† Friction factor, f, is related to Bazin's coefficient, K_B and Kutter's coefficient, K_k, by:

$$f = 0.01037 \left(1 + \frac{2K_B}{\sqrt{D}}\right)^2 = 0.00785 \left(1 + \frac{2bK_k}{\sqrt{D}}\right)^2 \tag{4.19a}$$

In some practical problems it may be of interest to compare the values of K_B, K_s and those of the absolute roughness, ε, to which corresponds, for a given diameter, the same value of the friction factor, f. Graph 25 shows the relationship existing between the three coefficients of roughness. For instance, for an absolute roughness $\varepsilon = 10^{-3}$ m, the values of K_B vary between 0.1 and 0.2 m$^{1/2}$ when the diameter varies from 10.0 m to 0.5 m. For the same variation of diameters, the values of K_s vary between 70 and 82 m$^{1/3}$/s.

It will therefore be seen that when speaking of K_B or K_s, it is necessary to bear in mind the diameter of the pipe. In fact, for the same absolute roughness, ε, there are different relative roughnesses, ε/D, and it is the relative roughnesses that condition the value of f.

4.8 Flow in hoses

In a hose, resistance to the flow is a function not only of the roughness but also of the absolute pressure, since for high pressures the diameter will increase considerably, owing to the elasticity of the material.

There seems to be the following ratio between the variation in energy losses and variations in diameters, for some types of hoses and for small variations in pressure [3].

$$\frac{\Delta i}{i} = -5\,\frac{\Delta D}{D} \tag{4.22}$$

Table 24 gives the values of coefficient C of Chézy's formula, the value of the term $K = Di/U^2$ and the value of $\Delta D/D$ and $\Delta i/i$ for an increase of 10 N/cm^2 of pressure.

Diameter D is the diameter of the hose, without this being subject to any water pressure. The value of C is the value of Chézy's coefficient for this case. For a given Q the value of i can then be determined; if the pressure increases by 10 N/cm^2, the diameter will undergo a relative increase, $\Delta D/D$, and the value if i will be reduced by $\Delta i/i$.

4.9 Head losses of compressible fluids

When studying the flow of a compressible fluid in a pipe, two extremes must in principle be considered; the pipe is perfectly isolated thermally and there is thus no heat exchange with the exterior — *adiabatic* flow; or the pipe is completely permeable to heat and the flow will be *isothermal*, taking the ambient temperature to be uniform.

From the practical point of view, flows of air and of lighting gas can generally be regarded as *isothermal*[†], [9].

It must not be forgotten, however, that the pressure variation influences the density[‡].

[†] Taking the flow to be adiabatic, for air, we have the following ratio:

$$T_1^\circ - T_2^\circ \approx \frac{1}{100}\,\frac{U_2^2 - U_1^2}{2g} \tag{4.24}$$

Consequently, if the initial velocity is 1 m/s, it must increase to 44.3 m/s in order for the temperature to fall by one degree. Moreover, if account is taken of the heat caused by friction, which will only serve to heat the gas, since the pipe is insulated, the temperature difference will be even less and the flow can therefore be considered isothermal.

[‡] Boyle–Mariotte law: $p_1\,\rho_1 = p_2\,\rho_2$.

Owing to the head loss along the pipe, the pressure will fall† and with it the density ρ will also diminish. According to the equation of continuity, the mass flow rate, $G = \rho\, UA$ remains constant; if A remains constant and ρ decreases, U increases.

The Reynolds number, however, remains practically constant, since:

$$\mathbf{R}_e = \frac{D}{\nu}\, U = \frac{\rho D}{\mu}\, \frac{4G}{\rho \pi D^2} = \frac{4G}{\mu \pi D} \approx const. \tag{4.23}$$

if the viscosity variations, $\mu = \rho \nu$, are disregarded. Thus f also remains constant.

The Bernoulli equation is applied along elementary lengths of piping, ΔL, and the terms z and $U^2/2g$ can be disregarded; it is thus possible, by successive lengths, to calculate the head loss, $\Delta p/\gamma$, over a distance ΔL. This lowering of pressure has a corresponding reduction in the specific weight, γ, and $\Delta p/\gamma$ should be calculated with the mean of the initial and final value of γ. In this case calculation by computer is advisable.

To sum up: if the flow of gas takes place at low velocities ($U \approx 10$ m/s) and with small fall in pressure (up to 1000 N/m², equivalent to 10 cm of water column), the thermal energy of the flow and the unit weight can be considered constant, and the calculations will be similar to those for incompressible fluids.

For calculating the head losses, the Moody diagram can be used.

There are also various practical formulae for determining the head losses of different gases, namely air and lighting gas.

Given below is the *Aubéry formula* for cast iron pipes:

For air…

$$\Delta h = \frac{3347\ Q^{1.85}}{D^{4.92}} \tag{4.25}$$

For lighting gas…

$$\Delta h = \frac{1625\ Q^{1.85}}{D^{4.92}} \tag{4.25}$$

in which Δh is the head loss in millimetres of water per kilométre of pipe; Q is the discharge in m³ per hour and D is the diameter of the pipe in centimetres.

4.10   Ageing of pipes

It is very difficult to assess the ageing of pipes, i.e. the increase in roughness and, consequently, reduction in the carrying capacity of the piping, at the end of T years.

Various factors may contribute to the increase in roughness of a pipe, in particular the characteristics of the water, the material of which the pipe is made, the type of internal lining, if such exists, and how it is applied, and the activity of rust-generating bacteria, etc.

From the practical point of view, the phenomenon of ageing can be regarded in two ways; either by trying to reduce or eliminate it, through correction of the characteristics of the water carried or application of a suitable internal lining; or by over-dimensioning the pipe, in order to take

† Even in very steeply sloping pipes the factor $h = p/\gamma$, which might compensate the pressure reduction is insignificant. For air, with $h = 100$ m, we have $p/\gamma = 0.13$ m of water column.

account of the reduction in its carrying capacity. In the great majority of cases, except perhaps that of provisional piping, the first way is probably the most economical.

In order to be able to take the right decision, it is necessary to have an idea of the ageing.

Table 26a summarizes the indications of Colebrook and White[†].

Price's studies are also considered, the results being given in Graph 26b, which is valid for metal pipes, with indication of the reduction in the discharge according to the age of the pipe, for various types of water.

More recently, F. Lamont, on the basis of a large number of experiments, gives the following indications for calculating ageing.

(1) Langelier's index, I, characterizes the encrusting power of the water:

$$I = pH + \log Ca + \log Alc - K - 9.3 \qquad (4.26)$$

in which Ca is the calcium content expressed in p.p.m. of Ca^{2+}; Alc is the total alkalinity expressed in p.p.m. of $CaCO_3$; K is a coefficient varying with the temperature and the dry residue, according to Table 27.

If only the pH and Alc are known, the value of I can be calculated with sufficient approximation by the expression:

$$I = pH - pH_s \qquad (4.26a)$$

pH_s being the saturation pH obtained according to Table 28b.

(2) The increase in roughness, α, in mm/year, is given by the following equation:

$$\log \frac{\alpha}{K_1} = -\frac{I}{2.6} \qquad (4.27)$$

in which K_1 is a coefficient that must, as far as possible, be based on the observation of experimental data referring to existing pipes of the same type and carrying water with the same characteristics as that which will be carried by the new pipes.

Failing data of this type, and if it can be assumed that corrosion will be kept within normal limits, the value of K_1 will vary between 0.01 and 0.05, a value between those limits being taken according to the greater or lesser tendency that is assumed for corrosion. If high corrosion is suspected, higher values will have to be taken, and these may be as high as 0.2.

From what has been said, it will be seen that on the basis of present

† Colebrook, C. F. and White, C. M., *The Reduction in Carrying Capacity of Pipes with Age*, *Journal of the Institution of Civil Engineers*, No. 1, 1937–38.

knowledge it is very hard to indicate a sure process for forecasting corrosion.

Observations by P. Lamont† have suggested that, with the possible exception of particularly aggressive water, *in cast iron or steel pipes internally lined with bituminuous products applied centrifugally*, and in *asbestos cement pipes*, the effect of ageing is not apparent. In *concrete piping* and also in *cast iron or steel piping internally lined with cement mortar applied by centrifugally*, there may be a reduction in carrying capacity but this will be negligible if the internal concrete walls or the cement mortar have been given some coating of a bituminuous product.

From the practical point of view, however, in designing such pipes, it is necessary to account for a 5% reduction in their carrying capacity, and it would be prudent to increase this to 10% in the case of water that is aggressive in relation to cement and when no bituminous protection has been envisaged.

4.11   Choice of the formula to be used

The large number of formulae presented and countless others that have not been mentioned cause difficulties for the user when choosing a formula for resolving a given case.

In the selection that has been given, efforts have been made to eliminate a certain number of old formulae (Prony, Dupuit, Levy, Tutton, etc.), which have been replaced by others to advantage, since they are concerned with more specific cases.

The final choice of formula is basically up to the designer; according to the kind of problem, it will thus be convenient to use underestimated or overestimated values, to forecast whether or not the piping will deteriorate, fix the degree of ageing, etc.

Given below are some general indications:

(a)   Whenever laminar flow is involved, this being rare in practial calculations of hydraulics, though more frequent in the lubrication of conveyance of very viscous oils, the Poiseuille formula must be used, or amounting to the same thing, Moody's diagram must be resorted to.

(b)   In the case of pipes with large diameter (larger than 0.5 or 1 m) that are very smooth, it is very likely that the flow will not be rough turbulent. In practice, such cases occur in large diversion tunnels of reservoirs, in large water supply pipes and others. Under these conditions the Moody diagram gives the best indications, and it must be used exclusively.

(c)   In the case of small diameters and in rough turbulent flows, the Moody diagram continues to give good results. It is, however, advisable to pick out, from the indications given by the various formulae, the one

† Lamont, P., *Formulae for Pipe-Line Calculations*, International Water Supply Association, Third Congress, London, 1955.

that is best suited to each specific case. Whenever a monomial formula is well suited to the case in study, it is best to use it owing to facility of calculation.

(d)   Should it happen that the same case can be studied by two formulae leading to slightly different results, the project designer must, in accordance with other conditioning factors, choose the value to be utilized.

4.12   Economic dimensioning of pumping conduits

Given a discharge to be carried over a certain distance, for example by means of pumping, the fixing of the diameter and consequent head loss will in the long run be decided by economic considerations, in order to achieve minimum total investment and operation costs. The following is concerned with pumping pipelines, since this is the case that is generally faced by users of this handbook. Similar lines of reasoning are, however, also valid for the penstocks of turbines.

In most cases the difference in costs of pumping plants has little influence compared with the cost of the pipelines. This means that only the investment in the pipelines need be considered, at least in a first approximation[†].

In a *first approximation*, the economic diameter can be calculated by means of the simplified formula:

$$D \approx 0.95 \ Q^{0.43} \tag{4.28}$$

D is expressed in metres and Q in m³/s; Q is the constant discharge to be conveyed, this being fixed by the conveyance requirements at the end of the project horizon, n years.

If effect, uncertainty as to the cost of the pipelines and energy, and even discharge variations, do not justify very rigorous calculations, at least in simpler cases and at a preliminary design level.

B.   LOCAL HEAD LOSSES

4.13   General expression. Equivalent length of pipe

Local head losses in turbulent flow can be reduced to the following type:

$$\Delta H = K \ \frac{U^2}{2g} = bKQ^2 \tag{4.29}$$

with

$$b = \frac{1}{2gA^2} \tag{4.29a}$$

† Lencastre, A., *Hidráulica Geral* (*General Hydraulics*). In Portuguese. Hidroprojecto, Lisbon, 1983.

in which K is a function of the geometrical characteristics, the roughness and the Reynolds number.

In most cases, the influence of roughness and Reynolds number is negligible.

Sometimes it is advantageous to equate the local head loss to a fictitious length of conduit that would cause the same head loss, known as *an equivalent length of conduit*.

It has been seen that continuous loss in turbulent flow could always be put in the form $\Delta H = aLQ^2$. The fictitious length will be such that $aL = bK$, so that:

$$L = \frac{b}{a} K \qquad\qquad (4.30)$$

Table 52 gives the values for equivalent lengths of conduits, for standardized accessories.

4.14 Losses in sudden expansions
In the case of a sudden expansion, head losses are mainly due to the separation originated by a positive gradient of pressure due to reduction in velocity (Section 2.27).

Deduction of head losses due to sudden expansion turbulent flow, is due to Borda†, who applied the theorem of momentum to the volume between sections 1 and 2 (Fig. 4.1), with the following simplifying hypotheses:

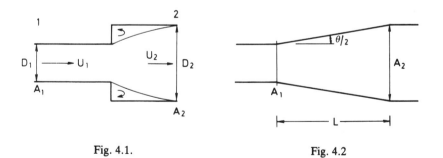

Fig. 4.1.   Fig. 4.2

uniform distribution of velocity in section 1, that is to say, coefficients of Coriolis, α, and of Boussinesq, β, equal to unity (see Sections 2.19 and 2.21); stagnation of the liquid in the separation zone; negligible friction losses at the walls. The well-known *Borda's formula* was deduced on the basis of these hypotheses:

† Borda, C. (1733–1799).

$$\Delta H = \frac{(U_1 - U_2)^2}{2g} = K \frac{U_1^2}{2g} \tag{4.31}$$

with:

$$K = (1 - \eta)^2 \text{ and } \eta = A_1/A_2. \tag{4.32}$$

Subsequently Idel'cik [11], in order to take account of the nonuniform distributions of velocities (α and β different from 1), proposed:

$$K = \eta^2 + \alpha - 2\eta\beta \tag{4.32a}$$

with α and β, being the coefficients of Coriolis and Boussinesq, respectively.

For the head loss at an exit of a pipe to a reservoir of large dimensions, $\eta = 0$, so that $\Delta H = \alpha U_1^2/2g$, which is equivalent to the total loss of kinetic energy.

In laminar flow, with $R_c < 10$ and for $\alpha = \beta = 1$, Karev[†] indicates $K = 26/R_e$; for values of $10 < R_e < 3500$, values for K are given in Table 29.

4.15 Gradual expansion. Diffusers

We have seen that the head loss due to sudden expansion is mainly attributable to separation of the flow. The best way of reducing such losses is to make a progressive widening, by means of a diffuser. The head loss in a diffuser consists of two parts: head loss due to friction and head loss due to the shape of the diffuser itself. We thus have:

$$\Delta H = K_d U_1^2/2g \tag{4.33}$$

in which:

$$K_d = K_f + K_a \tag{4.33a}$$

K_d being the coefficient of total head loss in the diffuser, K_f being the coefficient of head loss due to friction (see Sections 4.1 and 4.33) and K_a the coefficient of head loss due to expansion.

Head losses due to expansion can be determined by the following formula deduced by Idel'cik [11].

$$K_a = \psi.\phi(1 - \eta)^2 \tag{4.34}$$

It will be seen that in relation to the Borda-type head loss, represented by $(1 - \eta)^2$, a coefficient, ϕ, is introduced, this being a function of the greater or

[†] Karev, B. H. referred to in [11].

lesser expansion of the diffuser, and a coefficient, ψ, accounting for the non-uniform distribution of velocities.

Head loss coefficients for conical and rectangular diffusers are presented in Table 30.

In passing from the initial section A_1, to the final section, A_2, the smaller the angle of divergence of the diffuser, the smaller will be the losses from shock due to the expansion, but on the other hand the greater will be the losses from friction along the diffuser.

The angle of divergence that gives a minimum head loss from expansion and friction is given by:

$$\theta_{opt} = 0.43 \left(\frac{f}{\psi} \frac{1+\eta}{1-\eta} \right)^{4/9} \tag{4.35}$$

in which f is the friction factor (see Section 4.1).

For example, for $f = 0.015$, $\eta = 0.45$ and $\psi = 1.0$, we have $\theta_{opt} = 6°$. This formula is valid for a rectangular section. For a flattened section, the optimum angle is between $10°$ and $12°$.

It does not, of course, suffice to take into account the minimum head loss; what is important is the minimum cost of the diffuser, in terms of investment and operation, involving problems similar to those of calculating the most economical diameter of piping (see Section 4.12).

The head loss in the passage from a diffuser to a large flow cross section is given in Table 31.

In cases in which wide-angle diffusers must be used, their performance can often be markedly improved by use of auxiliary devices.

One of the means used consists of *boundary layer suction* in the separation zone (Fig. 4.3a),

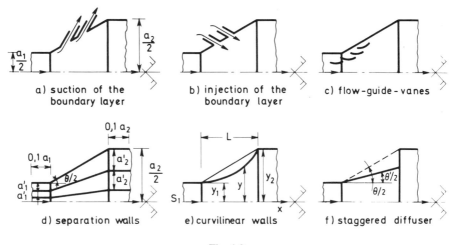

a) suction of the boundary layer

b) injection of the boundary layer

c) flow-guide-vanes

d) separation walls

e) curvilinear walls

f) staggered diffuser

Fig. 4.3

and by this means head losses can be reduced by 30 to 50%.

Instead of suction, *injection of fluid in the boundary layer zone* can be undertaken; this increases the velocity in that zone and supresses separation in the diffuser (Fig. 4.3b).

Flow guide-vanes can also be used, particularly for angles of divergence, θ, that are greater than 90°, angles which are rare in hydraulics but frequent in ventilation systems (Fig. 4.3c). Concerning the placing of such guide-vanes, reference [11] should be consulted.

In the case of wide-angles, the diffuser can also be divided by *separation walls* along the diffuser (Fig. 4.3d), this being equivalent to replacing a single diffuser with a wide angle of divergence by various diffusers with narrower angles; this process may be inefficient for small angles, since by increasing the walls of the diffuser the friction losses may be increased more than the expansion losses are reduced, apart from the fact that more costly works are involved.

Idel'cik recommends the following rules for dimensioning a diffuser of this type (Fig. 4.3d):

— the number of separation walls depends on the angle of the diffuser; for 30°; two walls, for 45° and 60°, four walls; for 90°, six walls; for 120°, eight walls;
— the intervals, a_1', between the separation walls, in the intake section, must be equal, as must the intervals, a_2', in the outlet section;
— the walls must be extended about 0.1 a_1 upstream of the diffuser and about 0.1 a_2 downstream of it, a_1 and a_2 being the linear dimension of the intake and outlet sections, respectively.

Another way of reducing the head loss in a diffuser, when the angle is greater than 20°, is to vary this angle. Fig. 4.3e gives an example of a *curved wall*, with the pressure gradient, dp/dx, constant along the diffuser, allowing head losses to be reduced (40% for angles between 25 and 90°). For angles of less than 15° or 20° there will be increased head loss, so that it is not appropriate to use diffusers with curved surfaces.

The equation for the generatrices in a diffuser, with dp/dx = const., is given by:

$$y = \frac{y_2}{\sqrt[n]{1 + \left[\left(\frac{y_2}{y_1}\right)^n - 1 \right] \frac{x}{L}}} \tag{4.36}$$

The letters have the significance given in Fig. 4.3e, in which $n = 4$ for diffusers of circular or square section; $n = 2$ for diffusers of flattened section (2 parallel faces).

For the coefficient of head loss in this case, within the interval $0.1 \leqslant (A_1/A_2) \leqslant 0.9$, Idel'cik proposed the following expression:

$$K = \phi_0 (1.43 - 1.3\eta)(1 - \eta)^2 \tag{4.37}$$

in which ψ_0 is given by Table 30–3 and $\eta = A_1/A_2$.

In a *staggered diffuser* (Fig. 4.3f), the continuous variation of section ends in a sudden expansion, thus attempting to achieve a compromise between the two types of head loss. This type is only justified in wide angles, which are mostly used in ventilation systems rather than being recommended in hydraulic structures. For its dimensioning [11] should be consulted.

4.16 Losses in contractions and entrances

Head losses in a contraction are mainly due to the losses by expansions in the passage from the contracted section, A_c, to the section A_2 (Fig. 4.4). From the practical point of view, $\Delta H = KU^2/2g$ is once again valid.

For *abrupt contractions*, the value of K is given by Table 32.

Losses from *gradual contraction* are a function of their shape. Given the stability of flow in accelerated systems, these losses are always small, and K can take values of about 0.01 or even 0.005; for this reason they are generally disregarded. In a gradual contraction, the prevention of cavitation or separation of the flow is more important than the actual energy loss.

Studies based on an electrical analogy [7] made it possible to establish the curves of Graph 33, for designing bell-mouth pipe entrances.

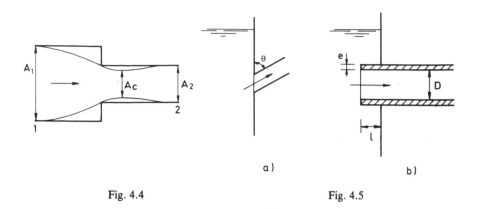

Fig. 4.4   Fig. 4.5

In the passage from a reservoir to a pipe, if it takes place over a square-edge and the pipe is not re-entrant (Fig. 4.5a), the coefficient of head loss is given by Weisbach's formula:

$$K = 0.5 + 0.3 \cos\theta + 0.2 \cos^2\theta \qquad (4.38)$$

in which θ is the angle between the axis of the pipe and the wall of the reservoir.

If the axis of the pipe is normal to the reservoir wall, $\theta = 90°$ and $K = 0.5$; if the pipe is re-entrant, with length l (Fig. 4.5b), the value of K depends on the ratio l/D and also on the ratio e/D, where e is the thickness of the conduit wall (see Tables 34 and 147).

In the case of non-circular pipes, the hydraulic diameter, $D = 4R$, is taken. Table 34 also gives the values of K in the passage from a reservoir to a conduit with a conical connection.

If the edges of a pipe entrance are rounded to produce a *streamlined bellmouth*, the value of K varies between 0.01 and 0.05 and can be disregarded. In this case it is again important to prevent cavitation; Graph 35 gives indications for the design of intakes without cavitation.

In the case of an entrance very close to a front wall, the values of K are given in Table 36.

4.17 Losses in valves and taps

In these cases the losses are again given by $\Delta H = KU^2/2g$, in which U is the mean velocity in the normal section of the piping.

From the point of view of head losses, valves can be divided into two basic groups, in which the classification criterion is broadly the form of flow. The first group includes those types of valve in which, when fully open, the flow does not undergo great changes of direction; this group includes *gate valves*, *butterfly valves*, *plug valves* and also *swing-check valves* and *ball-check valves*. The second group covers valves in which the flow is very sinuous, or those whose outlet section has a different direction from that of the intake section; this group includes *globe valves*, *angle valves* and '*Y*' *valves*.

The most important head loss in all these type of valves, when they are not fully open, is that of the *Borda-type* (see Section 4.15), owing to rapid expansion of the flow downstream of the valve. In the first category, this *Borda-type* loss constitutes nearly all of the head loss which really occurs [4].

In this case of partly open valves, of the first group, values for *K* are indicated in Table 37; for fully open valves of the second group, *K* values are presented in Table 52. For large gates in hydraulic development schemes, the appropriate values are given in Table 38; for Howell–Bunger Table 39 applies.

All of these values must be taken as a general indication, since the results sometimes differ according to the experimenter.

See also Section 8.23 and Graphs and Tables 144, 145 and 146.

4.18   Losses in pipe bends

A bend causes a disturbance in the flow owing to an increase in pressure and corresponding reduction in velocity in the outer part of the curve, and a diminution in pressure and corresponding increase in velocity in the inner part (Fig. 4.6). This difference in pressures, due to centrifugal forces on fluid

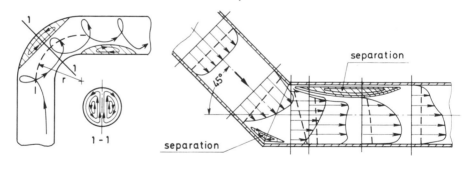

Fig. 4.6.   Fig. 4.7

particles, causes a modification in the flow field, and in the tansverse cross-section of pipe there appears a double spiral motion whose influence is felt over a considerable length downstream of the bend (about 50 times the diameter of the pipe).

This difference in pressure combined with the double spiral motion and the separation are the main factors responsible for head losses in bends. By increasing the width and reducing the height of the cross-section of the pipe, it is possible to reduce the intensity of the double spiral motion and thus the head loss.

Fig. 4.7 shows a flow scheme in a 45° bend, with the velocity diagrams and separation zones of the flow.

Table 40 gives coefficients of head loss in 90° bends, for pipes with rectangular cross-section and for various ratios of l/h, l being the width of the section and h the height.

In circular cross-sections of small diameter (in order of magnitude less than 0.50 m) and for 90° bends, the values of head loss are as indicated in Table 41.

Should the angle of the bend be different from 90°, the value of the coefficient of head loss can be obtained by multiplying the values given in Table 41, by the values of Table 42. Graph 44 should also be referred to.

For large industrial pipelines, the values indicated by these tables are exaggerated. The values given by Graph 43 must therefore be taken.

Table 45 gives the K coefficients for three special bends, which make it possible to compare the variation in head loss with modifications to the cross-section of the bend, the cross-sectional area remaining constant. The minimum for head loss in 90° bends is reached when the ratio between the inlet section and outlet section, A_1/A_2, falls between 1.2 and 2.

In the case of acute-angle changes of direction, the values in Table 46 are indicated.

Although a modification of section of the type shown in Table 45 leads to lower energy losses, it seems that the most effective way of reducing head loss in bends is to use suitably dimensioned guide vanes in order to orient the flow. Apart from bringing the head loss coefficient down to 0.15 or less, a better distribution of velocities is obtained.

The most favourable arrangement of the vanes is to place them concentrically, aligned according to the bisectrix of the elbow or bend (Fig. 4.8(b)) and (Fig. 4.9). A single vane is more

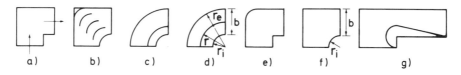

Fig. 4.8

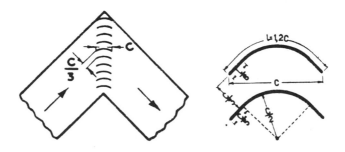

Fig. 4.9

effective next to the inner wall (Fig. 4.8(c)) than along the axis (Fig. 4.8(d)) or next to the outer wall†.

The shape that seems to be most advisable for the vaned bends is as shown in Fig. 4.9‡. In special conditions (air or water tunnels), the transverse cross-section of the vanes must be aerodynamic; in normal conditions it is not worthwhile. Other investigations lead to the conclusion that the influence of the *upstream extensions* of the vanes is small and that vanes without such extensions also give good results.

There does not, however, seem to be any advantage in increasing the external radius (Fig. 4.8e); if the inner wall makes an acute angle, a significant increase in the curvature of the outer wall increases the head loss since it causes a reduction in the cross-section.

On the other hand, it is advantageous to increase the inner radius (Fig. 4.8f); if the outer wall makes a right angle, i.e. without curvature, and if r_i is the inner radius the minimum head loss is obtained for $r_i/b = 1.2$ to 1.5. For greater values of r_i/b, the head loss increases, since there is a corresponding increase in the section of flow with a corresponding reduction in velocity, and this causes a marked increase in separation of the flow upstream of the bend.

In the case of bends with concentric walls (Fig. 4.8d), the minimum head loss is obtained when $r_e/b = r_i/b + 1$. This ratio is considered optimum, and is not difficult to obtain technically. It is also advantageous to suppress the 'dead spaces', where there is a tendency for vortices to develop, particularly if the bend is followed by a sufficiently gradual diffuser (Fig. 4.8g).

A modification of the A_1/A_2 ratio between the intake area and outlet area can reduce the head loss. In 90° bends the minimum head loss is obtained for A_1/A_2 between 1.2 and 2.0. If the exact value of the head loss unknown, it is reasonable to adopt the loss that would correspond to $A_1/A_2 = 1$.

For 180° combined bends containing two similar 90° bends, the head loss depends a great deal on the relative distance L/b (Fig. 4.10a). For $L/b \geqslant 4.5$, the head loss approaches double of that corresponding to a single 90° bend. The minimum value of the head loss is obtained for L/b between 0 and 1.0, with a head loss coefficient equal to 1.4 times that corresponding to a single 90° bend. These values refer to sharp right-angle curves. If the curves are rounded the head losses are lower when L/b is less than about 30.

In the case of Z bends, also called 0° combined bends, (Fig. 4.10b) with two similar 90° bends, the head loss is minimum for values of L'/b equal to 4.0, L' being the distance between the axes of the two straight lengths. The head loss grows rapidly as L'/b increases and for $L'/b = 11.0$ reaches maximum values of twice the value corresponding to a single 90° curve.

For a 90° combined mitre bend containing two similar 90° bends (Fig. 4.10c), for ratios of

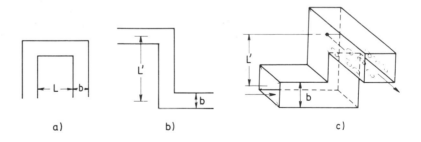

Fig. 4.10

$L'/b > 4$, the head loss, being practically the same for each bend, approaches double that corresponding to a single 90° bend; when L'/b decreases, K increases, reaching for $L'/b = 2$, a value equal to about 2.4 times the value corresponding to a single 90° curve.

† Tests carried out of Zurich Federal Polytechnic School, quoted in [9].
‡ Tests by Klein, Tupper and Green, quoted in [9].

If bends are separated by a spacer of more than 30 diameters interaction effects are not important†.

4.19 Losses at combining and dividing junctions

(a) In a combining junction, the flow takes the form shown in Fig. 4.11.

If the two flows have different velocities, there will be a shock and mixing zone, with

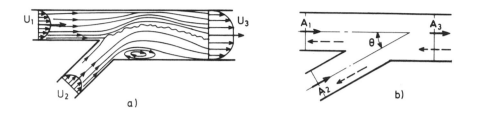

Fig. 4.11

transfer of momentum between the particles of different velocity. In the direction of the more rapid flow, there will always be head losses $(K > 0)$; as regards the lower velocity flow, its velocity increases when it mixes with the more rapid one, and energy may be communicated to it by the more rapid vein $(K < 0)$.

Combining junction losses are fundamentally due to the shock between flows of different velocity, the change of direction of one of the flows and expansion in the divergent section.

We thus have:

$$\Delta H_{1.3} = K_{1.3}\ U_3^2/2g \text{ and } \Delta H_{2.3} = K_{2.3}\ U_3^2/2g$$

Tables 47 and 48 give the values of $K_{1.3}$ and $K_{2.3}$.

One of the ways of reducing head losses in combining junctions is to introduce a readius or chamfer, r_1, or to diffuse the flow before the junction, or both together; in the case of a dividing flow, it is necessary to round the lesser angle at the expense of the radius r_2 (Fig. 4.12a). Even better results are obtained if the combining or dividing is made by

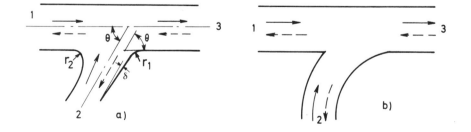

Fig. 4.12

means of a curve (Fig. 4.12b). Table 48 gives the values of K_{13} and K_{23} for some types of non sharp-edged combining junctions.

(b) Table 49 gives the values of $K_{3.2}$ for sharp-edged dividing junctions.

† For more details see Miller [13].

The value of $K_{3.1}$, for $A_3 = A_1$ and $U_1/U_3 < 1$, is given by the following formula:

$$K_{3.1} = 0.4 \, (1 - U_1/U_3)^2 \tag{4.39}$$

The value of $K_{3.1}$ for $A_3 = A_2$ is given in Table 49.

To reduce head losses, use is made of the same schemes given for combining flows (Fig. 4.12), the coefficients $K_{3.1}$ and $K_{3.2}$ being given in Table 50.

(c) *At three-branch symmetrical trifurcations* (Fig. 4.13a) — The following formula applies where the three combining flows meet;

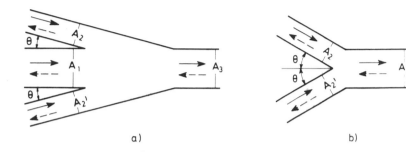

a)   b)

Fig. 4.13

$$K_{23} = A \, \frac{Q_2}{Q_3} + B \left[\left(\frac{Q_2}{Q_3} \right)^4 + \left(1 - \frac{Q_2}{Q_3} \right)^4 \right] - C \left(\frac{Q_3}{Q_2} \right)^2 - D \tag{4.39a}$$

The values of A, B, C and D are given in Table 51.

4.20   Trashracks

(a) *Head losses* — The trashrack is a fundamental element in protecting water intakes from solid bodies: intakes to pumps, turbines, channels, etc.

The rack, as a whole (Fig. 4.14a), consists of rectangular panels supported on a structure defined by transverse and longitudinal bars, this structure being set on the concrete or masonry of the intake. The structural elements of the trashrack are supported on the transverse bars and on intermediate support elements (Fig. 4.14b) in order to prevent vibrations, as is explained below.

From the point of view of head loss, the rack is defined by the distance between bars, a, their dimension, b, in the direction of the flow, their maximum width, e (Fig. 4.14c), and the geometrical shape of their transverse cross-section. Head loss at racks is given by the formula $\Delta H = K U^2/2g$, in which U would be the velocity at the cross-section of the rack in its absence. Values of K are given in Tables 54 and 55.

(b) *Stability of the bars* — The flow, when passing through the bars, shed

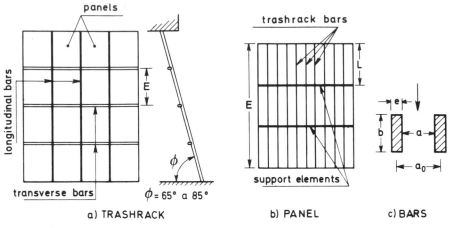

Fig. 4.14

alternate vortices of which it is necessary to know the vortex-shedding frequency, in order that the natural frequencies of the bars, f_b, may be kept away from the frequencies of the vortices, f_t. In this way it is possible to prevent the phenomenon of resonance in the bars, which lead to vibrations and in some cases has caused trashrack failures.

The dominant vortex-shedding frequency is given by:

$$f_t = S_t U/e \qquad (4.40)$$

U is the approach flow velocity; e is the thickness of the bar; S_t is the Strouhal number, a function of the cross-section of the bars and of the ratio $(a + e)/e$ (Fig. 4.14c). Graph 56(a) gives the value of the Strouhal number, S_t, for different shapes of single bars, with small angles of attack† the values of which must be increased by the coefficients of Table 56(a).

The natural frequency of the bars is given by:

$$f_b = M \frac{K}{L^2} \sqrt{\frac{gE_b}{\gamma_b + \frac{a}{e}\gamma}} \qquad (4.40a)$$

in which:

M – *fixity factor*, with the following values:
— *clamped bars*: $M = k/2\pi$ ($k = 22.4$ for the first mode);

† If the flow is turbulent.

— *hinged bars*: $M = k'\pi/2$ ($k' = 1$ for the first mode);

K — *radius of gyration* of the transverse cross-section of the bar in relation to an axis parallel to the velocity of the stream;

L — distance between the supports of the bars;

E_b and γ_b — modulus of elasticity and specific weight of the material of the bar;

γ — specific weight of the fluid.

This formula is only valid for $a < 0.7b$; if $a \geq 0.7b$, $a = 0.7b$ must be used in calculations.

The conditions of stability require that f_b shall be significantly different from f_t.

$$f_b \gg f_t$$

In practice it has been found that for natural frequencies of the bar that are about 1.5 times the vortex-shedding frequency, no dangerous resonances are noted.

Example: Check the stability of a trashrack consisting of bars of rectangular cross-section ($e = 10$ mm; $b = 100$ mm), $a = 30$ mm apart and welded to supports $L = 700$ mm apart. The approach velocity is $U = 1.0$ m/s. The bars are made of steel ($E_b = 2.1 \times 10^{11}$ N/m^2); $\gamma_b = 78\ 000$ N/m^3.

Solution:

Vortex-shedding frequency — For a rectangular cross-section with $b = 10e$, Table 56a gives the value of the Stouhal number, $S_t = 0.240$. For $(a + e)/e = 4$, the multiplying factor, in accordance with Table 56b, will be $c = 1.05$, so that $S_t = 1.05 \times 0.240 = 0.252$.

The vortex-shedding frequency will thus be:

$$f_t = S_t U/e = 0.252 \times 1/0.01 = 25.2 \text{ Hz}$$

Natural frequency of the bars

— $a/b = 30/100 = 0.3 < 0.7$; in the calculations it can therefore be taken that $a = 30$ mm.

  — since the bars are welded to the supports, they can be regarded as clamped: $M = k/2\pi = 3.57$;

  — the radius of gyration of a rectangular cross-section, in relation to an axis parallel to the flow, is (Table 14);

  — $K^2 = e^2/12 = 0.01^2/12 = 8.3 \times 10^{-6}$ m^2; thus $K = 2.9 \times 10^{-3}$ m.

$$\gamma_b + \frac{a}{e}\gamma = 78\ 000 + \frac{0.03}{0.01}\ 10\ 000 = 108\ 000 \text{ N/m}^3$$

$$f_b = M\frac{K}{L^2}\sqrt{\frac{gE_b}{\gamma_b + \dfrac{a}{e}\gamma}} = 3.57 \times \frac{2.9 \times 10^{-3}}{0.7^2}\sqrt{\frac{9.8 \times 2.1 \times 10^{11}}{1.08 \times 10^5}} = 92 \text{ Hz.}$$

Accordingly, $f_b = 4f_t$, which guarantees the stability of the bars. It must be noted, however, that if the approach velocity increases from 1 m/s to 4m/s, the vortex-shedding frequency will increase to $\simeq 100$ Hz and the danger of vibration will become important.

(c) *Dimensioning parameters* — The following approach velocities U, are advisable for trashracks†:

† According to {12}.

— Surface trashrack directly protecting
   hydraulic equipment $U = 0.8 - 0.9$ m/s
— Ditto, protecting a penstock $U = 0.9 - 1.0$ m/s
— Ditto, protecting a conveyance channel $U = 1.0 - 1.1$ m/s
— Trashrack at a medium depth intake (at a depth
   2 to 3 times its height) with automatic cleaning $U = 0.8 - 1.0$ m/s
— Trashrack at a deep intake with mechanical cleaning $U = 0.6 - 0.8$ m/s
— Trashrack at a very deep intake (depth 50 to 100 m) $U = 0.4 - 0.6$ m/s

The following spacing, a, between bars is advisable:

— Kaplan turbine, $n_s = 750$ to 1000, $Q = 150$ m³/s $a = 120\text{--}150$ mm
— Ditto, for $Q = 100$ m³/s $a = 100\text{--}120$ mm
— Francis turbine — high speed $a = 80\text{--}100$ mm
— Ditto — low speed $a = 25\text{--}50$ mm
— Pelton turbine $a = 25\text{--}50$ mm
— Small pumping plants (0.5 to 1 m³/s) $a = 20$ mm

4.21 Head losses in perforated plate
Head losses in perforated plate can be calculated by the expression [11]:

$$K = \frac{(0.71 \sqrt{1 - \eta} + 1 - \eta)^2}{\eta^2} \qquad (4.41)$$

in which $\eta = A_0/A$ is the ratio between the total area of the orifices and the total area of the screen. This expression is valid for a thin plate, that is to say, $e/d_0 < 0.015$ (e being the thickness of the plate and d_1 the diameter of the orifice) and for values of the Reynolds number of the orifice, $\mathbf{R}_{e_0} = V_0 d_0/\nu \geqslant 10^5$, V_0 being the velocity through each orifice.

4.22 Head losses in slots and expansion joints
A slot is characterized by the following elements: l — width in the direction of the flow; b — length normal to the flow; p — depth; θ — angle of the downstream wall with the horizontal.
   Taking λ to represent the ratio l/p, slots are considered to be short if $\lambda < 4$; medium for $4 < \lambda < 6$; long for $\lambda > 6$.
   We have $\Delta H = K U^2/2g$, with $U = Q/A$ the mean velocity in the section upstream and downstream of the slot.
   The values of K are given [12] by

(a) $\lambda < 4$

$$K = \left(\frac{\eta^{1.8} - 1}{1.43 \eta^{-1.8} + 1} \right)^2 \sin\theta \qquad (4.42)$$

in which:

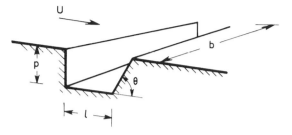

Fig. 4.15

$$\eta = \frac{A + 0.25lb}{A}$$ (4.42a)

(b) $4 < \lambda < 6$

The previous expression is valid with:

$$\eta = \frac{A + pb}{A}$$ (4.42b)

(c) $\lambda > 6$

The total head loss will be given by a coefficient:

$$K = K_1 + K_2$$

in which K_1 is the value corresponding to the previous case and:

$$K_2 = \left(1 - \frac{20}{e^{0.5\lambda}}\right)\left(1 - \frac{1}{\eta}\right)^2$$ (4.42c)

η being given by the expression 4.42b.

Table 55 gives the values of K, K_1 and K_2.

Example: Consider a circular conduit, diameter $d = 1.0$ m, in which there is an expansion joint with the following characteristics: $l = 0.5$ m; $p = 0.08$ m; $\theta = 90°$. The discharge is $Q = 4.2$ m³/s.

Solution:

$$\lambda = l/p = 6.25 > 6; \ b = \pi d = 3.14 \ m; \ A = \pi d^2/4 = 0.78 \ m^2$$

whence:

$$\eta = \frac{0.78 + 0.08 \times 3.14}{0.8} = 1.28$$

From Table 55a, for $\eta = 1.28$ we have $K_1 = 0.035$. For Table 55b we also have for $\eta = 1.28$ and $\lambda = 6.3$, $K_2 = 0.007$; whence $K = K_1 + K_2 = 0.035 + 0.007 \approx 0.042$.

Since: $U = 4.2/0.78 = 5.38$ m/s; $U^2/2g = 1.48$ m, and finally we have $\Delta H = 0.04 \times 1.48 = 0.06$ m.

4.23 Singularities in series

Downstream of a singularity, the turbulence of a flow increases; the flow characteristics return to the upstream conditions at a distance, L_0 that is given approximately, according to Levin [12], by:

$$L_0 = 0.075 \, D \sqrt[4]{\frac{K}{f}} \, \mathbf{R_e} \tag{4.43}$$

in which

$\quad K$ — coefficient of head loss of the singularity;
$\quad f$ — friction factor (Moody diagram);
$\quad \mathbf{R_e}$ — Reynolds number of the pipe;
$\quad D$ — diameter (hydraulic) of the conduit.

The head loss in the singularity thus takes place in distance L_0.

Should there be a further singularity at a distance $L < L_0$, the head losses that would correspond to each singularity considered separately should not be added to one another. Levin proposes a coefficient of reduction of head loss for the first singularity.

$$\Delta H_L = b\Delta H \tag{4.44}$$

with:

$$b = 1 - \frac{1}{e^{5.2\lambda}} \tag{4.44a}$$

$\lambda = L/L_0$. This is, however, a subject in which much remains to be studied.

C.   ENERGY LINE AND PIEZOMETRIC LINE

4.24 Design of the energy line and piezometric line

Consider a conduit with section A, in which flows a given discharge, Q, the total energy, per unit weight flowing in each section in relation to a horizontal datum is (Sections 2.18 and 2.19):

$$E = z + \frac{p}{\gamma} + \alpha\frac{U^2}{2g} = z + \frac{p}{\gamma} + \alpha\frac{Q^2}{2gA^2} \tag{4.45}$$

Considering two sections, 1 and 2, in which the total heads are (Fig. 4.16) E_1 and E_2, the difference $\Delta E_{12} = E_1 - E_2$ represents, according to Bernoulli's theorem, the total energy losses of the hydraulic flow: energy degraded into heat by friction of the particles against one another and of the particles against the walls of the conduit; or hydraulic energy converted into mechanical energy or inversely, by means of a hydraulic machine (turbine or pump).

The energy dissipated or transformed between sections 1 and 2 will be equal to:

$$\Delta W = \gamma Q . \Delta E_{12} \tag{4.46}$$

If there is no hydraulic machine in the circuit, this term corresponds only to head losses, which can be of two types, as has already been seen: *due to friction*, occurring along flowing full pipes, and *local*, due to singularities (contractions, expansions, changes of direction, etc.).

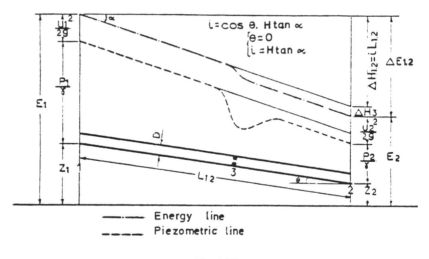

—·—·— Energy line
—————— Piezometric line

Fig. 4.16

The friction head losses between the generic sections a and b are represented by ΔH_{ab}. This head losses per unit length of pipe are, as has been seen, represented by i. With L_{ab} the distance between section a and b, we shall have:

$$\Delta H_{ab} = iL_{ab} \tag{4.47}$$

The local head losses in section c will be represented by ΔH_{c}.

The sum of the friction and local head losses between a and b will be represented by ΔE_{ab}.

In the case of horizontal or near-horizontal conduits (see Fig. 4.16), the energy line has an inclination, i, corresponding energy loss due to friction. Thus $i \approx \tan \alpha$†.

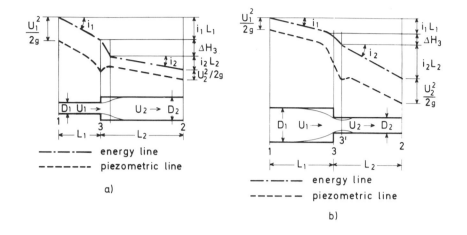

Fig. 4.17

If there are local losses due to singularities, the energy line goes down suddenly (Section 3 of Fig. 4.16 serves as an example).

If the section remains constant, the *piezometric line* is parallel to the energy line and separated from it by a distance equal to $U^2/2g$. If there is a localized increase in velocity (diminution of section), a corresponding lowering of the piezometric line occurs. If there is a diminution of velocity (increase of section), there is a corresponding rise in the piezometric line.

It is to be noted that the energy line always falls in the direction of the flow. The same does not occur with the piezometric line. For example, the details of the energy line and piezometric line, in the cases of a contraction and an expansion, are represented in Fig. 4.17.

4.25 Position of the piezometric line in relation to the conduit

Consider a conduit that connects two reservoirs, this being schematically represented in Fig. 4.18. The piezometric line, corresponding to the relative pressures, is approximately the line AA', that joins the free surfaces of the two reservoirs (excluding the local variations of the piezometric line at the intake and outlet, due to contraction and expansion, as shown in Fig. 4.17).

The piezometric line BB', corresponding to the absolute pressures, is

† Bearing in mind the angles α and θ, of Fig. 4.16, we have $i = \cos \theta \tan \alpha$. If the conduit is horizontal: $\theta = 0$; $\cos \theta = 1$; $i = \tan \alpha$. For small inclination of the conduit, we also have approximately $i = \tan \alpha$. As order of magnitude, it is to be noted that for $\theta = 2°$, which corresponds to an inclination of about 3.5%, $\cos \theta = 0.999$; for $\theta = 8°$, which corresponds to an inclination of about 14%, $\cos \theta = 0.996$.

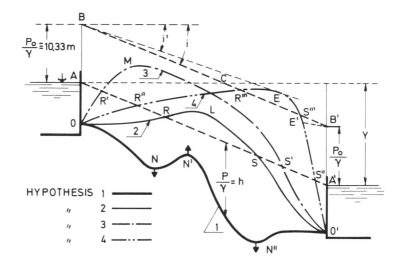

Fig. 4.18

parallel to this line and is vertically separated from it by a length equivalent to the atmospheric pressure; under normal conditions we shall have $p_0/\gamma =$ 10.33 m water column.

In the hypothesis of the whole conduit being situated lower than AA' (Hypothesis 1, line $ONN'N''O'$), the pressure is higher than atmospheric pressure in all sections, by a quantity $h = p/\gamma$. This hypothesis corresponds to the normal situation. Air escape valves must be provided at the high points (N'), for releasing accumulated air, and bottom outlets at the low points (N, N'') for emptying and cleaning, apart from judiciously located cut-off valves.

In the hypothesis of the conduit being represented by $ORLSO'$ (Hypothesis 2), depressions occur relative to atmospheric pressure, in the zone RLS situated above the relative piezometric line AA'. In general, particularly in waterline distribution networks, these depression zones must be avoided. An eventual crack will facilitate the entry of foreign bodies and lead to the contamination of the water. In the length RLS, there will be a release of air dissolved in the water and also production of water vapour. Placing of an air escape valve would be inadvisable since it would allow the entry of air and consequently there would be a reduction in the discharge. If the pressure reaches an absolute pressure equal to the saturated vapour pressure of water, cavitation will begin (at about 10 m of water column), and there is a tendency for the flow to become unstable.

If the conduit proceeds as represented by $OR'MS'O'$ (Hypothesis 3), and rises above the horizontal line that passes through A (relative energy line upstream, in the reservoir), there will only be flow if the whole conduit was filled beforehand with water (siphon effect). There continue to be depressions relative to atmospheric pressure in the length $R'MS'$.

If the conduit proceeds as represented by OR''R'''ES'''ES'''S''O' (Hypothesis 4), always remaining below the head plane (level in the reservoir), but rising above the line BB' (absolute piezometric line), flow begins without the need to resort to siphoning. The piezometric line will, however, cease being BB' and becomes $BEE'B'$. The absolute pressure will thus be null between E and E', and the flow will take place with the section partly full, i.e. as in a channel. The discharge will theoretically be that corresponding to the inclination i'. Owing to the release of air and water vapour in the zones of great depression, the flow is irregular, and this situation must be avoided.

If the conduit rises higher than line BC, though not exceeding the elevation of B, it would be necessary to establish a siphon, with partly full flow occurring in a very irregular manner in part of the conduit.

If the conduit at any point rises above elevation B that represents the absolute energy available, it will be impossible for flow to be started.

D. SPECIAL PROBLEMS

4.26 General expression
When solving prolems of steady pressure flows in pipes, it is necessary to take account of the general equations of hydraulics and the expressions of friction and local energy losses.

Friction losses are, as has been seen, always of the type $\Delta H = aLQ^2$. In general, a is a function of the velocity and, consequently, of Q.

Local losses, as has also been explained, are always of the type $\Delta H = bKQ^2$. K is also a function of the velocity and, therefore, of Q.

The general expression of energy losses will therefore be:

$$\Delta E = Q^2(\Sigma aL + \Sigma bK) \tag{4.48}$$

or

$$Q = \sqrt{\frac{\Delta E}{\Sigma aL + \Sigma bK}} \tag{4.49}$$

4.27 Examples
(1) Reservoir A, with constant level at elevation 600 m feeds reservoir B, also with constant level at elevation 520 m. The water supply conduit is made up of cast iron, consisting of the following: a passage from reservoir A to the pipe, through a sharp-edged entrance, a stretch of 200 m in length and 100 mm in diameter; a stretch of 300 m in length and 200 mm in diameter; a gate valve of 200 mm diameter; a stretch of 100 m in length and 200 mm in diameter; an entry into reservoir B through a sharp exit. Determine the discharge.

Solution: K_s is fixed for Strickler's coefficient of roughness. From Table 21, for $d_1 = 100$ mm and $K_s = 75$ m$^{1/3}$/s, we have, $a_1 = 394.17$ and for $d_2 = 200$ mm, $a_2 = 9.779$. For $d_1 = 100$ mm, we have $b = 826.38$ and for $d_2 = 200$ mm, $b = 51.65$.

The coefficient of local head losses are (Table 53):

> Sharp-edged entrance $K = 0.5$
> Expansion $K = 0.6$
> Gate valve $K = 0.1$
> Sharp exit $K = 1.0$

Accordingly:

$$\Sigma\ aL = 394.17 \times 200 + 9.779 \times 400 = 82,\ 746$$
$$\Sigma\ bK = 826.38(0.5 + 0.6) + 51.65(0.1 + 1.0) = 909 + 57 = 966.$$

$$Q = \sqrt{\frac{\Delta E}{\Sigma\ a\ L + \Sigma\ b\ K}} = \sqrt{\frac{80}{83712}} = 0.031\ \text{m}^3/\text{s}$$

(2) Lighting kerosene is pumped at 20°C into a galvanized iron pipe of 50 mm internal diameter (Fig. 4.19). The pipe is fitted with threaded accessories and discharges into a tank which

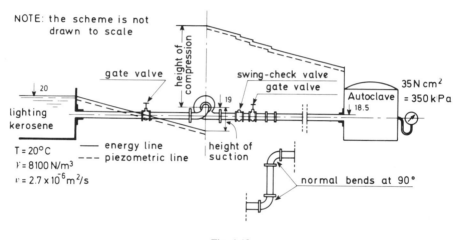

Fig. 4.19

pressure remains constant at 35 N/cm² = 350 kN/m², at the level of entry of the pipe.

Bearing in mind the scheme of Fig. 4.19, it is necessary to determine the suction and compression heads at the pump and the pressure in the outlet section of the pump. The discharge is 6.31 l/s. The length of the suction piping is 6.5 m; the length of the compression piping is 116 m.

Solution: For $d = 5$ cm we have $A = 19.63$ cm² = 0.196 dm². The mean velocity is:

$$U = Q/A = 6.31/0.196 = 32\ \text{dm/s} = 3.2\ \text{m/s}.$$

The velocity head is therefore $U^2/2g = 0.522$ m of kerosene column.

The kinematic viscosity of lighting kerosene at 20°C is $v = 2.7$ centistokes $= 2.7 \times 10^{-6}$ m²/s. The Reynolds number will thus be.

$$\mathbf{R}_e = UD/v = 3.2 \times 0.05/2.7 \times 10^{-6} = 59,000 = 5.90 \times 10^4$$

Table 16 gives $\varepsilon = 0.05$ mm.
So that $\varepsilon/D = 0.001$.
From Graph 17, $f = 0.0234$. The flow is turbulent, though not completely so.

The friction head loss will be $i = f \dfrac{1}{D} \dfrac{U^2}{2g} = 0.0234 \times \dfrac{1}{0.05} + 0.522 =$

$= 0.244$ m of kerosene column per metre of pipe.

The coefficients of local head losses are (Table 53).

Sharp-edged entrance $K = 0.5$
Gate valve $\qquad\qquad K = 0.17$
Swing-check valve $\quad K = 2.00$
Mean radius curve $\quad K = 0.8$
Sharp exit $\qquad\qquad K = 1.0$

(a) *Suction pipe*:

$$\Delta H = iL + \Sigma K \; \frac{U^2}{2g} = 0.244 \times 6.5 + (0.5 + 0.17) \times 0.522 = 1.59 + 0.35 = \quad 1.94 \;\; \text{m}$$

kerosene column.

The *suction head* will thus be:

$$H_a = (20 - 19) - 1.94 = -0.94 \text{ m kerosene column.}$$

(b) *Compression pipe*:

$$\Delta H = iL + \Sigma K \frac{U^2}{2g} = 0.244 \times 116 + (2.0 + 0.17 + 2 \times 0.8 + 1.0) \times 0.522 =$$
$$28.2 + 2.5 = 30.7 \text{ m of kerosene column.}$$

Pressure in the tank, expressed in metres of kerosene column, with density $\rho = 810$ kg/m^3, will be:

$$h = \frac{p}{\rho g} = \frac{350\ 000 \text{ N/m}^2}{9.8 \times 810 \text{ N/m}^3} = 44.09 \text{ m of kerosene column.}$$

The *compression head* will thus be:

$$H_c = (18.5 - 19.0) + 30.7 + 44.09 = 74.29 \text{ m of kerosene column.}$$

(3) For aeration of a hydraulic flow, in order to prevent the occurrence of dangerous negative pressures, a horizontal smooth concrete pipe with a rectangular cross-section of 0.90 m $\times$ 1.20 m and length of 150 m was used. The entrance of the pipe is rounded ($K = 0.10$).

The difference in pressure between the entrance of the pipe, in communication with the atmosphere, and the outlet, in contact with the flow that it is wished to aerate, is equal to 0.20 m of water column, with the air at a temperature of 15.6°C (at sea level). It is wished to find the velocity of the air in the pipe.

Solution: Cross-section of the pipe $A = 0.90 \times 1.20 = 1.08$ m^2; wetted perimeter $P = 2(0.90 + 1.20) = 4.20$ m; hydraulic radius: $R = A/P = 1.08/4.20 = 0.257$ m.

Table 16 gives $\varepsilon = 0.09$ mm, so that $\varepsilon/D = \varepsilon/4R = 0.09/(4 \times 257) = 0.00009$

Assume $\mathbf{R}_e = 2.5 \times 10^6$. From Graph 18 we have $f = 0.0126$

Accordingly:

$$\frac{p_1 - p_2}{\gamma} = \left(K + f\frac{L}{4R}\right)\frac{U^2}{2g} = \left(0.10 + 0.0126 \times \frac{150}{1.028}\right)\frac{U^2}{2g} = 1.93\frac{U^2}{2g}$$

The higher values of K are taken, since the accessories are threaded and the diameter of the pipe is small.

The specific weight of air at 15.6°C (Table 11) is $\gamma = 11.989$ N/m³.

We thus have:

$$\frac{p_1 - p_2}{\gamma} = \frac{0.20 \times 9810}{11.980} = 164 \text{ m} \qquad \text{air column}$$

Thus $U^2/2g = 164/1.93 = 84.97$ m; $U = 40.8$ m/s.

Since R_e was taken arbitrarily, it is necessary to verify it.

We have: $v = 1.47 \times 10^{-5}$ m²/s and $R_e = UD/v = U.4R/v = 40.8 \times 1.028/1.47 \times 10^{-5} = 3.5 \times 10^6$.

Graph 17 indicates that there is no appreciable change in the value of f.

4.28 Pipes in series

In the case of various pipes of different diameters or roughnesses, arranged in series together with their accessories, the application of the general expressions in accordance with 4.26 is sufficient for solving the problem.

If the cross-sections and discharge are known, determination of the head loss is made directly, by applying equation 4.48.

If Q is the unknown, an approximate value is estimated for the velocity, and this enables approximate values to be determined for a and K and, from equation 4.49, an approximate value for Q is obtained. From this value, new values are determined for a and K and a new value for Q, until the desired approximation is obtained.

4.29 Branched pipes

A system is called a branched system when the pipes that compose it subdivide successively from a common feed point, with no circuit ever being closed (Fig. 4.20).

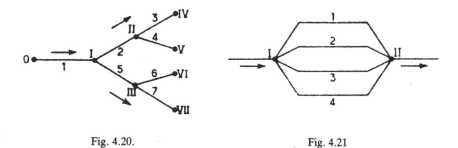

Fig. 4.20.   Fig. 4.21

Given the energy E in the initial section O, an equation of the following type can be written in relation to each terminal point, IV, V, VI and VII:

$$E_0 - E_{IV} = \Delta E_{0.IV} =$$
$$= Q_1^2 \, (a_1 L_1 + b_1 \Sigma K_1) + Q_2^2 \, (a_2 L_2 + b_2 \Sigma K_2) + Q_3^2 (a_3 L_3 +$$
$$+ b_3 \Sigma K_3) \tag{4.50}$$

In relation to each node (I, II and III), there will be an equation of continuity.

$$\Delta Q = O \tag{4.51}$$

If the head difference $\Delta E_{0.IV}$ and the diameters are given, and if it is wished to find the discharges in each branch, we shall have a system of 7 equations and 7 unknown quantities.

If the diameters and the discharges are known, the head at each terminal is determined directly by application of an equation of the type (4.50), since it will then be a case of conduits in series in which the discharges vary from one length of piping to another.

4.30 Pipes in parallel
Given a system in parallel (Fig. 4.21), two cases may occur: the head loss between I and II is known and it is wished to find the discharge; or the total discharge, Q, is known and it is wished to determine the head loss.

The first case is solved directly, since the head loss is known the discharge in each pipe length is known; the total discharge being the sum of the discharges in the various lengths.

The second case requires a calculation by successive approximations that can be carried out as follows:

(a)   A discharge is estimated for one of the conduits, for example Q_1' and the corresponding head loss is calculated, ΔH.
(b)   From the value of ΔH, the values of Q_2' and Q_3' are calculated for the other conduits. These discharges are summed, and Q' is obtained.
(c)   The total discharge, Q is divided, for each conduit, in the same proportion as the calculated discharges. We then have $Q'' = Q_i' \times Q/Q'$.
(d)   From these values, the head loss in each conduit is calculated. If the results are sufficiently accurate, the problem is considered to be solved; otherwise the calculation is repeated, on the basis of the value Q_i'', previously determined.

4.31 Several reservoirs connected to one another
In the case of several reservoirs connected to one another (Fig. 4.22), the equation of continuity to be satisfied will be $Q_1 = Q_2 + Q_3$ or $Q_1 + Q_2 = Q_3$; according to the relative position of the levels of the reservoirs. The procedure is thus the following:

(a)   the piezometric elevation at point I is estimated;

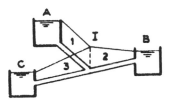

Fig. 4.22

(b) Q_1, Q_2 and Q_3 are calculated from that value;
(c) if the equation of continuity is satisifed, the problem is solved. Otherwise a new value is fixed for the piezometric elevation at I, the previous value being corrected from observation of the equation of continuity, until a value which satisfies that equation is obtained.

4.32 Complex systems. Pipe networks

A case of a complex system consists of calculating a distribution network with the help of meshes, in which it is wished to find a balanced solution for the discharges and head losses. Resolution of these systems must comply with the following laws:

(a) the algebraic sum of the successive head losses in each mesh must be zero;
(b) the discharges flowing to a node must be equal to the discharges leaving that node;
(c) on each side there must be compliance with the law of head loss between the extremes.

One of the ways of doing the calculation is the *Cross* method, which is as follows:

— choose a distribution of discharges that satisfies the equation of continuity;
— for each side of the mesh calculate the head loss $\Delta H = aLQ^2 = sQ^2$, with $s = aL$. In each circuite we shall have:

$$\Sigma\Delta H = \Sigma sQ^2 \qquad\qquad (4.52)$$

bearing in mind the sign of ΔH;
— ·for each circuit, make the sum;

$$\Sigma 2sQ \qquad\qquad (4.53)$$

without taking the signs into consideration:
— if $\Sigma\Delta H \neq 0$, a discharge of ΔQ is added to each circuit so that:

$$\Delta Q = \frac{\Sigma s Q^2 \text{ (taking the signs into consideration)}}{\Sigma 2sQ \text{ (not taking the signs into consideration)}} \qquad (4.54)$$

— the calculation is repeated until the necessary approximation is obtained.

Example [8]: It is wished to calculate the discharge in the system shown in Fig. 4.23, of which, for each pipe, the length, L and the diameter, D, are known.

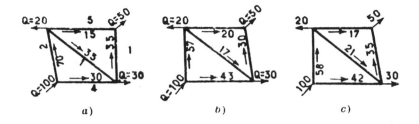

$a)$ $\qquad\qquad\qquad\qquad b)$ $\qquad\qquad\qquad\qquad c)$

Fig. 4.23

Suppose that the values of s for each pipe have been calculated by the formulae given previously and are as indicated on Fig. 4.23a.

Then the equations can be written, bearing in mind the appropriate signs:

	Circuit nr. 1 (Fig. 4.23, left)		Circuit nr. 2 (Fig. 4.23, right)	
1st approximation	$+70^2 \times 2 =$ 9800 $+35^2 \times 1 =$ 1225 $-30^2 \times 4 = -3600$ $\Sigma s Q^2 = +7425$	$2 \times 70 \times 2 = 280$ $2 \times 35 \times 1 = 70$ $2 \times 30 \times 4 = 240$ $\Sigma snQ = 590$	$+15^2 \times 5 =$ 1125 $-35^2 \times 1 = -1225$ $-35^2 \times 1 = -1225$ $\Sigma s Q^2 = -1325$	$2 \times 15 \times 5 = 150$ $2 \times 35 \times 1 = 70$ $2 \times 35 \times 1 = 70$ $\Sigma snQ = 290$
	$\Delta Q = \dfrac{+7425}{590} = +13$		$\Delta Q = \dfrac{-1325}{290} = -5$	
2nd approximation	$+57^2 \times 2 =$ 6500 $+17^2 \times 1 =$ 289 $-43^2 \times 4 = -7400$ $- 611$	$2 \times 57 \times 2 = 228$ $2 \times 19 \times 1 = 38$ $2 \times 43 \times 4 = 344$ 610	$+20^2 \times 5 =$ 2000 $-17^2 \times 1 = -289$ $-30^2 \times 1 = -900$ $+ 811$	$2 \times 20 \times 5 = 200$ $2 \times 17 \times 1 = 34$ $2 \times 30 \times 1 = 60$ 294
	$\Delta Q = \dfrac{- -611}{610} = -1$		$\Delta Q = \dfrac{+811}{294} = 6$	
3rd approximation	$+58^2 \times 2 =$ 6470 $+21^2 \times 1 =$ 441 $-42^2 \times 4 = -7050$ $- 131$	$2 \times 58 \times 2 = 232$ $2 \times 21 \times 1 = 42$ $2 \times 42 \times 4 = 336$ 610	$+17^2 \times 5 =$ 1444 $-21^2 \times 1 = -441$ $-33^2 \times 1 = -1089$ $- 86$	$2 \times 17 \times 5 = 170$ $2 \times 21 \times 1 = 42$ $2 \times 33 \times 1 = 66$ 278
	$\Delta Q = \dfrac{-131}{610} \approx 0$		$\Delta Q = \dfrac{-86}{278} \approx 0$	

In the first approximation the value of $\Sigma s Q^2$ is positive in circuit number 1, giving $\Delta Q = 13$ m³/s.

Then it is necessary to subtract 13 m³/s from each positive discharge and to add 13 m³/s to each negative discharge to obtain $\Sigma s Q^2 = 0$. In circuit number 2 the converse occurs: it is necessary to add 5 m³/s to the discharges which were considered as positive and to subtract 5 m³/s from the discharges which were considered negative. The common pipe is subject to both corrections.

We shall then have the new distribution shown in Fig. 4.23b. In the same way, the distribution shown in Fig. 4.23c was obtained, in which the values of ΔQ are practically negligible.

4.33 Conduits with constant discharge and variable diameter

In the case represented schematically in Fig. 4.24, the continuous head loss

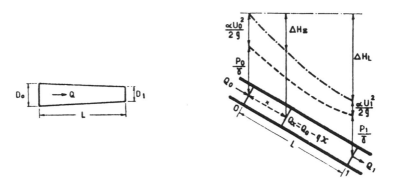

Fig. 4.24.   Fig. 4.25

that occurs between 0 and 1 is:

$$\Delta H_{0.1} = \int_0^L a_x Q^2 \, dx = Q^2 \int_0^L ax \, dx \tag{4.55}$$

By substituting and integrating, we have:

$$\Delta H_{0.1} = \frac{1}{4} a_1 Q^2 \, c^5 \left(\frac{1}{c^4} - \frac{1}{(c+L)^4} \right) \tag{4.56}$$

in which a_1 is the coefficient of head loss for the discharge Q at the diameter d_1 and:

$$c = L d_1 / (d_0 - d_1) \tag{4.57}$$

4.34 Conduits with constant diameter and uniformly varying discharge along their lengths

In the case represented schematically in Fig. 4.25, if Q_0 is the discharge that enters the upstream section and q the discharge leaving the conduit per unit length of the conduit, the discharge, at a distance x from the initial section, is:

$$Q_x = Q_0 - P_x = Q_1 + P_x(L - x) \tag{4.57a}$$

in which:

$$P_x = qx \tag{4.58}$$

The head loss over a length x is written:

$$\Delta H_x = \int_0^x aQ_x^2 \, dx \tag{4.59}$$

which gives approximately:

$$\Delta H_x = ax(Q_x + 0.55 \, P_x)^2 \tag{4.59a}$$

Over a distance L we have:

$$\Delta H_L = aL(Q_1 + 0.55 \, P)^2 \tag{4.60}$$

in which $P = qL$, Q_1 is known as the *end discharge* and the term $Q_1 + 0.55P$ is known as the *equivalent end discharge*.

We find that:

(a)   the piezometric line is a cubic parabola in x;
(b)   when there is no discharge along the length, the head loss is given by $\Delta H = aLQ^2$, as expected;
(c)   when $Q_1 = 0$, the head loss is about 1/3 of that experienced if the entire flow is discharged at the downstream end.

From the practical point of view, it is possible to proceed in the following ways: either by calculating the head loss corresponding to the initial discharge, Q_0, and taking a third of the value thus determined; or by calculating the head loss for the fictitious discharge, known as *equivalent end discharge*.

BIBLIOGRAPHY

[1] Bakmeteff, B. A., *Méchanique de l'Ecoulement Turbulent des Fluides*, Dunod, Paris, 1941.
[2] Dubin, C., *Recueil de Tables et Abaques pour le Calcul des Adductions d'Eau*, Compagnie Générale des Eaux.
[3] Forcheimer, P., *Tratado de Hidraulica*, Editorial Labor, Buenos Aires, 1939.
[4] Nece, R. E. and Dubois, E. R., *Hydraulic Performance of Check and*

Control Valves, Journal of the Boston Society of Civil Engineers, **42**, 3, 1955.

[5] Oniga, T., *Calcul des Tuyaux*, Matémine, Paris, 1949.

[6] Puppini, U., *Idraulica*, Nicola Zanichelli, Bologna, 1947.

[7] Rouse, H. and Hassam, M. M., *Cavitation Free Inlets and Contractions*, Mechnical Engineering, March, 1941.

[8] Scimemi, E., *Compendio di Idraulica*, Libreria Universitaria de G. Randi, Padua, 1952.

[9] Schlag, A., *Hydraulique Générale et Mécanique des Fluides*, Dunod, Paris, 1950.

[10] *La Houille Blanche*, May–June, 1947, and November, 1959.

[11] Idel'cik, I. E., *Memento des Pertes de Charges*, Eyrolles, Paris, 1969.

[12] Levin, L., *Formulaire des Conduites Forcées, Oléoducs et Conduits d'Aération*, Dunod, Paris, 1968.

[13] Miller, D. S., *Internal Flow Systems*, BHRA Fluid Engineering Series, Vol. 5, 1978.

5

Open-channel flows. Uniform flow

A. BASIC CONCEPTS

5.1 Reynolds number and Froude number

Open-channel flows, like closed-conduit flows, continue to be characterized by the Reynolds number, which is an inverse measure of the effect of viscosity. They are, however, also a function of the non-dimensional parameter that represents the influence of gravity and which, as previously explained (Section 1.18) is known as the Froude number. The expression of the Reynolds number, for free surface flows, is:

$$\mathbf{R}_e = \frac{UD}{v} \tag{5.1}$$

in which U is the mean velocity, D is the hydraulic diameter† equal to $4R$, where R is the hydraulic radius (see Section 1.15) and v is the coefficient of kinematic viscosity. For channels of unlimited width, $R = h$, in which h represents the depth of water. The Froude number for free surface flows is:

$$\mathbf{F}_r = \frac{U}{\sqrt{gh}} \tag{5.2}$$

This expression is also known as the *kinetic factor*, and represents the ratio of the velocity of the flow U to the velocity of propagation of the small disturbances (wave velocity), $\sqrt{gh}$. Some authors consider the square of this value.

---

† The hydraulic mean radius usually known as hydraulic radius, R, is also used as a geometrical parameter. The values thus obtained are 1/4 of the values given by (5.1).

5.2   Types of flow
(a) Influence of the Reynolds number, R_c
In free surface flows, laminar flow is found for values of the Reynolds number, defined in the equation (5.1), that are less than 2000. This flow only occurs in very small channels or with very low velocities, and its technical application is almost exclusively confined to the theory of lubrication. It may also occur in flows with reduced water depths, as sometimes happens in very flat fields when flooded. For values of the Reynolds number greater than 2000, the flow becomes turbulent.†

(b) Influence of Froude number, F_r
The velocity of propagation of small disturbances, in a rectangular channel of undefined width, is:

$$V_c = \sqrt{gh} \tag{5.3}$$

This velocity is known as *critical velocity* or *wave velocity*. In a channel of unlimited width, if the mean velocity of the stream is greater than this value, i.e. if the Froude number, F_r, is greater than 1, the small disturbances are not propagated upstream and the flow is known as *supercritical*; if $U < \sqrt{gh}$, that is to say, $F_r < 1$, the small disturbances are propagated upstream and the flow is known as *subcritical*. If $U = \sqrt{gh}$ and therefore $F_r = 1$, the flow is known as *critical*. In Chapter 6, on the basis of considerations of energy and momentum, new definitions will be given for these types of flow.

5.3   Distribution of pressures and velocities
The pressure at a depth y (measured normal to the bottom), in uniform flow, is:

$$p = \gamma y \cos \theta \tag{5.4}$$

in which θ is the angle of the bottom of the channel with the horizontal (Fig. 5.1).

For $y = 0$ have $p = 0$, that is to say, the relative piezometric line always coincides with the free surface.

Since generally $\theta \approx 0$ and $\cos \theta = 1$, we have:

$$p \approx \gamma y \tag{5.4a}$$

---

† From the qualitative point of view, the phenomenon is analogous to that in closed conduit flows. From the quantitative point of view, it has been less studied, since it is more difficult to determine energy losses in free surface flows.

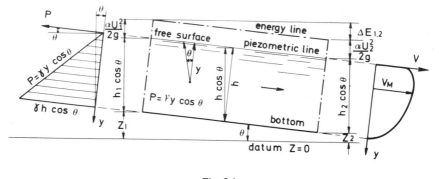

Fig. 5.1.

This expression is usually used† and it corresponds to a hydrostatic pressure distribution.

Should the bottom of the channel be concave (Fig. 5.2a) or convex (Fig. 5.2b), then the flow is not uniform in the longitudinal direction and there will be an increase or decrease of pressures owing to the effect of the centrifugal force originating in the curve of the streamlines. The distribution is not longer hydrostatic and the mean pressure in water column, at depth y, will not be equal to y but will have a value of $p/\gamma = y \pm a$, where:

$$a = \pm \frac{h}{g} \frac{U^2}{r} \tag{5.4b}$$

in which h is the water depth, r is the radius of curvature of the bottom and U, in practical terms, is the mean velocity in the section (Fig. 5.2). The positive sign is used if the bottom is concave and the negative sign if it is convex.

In the case of mild curvatures, i.e. if r is very large, this corrective term has no significance and a hydrostatic pressure distribution can be accepted.

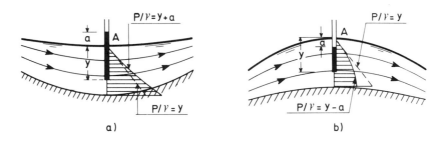

Fig. 5.2.

† In order to justify this simplification, the following values are given:

θ	10′	60′ = 1°	5°	10°	20°	30°	45°
$I \approx \tan\theta$	0.00291	0.01746	0.08749	0.17633	0.36397	0.57735	1.00000
$\cos\theta$	1.00000	0.99985	0.99619	0.98481	0.93969	0.86603	0.70711

The distribution of mean velocities in the section varies a great deal according to the shape of the section. Generally speaking, in artificial canals of regular shape, the distribution of velocities follows an approximately parabolic law, with values that decrease with depth (Fig. 5.1).

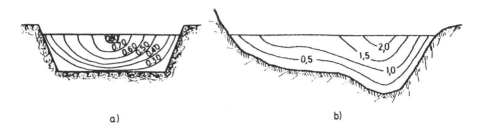

Fig. 5.3.

Fig. 5.3(a) is a typical example of the shape of the isotachs in trapezoidal canals. The maximum velocity occurs a little below the free surface, in a flow zone known as *filament*.

Fig. 5.3(b) shows the distribution of velocities in an irregular channel.

With U representing the mean velocity in the section, V_M the maximum velocity in the free surface, V_f the velocity next to the bottom, V_y the velocity at depth y and h the depth of water in the channel, the following approximate relationships can be used:

Author	U	V_f	V_y
Prony	$0.8\,V_M$	$0.6\,V_M$	—
Fargue	$0.842\,V_M$	—	—
Bazin	$V_M - 14\,Ri$	—	$V_M - 20\left(\dfrac{y}{h}\right)^2\sqrt{hi}$

B.   FLOW RESISTANCE. ENERGY LOSSES

5.4   Uniform flow formulae

In a uniform flow, energy losses, ΔE_{12}, in relation to a datum (see Fig. 5.1) are fully compensated by the lowering of the bottom of the channel.

In open-channel flows, it is convenient to consider the energy in relation to the bottom of the channel, H, often termed the *specific energy*.

$$H = \frac{p}{\gamma} + \frac{\alpha U^2}{2g} = h + \frac{\alpha U^2}{2g} \tag{5.5}$$

In uniform flow, the specific energy, H, and the depth of the water, h, remain constant; there is a parallelism between the bottom of the channel, the water surface and the energy line (or head line).

Thus (Fig. 5.1)

$$I = J = i \tag{5.6}$$

in which $I = \sin \theta \approx \tan \theta$ is the slope of the bottom of the channel, θ being the angle to the horizontal; J = slope of the free surface; i = energy loss per unit weight discharged and per unit length of the channel, also called *energy slope* or *friction slope*.

Generally speaking, the precision obtained in open-channel flows is less than that obtained in pipes. In effect, in the latter the section is more constant and it is easier to obtain uniform flow. Thus, there is higher inaccuracy in the calculation of open-channels flows. Broadly speaking, at least from a qualitative point of view, friction head losses are given by expressions analogous to those shown for closed-conduit flows (see Sections 4.5 and 4.6).

If the flow is laminar, which rarely occurs in cases of hydraulic application, the *Poiseuille formula* is valid, provided that the water surface is considered as the lower half of a pipe having the free surface as plane of symmetry.

In the case of turbulent flow, which is usual in hydraulic phenomena, the most commonly used formulae are those of Chézy and Manning.

(a) Chézy formula (see equations (4.18), (4.19) and (4.20) and Section 4.6)
The values of the coefficients K_B and K_k represent the roughness of the surface and are given in Tables 58 and 59, respectively, for channels.

(b) Manning–Strickler formula (see Section 4.6 and equation 4.14)
The values of K_s are given by Table 60 for regular canals and by Tables 61 and 62 for natural watercourses. Between coefficient K_s and coefficient C, of the Chézy formula, there is the relationship $C = K_s R^{1/6}$.

Roughness of channels, in non-cohesive material, is a function of the mean diameter of the particles and of the hydraulic radius. We thus have, approximately:†

$$K_s = \frac{1}{n} \approx 26 \left(\frac{1}{d_{65}} \right)^{1/6} \text{m}^{1/3}/\text{s} \tag{5.7}$$

in which d_{65} is the diameter in metres such that 65% of the material is of a lesser diameter. This expression, deduced experimentally by Strickler, is

---

† Various expressions of this type, all approximate, have been presented by Strickler (1923), Meyer–Peter (1948), Keulegan (1949), etc.

valid for $4 < R/k < 4000$, in which R is the hydraulic radius and k the equivalent roughness, which can be taken as equal to d_{65}.

When designing a channel, attention should be paid to the fact that the finer elements will be transported (see Section 5.9); thus the roughness must be selected in relation to the material which, according to the tractive forces, or the assumed velocities, will not be transported (threshold conditions).

The advantage of the Manning formula over the Bazin or Kutter formulae is its logarithmic form and its easier to use in calculations.

5.5   Determination of the uniform depth: discharge capacity curves

Chézy and Manning formulae can be written, respectively:

$$\frac{Q}{\sqrt{i}} = CA\sqrt{R}; \qquad \frac{Q}{\sqrt{i}} = K_s AR^{2/3} \tag{5.8}$$

The second members of these equalities are functions only of the nature of the surface boundaries (by means of C or K_s) and of the geometrical nature of the cross-section; once the nature of the surface is fixed, they are functions of the wetted section and, consequently, of the depth of water.

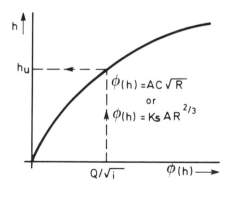

Fig. 5.4.

When the function $\psi(h) = CA\sqrt{R}$ or $\phi(h) = K_s AR^{2/3}$ has been plotted (Fig. 5.4), the value of h, with a corresponding value of $\phi(h) = Q/\sqrt{i} = Q/\sqrt{i}$, is the value corresponding to the uniform depth, h_u. The function $\phi(h)$ is known as the *discharge capacity curve* of the cross-section. The curve Q/K_s $\sqrt{i}$ could also be used.

Whereas in open sections the discharge always grows with an increase in the water depth, this does not occur in closed sections. In the latter case at deeper depths the wetted perimeter increases more rapidly than does the cross-sectional area, causing a reduction in the hydraulic radius and there-fore of the discharge capacity (see Graphs 65, 69, 71, 75, 78 to 82). In general, the flow corresponding to the maximum discharge in a closed section is characteristically unstable and so the discharge must not be considered higher than that of the section running completely full.

Graphs and Tables 63 to 89 give geometric and hydraulic data for the resolution of various problems in uniform flow, for different types of section (see Table 57, Index).

Example:
(1) A canal with trapezoidal cross-section has the following characteristics: width at the base $l = 4$ m; slope of the sides $m = 1/1$; roughness of the surfaces $K_B = 0.16$ m$^{1/2}$; slope of the bottom $I = 0.30$ metres per kilometre; depth of water $h = 1.60$ m. Calculate U and Q.

Solution:

$$A = h(l + mh) = 1.60 \ (4 + 1.60) = 8.96 \text{ m}^2; \ P = l + 2h\sqrt{2} = 4 + 2\sqrt{2} \times 1.60 =$$
$$= 8.52 \text{ m}; \ R = \frac{A}{P} = \frac{8.96}{8.52} = 1.05 \text{ m}$$

Thus:

$$U = C\sqrt{Ri} = 75.2\sqrt{1.05 \times 3 \times 10^{-4}} = 1.33 \text{ m/s}$$

$$Q = UA = 1.34 \times 8.96 = 11.9 \text{ m}^3/s$$

(2) A canal of rectangular section has a width of 4 m; roughness of the canal surfaces, according to Kutter, is $K_k = = 0.25$ m$^{1/2}$ and the slope of the bottom is $40^0/_{00}$. Determine the water depth at which, in uniform flow, there will be a discharge of 170 m^3/s.

Solution:

$$Q/\sqrt{i} = 170/\sqrt{0.04} = 850 \ m^3/s$$

In the table given below the function $\psi(h) = CA\sqrt{R}$ is calculated.

h m	A m^2	P m	R m	$\sqrt{R}$ m$^{1/2}$	$C = \dfrac{100\sqrt{R}}{0.25 + \sqrt{R}}$	$\psi(h) = CA\sqrt{R}$ m^3/s
2.40	9.6	8.8	1.09	1.044	80.7	809
2.45	9.8	8.9	1.10	1.049	80.8	831
2.50	10.0	9.0	1.11	1.054	80.8	852
2.55	10.2	9.1	1.12	1.058	80.9	873
2.60	10.4	9.2	1.13	1.063	81.0	895

By interpolation or by graphically plotting the curve $\psi(h)$, we have $h_u \approx 2.50$ m.
(3) For the preceding example, determine the slope which, with uniform flow, corresponds to a water depth of 2.52 m.

Solution:
By interpolation or by graphically plotting $\psi(h)$, for $h = 2.52$ m, we have $\psi(h) = 860$ m³/s. Thus $\sqrt{i} = Q/\psi(h) = 0.1977$, whence $I = 39.1\%$ since $i = I$.
(4) Other examples are given with the tables for uniform flow.

5.6 Maximum discharge cross-section†

Sometimes it is necessary to determine, for certain geometrical shapes, what cross-section, of equal area, has greatest discharge capacity. For the same area, A, the discharge will of course be maximum when the hydraulic radius, R, is maximum, and consequently, since A is constant, when the wetted perimeter P is minimum.

The ideal cross-section would, therefore, be a semicircle. The water depth will be equal to the radius of the circle and the hydraulic radius will be $R = h/2$.

The isosceles trapezoidal cross-section that in equality of area corresponds to the maximum discharge is, for each slope of the sides m, that which can be circumscribed to a semicircumference whose diameter coincides with the free surface (Fig. 5.5).

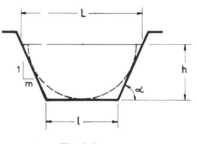

Fig. 5.5.

The rectangular cross-section can be regarded as the boundary case of the previous section, when m tends to zero, which gives a width equal to twice the depth.

It is easy to ascertain that, of the three profiles indicated, the semicircle is the one which, in order to allow the same discharge, requires lesser dimensions. Since, however, it is difficult to construct, the trapezoidal cross-section is more frequently used.

Within the trapezoidal profile there is sometimes a departure from that corresponding to the maximum discharge, when this profile calls for great depths.

It should be noted that when the canal is lined, the minimum cost may not correspond to the minimum amount of excavation. In practice, from the various dimensions that may satisfy requirements, a cross-section with minimum cost must be sought.

For studying trapezoidal cross-sections corresponding to maximum discharge, a parameter M is defined.

$$M = 2\sqrt{1 + m^2} - m \qquad (5.9)$$

The following ratios occur:

Wetted section $\qquad A = h^2 M \qquad\qquad\qquad\qquad\qquad (5.9a)$

Mean radius $\qquad R = \dfrac{h}{2} = \dfrac{\sqrt{A}}{2\sqrt{M}} \qquad\qquad\qquad (5.9b)$

Width at bottom $\qquad l = h\,(M - m) = \sqrt{A}\left(\sqrt{M} - \dfrac{m}{\sqrt{M}}\right) \qquad (5.9c)$

† Also called *most efficient channel section*.

5.7 Compound cross-sections

In the case shown in Fig. 5.6, consisting of a minor bed and a major bed, the discharge must be calculated by adding the discharge corresponding to the whole central section A_c, defined by the points C'CDEFF' and corresponding to the wetted perimeter $\overline{CD} + \overline{DE} + \overline{EF}$; and the value corresponding to the two lateral stretches ABCC' and HGFF', corresponding to the wetted perimeters $\overline{AB} + \overline{BC}$ and $\overline{HG} + \overline{GF}$, respectively.

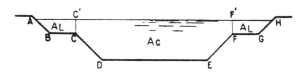

Fig. 5.6.

In the transverse cross-section of a flow, if there are various types of roughness, it is possible, in accordance with *Einstein's formula*, to consider a Strickler's coefficient of roughness, $\overline{K}_s$, for the whole, given by:

$$\overline{K}_s = \left[\frac{P}{\Sigma \Delta P_i / K_{s_i}^{3/2}} \right]^{2/3}$$

(5.10)

in which P is the total wetted perimeter and ΔP_i is the length of the wetted perimeter which has a corresponding coefficient K_{s_i}.†

5.8 Freeboard in channels

As has already been mentioned, the calculation of head losses in open-channel flows cannot be solved as accurately as those for closed-conduit flows. An unforeseen head loss causes a rise in the free surface and the risk of an overflow.

Accordingly, it is always necessary to provide freeboard in a channel, above the water line calculated, in order to take account of the difficulty in calculating continuous and localized losses, surcharge at bends and curves,

---

† The method sometimes used, which consists of defining a Strickler's coefficient as the mean weighted in proportion to the lengths, of the various values of K_{s_i}, is *not acceptable*. In other words, the expression

$$P\overline{K}_s = \Sigma \Delta P_i K_{s_i}$$

is not valid. (Consult: Remenieras, G., and Bourguignon, P., *Prédétermination des pertes de charge d'une canalisation d'eau par circulation d'air*. Le Génie Civil, Nos. 5, 7, 6 and 9, 1953).

waves generated by the wind, variation in atmospheric pressure, the accumulation of solid debris, growth of vegetation, etc.

Generally the freeboard varies between 0.30 m for small channels and 0.60 to 1.20 m for large channels. As a rule, it is reasonable to consider 1/4 of the depth. It is essential, however, to bear in mind the circumstances that may lead to an increase or decrease of such values.

C. STABILITY OF UNLINED CHANNELS

5.9 Dimensioning criteria for noncohesive materials

When dimensioning an unlined channel in which the banks and bottom consist of noncohesive materials, it is necessary to ensure its stability in relation to the hydrodynamic forces generated by the flow. In fact, if the hydrodynamic forces acting on the particles of the noncohesive material of the banks and bottom are small enough, those particles remain stable. However, as the hydrodynamic forces increase, a situation may be reached in which the particles are dislodged from their initial position and move with the flow. Motion is not, however, instantly realised for all the particles when a gradual increase of the hydrodynamic forces occurs. In effect, the random nature of the forces, which is typical of turbulent flows, determines that the start of the process of displacement of the particles is similarly random. Nevertheless, the lighter particles are, on average, dislodged from their initial positions more rapidly. If the weight of the particles constituting the noncohesive material varies considerably, a situation may occur in which only the lighter particles move.

The conditions in which the particles of noncohesive material constituting the bottom and banks of a channel start moving, known as critical conditions or threshold of movement, can be expressed in terms of the velocities of the flow, as *critical velocities*, or the shear stresses exerted on the particles, as *critical shear stresses*. These two criteria, given below, have been used in analysing the stability of channels.

5.10 Critical velocities

In most practical applications it is impossible to determine, with sufficient precision, the critical velocity at the channel bed $(V_f)_{crit}$. For this reason, the analysis of the stability of the bottom of channels is traditionally based on the mean velocity of the flow. As a general indication, Table 88 gives the *mean critical velocities* for various materials, together with the corrective factors that must be used in the case of non-rectilinear channels.

Graph 89 shows the results of *Hjulström's†* experiments for particles of uniform diameter.

Neill [10] presented the following conservative equation for dimensioning channels with noncohesive and uniform bottom material:

† Hjulström (1953) quoted by [12].

$$\frac{U_{crit}^2}{(\gamma_s/\gamma - 1)d} = 2.5 \times 10^{-4} \left(\frac{d}{h}\right)^{-0.20} \tag{5.11}$$

in which U = mean velocity of the flow (m/s); d = mean diameter of the bottom material (mm); h = mean depth of the flow (m).

The use of criteria for dimensioning stable channels, based on the critical velocity, has been rightly critized by various researchers (e.g. Lane [5]). In fact, the analysis of stability is based on the velocity of the flow at the bottom and not on the mean velocity of the flow. For two flows with the same mean velocity, with the same bottom material, but with different depths, the velocity at the bottom is higher for the flow with less depth. In analysing the stability of channels, using a criterion based on mean critical velocities, it is always necessary to take account of the assumption on which the criteria for motion were based, particularly in relation to the depth of flow (Table 88(b)). There are, however, in practice various situations in which only the mean velocity of the flow is known, so that the criteria based on mean velocities are the only ones allowing an analysis of the problem, at least as a first approach.

5.11 Critical shear stress
(a) Stability of the material at the bottom
The drawbacks in using the criterion of critical velocities when dimensioning stable channels, have led to the development of the criteria of *critical shear stress*. For a two-dimensional flow (in a rectangular channel of undefined width) the shear stress exerted by the flow at the bed, τ_0, is given by

$$\tau_0 = \gamma h i \tag{5.12}$$

The dimensions of τ are $ML^{-1}T^{-2}$; in the SI system it is expressed in N/m^2.

In the general case, the maximum shear stress at the bed of a channel is given by:

$$\tau_0 = \gamma R i \tag{5.12a}$$

Table 90 shows the distribution of the shear stress, τ, at the bed and on the banks of trapezoidal and triangular channels, with various bank slopes.

For practical purposes, the maximum values of the shear stress at the bed of a channel can be taken as $\tau_0 = \gamma h i$, and on the banks as $\tau_0' = 0.76\, \gamma h i$ [7].

A non-dimensional parameter, τ^*, known as *Shields parameter* can be defined, τ^*, as:

$$\frac{(\tau_0)_{crit}}{(\gamma_s - \gamma)\, d} = \tau^* \tag{5.13}$$

Shields found that the parameter τ^* was not constant, and analysed the dependence of τ^* on R_e^*, in which R_e^* is a non-dimensional parameter defined by

$$R_e^* = \frac{u^* d}{\nu} \tag{5.14}$$

where ν is the viscosity of the water and u^* is the friction velocity at the bed, defined by $u^* = \sqrt{\tau_0/\rho}$.

In view of the analogy between equation 5.14 and the equation that defines the Reynolds number, the parameter R_e^* is known as the *boundary Reynolds number*. The functional relationship between τ^* and R_e^*, derived experimentally by Shields, is represented in Graph 94 and is widely known as *Shields curve*.

The curve shows that, for $R_e^* > 400$,

$$\frac{(\tau_0)_{\text{crit}}}{(\gamma_s - \gamma)d} = 0.06 \tag{5.14a}$$

Shields curve was established for flows in sandy beds, with grains of uniform dimensions. For noncohesive material with grains of various dimensions, d represents the median of the grain-size curve. It is found that in the case of sand with grains of various dimensions, the Shields criterion is conservative.

The principal inconvenience of using Shields curve in dimensioning stable channels is due to the fact that the shear stress of the flow, τ_0, appears simultaneously on the abscissae (in the parameter R_e^*) and on the ordinate (in the parameter τ^*). This means that in calculation it becomes necessary to use an iterative method. In order to obviate this drawback, Graph 91.b shows the ratio between the critical shear stress and the mean diameter of the material, extracted from the Shields curve.

An approximation to the value of the critical shear stress, for coarse non-cohesive materials, is obtained by means of the *Lane criterion* [5], which can be expressed by

$$(\tau_0)_{\text{crit}} \approx 8 \times d_{75} \tag{5.15}$$

with $(\tau_0)_{\text{crit}}$ in N/m^2 and d_{75} in cm. This equation is valid for rectilinear channels and for materials with specific weight $\gamma = 26 \times 10^3$ N/m^3. For materials with different specific weight, γ', the value $(\tau_0)_{\text{crit}}$ must be multiplied by the factor $c = \gamma'/\gamma$.

For application of Lane criterion, consult Table 92, which also gives indications for fine noncohesive materials and cohesive materials.

(b) Bank stability

An analysis of the stability of channel side slopes is conceptually analogous to the analysis mentioned for the bed material.

For rectilinear flows, the critical shear stress for a particle on the slope of the bank can be expressed in terms of the critical shear stress for a bed particle by means of the following equation:

$$(\tau_0^t)_{crit} = K \, (\tau_0)_{crit} \tag{5.16}$$

The value of K can be obtained directly from Table 94.

A first condition of equilibrium of the material on the channel side slopes is, of course, expressed by $\phi < \psi$, that is to say, the angle of the side slope to the horizontal must be less than the angle of repose of the bed material (Graph 93b). It must also be noted that the maximum shear stress exerted by the flow on the side slopes is less than the maximum shear stress exerted on the bottom (see Table 90).

Table 93 gives indications of the angles of repose that in principle are advisable for various cohesive and noncohesive materials.

5.12 Example

Determine the dimensions of a trapezoidal channel, unlined, to carry a discharge of 20 m³/s of clean water. The slope of the channel is 0.0003. The bed material and that of the channel side slope has a mean diameter of 8 mm and a characteristic diameter, d_{65}, of 12 mm.

(1) *Criterion of critical shear stresses*
  (1a) *Manning–Strickler coefficient of roughness* (equation (5.7)):

$$K_s = \frac{26}{(12 \times 10^{-3})^{1/6}} \, m^{1/3}/s = 54 \, m^{1/3}s$$

  (1b) *Angle of repose of the material* (Graph 95b) — for $d = 8$ mm, $\psi \simeq \approx 31°$ (assuming that the material is moderately angular).
  (1c) *Angle of the side slopes* — for reasons of stability, it is necessary that $\phi < \psi$; assuming a side slope equal to $2H{:}1V$, $\phi = 26.6°$.
  (1d) *Bottom shear stress* (equation (5.18)), assuming that $R_e^* > 400$;

$$(\tau_0)_{crit} = 0.06 \times (\gamma_s - \gamma)d = (0.06 \times 1650 \times 9.8 \times 8 \times 10^{-3}) \, N/m^2 =$$
$$= 7.76 \, N/m^2$$

Thus:

$$R_e^* = \frac{u^* d}{v} = \frac{\sqrt{\tau_0/\rho.d}}{v} = \frac{(7.76/1000)^{1/8} \times 8 \times 10^{-3}}{1.01 \times 10^{-6}} = 698 > 400$$

The application of equation 5.14 is therefore valid. The value of $(\tau_0)_{crit}$ could have been read directly from Graph 91b.
  (1e) *Critical shear stress on channel side slopes* (equation (5.16))

$$(\tau_0^t)_{crit} = (\tau_0)_{crit}\left(1 - \frac{\sin^2 \phi}{\sin^2 \psi}\right)^{1/2} = 7.76 \left(1 - \frac{\sin^2 26.6}{\sin^2 31.0}\right)^{1/2}$$
$$= 3.83 \, N/m^2$$

This value could have been obtained with the help of Graph 94.
  (1f) *Ascertaining the hydraulic radius* — In a first approximation that can be checked later,

assuming a ratio $l/h = 5$, Table 90 indicates that the maximum shear stress caused by the flow is $0.98\ \gamma hi$ at the bottom and $0.77\ \gamma hi$ at the sides.

In order to ensure stability of the bottom, the following is required:

$$h < \frac{(\tau_0)_{crit}}{0.98\gamma i} = \frac{7.76}{0.98 \times 9800 \times 0.0003} = 2.69 \text{ m}$$

In order to ensure stability of the sides, the following is required:

$$h < \frac{(\tau'_0)_{crit}}{0.77\gamma i} = \frac{3.83}{0.77 \times 9800 \times 0.0003} = 1.69 \text{ m}$$

Taking $h = 1.5$ m, the critical conditions are satisfied.

(1g) *Calculation of the geometry of the cross-section* — Taking $b = 10$ m, we have $A = 19.5$ m^2 and $R = 1.17$ m.

The discharge calculated from application of the Manning–Strickler equation is:

$$Q = K_s\ R^{2/3}\ Ai^{1/2} = 54 \times 1.17^{2/3} \times 19.5 \times (0.3 \times 10^{-3})^{1/2} \text{m}^3/\text{s} = 20 \text{ m}^3/\text{s}$$

The mean velocity of the flow is 1.03 m/s.

(2) *Criterion of critical velocities*

(2a) *Using Hjulström's curve* — for $d = 8$ mm, Graph 89 gives us $U_{crit} \approx 1$ m/s, which agrees with the criterion of the critical shear stress. This value of U_{crit} is also obtained with the help of Table 88.

(2b) *Using Neill's criterion* — from equation (5.11) we have

$$U_{crit}^2 = 1650 \times 8 \times 2.5 \times 10^{-4} \times (8/1.3)^{-0.20} \text{ (m/s)}^2 = 2.3 \text{ (m/s)}^2$$

whence $U_{crit} = 1.5$ m/s

5.13 Example

Determine the maximum slope that can be given to a channel with a trapezoidal cross-section, width at base $l = 8$ m and water depth $h = 2$ m, the characteristic diameter, d_{65} of the noncohesive bed material being 15 mm and the mean diameter 10 mm. The channel side slope is 2H:1V.

Solution:

$l/h = 4$. The maximum shear stress (Table 90) is: at the bed $\tau_M = 0.97\ \gamma hi = 19 \times 10^3$ N/m^2;

at banks $\tau'_M = 0.77\ \gamma hi = 15.1 \times 10^3\ i$ N/m^2.

The critical shear stress, given by the Lane criterion, is $\tau_0 = 0.8 \times \times d_{75} = 0.8 \times 15 = 12$ N/m^2.

The angle of repose is (Graph 93) $\psi \approx 32°$, therefore a side slope angle of 30° is taken. From Graph 94, for $\psi = 32°$ and $\phi = 30°$, we have $K = 0.3$.

The critical shear stress is thus $\tau'_0 = K\tau_0 = 0.3 \times 12 = 3.6$ N/m^2.

For stability of the bottom, it is necessary for $\tau_M \leqq \tau_0$, that is to say, $19 \times 10^3 \times i \leqq 12$, whence $i \leqq 0.00063$.

For stability of the sides, it is necessary that $\tau'_m < \tau'_0$, that is to say, $14.9 \times 10^3 \times i < 3.6$, whence $i < 0.00024$. It is therefore necessary that $i < 0.00024$.

The mean velocity of the flow is given by $U = K_s R^{2/3} i^{1/2}$. With $K_s = 52$ m$^{1/3}$/s and $R = 1.4$ m, we have $U = 1.0$ m/s. The critical velocity given by Hjulström criterion is 1.6 m/s.

5.14   Most stable shape of the cross-section

As we have seen, in a tapezoidal channel the value of the fluid shear stress, τ_0, is not the same on the entire perimeter. This means that the transport capacity of the channel is limited by a small zone in which τ_0 reaches maximum values.

For a given bed material, there will be a shape of cross-section corresponding to a uniform distribution of the tractive force. This cross-section is known as the *optimized stable cross-section*. It is easy to deduce the equation $y(x)$ of this section,[†] and a cosinusoidal curve is obtained:

$$y = h_u \cos \left(\frac{\tan \psi}{h_u} x \right) \tag{5.17}$$

in which: h_u is the depth of water corresponding to the middle of the *channel*, considering that flow is established and ψ is the angle of repose of the bed material (Fig. 5.7(a)).

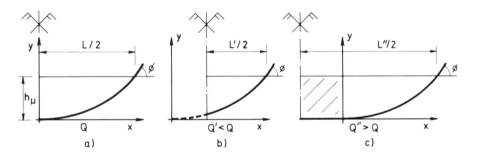

Fig. 5.7.

According to data of the US Bureau of Reclamation, it is necessary to take:

$$h_u = \frac{(\tau_0)_{crit}}{0.97 \, \gamma \, i} \tag{5.18}$$

$\tau_{0(crit)}$ is the critical shear stress of the bed material in metric units. The mean velocity, U, will thus be defined for the section as follows:

$$U = K_s \, (0.91 - 0.8 \, \tan\psi) h_u{}^{2/3} \, i^{1/2} \tag{5.19}$$

and the wetted area, A, of the cross-section, will be:

$$A = \frac{2.04 \, h_u{}^2}{\tan\psi} \tag{5.20}$$

The resulting discharge will be $Q = UA$.

If the effective discharge is $Q' < Q$, it will be necessary to remove a central part of the channel, until the surface width is (Fig. 5.7(b)):

† See, for example, [7].

$$L' = 0.96 \left(1 - \sqrt{\frac{Q'}{Q}}\right) L \qquad (5.21)$$

If the effective discharge is $Q'' > Q$, it will be necessary to add a portion in the centre of the channel so that the surface width is (Fig. 5.7(c)):

$$L'' = \frac{Q'' - Q}{h_u^{5/3} i^{1/2} K_s} \qquad (5.22)$$

Example: Determine the optimized stable cross-section of a channel excavated in material with $d_{65} = 15$ mm, $d_{75} = 20$ mm, $\psi = 35°$, in which $i = 0.0016$ and $Q = 10$ m³/s.
Solution: From the Lane criterion we have $\tau = 8 \times 2 = 16$ N/m².
Formula 5.18 gives

$$h_u = \frac{(\tau_0)_{crit}}{0.97 \, \gamma i} = \frac{16}{0.97 \times 9800 \times 0.0016} = 1.56 \text{ m}$$

According to formula (5.17), the equation of the shape of the cross-section is

$$y = 1.06 \cos\left(\frac{\tan 35.0}{1.06} x\right) = 1.06 \cos(0.30 \, x)$$

It must be noted that $0.30 \, x$ is expressed in radians.
The mean velocity will be (equation 5.19), taking $K_s = 26/d_{65}^{1/6} = 26/(0.015)^{1/6} = 26/0.4966 = 52m^{1/3}$/s:

$$U = K_s(0.91 - 0.8 \tan\psi)h_u i^{1/2} = 52(0.91 - 0.8 \times 0.7) \times 1.06^{2/3} \times 0.0016^{1/2} = $$
$$= 0.76 \text{ m/s.}$$

The cross-section will be (equation (5.20)):

$$A = \frac{2.04 \, h^2}{\tan\psi} = \frac{2.04 \times 1.06^2}{0.7} = 3.3 \text{ m}^2$$

Whence $Q = UA = 3.3 \times 0.76 = 2.5$ m³/s.
For a discharge of 10 m³/s, it will be necessary to have a horizontal length at the bottom of L'', given by

$$L'' = \frac{Q'' - Q}{h_u^{5/3} i^{1/2} K_s} = \frac{10 - 3.3}{1.06^{5/3} \times 0.0016^{1/2} \times 52} = 3.27 \text{ m}$$

5.15 Channels with vegetation

Vegetation is an effective means of protecting channels against scour, though it means an increased roughness and a certain slowing down of the flow.

In dimensioning a channel lined with vegetation, it is necessary to bear in mind the state of development of the vegetation. The calculation must be made in two stages: in the first, the calculation is made for a development of vegetation that ensures the stability of the channel; in the second, it is assumed that the vegetation has developed more than is necessary for stability against scour, with the corresponding increase in roughness: it is necessary to ensure that under such circumstances there is the maximum discharge.

Table 95 summarizes the criteria laid down by the US Soil Conservation Service for dimensioning channels lined with vegetation (quoted by [7]).

Given the discharge Q, the slope, i, of the channel, and its geometrical shape, and after choice of the vegetal cover, the procedure is as follows:

(a) *Calculation to ensure stability against scour*

  (a1) Given the species that constitute the vegetal cover, an analysis is made, in the light of the local conditions, of the extreme values of the height and density of vegetation that can be foreseen. On the basis of Table 95, the vegetation is classified according to one of the five types, identified from A to E.

  (a2) A value of K_s is assumed and, on the basis of the curve of Graph 95, corresponding to the least foreseeable vegetal cover that can occur with discharge Q (curves D or E, for example), the value of UR is extracted.

  (a3) From Table 95, the admissible velocity, U, is determined; since the geometrical shape of the cross-section is given, the value of R is known, corresponding to a wetted area $A = Q/U$; from the Manning formula the value of $UR = K_s R^{5/3} i^{1/2}$ is calculated; this value is compared with that obtained in (a2).

  (a4) The calculation is repeated with other values of K_s, until the value of UR calculated in (a3) coincides with the value of UR determined in (a2).

(b) *Checking the cross-section for greater development of the vegetation*

  (b1) A water depth, h, is chosen and the corresponding values of A, R, $U = Q/A$ and UR are calculated.

  (b2) For the maximum possible development of the vegetation as analysed in (a1), and with the value of UR calculated in (b1), the value of K_s is obtained from Graph 95.

  (b3) $U = K_s R^{2/3} i^{1/2}$ is calculated and compared with the value of U obtained in (b1).

  (b4) New values of h are selected until the value of the velocity calculated in (b3) is equal to that obtained in (b1).

BIBLIOGRAPHY

[1] Arghyropoulos, P., *Calcul de l'Ecoulement en Conduites sous Pression ou à Surface Libre*. Dunod, Paris, 1957 (tables pour le calcul d'après la formule de Manning–Strickler).

[2] Bakhmeteff, B. A., *Hydraulics of Open Channels*. McGraw-Hill, New York, 1932.

[3] Crausse, E., *Hydraulique des Canaux Découverts en Régime Permanent*. Eyrolles, Paris, 1951.

[4] Lencastre, A., *Alguns Aspectos do Transporte Sólido em Problemas Hidráulicos*. Technical Memory No. 64, *L.N.E.C.* and Técnica No. 247, December, 1954.

[5] Lane, E. W., *Progress Report on Results of Studies on Design of Stable Channels*. U.S. Bureau of Reclamation Hydro Lab. Report Hyd-352, 1952.

[6] Inspection Fédérale Des Travaux Publics, *Abaques pour l'Ecoulement Uniforme dans les Canaux à Profil Rectangulaire ou Trapézoidal*. Berne, 1956.

[7] Chow, V. T., *Open-Channel Hydraulics*. McGraw-Hill, New York, 1959.

[8] Mavis, F. T., Liu, T., Soucek, E., *The Transportation of Detritus by Water — II*. University of Iowa Studies in Engineering No. 341, 1937.

[9] Mavis, F. T., Laushey, L. M., *A Reappraisal of the Beginning of Bed Movement — Competent Velocity*, Proc. 2nd IAHR Congress, Stockholm, 1948.

[10] Neill, C. R., *Mean Velocity Criterion for Scour of Coarse Uniform Bed Material*. Proc., 12th Congress of IAHR, Fort Collins, 1967.

[11] Shields, A., *Anwendung der Aznlichkeitsmechanik und der Turbulenzforschung auf die Gerchiebebewegung*. Mitteil. Preuss. Versuchsants, Wasser and Schiffsbau, Berlin, No. 26, 1936.

[12] Graf, W. H., *Hydraulics of Sediment Transport*. McGraw-Hill, New York, 1971.

6

Open-channel flow. Steady flow

A. GENERAL EQUATIONS

6.1 Types of flow

In a channel that is sufficiently long, and has constant slope, cross-section, roughness and discharge, a *uniform flow* is always eventually established, and friction losses in this flow are entirely compensated by the slope of the bed. The existence of any singularity (contraction, expansion, discontinuity of the bottom, etc.) causes, as in closed-conduit full flows, a local energy loss and consequently an alteration of the free surface. The flow is no longer uniform, and is therefore known as *nonuniform* or *varied* since the depth of the liquid continually varies from one section to another.†

When the velocities increase in the flow direction, the flow is said to be *accelerated*; when they decrease, the flow is said to be *retarded*.

Nonuniform flows may be divided into two major groups: *gradually varied* flows, in which the hydraulic parameters only change very slowly from one section to another, and therefore sometimes extend for consider-able distances, giving rise to *different types of surface profiles*; and *rapidly varied* flows in which there is a rapid and sometimes discontinuous evolution of the characteristics of the flow; such flows usually occupy a relatively small zone, the most important being the *hydraulic jump*, the *abrupt drop* (of special note being spillways) and *contractions*.

The basic equations that allow a study of steady flow are the general equations of Chapter Two: the equation of continuity, Bernouilli's theorem associated with the energy loss equations, and Euler's theorem.

---

† *Nonuniform* or *varied* must not be confused with *unsteady* flow. Nonuniform flow, among the steady flows, is the opposite of uniform flow; unsteady flow is the opposite of steady flow, i.e. a flow whose characteristics, in a given cross-section, change with time.

6.2   Energy losses

In uniform flow, the energy loss, i, per unit weight discharged and per unit length of the channel, can be calculated, as was seen in Chapter 5, by formulae such as Chézy's or Manning–Strickler's.

In any section, the value of i is constant and the energy line is parallel to the bottom of the channel.

In nonuniform flow, since the hydraulic radius varies from one section to another, the energy loss also varies. *In the case of gradually varied flow*, it is accepted that in a sufficiently small reach of channel, the value of i will be the same as what would be obtained if the same discharge were taking place in uniform flow, with a depth of water equal to that of the mean cross-section of that reach.

In rapidly varied flow, the slope of the trajectories of the various particles modifies the characteristics of the flow in such a way that the equations governing uniform flow are no longer applicable, and the total energy loss has to be calculated between the two extreme sections.

6.3   Energy in relation to a horizontal datum

It will be remembered that the total energy, E, in a cross-section (see Section 2.16) in relation to a horizontal datum, per unit weight of the flow, is the sum of three terms: elevation head, piezometric head and kinetic head. Expression (2.26), applied to free surface flows, is then written as follows:†

$$E = z + h\cos\theta + \alpha\frac{U^2}{2g} \simeq z + h + \alpha\frac{U^2}{2g} \qquad (6.1)$$

The total energy line always descends in the direction of the flow.

Between two cross-sections, 1 and 2, the energy E undergoes a variation, $\Delta E_{1.2} = E_1 - E_2$, corresponding to friction losses.

In uniform flow, the energy line is straight and parallel to the free surface and to the bottom (see Fig. 5.1). Taking i to represent the slope of the energy line, J the slope of the free surface and I the slope of the bottom, it is found that in uniform flow $i = J = I$.

In gradually varied flow, the energy line is curved (see Fig. 6.1). If the

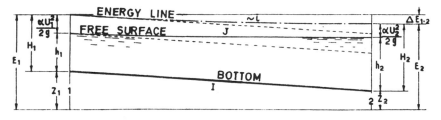

Fig. 6.1.

† In gradually varied flow it is generally assumed that the energy line, plotted for the points of the free surface, is valid for the whole liquid mass.

bed slope is small, and if the variations in kinetic energy are negligible compared to the variations in the water depth, the energy line and free surface are also nearly parallel, and the inclination of the energy line may be regarded as equal to the slope of the free surface: $I = J$.

6.4   Energy in relation to the bottom

In Section 5.4, specific energy was defined as the energy referred to the channel bed as datum, i.e.,

$$H = h\cos\theta + \frac{\alpha U^2}{2g} \simeq h + \frac{\alpha U^2}{2g} = h + \frac{\alpha Q^2}{2gA^2} \tag{6.2}$$

Whereas the total energy E always decreases in the direction of the flow, in relation to a horizontal datum, the specific energy, H, may remain constant, as occurs with uniform flow, and may increase or decrease according to the flow conditions in nonuniform flow.

Equation (6.2) defines, for a *given section*, a relationship between H, h and Q that is valid for any type of open-channel flow.

By making $Q = Q_1$ (constant) in this expression, a curve $H = f(h)$ is obtained, as shown in Fig. 6.2(a), which gives the water depths h according

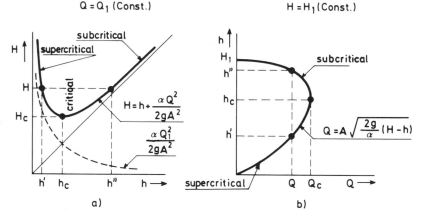

Fig. 6.2.

to the specific energy H.

For $h = 0$, $A = 0$ and $H = \infty$, i.e. the axis OH is an asymptote of the curve; for $h = \infty$, $A = \infty$ and $H = h$, i.e. the curve has a second asymptote at 45°.

It will be seen that the same discharge, with the same specific energy, H, can occur with two different depths: h', corresponding to the supercritical flow (see 5.2) and h'', corresponding to the subcritical flow. These depths, h'

and h'', are known as *alternate depths of equal specific energy H*. The point on the curve defined by (H_c, h_c) represents the *critical flow*; the corresponding energy H_c, known as *critical specific energy*, represents the minimum specific energy for which a given discharge can take place at a cross-section. The water depth h_c, corresponding to the critical flow, is known as *critical depth*.

The minimum point of the curve is obtained by differentiation.

$$\frac{\mathrm{d}H}{\mathrm{d}h} = 1 - \frac{Q^2}{gA^3}\frac{\mathrm{d}A}{\mathrm{d}h} = 0 \tag{6.2a}$$

with L representing the width of the surface, $\mathrm{d}A = L\mathrm{d}h$. By substitution we have, as the condition of minimum:

$$\frac{Q}{\sqrt{g}} = A\sqrt{\frac{A}{L}} \tag{6.3}$$

Table 97 gives the formulae for the critical flow in various sections. Table 98 and Graph 99 enable the critical depth to be determined for rectangular, triangular, trapezoidal and circular cross-sections. Tables 100 and 101 give the values of h' and h''.

Examination of curve $H(h)$ in Fig. 6.2(a) shows that in subcritical flow H and h vary in the same direction, that is to say, they simultaneously increase or decrease; in supercritical flow the opposite occurs. Analysis of the same curve shows that in the vicinity of the critical flow a small variation in energy H leads to appreciable variations in the water depth h. This means that in any flow close to the critical flow, small irregularities are sufficient to cause an appreciable undulation of the free surface.

If in expression (6.2) we make $H = H_1 = $ constant, we obtain the following equation:

$$Q = A\sqrt{\frac{2g}{\alpha}(H_1 - h)} \; . \tag{6.4}$$

According to this expression, for $h = 0$ $(A = 0)$ and for $h = H$, the discharge is zero. By mere mathematical transformation, we arrive at the same condition that was deduced previously (equation (6.3)).

For a constant energy $H = H_1$, the same discharge can take place with two depths: h', corresponding to supercritical flow and h'', corresponding to subcritical flow (Fig. 6.2(b)).

Point (Q_c, h_c) corresponds to the maximum discharge in the cross-section that can take place with energy H_1. This point represents the critical flow which coincides with the critical flow defined above.

If the energy H remains constant, an increase in discharge means an increase in the water depth for a supercritical flow and a decrease in the water depth for a subcritical flow.

6.5 Momentum function

Applying Euler's theorem (see Section 2.22) to two cross-sections, 1 and 2, of a free surface flow, let us suppose a channel with relatively mild slope, such that $\cos \theta \approx 1$. The resultants of the pressure forces in cross-sections 1 and 2 are $I_1 = \gamma y_1 A_1$ and $I_2 = \gamma y_2 A_2$, respectively (see equation (3.7)), in opposite directions and parallel to the axis of the channel. The momentum flux across the two cross-sections (see equation (2.50)) is $M_1 = \beta \rho Q U_1$ and $M_2 = \beta \rho Q U_2$.

Taking F_H as the resultant of the forces opposing the flow, also parallel to the axis of the channel, by applying equation (2.31) the resultant along that axis will be:

$$F_H = (\gamma y_1 A_1 + \beta \rho Q U_1) - (\gamma y_2 A_2 + \beta \rho Q U_2) = \mathcal{M}_1 - \mathcal{M}_2 \quad (6.5)$$

In this equation the terms in parentheses represent the sum of the pressure force and momentum flux at each section.

That is to say, the resultant F_H of the forces acting in the direction of the flow between the two cross-sections is equal to the difference of the momentum function value, $\mathcal{M}_1$ and $\mathcal{M}_2$.

The momentum function is defined by:

$$\mathcal{M} = \gamma y A + \beta \rho Q U = \gamma y A + \beta \rho \frac{Q^2}{A} \quad (6.6)$$

For a given cross-section, this expression defines a ratio between $\mathcal{M}$, h and Q.

By making $Q = Q_1$ (constant), a curve is obtained as shown in Fig. 6.3.

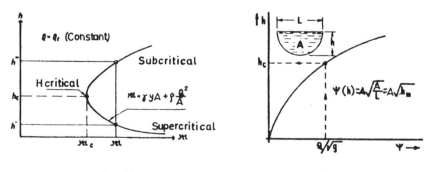

Fig. 6.3.   Fig. 6.4

The same value of the momentum function has two corresponding depths, h' and h'' known as *conjugate depths of constant momentum*. One of these depths corresponds to supercritical flow, the other to subcritical flow. Point $(h_c, \mathcal{M}_c)$ corresponds to the critical flow. The critical depth h_c thus determined coincides with the critical depth determined from the energy curves. Tables 64(d) and 84(d) give the depth from the surface to the centroid y, in circular and trapezoidal cross-sections.

6.6 Critical flow
(a) Graphical evaluation of critical flow
The critical flow, as defined in section 6.4, corresponds to the minimum specific energy for a given discharge in a cross-section, or to the maximum discharge, for a given specific energy, in that same cross-section.

The general equation for the critical flow, as we have seen, is:

$$\frac{Q}{\sqrt{g}} = A \sqrt{\frac{A}{L}} \tag{6.7}$$

Generally speaking, the second term is only a function of h, so that the equation may also be written as $Q/\sqrt{g} = \psi(h)$.

By plotting the curve $\psi(h)$ characteristic of the cross-section, the critical depth for discharge Q can be determined graphically, by noting $\psi = Q/\sqrt{g}$ and reading off h_c (Fig. 6.4). This method is valid for any cross-section.

Table 64(c) gives the value of $h_m = A/L$ in circular cross-section. (See also Tables 97 to 99.)

Examples

(1) Given a channel of trapezoidal cross-section with the following characteristics: width at base $l = 4$ m; side slopes $1\,H : 1\,V$, determine the critical depth for $Q = 6\,\text{m}^3/\text{s}$.

Table 101 could be used. This would be the simplest method. For the purpose of this exercise, however, the curve $\psi(h) = A\sqrt{A/L} = A\sqrt{h_m}$ is plotted in accordance with the table shown below.

h	$A = h(l + mh)$ $= h(4 + h)$	$L = l + 2mh$ $= (4 + 2h)$	$h_m = \dfrac{A}{L}$	$\sqrt{h_m}$	$A\sqrt{h_m}$
(m)	(m^2)	(m)	(m)	(m$^{1/2}$)	(m$^{5/2}$)
0.45	2.00	4.90	0.41	0.64	1.28
0.50	2.25	5.00	0.45	0.67	1.51
0.55	2.50	5.10	0.49	0.70	1.75
0.60	2.76	5.20	0.53	0.73	2.00
0.65	3.02	5.30	0.57	0.76	2.29

Therefore $Q/\sqrt{g} = 6/3.13 = 1.92\,\text{m}^{5/2}$. By interpolation, the corresponding depth for $A\sqrt{h_m} = 1.92\,\text{m}^{5/2}$ is about $h_c = 0.58$ m.

(2) In the case of example (2) of Section 5.5, determine the critical depth. By direct application of formula

$$h_{\mathrm{c}} = \sqrt[3]{\frac{1}{g}\left(\frac{Q}{L}\right)^2}$$

(Table 97), we have:

$$h_{\mathrm{c}} = 0.47\left(\frac{Q}{L}\right)^{2/3} = 0.47 \times \left(\frac{6}{4}\right)^{2/3} = 0.47 \times 1.31 = 0.62\,\mathrm{m}$$

(b) Critical, mild and steep slopes

The critical slope of a channel, for a given discharge, is the slope for which that discharge take place under critical uniform flow conditions or, in other words, the slope for which the discharge occurs with minimum energy.

For a given discharge, the general formula for the critical slope is:

$$I_{\mathrm{c}} = \frac{gA_{\mathrm{c}}/L_{\mathrm{c}}}{C^2 R_{\mathrm{c}}} \qquad \text{or} \qquad I_{\mathrm{c}} = \frac{gA_{\mathrm{c}}/L_{\mathrm{c}}}{K_{\mathrm{s}}^2 R_{\mathrm{c}}^{4/3}} \qquad (6.8)$$

in which $A_{\mathrm{c}}/L_{\mathrm{c}}$ is the mean-depth, C is Chézy's coefficient and K is Strickler's coefficient.

In a channel, each discharge Q has a corresponding critical depth h_{c}, determined from equation (6.7), and a critical slope, I_{c}, determined from equation (6.8), knowing h_{c}.

The curve of equation (6.8) therefore defines a relationship between Q and I_{c} and is known as the *critical slope curve* $I_{\mathrm{c}} = f(Q)$, shown as in Fig. 6.5.

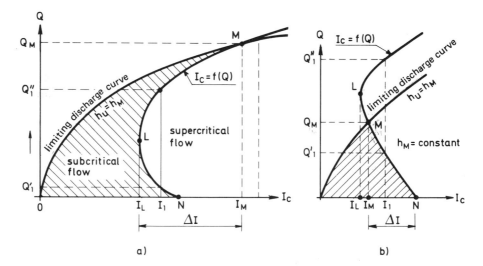

Fig. 6.5.

This curve is asymptotic to the axis of the abscissa. In practice, however, it so rapidly approaches that axis that it is common with it, in effect from a point *N*.

A *mild slope* is one on which uniform flow is subcritical; on a *steep slope* uniform flow is supercritical.

A channel with slope I_1 (see Fig. 6.5(a)) will be a *mild slope* channel for a discharge of more than Q''_1 or less than Q'_1; it will be a *steep slope* channel for discharges of between Q'_1 and Q''_1.

Discharges Q'_1 and Q''_1 are known as *characteristic discharges* of that channel, with a corresponding slope of I_1.

(c) *Limiting slope* I_L. *Limiting discharge curve*

Limiting slope, I_L, is defined as the slope below which the flow in a channel is always subcritical, whatever the discharge. The limiting slope I_L corresponds to the minimum on the critical slope curve $I_c = f(Q)$, and for each channel there is only one value of I_L.

If the maximum depth H_M of the flow in the channel is fixed, it will have a corresponding *limiting discharge curve*, i.e. a curve of the possible maximum discharge without exceeding h_M, this curve being a function of the slope and being determined by uniform flow formula.

Point *M*, the intersection of the limiting discharge curve and curve $I_c = f(Q)$, corresponds to the maximum discharge which, with its limiting depth h_M, can take place in critical flow.

Points (I, Q) situated within the area contained between the limiting curve and curve $I_c = f(Q)$ (shaded area of Fig. 6.5) correspond to subcritical flow; points situated to the right of curve $I_c = f(Q)$ correspond to supercritical flows. Point *L* may be below or above the limiting discharge curve (Fig. 6.5(a) and 6.5(b)); it will be infinite for a rectangular cross-section of undefined width. ΔI is taken to represent the interval between the value of I_M corresponding to point *M* and (in the case of Fig. 6.5(a)) the value of I_L corresponding to point *L*; or (for the case of Fig. 6.5(b)) the value of I_N corresponding to point *N*. For slope values *I* falling within interval ΔI, the channel behaves as a steep slope channel or a mild slope channel depending on the discharge. For values of *I* outside interval ΔI, the channel behaves, whatever the discharge compatible with maximum depth h_M, always either as a mild slope channel (for $I < I_L$) or as a steep slope channel (for $I > I_M$ in the case of Fig. 6.5(a)) or for $I > I_N$ in the case of Fig. 6.5(b)).

If the limiting curve passes below *L*, it means that there is no limiting slope for the depth of flow imposed (Fig. 6.5(b)).

6.7 The kinetic factor: Froude number

Froude number, defined in Section 5.1, is the ratio of the flow velocity, U, to the celerity of small gravity waves, $\sqrt{gh}$; $\mathbf{F}_r = U/\sqrt{gh}$. Froude number is thus a measure of kinetics of the flow, i.e. of its rapidity or slowness, and is therefore sometimes known as the *kinetic factor*.

The specific energy (i.e. the energy relative to the channel bed) can be written):

$$H = h\left(1 + \frac{1}{2}\mathbf{F}_r^2\right) \qquad (6.9)$$

In a rectangular channel, critical flow is defined by $\mathbf{F}_r = 1$ and consequently $H_c = 1.5\,h_c$. Subcritical flow is characterized by $\mathbf{F}_r < 1$ and $H < 1.5\,h$; and in supercritical flow $\mathbf{F}_r > 1$ and $H > 1.5\,h$.

If in the expression of Froude number the depth h is substituted by the mean depth h_m, we have:

$$\mathbf{F}_r = \frac{U}{\sqrt{gh_m}} \qquad (6.10)$$

Accordingly, whatever the shape of the channel, the previous considerations can be generalized as follows: for $\mathbf{F}_r > 1$, the flow is supercritical, for $\mathbf{F}_r = 1$, the flow is critical and for $\mathbf{F}_r < 1$, the flow is subcritical.

B.   GRADUALLY VARIED FLOW. SURFACE PROFILES

6.8   General equation of gradually varied flow

As was seen in Section 6.3, in uniform flow, that is to say, if $h = h_u$, then $i = I = J$. If $h > h_u$, the energy losses are less than the slope of the bottom, and $i < I$. If $h < h_u$, the energy losses are higher then the slope of the bottom and $i > I$.

In a short reach, Bernoulli's equation is written: $dE = ids$ or $d(H + z) = ids$, so that we have:

$$\frac{dH}{ds} = i - \frac{dz}{ds} = i - I \qquad (6.11)$$

Bearing in mind that:

$$\frac{dH}{ds} = \frac{\partial}{\partial H} \cdot \frac{\partial H}{\partial s} \qquad \text{where} \qquad H = h + \frac{\alpha U^2}{2g} = h + \frac{\alpha Q^2}{2gA^2} \qquad (6.11a)$$

We have:

$$\frac{dH}{ds} = \frac{\partial}{\partial h}\left(h + \frac{\alpha Q^2}{2gA^2}\right)\frac{dh}{ds} = \left(1 - \frac{\alpha Q^2}{gA^3}\frac{dA}{dh}\right)\frac{dh}{ds} =$$

$$= \left(1 - \frac{\alpha Q^2 L}{gA^3}\right)\frac{dh}{ds} \qquad (6.12)$$

where $dA/dh = L$ (surface width).

Making a comparison with (6.11) we have:

$$\frac{dh}{ds} = (I - i) \bigg/ \left(1 - \frac{\alpha Q^2 L}{gA^3}\right) \tag{6.13}$$

in which, as has been said, s is the length of the channel from an initial cross-section, h is the water depth, Q is the discharge, L is the surface width, A is the wetted area, I is the slope of the bottom of the channel and i is the slope of the energy line.

Thus:

$$i = \frac{bQ^2}{RA^2}$$

in which the coefficient b is given by:

$$b = \frac{1}{C^2} \text{ (Chézy)} \quad \text{or} \quad b = \frac{1}{K_s^2 R^{1/3}} \text{ (Manning)} \tag{6.14}$$

Calculation of the *water surface longitudinal profile, h(s)*, or more usually, *surface profile*, consists of integration of equation (6.13), for which some methods are given below.

However, before analysing the integration methods, a qualitative analysis must be made of the forms of the *surface profiles*.

6.9 Classification of surface profiles
In gradually varied flow, the slopes and curvatures of free surface profiles are very slight, and it may be assumed that, as in uniform flow, there is a hydrostatic pressure distribution.

Fig. 6.6 shows the twelve possible types of the surface profiles, for various cases of bottom slope.†. The locus of the representative point of the flow in the (h, H) plane is also shown in the diagram.

The profiles are further classified according to the depth of the flow: the depth may be greater or smaller than the uniform depth, h_u, and greater or smaller than the critical depth, h_c. If the depth is greater than both the uniform depth and the critical depth, the profile is of type 1; if the depth is between h_u and h_c the profile is of type 2; and if it is less than both the profile is of type 3:

$h > h_u$ and	$h > h_c$	Type 1
$h_u > h > h_c$		Type 2
$h_u > h$ and	$h_c > h$	Type 3

† For further illustrations see [1], [3], [9], for example.

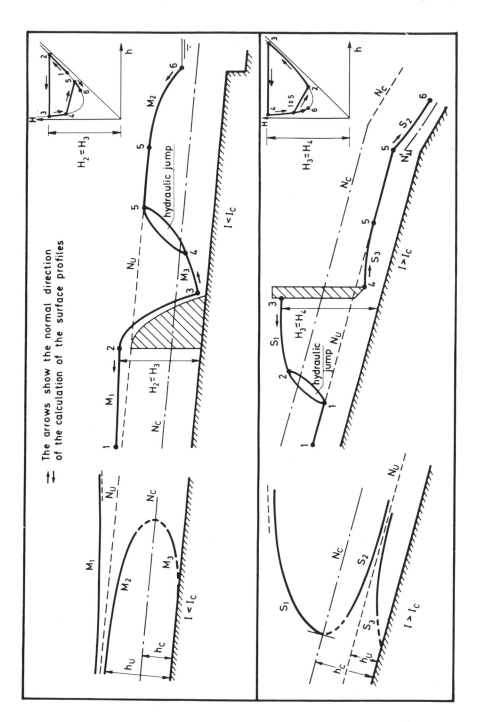

Fig. 6.6.

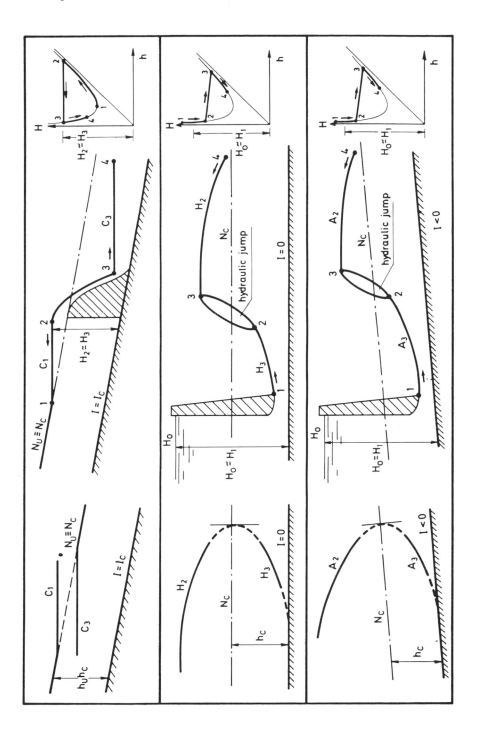

Fig. 6.6a.

(a) *Mild slope channel:* $(I < I_c; h_u > h_c)$. *Curve M*

M_1 profile: $(h > h_u)$

The numerator and denominator of (6.13) are positive: thus $dh/ds > 0$.

When h tends to h_u, i tends to I and dh/ds tends to zero, that is to say, the surface profile tends asymptotically to the uniform depth. When h tends to infinity i tends to zero; Q^2L/gA^3 also tends to zero, even in closed cross-sections in which L/A tends to zero, which shows that in such a situation the curve has a horizontal asymptote.

To sum up: the surface profile is concave and ascendent. Upstream, it tends asymptotically to uniform depth; downstream, it tends asymptotically to horizontal. This profile is found particularly upstream of a dam or of the piers of a bridge, or in certain cases of sudden change of slope. From a practical point of view, it is the M_1 curve that is of most interest.

M_2 profile: $(h_c < h < h_u)$

By simple reasoning similar to that for the M_1 curve, it is easily seen that the curve is convex descendent. Upstream, it tends asymptotically to uniform depth; downstream, it reaches the critical level perpendicularly (sudden fall). It is found in certain cases of change of slope, upstream of a expansion, sudden drop, etc.

M_3 profile: $(h < h_c)$

Here also it can easily be seen that the curve is ascendent. It leads to the hydraulic jump, near the critical depth. It occurs when superficial flow is created by some upstream control, such as sluice gate and at certain changes of slope.

(b) *Steep slope channel* — $(I > I_c; h_u < h_c)$. *Curve S*

S_1 profile: $(h > h_c)$

The curve is convex ascendent. Upstream it originates perpendicularly to the critical depth line, downstream, it tends asymptotically to the horizontal. Such profiles occur upstream of dams and contractions and at certain changes of slope.

S_2 profile: $(h_u < h < h_c)$

The curve is convex descendent. Upstream, it again originates perpendicularly to the critical depth line; downstream, it tends asymptotically to the uniform flow depth. Generally, the extent of this curve from a practical point of view is very small, i.e. it tends rapidly to uniform flow. It occurs, as a transition, between a sudden fall and uniform flow, or at an increase in slope in steep channels.

S_3 profile: $(h < h_u)$

The curve is ascendent and tends asymptotically to uniform flow. It is found downstream of gates, at the base of spillways, etc.

(c) *Critical slope channel:* $(I = I_c; h_u = h_c)$. *Curve C*

Both C_1 and C_3 profiles must be intermediate between the M_1 and S_1 profiles and the M_3 and S_3 profiles respectively. Some of these are concave and other convex; C_1 and C_3 profiles are thus horizontal or practically horizontal, according to whether Chézy's or Manning's formula is used. The C_2 profile which would correspond to water depths between uniform flow and critical flow, does not exist, since $h_u = h_c$.

C_1 profiles are found in cases similar to those of curves M_1 and S_1 and C_3 profiles occur in cases that are analogous to those of curves M_3 and S_3.

(d) *Horizontal channel:* $(I = 0; h_u = \infty)$. *Curve H*

In a horizontal channel uniform flow cannot be established. The critical depth, however, is defined, this being solely a characteristic of the geometry of the cross-section and of the discharge (see Section 6.6).

Surface profiles are the limiting case of the M curves when the slope tends to zero. The profile that would correspond to M_1 moves to infinity, so that only profiles H_2 and H_3 occur in situations that are analogous to those of M_2 and M_3 profiles. Curve H has the general form of a parabola.

(e) *Adverse slope channel:* $(I < 0)$. *Curve A*

As in the case of the horizontal channel, uniform flow is not defined, although the critical depth exists. The curve A is similar to a parabola; the two profiles, A_2 and A_3, correspond to H_2 and H_3 profiles respectively and occur in similar situations.

(f) *Partly full closed cross-sections*

As has been seen, in a partly full closed-conduit, e.g. a circular conduit (see Graph 65), the discharge in uniform flow grows to a maximum value that does not correspond to the cross-section being completely full, and decreases because the increase in the wetted perimeter is not compensated by the increase of the cross-section. There is thus a zone in which, for the same discharge, it is possible for two uniform depths to be established (Fig. 6.7(a)).

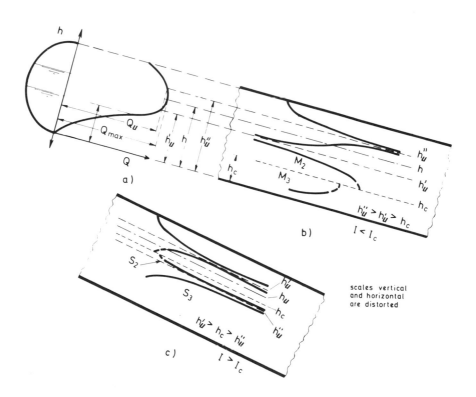

Fig. 6.7.

For the purpose of plotting the longitudinal surface profiles, there are therefore two uniform depths to be taken into account: h'_u and h''_u, situated below and above the value of h_M, corresponding to the maximum discharge.

Figs. 6.7(b) and 6.7(c) provide clear examples: the surface profiles, which tend asymptotically to uniform flow, may do so in two ways, since there are two possibilities of uniform depth; the other profiles remain qualitatively similar.

6.10   Reference cross-section

Integration of equation (6.13) leads to an indefinite integral; it is thus necessary to know the characteristics of the flow at a *reference cross-section* that will be represented by 0.

This reference cross-section *is located downstream*, for profiles M_1, S_1, C_1, M_2, H_2 and A_2, that is to say, whenever the flow is subcritical ($h > h_c$); under these conditions the surface profile must be calculated progressively upstream.

The reference cross-section is located upstream, for profiles S_2, M_3, S_3, C_3, H_3 and A_3, that is to say, whenever the flow is supercritical ($h < h_c$). In this case the surface profile must be calculated progressively downstream.

If the surface profile is caused by an overflow dam (backwater curve), the datum cross-section is located immediately upstream of the dam in a zone where the effect of local lowering of the free surface, caused by the increased velocity at the overflow is no longer felt. The flow conditions in the datum cross-section are determined from knowledge of the head required for a given discharge to be effected by the spillway.

Similar considerations are made for the case of a gate, contraction or expansion, sudden drop, or any other singularity.

6.11   Graphical method

The second member of the general equation, if the cross-section is constant, defines a function of the depth, h. In effect, given the geometrical and roughness characteristics of the cross-section, the variables that come into expression (6.13) are only a function of h. The value of the function $f(h)$ is then calculated for various values of h, within the zone necessary for a solution of the problem.

Since $ds/dh = f(h)$, we shall have, integrating between one cross-section, at a distance s_0 from an origin and another generic cross-section at a distance s

$$s - s_0 = \int_{h_0}^{h} f(h)\,dh \tag{6.15}$$

That is to say, the area between the curve $f(h)$, the axis $0h$ and the vertical straight lines of abscissa h_0 and h, will give the value of $s - s_0$.

Accordingly, from an abscissa cross-section, s_0 and water depth h_0, when known, it is possible to determine any intermediate point. This method, without any simplifying hypothesis that might compromise the precision of its results, is valid in any prismatic channel, whatever the geometrical shape of its cross-section.

Example (Ref. [3]): A trapezoidal channel of base width 10 m and side slopes 1 H: 1 V is laid on a slope of 0.0003 and carries a discharge of 15 m³/s. The roughness of the sides corresponds to a Strickler's coefficient $K_s = 100\,\text{m}^{1/3}$. An overflow dam causes the discharge to take place at a depth $h_0 = 2.5\,\text{m}$ in the cross-section just upstream of the dam. Compute the surface profile.

Solution: The uniform depth, obtained as explained in Section 5.5, is $h_u = 0.92$ m. The critical depth, obtained as indicated in Section 6.6, is $h_c = 0.60$ m. Since $h_u > h_c$, the channel has a mild slope and a surface profile of type M. Since h_0 is greater than h_u, the surface profile will be of type M_1. The function $f(h)$ is calculated in the following table and represented in Fig. 6.8.

h (m)	2.50	2.25	2.10	1.70	1.30	1.10	1.00
A (m^2)	31.2	27.5	25.4	19.9	14.7	12.2	11.0
P (m)	17.1	16.3	15.9	14.8	13.7	13.1	12.8
R (m)	1.83	1.68	1.57	1.34	1.07	0.93	0.81
$R^{4/3}$ (m$^{4/3}$)	2.24	1.99	1.96	1.49	1.10	0.91	0.75
l (m)	15.0	14.5	14.2	13.4	12.6	12.2	12.0
$\dfrac{Q^2L}{gA^3}$ $(\times 10^{-2})$	1.10	1.57	1.98	3.82	9.11	15.4	20.7
$N = 1 - \dfrac{Q^2L}{gA^3}$	0.989	0.984	0.978	0.961	0.909	0.846	0.793
$\dfrac{Q^2}{K_s^2}R^{4/3}a^2$ $(\times 10^{-5})$	1.06	1.49	1.90	3.93	9.46	16.7	14.8
$D = 1 - \dfrac{Q^2}{K_s^2 R^{4/3}A^2}$ $(\times 10^{-4})$	2.89	2.85	2.81	2.61	2.05	1.33	0.52
$f(h) = \dfrac{N}{D}$	3420	3460	3500	3680	4450	6360	15260
$s_0 - s$ (m)	0	860	1370	2790	4390	5455	6355

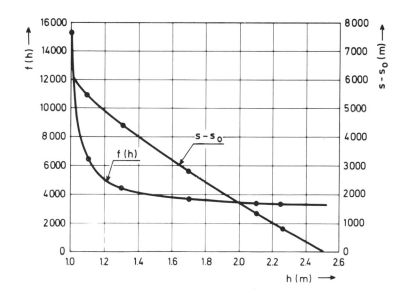

Fig. 6.8.

The area between the curve $f(h)$ and the axis of the abscissae gives the distance $s_0 - s$ for various depths, thus defining the backwater curve. Accordingly, at a distance of 860 m the water depth will be 2.25 m and at a distance of 6355 m it will be 1.00 m.

6.12   Step-by-step method

This method is a direct application of Bernoulli's theorem combined with the assumption that the head loss for nonuniform flow is equal to the head loss for a uniform flow at the same depth and discharge.

Between two cross-sections 0 and 1, at a distance Δs, assuming as is most frequent, that the datum cross-section 0 is downstream of 1, we can write:

$$\frac{U_1^2}{2g} + h_1 + z_1 = \frac{U_0^2}{2g} + h_0 + z_0 + i\Delta s \qquad (6.16)$$

in which i is the energy slope along the reach considered. If we represent the bottom slope by I, we have:

$$z_1 - z_0 = I\Delta s \qquad (6.17)$$

The expression then takes the following form:

$$\left(\frac{U_1^2}{2g} + h_1\right) - \left(\frac{U_0^2}{2g} + h_0\right) = (i - I)\Delta s \qquad (6.18)$$

or

$$H_1 - H_0 = (i - I)\Delta s \qquad (6.19)$$

Should the datum cross-section 0 be upstream of 1, that is to say, if the surface profile is calculated proceeding downstream (see Section 6.9), s must be given a negative sign. I is considered negative if we have an adverse-slope channel.

The method consists of applying this expression to sufficiently short reaches. Generally speaking, the following procedure can be used:
(1) The longitudinal profile is divided into lengths which are in accordance with the precision required.
(2) The energy curve H is plotted according to the depth, h (equation 6.2).

$$H = h + \frac{\alpha Q^2}{2gA^2} = f(h) \qquad (6.20)$$

If the cross-section is variable, several curves of this type must be plotted.
(3) The curve is plotted for the energy slope i, as a function of the depth, h, for each discharge Q:

$$i = \frac{Q^2}{C^2 A^2 R} \qquad \text{or} \qquad i = \frac{Q^2}{K_s^2 A^2 R^{4/3}} \qquad (6.21)$$

according to whether Chézy or Manning formulae are used.
(4) From h_0 known at the datum cross-section, and the corresponding values H_0 and i_0 taken from the curves (equation 6.20 and 6.21) the approximate value H_1 in the cross-section is determined, at a distance Δs from 0.

We thus have:

$$H_1' = H_0 + (i - I)\Delta s \tag{6.22}$$

(5) h_1' is determined from H_1' equation (6.20); and from h_1', i_1' is determined, equation (6.21). The previous calculation is repeated, taking for i the value:

$$i = \frac{i_1' + i_0}{2} \tag{6.23}$$

giving new values H_1'' and h''_1 and so on, successively, until for different values of h_1 reaches are obtained with the required precision.

(6) After definite values, H_1 and h_1 have been obtained, the next point at cross-section 2 is calculated in an analogous way to the previous one.

It is to be noted that this method is valid even when the geometrical cross-section and slope of the bottom are variable, as usually happens in natural watercourses, provided that there is judicious choice of the reaches Δs (see Section 6.15).

In the case of marked variations in the cross-section and, consequently, of the velocity, it will be necessary to take into account the local losses to be added to the friction losses (see Section 6.22).

Equation (6.22) is then written as follows:

$$(H_1' - H_0) = (i - I)\Delta s + \Delta E \tag{6.24}$$

The additional term ΔE represents the local head loss.

6.13   Bakhmeteff's method

This method if based on the empirically established rule that the function $\phi(h) = CA\sqrt{R}$ or $\phi(h) = K_s AR^{2/3}$ of the discharge capacity of the cross-section, in normal cases satisfies the equation:

$$[\phi(h)]^2 = A^2C^2R = \text{constant} \times h^n \tag{6.25}$$

or

$$[\phi(h)]^2 = K_s^2 A^2 R^{4/3} = \text{constant} \times h^n \tag{6.25a}$$

The exponent n is known as the *hydraulic exponent*.

Broadly speaking, the procedure is as follows:

(1) The value of n is determined, writing the equation $\phi(h) = CA\sqrt{R}$ or $\phi(h) = K_s AR^{2/3}$ in logarithmic form. For various values of h the value of $\phi(h)$ is calculated and the values *log h* and *log* ϕ are recorded in a system of axes. These points are joined by a straight line. Since α is the angle that this straight line makes with the axis of *log h*, $n = 2 \tan \alpha$.

For special cases, and if greater precision is required, one value of n can be used for one water depth zone and a different value for another zone. Analytically, between two extreme points of the limited zone, we shall also have:

$$n = 2\frac{\log[\phi(h)/\phi(h_0)]}{\log(h/h_0)} \tag{6.26}$$

In wide rectangular cross-sections (studied by *Bresse*), $n = 3$; in parabolic cross-section, with sufficiently small water depths, so that the wetted perimeter is practically the same as the surface width (studied by *Tolkmitt*), $n = 4$.

The extreme values of n, correspond to a triangular cross-section ($n = 5.3$ to 5.5) and to a very narrow rectangular cross-section ($n = 2$).

Table 104 gives the values of n in rectangular, trapezoidal and circular cross-sections.

(2) The uniform depth, h_u and critical slope, I_c (see Sections 5.5 and 6.6) are calculated. Parameters are defined:

$$\beta = \frac{I}{I_c} \qquad \text{and} \qquad \eta = \frac{h}{h_u} \qquad\qquad \text{(6.27 and 6.28)}$$

(3) The values $\eta_0 = h_0/h_u$ and $\eta_1 = h_1/h_u$ are calculated, corresponding to the cross-sections in which the water depths are h_0 and h_1.

(4) From Table 102, according to whether $\eta > 1$ or $\eta < 1$, we can obtain the value of the function:

$$B(\eta) = \int_0^\eta \frac{d\eta}{\eta^n - 1} \qquad\qquad \text{(6.29)}.$$

for the values of η_0 and η_1.

(5) The distance between the two cross-sections 0 and 1 will be given by the expression:

$$\Delta s = s_1 - s_0 = \frac{h_u}{I}(\eta_1 - \eta_0) - (1 - \beta)[B(\eta_1) - B(\eta_0)] \qquad\qquad \text{(6.30)}$$

In this method, there is no need to calculate intermediate points between the extreme values chosen for h, provided that the hydraulic exponent and the coefficient β remain constant. See example of Table 102.

6.14   Approximate Bakhmeteff's solution

Whenever the value of β is close to 0, that is to say, whenever the bottom slope is very small compared with the critical slope and the energy of the stream is also very small, as occurs upstream of large dams, equation (6.30) can be written:

$$s_1 - s_0 = \frac{h_u}{I}\{(\eta_1 - \eta_0) - [B(\eta_1) - B(\eta_0)]\} =$$

$$= \frac{h_u}{I}[\phi(\eta_1) - \phi(\eta_0)] \qquad\qquad \text{(6.31)}$$

in which:

$$\phi(\eta) = \eta - B(\eta) \qquad\qquad \text{(6.32)}$$

The values of ϕ are given in Table 103.

This simplified method must only be used for determining type M_1 curves and for small values of Froude number.

6.15 Surface profiles in natural watercourses

The methods explained previously for determining profiles presuppose regular, uniform cross-sections, and are thus not applicable to natural watercourses.

In natural watercourses, calculation of the surface profiles must always be undertaken by dividing the channel into elementary reaches. Division cannot be arbitrary but the reaches must be such that it is possible to consider mean values for the slope, cross-section and coefficient of roughness.

— The mean slope in a reach is considered equal to the mean slope of the bottom, corresponding to that reach.
— The mean cross-section in a reach can be taken to be the cross-section corresponding to the middle of the reach. A better way consists of plotting various cross-section of that reach and finding a mean line of those cross-sections (line *mn* of Fig. 6.9).

Fig. 6.9.

— Selection of the mean coefficient of roughness presents certain difficulties. It can be determined from the water depth for a given discharge in that reach, with the help of the formulae of uniform flow. If these are not available, it must be determined from knowledge of the nature of the bed, always bearing in mind that any increase in turbulence will cause an increase in energy losses.

Bearing in mind the definition of mean values, it is possible to apply to a natural watercourse step-by-step method, as explained in Section 6.12, without forgetting local losses. Automatic calculation is particularly indicated in such cases. However, some simplified methods can be used, e.g. Grimm method, Leach diagram, Ezra method. See [9].

C. RAPIDLY VARIED FLOW. HYDRAULIC JUMP

6.16 Definitions

A hydraulic jump consists of a sudden rise in the free surface of a steady flow, occurring in the passage from supercritical to subcritical flow and occupying a fixed position, being accompanied by violent turbulence, eddying, air entrainment, surface undulations and high energy losses.

We shall take cross-section 1 as the upstream cross-section or beginning of the jump, and cross-section 2 as the downstream cross-section or end of the jump (Fig. 6.10(a)); sometimes these cross-sections are not perfectly

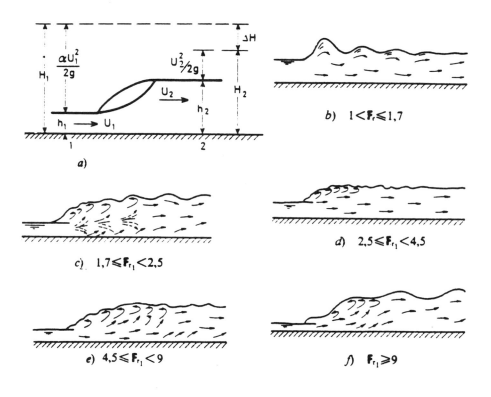

Fig. 6.10.

defined. The depths h_1 and h_2 are known as *conjugate depths* of the jump.[†] The distance between cross-sections 1 and 2 is known as the *length of the jump*. The energy loss is represented by H. The jump is generally characterized by the upstream Froude number:[‡]

$$\mathbf{F}_{r_1} = U_1/\sqrt{gh_1} \tag{6.33}$$

For values of $\mathbf{F}_{r_1}$ that are 1 or less, the flow is critical or subcritical and

---

[†] Not to be confused with the alternate depths of equal energy, H, defined in Section 6.4. What is referred to here are alternate depths of equal momentum as defined in Section 6.5.

[‡] Sometimes the square of this value is taken for Froude number.

there is no jump.† For values of Froude number between about 1 and 1.7 there is very little difference between upstream and downstream conjugate depths and the jump is characterized by a slight rippling of the free surface that differs little from what occurs in critical flow — *an undular jump* (Fig. 6.10(b)).

For values between about 1.7 and 2.5 the same phenomenon occurs, though it is more marked, and in this case small surface vortices appear. Up to these values of F_{r_1} the free surface is reasonably smooth and the distribution of velocities is regular — *a weak jump* (Fig. 6.10(c)).

For values between 2.5 and 4.5 the flow is pulsating in character; greater turbulence occurs, sometimes near the bed, at other times at the surface — *an oscillating jump* (Fig. 6.10(d)). Each pulsation causes a wave of irregular period that in nature may be propagated downstream for some kilometers and cause damage on the banks.

For Froude numbers between about 4.5 and 9, the jump is stable and a good dissipator of energy — *a steady jump* (Fig. 6.10(e)).

For values of F_{r_1} that are over 9, masses of water roll down at the start of the jump and fall into the upstream circuit in an intermittent manner, causing new undulations downstream — *a strong jump* (Fig. 6.10(f)).

These figures refer only to channels of rectangular section. In other channels the face of the jump is often complicated.

6.17   Determination of the conjugate depths of the jump

To determine the conjugate depths of the jump, it is not possible to apply Bernoulli's theorem between cross-sections 1 and 2, since the term ΔE, representing the energy loss, is unknown, and because the formulae corresponding to the uniform flow are not applicable. It is Euler's theorem that enables us to solve the problem. Consider a mass of liquid flowing in a horizontal channel; the friction losses at the end are disregarded. Application of equation (2.31), on an axis parallel to that of the channel, leads to:

$$\Pi + M_1 - M_2 = 0 \tag{6.34}$$

in which:

$$\Pi = \gamma y_1 A_1 - \gamma y_2 A_2 \tag{6.34a}$$

We therefore have:

$$\rho Q U_1 + \gamma y_1 A_1 = \rho Q U_2 + \gamma y_2 A_2 \tag{6.34b}$$

or

---

† The nomenclature followed hereafter corresponds to that in use by the U.S. Bureau of Reclamation.

$$\rho \frac{Q^2}{A_1} + \gamma y_1 A_1 = \rho \frac{Q^2}{A_2} + \gamma y_2 A_2 \tag{6.34c}$$

i.e. (see equation (6.5))

$$\mathcal{M}_1 = \mathcal{M}_2 \tag{6.34d}$$

If, for a given type of cross-section and a given discharge, the curve $\mathcal{M}$ is plotted, we obtain, by means of horizontals, different pairs of conjugate depths (Fig. 6.11). By making h_1, we can determine h_2 and vice versa as

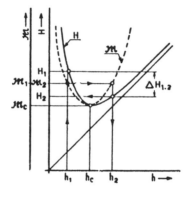

Fig. 6.11.

indicated in the figure.

The preceding equation can also be written, after dividing by γ and arranging the terms, as follows:

$$y_1 A_1 - y_2 A_2 = \frac{Q^2}{g} \left(\frac{1}{A_2} - \frac{1}{A_2} \right) \tag{6.34e}$$

By expressing the depth y of the centre of gravity as a fraction of the water depth h, that is to say, by making $y = \theta h$ and introducing the Froude number, $\mathbf{F}_{r_1}$, of the upstream cross-section, the equation takes the following form:

$$\theta_2 \frac{A_2}{A_1} \cdot \frac{h_2}{h_1} - \theta_1 = \mathbf{F}_{r_1}^2 \left[1 - \frac{A_1}{A_2} \right] \tag{6.34f}$$

The form of this equation, in certain particular cases† is analysed below:

(a) *Rectangular channel.* In a rectangular channel it is found that: $\theta_1 = \theta_2 = 1/2$ and $A_1/A_2 = h_1/h_2$ so that

$$\frac{h_2}{h_1} = \frac{1}{2}[-1 + \sqrt{1 + 8F_{r1}^2}]$$ (6.34g)

(b) *Triangular channel.* In a triangular channel $\theta_1 = \theta_2 = 1/3$ and

$$\frac{A_2}{A_1} = \left(\frac{h_2}{h_1}\right) \text{ giving the equation:}$$

$$\left(\frac{h_2}{h_1}\right)^2 = 1 + 3F_{r1}^2\left[1 - \left(\frac{h_1}{h_2}\right)^2\right]$$ (6.34h)

(c) *Trapezoidal channel.* For this case it is possible to obtain curves for various values of the parameter:

$$K = l/(mh_1) = 2l/(l - m)$$

in which l is the width of the bottom, m the slope of the banks (horizontal/vertical) and h the water depth upstream. Values of $K = 0$, and $K = \infty$ correspond to triangular and rectangular channel respectively.

(d) *Parabolic channel.* In a parabolic channel we have:

$$\theta_1 = \theta_2 = 2/5 \; ; \qquad \frac{A_2}{A_1} = (h_2/h_1)^{3/2}$$

whence

$$\left(\frac{h_2}{h_1}\right)^{5/2} = 1 + \frac{5}{2} \times F_{r1}^2\left[1 - \left(\frac{h_1}{h_2}\right)^{3/2}\right]$$ (6.34i)

(e) *Circular channel.* It is necessary to distinguish between the case of the channel being full and the case of it being completely full.‡ In the former case ($h_2 < D$), the points representing h_2/h_1, as a function of F_{r1}, occupy a band; in the latter case ($h_2 > D$), various lines are obtained, depending on the ratio h_1/D.

Graph 105 enables these various equations to be solved.

In a channel with vertical sides and sloping bottom, the conjugate depths are related by the expression:

---

† This closely follows Silvester, R., Hydraulic Jump in all Shapes of Horizontal Channels. *Proceedings ASCE, Journal of the Hydraulics Division*, Jan. 1964.

‡ It is demonstrated that in this case the piezometric head lines that occur represent the profile of the jump: Lane, E. W., Hydraulic Jump in Enclosed Conduits. *Engineering News Record*, **121**, (1938).

$$\frac{h_2}{h_1} = \frac{1}{2\cos\theta}\left[\sqrt{\frac{8\mathbf{F}_{r_1}^2\cos^3\theta}{1-2K\tan\theta}+1}-1\right] \tag{6.34j}$$

in which $\mathbf{F}_{r_1} = U_1/\sqrt{gh_1}$ and K is an experimental coefficient whose value depends on the slope of the channel; it is given in Table 110(b).

6.18 Determination of energy loss

If, together with the momentum function, the energy function H is also plotted, then the energy loss, $\Delta H = H_1 - H_2$, occurring in the jump can be determined. The energy loss could also be determined analytically by means of Bernoulli's theorem. Taking $z_1 = z_2$ we have:

$$\Delta E_{12} = \Delta H_{12} = \left(\frac{U_1^2}{2g}+h_1\right)-\left(\frac{U_2^2}{2g}+h_2\right) \tag{6.35}$$

The ratio H_2/H_1 is known as *jump efficiency*. The relative energy loss is given by:

$$\frac{H_1-H_2}{H_1} = 1-\frac{H_2}{H_1} \tag{6.35a}$$

Graph 109 gives the value of this energy loss for various cross-sections and various values of $\mathbf{F}_{r_1}$[†].

6.19 Location and length of the jump

As regards location of the hydraulic jump, it is clear from what has been said that it always takes place in the transition from supercritical flow to subcritical flow, and in cross-sections with the same value of the momentum function.

Assuming that the free surface forms of the upstream supercritical flow and downstream subcritical flow are known, the value of $\mathcal{M} = \rho Q^2/A + \gamma yA$ is determined along the longitudinal profile, corresponding to the various water depths, for both supercritical and subcritical flow. The meeting point of these two curves defines the position of the jump.

In the same way, it is possible to use the notion of conjugate depth, particularly in rectangular channels. Along the longitudinal profile, the conjugate depths corresponding to the different water depths known from the surface profile are determined. The hydraulic jump is located where the line of conjugate depths of the supercritical flow meets the free surface of the subcritical flow, or vice versa.

† Posey, C. J. and Hsing, P. S., Hydraulic Jump in Trapezoidal Channels. Proc. ASCE, **HY1**, Jan. 1964.

The length C is the most difficult characteristic of the jump to determine. In fact, as a rule it is difficult to define the end of the hydraulic jump.

From a practical point of view, when dimensioning stilling basins, the end of the hydraulic jump must be considered to be the place where concrete lining is no longer required.

Graph 107 should be consulted for determining the length of the jump in rectangular channels.

For trapezoidal channels, the following very approximate formula is given, for a value of the length C:

$$\frac{C}{h_2} = 5\left(1 + 4\sqrt{\frac{L_2 - L_1}{L_1}}\right) \tag{6.36}$$

in which h_2 is the depth downstream of the jump: L_1 and L_2 are the widths of the free surface upstream and downstream of the jump.

6.20 Submerged hydraulic jump
Whenever the water depth downstream is greater than the conjugate depth of the jump, h_2, a submerged hydraulic jump is formed (Fig. 6.12).

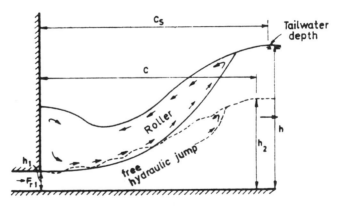

Fig. 6.12.

The degree of submergence is defined by the ratio

$$S = \frac{h - h_2}{h_2} \tag{6.37}$$

The length of the submerged jump is given by the approximate expression:

$$C_s = (4.9S + 6.1)h_2 \tag{6.38}$$

6.21 Hydraulic jump in rectangular channels

Consider a rectangular channel, taking the energy, H_1, upstream of the jump as unity. The various parameters defining the flow are called *reduced parameters* and are expressed according to the reduced upstream depth, $h'_1 = h_1/H_1$, by means of the following expressions:

$$U'_1 = U_1/\sqrt{H_1} = \sqrt{2g(1 - h'_1)} \tag{6.39}$$

$$q' = q/H_1\sqrt{H_1} = h'_1\sqrt{2g(1 - h'_1)} \tag{6.39a}$$

$$\mathbf{F}_{r_1} = 2\frac{1 - h'_1}{h'_1} \tag{6.40}$$

$$h'_c = \frac{h_c}{H_1} = h'_1\sqrt[3]{2\frac{1 - h'_1}{h'_1}} \tag{6.41}$$

$$h'_2 = \frac{h_2}{H_2} = \frac{h'_1}{2}\left[-1 + \sqrt{\frac{16}{h'_1} - 15}\right] \tag{6.41a}$$

$$U'_2 = U'_1(h'_1/h'_2) \tag{6.42}$$

$$\Delta H' = \frac{\Delta H}{H_1} = \frac{H_1 - H_2}{H_1} = 1 - H'_2 \tag{6.43}$$

It can be seen that the maximum value, and consequently the maximum reduced depth reached downstream occurs for $h'_1 = 0.4$, corresponding to $h'_2 = 0.8$.

The maximum value of the difference $(h'_2 - h'_1)$ is 0.507 for $h'_1 = 0.206$.

Should devices be used for locating the hydraulic jump and controlling energy dissipation, such as dentated sills, baffle piers or blocks, the characteristics of the hydraulic jump would vary considerably. See Section (6.37)

D. SINGULARITIES

6.22 Borda losses — Escande's formula

In open-channel flows, a singularity causes a change in uniform flow, associated with a local energy loss.

In closed conduit pressure flows, a local energy loss causes a definite lowering of the energy line along the pipe; in an open-channel flow in a sufficiently long channel, on the other hand, a sudden transition gives rise to a disturbance that is propagated along a more or less extensive reach, but outside that reach uniform flow is maintained, with energy characteristics that are analogous to those that would occur if the sudden transition did not exist.

Therefore, if the channel is sufficiently long for uniform flow to be re-established upstream and downstream of the singularity, the total energy loss between the extreme cross-sections is the same as that which would occur if the flow remained uniform, i.e. if there were no singularity.

The expression for the local energy loss has a somewhat different significance and is usually harder to determine. Sometimes cases that seem similar lead to very different flow conditions and energy losses. Whenever the nature of the problem so justifies, it is therefore advisable to carry out hydraulic model studies.

In the case of an abrupt *enlargement* there occurs *a Borda head loss*, which is studied in closed conduit pressure flows (see Section 4.14).

On the basis of Bernoulli's equation and Euler's theorem, Escande[†] proposed the following expression for losses in an abrupt enlargement:

$$\Delta E = K \frac{(U_1 - U_2)^2}{2g} - (h_2 - h_1) \tag{6.44}$$

If h_2 is close to h_1, the second term can be disregarded, so that Escande's equation becomes equal to Borda's equation.

6.23 Passage from a reservoir to a channel
The following cases are considered:

(1) When $I > I_c$ (Fig. 6.13): the flow is critical at a cross-section near the

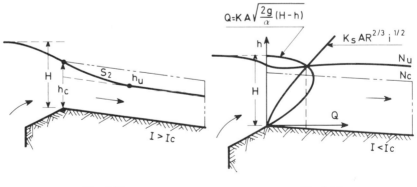

Fig. 6.13.   Fig. 6.14

intake of the channel, and quickly becomes uniform describing an S_2 type curve. Expression (6.4) can be used to determine the discharge, taking

† Escande, I., *Hydraulique Générale*, Université de Toulouse.

$h = h_c$. The flow is similar to that which takes place over a spillway, and the discharge can be given by:

$$Q = \mu L\sqrt{2g}H^{3/2} \tag{6.45}$$

in which L is the width, H the total energy over the spillway crest and μ the coefficient of discharge.

Spillway hydraulics is discussed further in Section 6.31 *et seq.*, including the appropriate value of μ.

(2) When $I = I_c$: the preceding considerations are valid for any discharge taking place. The flow, in this case, is always uniform.

(3) When $I < I_c$: the discharge in this case is that compatible with the energy conditions and the uniform flow in the channel (Fig. 6.14). It is obtained after solving, with the help of a graphical construction, the system of equations (6.4) and (5.7) or (5.8).

$$Q = KA\sqrt{\frac{2g}{\alpha}(H-h)} \qquad (H = \text{constant})$$
$$\tag{6.45a}$$
$$Q = K_sAR^{2/3}i^{1/2}$$

These two equations give Q as a function of h. When the two functions are drawn on the same system of axes (Q, h), the intersection defines the discharge Q that takes place at a water depth h.

The coefficient K introduced in equation (6.4) takes account of the energy losses at the intake; for a well designed intake, $K = 0.90$ would be appropriate.

6.24 Passage from a channel to a reservoir
The following cases are considered:

(1) When $I > I_c$: If the level N in the reservoir is lower than the lower level N_u at the outlet from the channel (Fig. 6.15), there will be a sudden

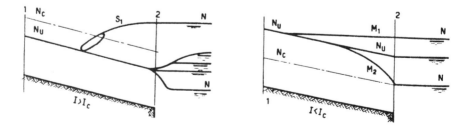

Fig. 6.15.   Fig. 6.16

drop (free overfall) in the free surface at the outlet. If N is between N_u (corresponding to the uniform depth) and N_c (corresponding to the critical depth), there will continue to be no change in the surface profile inside the channel and the flow entering the reservoir will have a curved surface profile that can be called a *false hydraulic jump*.

If N is higher than N_c, then in a steep channel, a hydraulic jump will be formed followed by a rising surface profile, the flow being subcritical (curve S_1).

In any of these cases the discharge will not be influenced since in supercritical flow the downstream conditions have no influence upstream unless the curve S_1 reaches the start of the channel.

(2) When $I = I_c$: If N is lower or equal to N_c, there will be uniform flow at a depth equal to the critical depth.

If N is higher than N_c, the free surface in the channel will be a surface profile in a critical slope channel and this, of course, is a horizontal (curve C_1).

As previously, the discharge will only be influenced if the surface profile reaches the beginning of the channel.

(3) When $I < I_c$: If N is lower than N_c, a surface profile of the M_2 type occurs, with a corresponding increase in mean velocity (Fig. 6.16). The discharge will be slightly higher than that a uniform flow, though this difference is only possible in very short channels or if they have an extraordinarily small slope. If N is equal to N_u, the discharge will be that corresponding to uniform flow. If N is higher than N_u, there will be a surface profile of the M_1 type in the channel, with a corresponding reduction of the discharge, such that as the profile approaches the beginning of the channel, the greater will be that reduction.

6.25   Raising of the bottom

The study of a disturbance caused by an obstacle in the bottom of a channel or watercourse (overflow spillway for example) can easily be undertaken using the energy curves appropriate to the water depth. Typical cases are presented in Fig. 6.17 where curves H and H' corresponding to the original channel and to that with a raised bed are plotted. Curve H' is obtained from curve H, after raising axis h by an amount equal to the height a of the obstacle.

Let h_u and H_u represent the water depth and the energy in relation to the bottom, in uniform flow; h_c and H_c represent the critical depth and critical energy in the channel; h'_c and H'_c represent the critical depth and critical energy at the top of the obstacle. The shape of the free surface and the path of the point that represents the flow on plane (h, H) are presented in Fig. 6.17, for various cases of slope and available energy.

6.26   Local constrictions; bridge piers

An analysis of disturbance caused by constrictions can also be easily be done with the help of the H curves for the full cross-section of the channel and H' for the cross-section constricted by the piers.

Analysis of the phenomenon, with a long reach upstream and down-

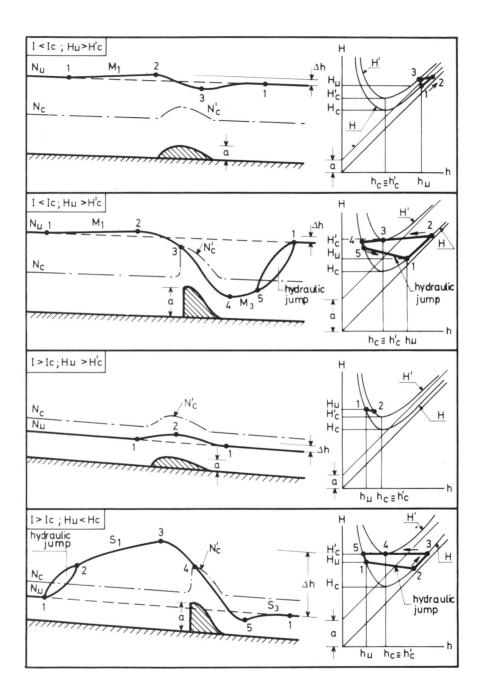

Fig. 6.17.

stream, is given for the most typical cases in Fig. 6.18, showing the shape of the free surface and the path of the point representing the flow (h, H).

It is necessary to analyse, in more detail, what occurs in the vicinity of a contraction, particularly in the case of channels with slope $I < I_c$, which corresponds to the most usual watercourses.

The flow is called *drowned* or *submerged* whenever $H_u > H'_c$, that is to say, when the upstream energy is enough to overcome the obstacle, as occurs in the first example of Fig. 6.18; it is said to be *undrowned* when $H_u < H'_c$, that is to say, when the energy is insufficient, and the flow becomes critical followed by a hydraulic jump (as occurs in the second example of Fig. 6.18).

In the case of nearly drowned flow, the shape of the flow is represented in greater detail in Fig. 6.19(a).

The constriction causes a rise in the free surface in cross-section 1 which may be located at about a distance l from the middle of the contraction, l being the width of the constriction. The transition between the flow in cross-section 1 and cross-section 0, upstream of the contraction corresponding to the undisturbed flow in which $h_0 = h_u$, is described by an M_1 type curve.

Between cross-section 1 and cross-section 3, which is the final cross-section of the constriction, as the flow is accelerated, the water surface profile falls rapidly and the jet stream contracts to a width somewhat less than the width of the opening; rapidly varied flow occurs downstream of cross-section 3 until uniform flow is reached in cross-section 4.

The contracted cross-section 2, situated between 1 and 3, controls the flow. It is in fact downstream of the minimum-width cross-section, when it widens until it reaches the total cross-section of the channel, that there are the highest energy losses, owing to the vortices which are generated due to the flow expansion.

Theoretically the calculation of the discharge should be related to cross-section 2, but from a practical point of view, it is related to cross-section 3, by the expression:

$$Q = CA_3 \sqrt{2g\left(\Delta h - h_f + \alpha_2\frac{U_1^2}{2g}\right)} \qquad (6.46)$$

in which $A_3 = lh_3$, l being the width of the contraction opening; h_f is the friction head loss between cross-sections 1 and 3; U_1 is the mean velocity in cross-section 1 and $\Delta h = Z_1 - Z_3$ is the fall in the free surface between 1 and 3.

C is a coefficient that takes into account the fact that cross-section 3 is considered instead of the contracted cross-section 2; the losses due to contraction between 1 and 2, are practically negligible, the losses due to the expansion downstream of 2 being the most important.

The value of C depends on various parameters and, taking into account only the main parameters for various shapes of contraction that are shown below, it could be written as follows[†] (see Graph 111):

$$C = C'K_F K_W K_r K_\phi K_y K_x K_e \qquad (6.46a)$$

in which: C' — discharge coefficient standard value, a function of the contraction ratio $\sigma = 100(L - l)/L$ (in which l is the width of the contraction opening and L the width of the channel) and of the ratio a/l, a being the length of the contraction, i.e. the dimension in the direction of the flow;

       K_F — function of Froude number, $\mathbf{F}_{r3} = Q/(A_3\sqrt{gh_3})$. In order to calculate $\mathbf{F}_{r3}$ it is necessary to assume a discharge, Q, and check the assumed value later. To

† Kindsvater, *Computer of peak discharge at contraction*, U.S. Geological Survey, Circular No. 289, page 1953. Quoted by [9].

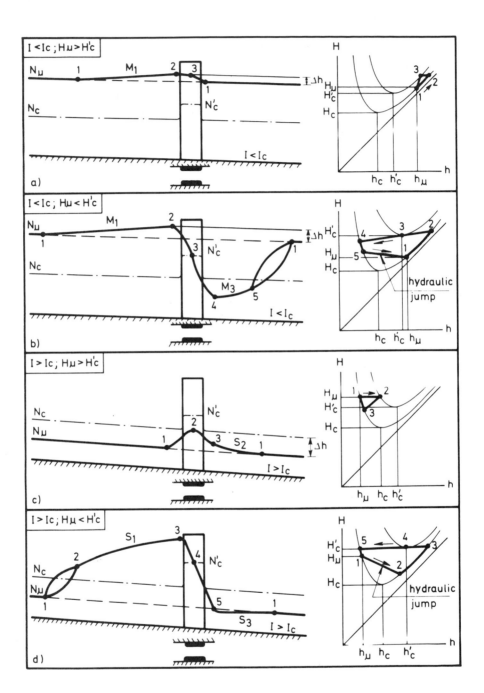

Fig. 6.18.

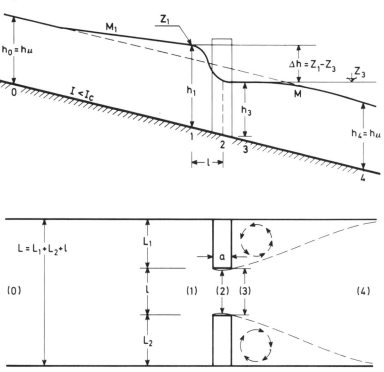

Fig. 6.19.

ensure the subcritical flow, it is necessary that $\mathbf{F}_{r_3} < 1$; in practice, for reasons of safety, it is better that $\mathbf{F}_{r_3} < 0.8$;

K_w — characterizes the abutment type, and is a function of the ratio between the length of wingwalls or chambers and, w, width l of the contraction opening (see Graph 108). In this case we have $K_r = 1$;

K_r — substitutes for K_w when the transition takes place with entrance rounding; r being the radius of the corner. In this case we have $K_w = 1$;

K_ϕ — function of the angle of the contraction measured in relation to the normal to flow direction;

K_y — function of $(h_a + h_b)/2\,l$ in which h_a and h_b are the water depths at the foot of each abutment;

K_x — represents the effect of side slope at abutments: $K_x = x/l$. The meaning of x is given in Graph 108(c);

K_e — function of the eccentricity of the opening, e (ratio between the discharge capacity of the upstream reach, to the right and to left of the contraction). In rectangular channels, $e = L_1/L_2$, with $L_2 > L_1$.

Graph 108 gives the various coefficients, for several values of the fill slope and angles of the wingwall abutments, for eccentricity values e, between 0.2 and 1.0, ($K_e = 1$). In the case of $e < 0.2$ the values of K_e are less than 1, and at the limit, for $e = 0$, the value is $K_e = 0.95$.

For determining the *rise of the free surface*, Δh, in drowned flow which frequently occurs in bridge piers[†], Rehbock's formula is given, since it is easier to apply:

† See [7] which gives a synthesis of the formulae that can be used in the various cases.

$$\Delta h = [\delta - \sigma(\delta - 1)](0.4\sigma + \sigma^2 + 9\sigma^4)(1 + \mathbf{F}_{r_3}^2)\frac{U_3^2}{2g} \qquad (6.47)$$

The values of δ, function of piers fairing, determined from Yarnell's tests are given in Graph 109, which also relates $\sigma = (L - l)/l$ to $\mathbf{F}_{r_3}$ and shows the region of validity of that formula.

6.27 Extensive contraction
Extensive contraction implies a decrease in the channel's cross-section for a sufficiently long reach so that gradually varied flow may occur in that reach.

Some typical cases of contraction are represented in Fig. 6.20 and 6.21, which show the shape of the free surface and the path of the point representing the flow in the plane (h, H).

6.28 Sudden slope changes
Fig. 6.22 shows the surface profiles in some typical cases of sudden changes of channel slope.

(a) *Increasing slope, always smaller than the critical slope.* In the second reach, the flow is uniform; in the first reach, there will be an M_2 curve that tends asymptotically to the uniform depth corresponding to h_1; the datum point for determining this curve is point 2.

(b) *Decreasing slope, always smaller than the critical slope.* This case is similar to the preceding one, the M_2 curve being replaced by an M_1 curve.

(c) *Steep slope channel connected to a mild slope channel.* There will be a hydraulic jump when the flow goes from supercritical to subcritical flow. One of the problems is to discover whether the jump is located in the steep slope reach or in the mild slope reach.

In this case h_1 and h_2 are taken as the uniform depths corresponding to the first and second reaches, respectively, and h_1' and h_2' as the conjugate depths of h_1 and h_2.

Two cases may occur: in the case of $h_1' > h_2$, the supercritical flow will enter the mild slope cross-section describing an M_3 curve, the datum point being at 1 and the hydraulic jump is located at the cross-section where the second conjugate water depth, h_2, occurs; in the case of $h'_1 < h_2$, the hydraulic jump occurs in the steep slope reach, being followed by an S_1 curve, 2 being the datum point. The hydraulic jump cross-section will be that where the conjugate water depth, h_2 has the same value of water depth as the surface profile.

(d) *Adverse or null slope channel followed by a steep slope channel.* In this case the upstream curve is necessarily A_2 or H_2 and the downstream curve S_2, tending asymptotically to uniform depth; the initial datum point for calculating both curves is point 1, defined by the critical depth:

(e) *Change from a mild slope channel ($I < I_c$) to a steep slope channel ($I' > I_c$).* Critical flow occurs at the change of slope; the flow is uniform

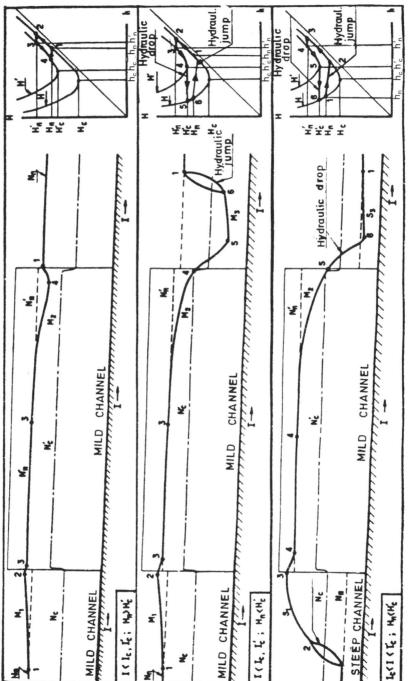

Fig. 6.20.

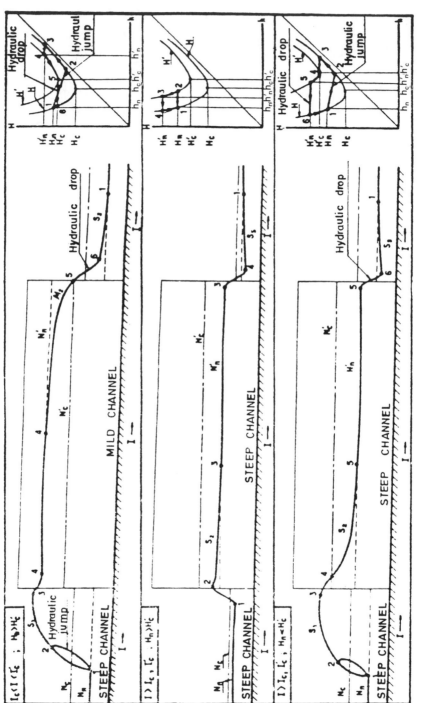

Fig. 6.21.

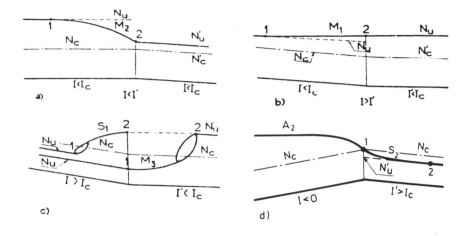

Fig. 6.22.

upstream describing an M_2 curve; downstream, uniform flow is reached and describes an S_2 curve.

6.29   Sudden drop

In a sudden drop, or free overfall, if the channel has a mild slope, $I < I_c$, critical flow occurs at the cross-section where the flow drops (Fig. 6.23(a)),

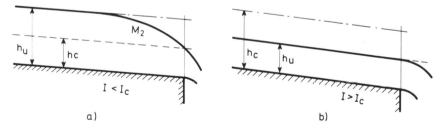

Fig. 6.23.

the backwater curve being describerd by an M_2 curve. In the case of $I = 0$ or $I < 0$, the flow also passes through critical, with H_2 or A_2 curves, respectively, upstream.

If the channel has a steep slope, $I > I_c$, the flow remains uniform until the cross-section where the free overfall occurs (Fig. 6.23(b)).

6.30 Flow at bends

Bends in watercourses and channels introduce various disturbances into the flow that may cause local head losses, formation of secondary currents and changes in the shape of the free surface. The secondary currents that are formed at bends cause a circulation of the liquid particles according to helical trajectories in the longitudinal direction of the flow, being responsible for an increase in the friction forces of the flow, and in the friction forces at the walls of the channel, as well as changing the shape of the water free surface. In flows with a mobile bed, typically alluvial beds, secondary currents are also responsible for a change in the shape of the alluvial bed, tending to scour sediment from the outside of the bend and deposit it on the inside. In the case of supercritical flows the bends, as well as other changes in the transverse cross-section of the flow, cause cross waves that are sometimes of sufficient amplitude to be of importance in the design of channels and which, as a rule, persist downstream for considerable distances.

In flow at bends it is necessary to take into account the head loss, ΔH, corresponding to an elevation of the free surface, Δh, these bing related to one another by equation (6.2a); apart from this head loss, there is a transverse rise of the free surface, ΔZ, that is due to the effects of centrifugal force which causes greater water depths at the outside and lesser depths on the inside of the bend.

(a) *Determination of head loss, ΔH*

The head loss introduced into the flow by a bend corresponds to one of the three cases represented schematically in Fig. 6.24.

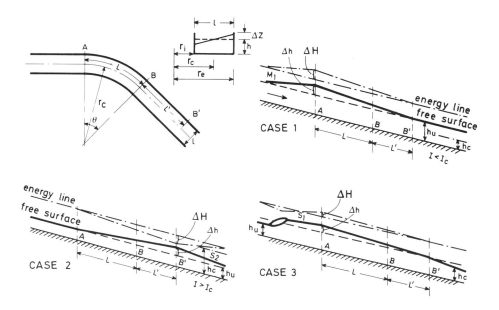

Fig. 6.24.

— *Case* 1 (subcritical flow): the head loss, ΔH, introduced by the bend, causes an elevation, Δh, of the free surface in the upstream cross-section A, of the bend. Uniform flow is reached in a cross-section B', with BB' being the zone of influence of the circulation caused by the bend.

— *Case* 2 (supercritical flow): the head loss introduced by the bend, ΔH, causes a gradual elevation, Δh, of the free surface downstream of cross-section A, until it reaches a maximum in cross-section B', with an S_2 curve being formed downstream of cross-section B'.

— *Case* 3 (supercritical flow changing to subcritical): when the energy of the supercritical flow is close to critical energy, the flow becomes critical at cross-section B' and induces a maximum head loss at cross-section A, with a corresponding elevation of the free surface. Upstream of cross-section A, an S_1 curve and a hydraulic jump are formed.

The head loss ΔH introduced into the flow is of the type:

$$\Delta H = \frac{KU^2}{2g}$$

in which U is the mean velocity corresponding to the undisturbed flow and K is a coefficient that can be determined as shown below (see Fig. 6.24).[†]

The ratios r_c/l and h/l are calculated and from Graph 113 the value of the base coefficient K_0 is obtained, this being valid for $\mathbf{R}_{eo} = UR/v = 31\,500$ (R is the hydraulic radius) and $\theta_0 = 90°$; for values of θ different from θ_0, the corresponding curve gives the values a and a_0 which correspond to θ and θ_0 respectively.

For values of $\mathbf{R}_e$ different from $\mathbf{R}_{eo}$, the respective curves of the same graph give the value of b and b_0 corresponding to the values of $\mathbf{R}_e$ and $\mathbf{R}_{eo}$, respectively.

The value of K_0 is then modified as follows:

$$K = K_0 \, \frac{a}{a_0} \frac{b}{b_0} \tag{6.48}$$

Example

Determine the head loss introduced in the flow by a 45° bend ($\theta = 45°$) and radius $r_c = 15$m, in a rectangular channel of width $l = 10$m, depth $h = 2.5$m; the mean velocity is $U = 3$m/s.

Solution

The hydraulic radius is $R = A/P = l.h/(l + 2h) = 25/15 = 1.67$;
The Reynolds number will be $\mathbf{R}_e = UR/v = 3 \times 1.67/(1.01 \times 10^{-6}) = 4.69 \times 10^6$.
Also $r_c/l = 15/10 = 1.5$ and $h/l = 2.5/10 = 0.25$.

---

† Shurkry, A., Flow around bends in an open flume. *Transactions, American Society of Civil Engineers*, **115**, (1950). Quoted by [9].

From Graph 110 we have $K_0 = 0.15$; from Graph 110, for $\theta_0 = 90°$, we have $a_0 = 0.30$ and for $\theta = 45°$ we have $a = 0.03$; from Graph 110 we have $b_0 = 0.15$ and $b \approx 0.20$. Thus;

$$K = K_0 \frac{a}{a_0} \frac{b}{b_0} = 0.15 \times \frac{0.03}{0.30} \times \frac{0.20}{0.15} = 0.002$$

$$\Delta h = KU^2/2g = 0.002 \times U^2/2g \approx 0$$

Example
In the previous case take $r_c = 10$ m and $\theta = 135°$.

Solution
We shall have $r_c/l = 1$; $K_0 = 0.30$; $a_0 = 0.30$; $a = 0.38$; $b_0 = 0.22$; $b = 0.35$
so that:

$$K = K_0 \frac{a}{a_0} \frac{b}{b_0} = 0.35 \times \frac{0.38}{0.30} \times \frac{0.35}{0.22} = 0.7$$

$$\Delta h = KU^2/2g = 0.7U^2/2g = 0.31 \text{ m}$$

This example, and an inspection of the graphs show that head losses at any bends will be negligible for $\theta < 45°$ and for $r_c/h < 2.0$.

(b) *Rise of the surface,* ΔZ

Centrifugal force, which acts on the flow at a bend, causes a transverse inclination of the free surface with a corresponding increase in the depth flow on the outside of the curve.

Assuming that the velocity has only tangential component and by applying Euler's equation radially, we have:

$$\frac{\partial}{\partial n}(z + p/\gamma) = -\frac{1}{g}\frac{V^2}{r} \tag{6.49}$$

Assuming $Z = z + p/\gamma$, the elevation on the outside of the bend will be:

$$\Delta Z = \frac{1}{g} \int_0^l \frac{V^2}{r} dn \tag{6.50}$$

in which l is the width of the bend measured along the radial.

As an approximate hypothesis for practical application, it may be assumed that the velocity of the streamlines at the bend is equal to the mean flow velocity, that all streamlines have the same radius of curvature, r_c, and that the shape of the free surface is a straight line, so that:

$$\Delta Z \approx \frac{U^2 l}{2gr_c} \tag{6.51}$$

In supercritical flow the effect of waves must also be considered.

Example

Calculate the elevation of the free surface in the channel mentioned in the previous example.

Solution

By applying formula (6.53) we have:

$$\Delta Z = \frac{9 \times 10}{9.8 \times 15} = 0.61 \, \text{m} \; .$$

In supercritical flow standing wave effects are superimposed on the centrifugal effects, whose maximum values are of the same order of magnitude as the latter and at the maximum ΔZ may reach the value $U^2 l/g r_c$.

E.   SUDDEN DROP. SPILLWAYS

6.31   General considerations

Consider a *sharp-crested weir* over which a discharge q, per unit width, takes place (see Section 8.29).

If the nappe is suitably aerated, the flow behaviour is similar to a projectile launched at a velocity V_0 and an angle θ; it can therefore be assumed that the horizontal component of the velocity is constant and that the only force which is acting is that of gravity. Thus a particle G (Fig. 6.25), on the lower face of the nappe, covers a distance x and a distance y in the horizontal

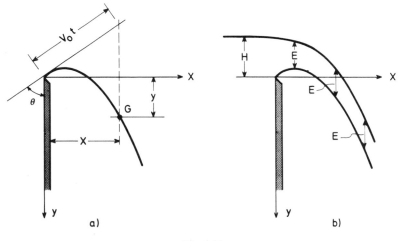

Fig. 6.25.

and vertical directions respectively, given by:

$$x = V_0 t \cos \theta$$
$$y = V_0 t \sin \theta + \frac{1}{2} g t^2 + c' \tag{6.52}$$

By eliminating t and dividing by the head, H, over the crest, we have:

$$\frac{y}{H} = A \left(\frac{x}{H} \right)^2 + B \frac{x}{H} + C \tag{6.53}$$

Since the horizontal component of the velocity is constant, the thickness E of the nappe will also be constant (Fig. 6.25(b)). By adding the term $D = E/H$ to the equation we obtain the equation of the upper face of the nappe.

There are several expressions for determining the constants, A, B, C, D, but in practice the problem is solved as explained in the following section, see [9].

6.32   Broad-crested weirs. High overflow spillways. The WES (Waterways Experiment Station) crest profiles

(a) Geometrical definition

Reference has been made to the nappe profile, in free fall, over a sharp-crested weir. For the purpose of high discharges (overflow spillways), broad structures are used, these being such that their downstream face has a profile which conforms to the lower face of the nappe from a sharp-crested weir. The coordinate axes of the preceding equations refer to the sharp-crested weir, but from a practical point of view this creates some difficulties.

The spillway crest shapes proposed by the Waterways Experiment Station (WES), also based on the lower surface of the nappe, are far more practical, since the origin of coordinate axes is the highest point of the curved crest; the shape of the profiles depends upon the head, H_d, and the inclination of the upstream face, that is sometimes imposed for reasons of structural stability of the dam. For this reason, and since this type of crest has been most commonly used, the following analytical definitions apply to these crest shapes.

As has been mentioned, taking the highest point of the curved crest (Fig. 6.26), as origin of the coordinates, the general expression for the *down-*

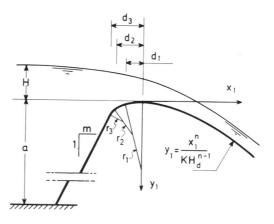

Fig. 6.26.

stream quadrant of the profile, according to WES, is as follows:

$$y_1 = \frac{x_1^n}{KH_d^{n-1}} \tag{6.54}$$

in which H_d represents the design head.

The *upstream quadrant* is defined by means of compounded circular curves, with radii expressed in terms of design head and of upstream face slope.

The values of the constants n and K, the radius of the circumference arcs r_1, r_2 and r_3, and the distances from the points of tangency at the sill d_1, d_2 and d_3 (Fig. 6.26), depend on the upstream face slope m and are given by Table 111.

For this type of crest profile those crests whose upstream face is formed by more than one straight alignment should also be considered: *crest with an offset and riser* (Fig. 6.27(a)), *crest with the upstream face partly vertical and*

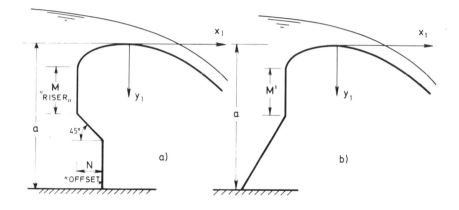

Fig. 6.27.

partly sloping (Fig. 6.27(b)), for which the parameters given in Table 114 for vertical faces are valid, provided that the values of M, N and M' represented in the figures are such that $M \geq 0.6H_d$; $M/N > 0.5$ and $M' \geq H_d$.

Crests of the type shown in Fig. 6.27(a) are generally used in gravity arch dams in which the stability profile is relatively sharp. Those of type shown in Fig. 6.27(b) are generally used, for stability reasons, in dams whose overflow spillway is not very high.

(b) Discharge capacity
The discharge over this type of spillway is given by the formula:

$$Q = \mu L \sqrt{2g}\, H^{3/2} \qquad\qquad (6.55)$$

in which L = effective length of the crest (m); H = functioning head, defined by the difference between the elevation of the reservoir level and the spillway crest (m); g = acceleration of gravity (m/s²); μ = coefficient of discharge; Q = discharge (m³/s).

The coefficient of discharge varies with the slope of the upstream face and with the ratio between the functioning and design heads. Table 111 gives the values of μ corresponding to different ratios of H/H_d and to different slopes of the upstream face.

The downstream face of the overflow crest, given by equation 6.54 is frequently continued by means of a tangent plane surface, but the location of the tangency plane may influence the discharge capacity of the overflow spillway. Table 114(c) gives that influence for a crest profile with a vertical upstream face.†

It is found that for $d \geqq 0.28 H_d$ (Table 111(c)), there is no reduction in the coefficient of discharge when the spillway operates with a discharge corresponding to the design head, H_d.

(c) *Free surface profile*
Knowledge of the profile of the free surface is of interest for placing the gates, when they are tainter gates, for dimensioning the height of the walls which laterally confine the spillway crest and for calculating the freeboard required for the flow when the crest of the dam is also designed to carry a road or railway. As regards tainter gates, proper clearance of the water surface for structural members or gate trunnions is important.

Table 111(d) shows the profiles of the free surface for various slopes of the upstream face and for H/H_d = 1.00 and 1.25.

(d) *Cavitation*
Cavitation is a dynamic phenomenon that may occur in high velocity flows and consists of the formation and subsequent collapse of pockets or cavities which are largely filled with water vapour. These cavities form in zones where, for any reason, the local pressure is below the inherent vapour pressure temperature limits of water and they begin to collapse further downstream, when they are carried by the flow to a region where the local pressure is higher than the vapour pressure. Collapse of these cavities can produce local destruction of the surface in contact with flow.

Since concrete is a material with low resistance to the action of cavitation, it is necessary to limit negative pressures and provide spillway faces with a carefully prepared finish. As an order of magnitude, mean local depressions must not exceed − 6 m water column.

The design head and location of the gate seat, the trunnion elevations and gate radii must be so planned that there occur no depressions in excess of that indicated above.

Table 111(e) shows, for spillway crest shapes with different upstream slopes, the values of $(p/\gamma H_d)_{min}$ as a function of H/H_d.

---

† Experiments have shown that these results are approximately valid for spillways with upstream face slopes of 1/3, 2/3 and 3/3 [27].

(e) Separation of the flow

It was explained in Section 2.27 that, in a flow with velocity V, if there is a positive gradient of pressures, $dp/dx > 0$, conditions are created for separation of the boundary layer to take place.

In spillways of this type, separation takes place whenever the Euler number is such that:[†]

$$E_u = -\frac{s}{\rho V^2}\frac{dp}{ds} \leqslant -0.25 \qquad (6.56)$$

in which s is a length along which the pressure gradient is constant. It is therefore necessary for:

$$\frac{dp}{ds} \leqslant -0.25\rho V^2/s \qquad (6.56a)$$

Table 111(f) shows the operating conditions likely to cause separation of the flow in normal spillways without gates, represented by the maximum value of H/H_d.

(f) Placing of gates

The possibility of separation is greater when the spillway has to operate with the gates partly open. In fact, it must be explained that for certain positions of the gate seat locations in relation to the highest point of the curved crest (crest axis), the jet issuing from under the gate tends to spring away from the face, resulting in regimes of pressures whose values may cause cavitation, or whose gradients are incompatible with the adhesion conditions and may cause separation of the flow (detachment of the jet).

When studying the positioning of spillway gates, it is necessary to avoid gate seat and trunnion locations that may lead to any of these phenomena (cavitation and detachment of the liquid filament).

Reference [27] gives criteria to satisfy these requirements.

6.33   Low overflow spillways

(a)   Crest shape. Analytical definition

In low overflow spillways, the approach velocity is usually fairly high. As a result, the profile of this type of crest depends not only on the slope of the upstream face and the design head, but also on the height of the overflow crest axis above the floor of the entrance channel (which influences the velocity of approach to the crest).

The profile proposed by the US Bureau of Reclamation takes account of the influence of the approach velocity, and it is defined by a shape analogous

† Schlichting, H., *Three-dimensional Boundary Layer Flow*, 9th IAHR Congress, Dubrov-nitk, 1961. Quoted by [28].

to that of high overflow spillways. Taking the highest point of the curved crest (crest axis) to be the origin of the coordinates, the general expression of the downstream profile is as follows (equation 6.54):

$$y_1 = \frac{x_1^n}{KH_d^{n-1}}$$

in which K and n are constants whose values depend on the slope of the upstream face and on the velocity of approach. The influence of the velocity of approach is quantified by the value of the velocity head in the approach to the crest defined by:

$$h_a = \frac{Q^2}{2g(a + H_d)^2 L^2} \tag{6.57}$$

in which a represents the height of the spillway crest; the other symbols have been previously defined. The upstream quadrant of the crest shape is defined by means of circular arcs. The values of the constants n and K of the radii of the circular arcs and of the distance of the tangency points from the crest axis (Fig. 6.28) are shown in Table 112(a), as functions of the slope of the upstream face.

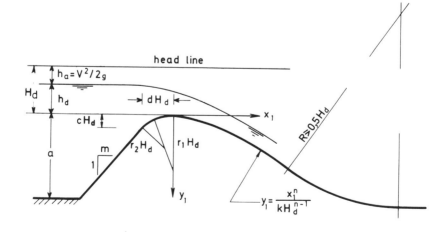

Fig. 6.28.

(b) Discharge capacity
In free flow the discharge is given by the expression:

$$Q = \mu L \sqrt{2g} H^{3/2} \tag{6.58}$$

The coefficient of discharge, μ, varies with the velocity of approach represented by the ratio a/H_d, with the ratio between the functioning head and design head, H/H_d, and with the slope of the upstream face, m (see Table 112(b)).

An analysis of this table makes it clear that ratios of $a/H_d < 0.4$ must not be used, because the coefficient of discharge in these spillways is low. It is also to be noted that it is advantageous to have upstream faces with slopes of 2:3 and 3:3, when $a/H_d < 0.6$. If $0.6 < a/H_d < 1.0$, it is advantageous for the spillway to have an upstream slope of 2:3.

(c) Discharge capacity in submerged flow

From the points of view of precision and practical application, the experiments of the US Bureau of Reclamation [23] for a vertical face are of particular interest, the results being represented in Graph 113.

The main variables of the graph are: on the horizontal axis, the ratio $\alpha = (y + h)/H$ representing the position of the downstream apron; y being the difference in level between the free surface upstream and the free surface downstream, h representing the water depth and H the functioning head over the spillway.

The percentage decrease in the coefficient of discharge is indicated by the continuous lines. These lines have an initial sector and a final horizontal sector. The vertical sector corresponds to the variations that are mainly caused by the position of the downstream apron; in the horizontal part a decrease in the coefficient of discharge is caused by the degree of submergence.

The graph is divided by dashed lines into four zones, corresponding to the various types of flow described below.

Type I — When a decrease in the coefficient of discharge occurs in this zone it is not caused by submergence but by the position of the downstream apron.

Type II — If the bed extends further downstream or the value of β is decreased a hydraulic jump forms and the flow is supercritical upstream and subcritical downstream.

Type III — If the value of β continues to diminish, even though the nappe continues to follow the face of the overflow, the water depth downstream is such that drowned jump with a flutuating path will occur except for very low values of α.

Type IV — For even lower values of β the flow becomes definitely submerged (lower zone of the graph). The submergence is very unstable except for very low values of α and it is in this zone that a decrease in the coefficient of discharge is most accentuated.

(d) Positioning of the gates

The general criteria given in item (f) of 6.32 are valid.

6.34   Distribution of pressures on the upstream face

On the upstream face the distribution of pressures is not hydrostatic. The shape of the distribution is represented in Fig. 6.29.

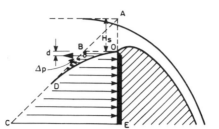

Fig. 6.29.

The hydrostatic pressure is represented by the area of the trapezium *OBCE*. The reduction in pressure Δ*p* in relation to hydrostatic pressure is represented by the shaded area defined by *OBD*.

The distance from the point of application of Δ*p* (centre of gravity of the area *OBD*) to point O will be represented by *d*.

Table 114 gives the value of $\Delta p/H_s^2$ and of d/H_s as function of h_a/H_s. The value of Δ*p* is expressed in surface units. Multiplying Δ*p* by the specific weight of the liquid gives the force per metre of crest length.

6.35   Circular spillway crest in plan. Shaft or morning glory spillways[†]
(a) Geometrical characteristics. Analytical definition
The main use of circular crests in plan is in shaft spillways. In this type of crest the flow is radial provided that there is no interference from nearby boundaries.

In general, shaft spillways are set on a platform, the shape of this platform being designed to make the flow conditions radial or nearly so. A shaft spillway installed far away from the hill slopes generally has no platform next to it and the upper part of the shaft has the shape of a tulip, so that it is usually known as a *morning glory spillway*.

The general shape of the crest profile is considered to coincide with the lower face of the liquid filament discharged by a sharp-edged, cylindrical, circular spillway. Table 117 gives the profile coordinates of the crest for various ratios of *a*/*r* and H_s/r.

H_s represents the head over the virtual sharp-edged, cylindrical-crested weir from which the nappe originates (Fig. 6.30).

H will continue to represent the head over the broad crest (functioning head) and H_d the head corresponding to the design discharge (design head).

The coordinates (x, y) thus refer to the sharp-edged, cylindrical crest, unlike the coordinates (x_1, y_1) used previously, which refer to the rounded spillway crest. Table 115 gives the relationships between the head on the sharp-crested weir H_s and the head on the rounded spillway crest H_d.

† A shaft spillway consists basically of a spillway crest and a shaft of which the diameter is generally variable, followed by a gallery. There is a transition zone between the crest and the shaft, and between the shaft and the gallery. For documentation on shaft spillways, see [29] and [31].

These data are based on the experimental tests carried out by Wagner.

Table 118 makes it possible to plot the free surface (upper profile of fully aerated nappe), determine the position of the *crotch* (Fig. 6.30) that defines the zone where the flow begins to occupy the whole transverse cross-section, the highest point of the *boil* (mass of water not concerned in the flow and situated above the *crotch* zone) and the point where the *boil* meets the nappe profile.

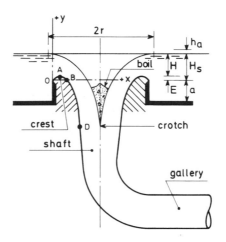

Fig. 6.30.

(b) Discharge capacity

When considering this type of spillway in operation, two cases may occur:

— the spillway is designed for pressure flow conditions at the maximum discharge;
— the spillway always functions with crest-control or free-flow conditions.

In the first case, the discharge is controlled by the spillway crest up to the design head and by the spillway throat when the head exceeds the design head. In the second case, the discharge is always controlled by the crest.

In both cases, the spillway crest is usually dimensioned for a design head, H_d, for which the discharge is still controlled by the crest and is obtained by the expression:

$$Q = \mu_d(2\pi r)\sqrt{2g}\,H_d^{3/2} = 27.83\mu r H_d^{3/2} \tag{6.59}$$

in which r is the radius of the circumference that defines the projection of the outer face of the crest (Fig. 6.30).

The coefficient of discharge of a shaft spillway differs, of course, from that of a rectilinear crest spillway, owing to the effects of convergence of the fillets and to change of control from the crest to spillway throat when the head exceeds the design head. The value of μ_d therefore depends, for each value of a, on the values of H_d and r. The value of μ_d is given by Table 116.

The coefficients of discharge that are given are only valid if the profile of the crest accords with the lower profile of a fully aerated nappe.

Inspection of Table 116 leads to the conclusion that, contrary to what occurs with spillways with a rectilinear crest in plan, the coefficients of discharge increase as the value of a decreases.

The values of μ for heads that are different from the design head are obtained by means of Table 116. In this table μ_d represents the coefficient of discharge for a head equal to the design head.

In order to prevent depressions in the shaft, the velocity at each of its cross-sections will have an upper limit so that the velocity head is less than the available head at that cross-section (distance from the cross-section to the free surface, less the friction and local head losses, as far as the cross-section).

Example

Dimension a shaft spillway so that a discharge of $200\,m^3/s$ takes place with a design head $H_d = 2\,m$. Consider that the approach velocity is neglible.

Solution

As a first approximation take $\mu = 0.45$. From formula 6.59 we then have:

$$r = \frac{Q}{27.83\mu H_d^{3/2}} = \frac{Q}{27.83 \times 0.45 \times 2.83} = 5.64\,m$$

Thus $H_d/r = 2/5.64 = 0.354$, corresponding to (Table 116) $\mu = 0.455$.
For $\mu = 0.455$, in a second approximation, we have $r = 5.59\,m$. Thus $H_d/r = 2/5.59 = 0.358$, corresponding to (Table 116) $\mu = 0.453$.
For $\mu = 0.453$, in a third approximation, we have $r = 5.61\,m$. Thus $H_d/r = 2/5.61 = 0.357$, a value that practically coincides with the previous one.
Hence $r = 5.61\,m$, corresponding to (Table 115) $H_s/H_d = 0.078$ and $H_s = 2.16\,m$.
Table 117 gives the coordinates of the spillway crest (lower profile of the fully aerated nappe as a function of H_s and r). Table 118 gives the coordinates of the upper nappe profile.

6.36 Coefficient of contraction in spillways with piers

The existence of piers in spillways causes a contraction which, from a practical point of view, corresponds to a reduction in the effective length of each opening. Accordingly, the length of the spillway crest to be considered in calculations will be:

$$L = L' - KnH_0 \tag{6.60}$$

in which L' is the real total opening between piers, K is the *coefficient of contraction of the piers*, n is the number of contractions (two per opening) and H_0 is the energy head on the spillway crest, including the kinetic head associated with the approach velocity.

The value of K is given in Table 119 for some pier shapes.

The shape of the upper surface of the liquid is deformed by the presence of the piers, and it has different characteristics at the middle of the opening compared with those next to the piers (see Table 120). The greater the contraction, the greater will be this deformation.

Experiments of Lemos [25], using piers with semi-elliptical and 1:2 and 1:3 ratios of semi-axes, have been satisfactory. The coefficients of contraction are of the same order of magnitude as those shown by semicircular piers, but the deformations of the free surface are less.

F.   ENERGY DISSIPATION

6.37   General

In order to minimize the effect of disturbances introduced into natural flow by the construction of a hydraulic development, restitution of the discharge must be carried out in conditions that are as close as possible to the natural conditions. It is therefore necessary for the excess energy created by the hydraulic scheme to be dissipated,† without the occurrence of significant scour in the river bed downstream of the works, which might endanger their stability.

In order to achieve this objective, the energy dissipators most frequently used are:

(a) *hydraulic jump* stilling basins;
(b) *roller buckets*;
(c) *impact type* stilling basins;
(d) *baffled apron drop spillways*.

Apart from these types, structures such as flip buckets,‡ free fall and crossed jets are also used.

In choosing the type of energy dissipator, for a design, a wide range of factors has to be considered, particularly the topography, geology, hydrology, type of dam and, of course, economy.

Only the types of dissipators mentioned in (a), (b), (c) and (d) above will be discussed, with indication of the geometrical characteristics of some

---

† This dissipation of energy is a transformation of part of the mechanical energy of the water into turbulent energy and finally into heat, owing to the internal friction of the flow and friction of the flow with the boundaries.

‡ Also termed ski-jump spillways.

typical stilling basins that have been tested, standardized and are most frequently in use.

The data given are intended mainly for works of small dimension or for preliminary design calculations. Study of the best structures for energy dissipation must be complemented with hydraulic model tests.

6.38   Hydraulic jump stilling basins

(a) Basins with rectangular plan and horizontal bottom

These will have the dimensions necessary for confining the jump formed at the design discharge. Their behaviour for lower discharges must be ascertained. Determination of the conjugate depths of the jump has been explained in Section 6.17; determination of the energy loss in Section 6.18 and of the length and location in Section 6.19.

The height of the lateral walls depends on the characteristics of the jump, particularly on the oscillations of the free surface. The value $0.25\,h_2$ can be taken to be the order of magnitude of the freeboard to be adopted.

This type of stilling basin is used for heads of over 60 m and discharges of $q > 45\,\mathrm{m^2/s}$ per unit width. The degree of submergence of the jump in this type of basin, given by equation (6.37), must have the value 0.1. The jump is very sensitive to any lowering of the downstream levels, which must not be less than the second conjugate depth of the jump under any circumstance.

Various accessories are used as baffles and sills to increase stability and shorten the length, and some of them are discussed below.

(1) *Basin with chute blocks and dentated sill near the downstream end*
   (Type II USBR basin†)
This can be used for heads of less than 65 m and discharges of $q < 45\,\mathrm{m^2/s}$ per unit width (Fig. 6.31).

With this type of basin it is possible to reduce its length to 70%, in relation to the length of a simple basin. It must not, however, be used for $\mathbf{F}_{r1} < 4.5$.

Its dimensioning is carried out according to Fig. 6.31, in which h_1 and $\mathbf{F}_{r1}$ represent the water depth and Froude number corresponding to supercritical flow immediately upstream of the basin; h_2 represents the second conjugate depth of the jump, as given by the formulae or by the graph.

A minimum submergence of the jump of 0.05 is advisable, i.e. the downstream water depth must be $1.05\,h_2$. The length L of the basin can be equal to 0.7 of the length of the jump C, that is to say, 0.7 of the value given by Graph 107(a) for a channel slope $I = 0$.

(2) *Basin with chute blocks, baffle blocks and continuous end sill (Type III USBR basin)*
These basins are used when upstream of the jump $U_1 \leqq 18\,\mathrm{m/s}$ and $q \leqq 18\,\mathrm{m^2/s}$. For higher velocities, cavitation may occur at the baffle blocks placed as shown in Fig. 6.32.

---

† Type I USBR (United States Bureau of Reclamation) basin corresponds to a stilling basin without any additional devices.

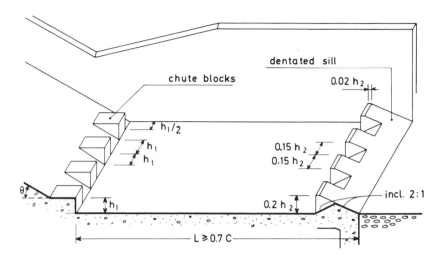

Fig. 6.31.

The chute blocks (first row going downstream) are the same as those of the Type II basin. The baffle blocks and end sill are dimensioned according to Graph 121.

If the configuration and spacing of the baffle blocks are modified, they can be used for higher heads [26].

With this type of basin it is possible to reduce the length of the basin by 45%, in relation to that of a simple basin. It must not be used for $F_{r1} < 4.5$.

The minimum downstream level compatible with containment of the jump, corresponds to $0.83 \, h_2$

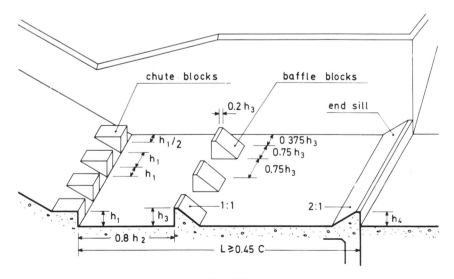

Fig. 6.32.

(3) *Basin with deflector blocks and continuous end sill (Type IV USBR basin)*

These basins are especially suitable for cases in which the jump is oscillating, which happens when the Froude number, at the cross-section of the first conjugate depth, is between 2.5 and 4.5.

Its efficiency in this range of values lies in the effect of the wave suppressor which significantly attenuates undulations.

Its dimensioning is carried out in accordance with Fig. 6.33.

The length L of the structure is that of the simple rectangular basin. $(L = C)$.

The degree of submergence of the jump must be 0.1.

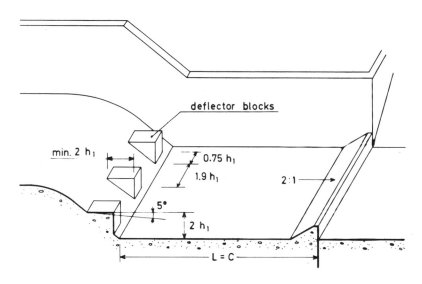

Fig. 6.33.

(b) Rectangular basins with vertical walls and sloping bottom

The conjugate depths of the jump are given by Graph 105 and the length of the basin is given by Graph 107(a).

When dimensioning stilling basins of this type, attention must be paid to the considerations set out above. It is always convenient at the end of the basin to have a continuous end sill, with a depth of about 0.05 to $0.10 h_2$ and an upstream face sloping at 3:1 to 2:1.

6.39 Roller buckets

(a) General characteristics

When the downstream water depth is substantially greater than the conjugate depth of the hydraulic jump that would be formed in a stilling basin with the bottom practically at the elevation of the bed (submerged jump — see 6.20), it is advisable to use roller bucket energy dissipators.

Two types of roller buckets have often been used: the *solid (roller) bucket* (Fig. 6.34(a)) and the *slotted bucket* (Fig. 6.34(b)).

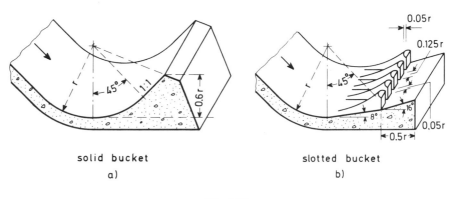

solid bucket   slotted bucket

a)   b)

Fig. 6.34.

In the solid bucket, all the flow is directed upward by the bucket lip, to create a boil on the water surface and a violent ground roller moving in the contrary sense downstream from the roller bucket.

In the case of slotted buckets part of flow passes through the slots, spreads laterally and is lifted away from the bottom by the apron. Thus the flow is dispersed and distributed over a greater area, providing a less violent surface boil and better dispersion of flow in the region above the ground roller.

For very low downstream levels, ejection of the roll wave may occur. The level from which the roll wave begins to be ejected is known as the *ejection level*, N_e. For exaggerated downstream levels, the buckets do not function well and produce inconvenient scour at the foundation of the structure.

(b) Dimensioning

Given the value of the Froude number upstream, $\mathbf{F}_{r_1} = U_1/\sqrt{gh_1}$, the value of the minimum radius, r, is determined from Graph 126. When the value of r is fixed, the maximum and minimum admissible tailwater levels N_{max} and N_{min} and the ejection level N_e are determined. It is to be noted that these levels are measured in relation to the bottom of the bucket and not in relation to the river bed.

If the natural level is higher or lower than N_{max} or N_{min}, respectively, the value of the radius r will have to be modified.

The difference between N_{min} and N_e gives an idea of the safety in relation to the ejection.

All the results given are valid for buckets with the geometrical characteristics indicated in Fig. 6.34.

Example

Dimension a roller bucket, considering that the maximum discharge will be $130\,m^3/s$ and that the basin has a width of $10\,m$. The characteristics of the flow at the entrance cross-section of the basin give a water depth h_1 of $0.60\,m$ and a velocity U_1 of $21.0\,m/s$.

Solution

We therefore have:

$$\mathbf{F}_{r1} = \frac{U_1}{\sqrt{gh_1}} = \frac{21.0}{\sqrt{9.81 \times 0.6}} = 8.7$$

This value of Froude number, at the entrance cross-section of the basin, corresponds to (Graph 123):

$$r_{min}/\left(h_1 + \frac{U_1^2}{2g}\right) = 0.15 \quad \text{and since} \quad h_1 + \frac{U_1^2}{2g} = 0.6 + \frac{21.0^2}{2 \times 9.81} = 23.1\,m$$

we have $r_{min} = 3.46\,m$.

If the value corresponding to r_{min} is taken for the bucket radius, we have (Graph 123), since

$$r/\left(h_1 + \frac{U_1^2}{2g}\right) = 0.15$$

and $\mathbf{F}_{r1} = 8.7$, that $N_{min}/h_1 = 12.7$ and $N_{max} = 18.0$ (Case II), that is to say $N_{min} = 7.6\,m$ and $N_{max} = 10.8\,m$.

If the downstream water depth, measured in relation to the bottom of the stilling basin, is less than the N_{min} or greater than the N_{max} that have been calculated it will be necessary to increase the value of the radius and repeat the calculation until this condition is met.

Once the value of r to be used has been determined the bucket behaviour must be checked for a discharge less than the maximum discharge.

The final geometry of the bucket is defined according to Fig. 6.34.

6.40   Impact-type stilling basins

(a) General characteristics

These are structures of small dimensions developed by the US Bureau of Reclamation, and they are particularly useful in bottom outlets and drainage structures. These energy dissipators do not depend upon tailwater.

(b) Dimensioning

Their dimensions are determined according to the scheme given in Fig. 6.35.

The incoming velocity must not exceed $9\,m/s$ and the diameter of the

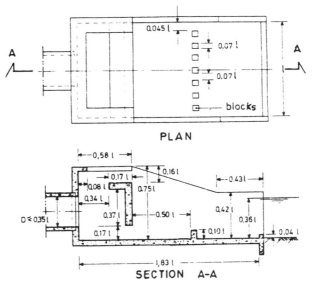

PLAN

SECTION   A-A

Fig. 6.35.

conduit may be as large as 1.80 m. With this structure riprap is not necessary. An alternative structure of the same type, but without the accessories (blocks, Fig. 6.36) has been developed. In this case riprap should be provided along the bottom and sides adjacent to the structure. The dimension of the riprap, d_{50} in this case, is a function of the diameter of the conduit, as shown in Table 125.

The thickness of the riprap must be at least equal to $1.5\,d_{50}$ and its length is equal to the width of the basin, l.

Fig. 6.36 shows this type of basin and its dimensions. The width, l, depends on the Froude number and diameter of the conduit, via the relationship:

$$\frac{l}{D} = 3.0\,\mathbf{F}_r^{0.55} \tag{6.61}$$

in which:

$$\mathbf{F}_r = \frac{U}{\sqrt{gD}} \tag{6.62}$$

The other dimensions are as follows:

$$L = 1.33l;\ \ f = 0.17l;\ \ e = 0.08l;\ \ a = 0.5l;\ b = 0.37l;\ \ g = \\ = 0.08l$$

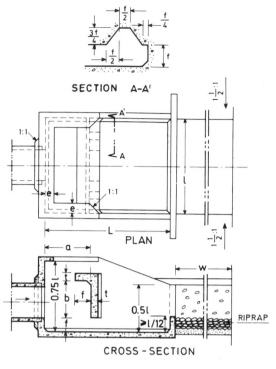

SECTION A-A'

PLAN

CROSS-SECTION

Fig. 6.36.

6.41 Baffled apron drop spillways

This type of structure is used to dissipate energy of the flow where it is desirable to avoid a stilling basin. Its use is confined to small discharges per unit width and in regions where there is no ice formation and no floating debris is expected. Limitation of the maximum admissible discharge is due to the possibility of cavitation occurring at the blocks.

The hydraulic design of this structure is undertaken in accordance with the scheme shown in Fig. 6.37. The discharge per unit width, q, must not exceed $5.6\,\text{m}^2/\text{s}$ and the approach velocity, U_a, must be less than the critical flow velocity.

$$q < 5.6\,\text{m}^2/\text{s} \tag{6.63}$$

$$U_a < \sqrt[3]{gq} \tag{6.64}$$

The ideal conditions occur when $U_a = 0.5\sqrt[3]{gq}$.

The minimum height, a, of the blocks must be practically equal to $0.8\,h_c$ $= 0.8\sqrt[3]{q^2/g}$.

Fig. 6.37 shows details of the other blocks dimensions and their spacing.

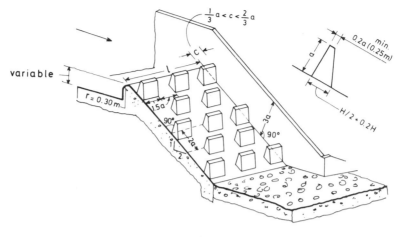

Fig. 6.37.

Good operating conditions of prototypes, in which the discharge value, even though for only short periods, was twice that of the established limit are given in [18].

BIBLIOGRAPHY

[1] Bakmeteff, B. A., *Hydraulics of Open-Channels*, McGraw-Hill, New York, 1932.

[2] Chabert, J., *Calcul des Courbes de Remous*, Collection du Laboratoire National d'Hydraulique, Eyrolles, 1955.

[3] Crausse, E., *Hydraulics des Canaux Découverts*, Eyrolles, 1951.

[4] Ippen, A. T., *Channel Transitions and Controls*, in Rouse, H. (editor): Engineering Hydraulics, John Wiley, New York, 1950.

[5] Lencastre, A., *Perdas de Carga Provocadas nos Pilares de Pontes e Pontões*, Tech. Memo. No. 53, LNEC, Lisbon, 1954.

[6] Manzanares, A., *Escoamento em Superfície Livre, Regime Perma-nente*, I. S. Técnico, Lisbon, 1947.

[7] Silber, R., *E'tude et Tracé des Ecoulements Permanents en Canaux et Riviéres*, Dunod, Paris, 1954.

[8] Allin, J. and Chee, S. P., *The Resistance to the Flow of Water Round a Small Circular Bend in Open Channel*, Proc. Inst. Civil Engrs., **23**, November, 1962, p. 423.

[9] Chow, V. T., *Open Channel Hydraulics*, McGraw-Hill, New York, 1959.

[10] Henderson, F. M., *Open Channel Flow*. Macmillan, New York, 1966.

[11] Mockmore, C. E., *Flow Around Bends in Stable Channels*. Trans. ASCE, **109**, 1944, pp. 593–618.

[12] Poggi, B., *Correnti Veloci nei Canali in Curve*. L'E'nergie Electrica, **33**, 5, May 1956, p. 465.

[13] Shukry, A., *Flow around Bends in an Open Flume*, Trans. ASCE, **115**, 1950, pp. 751–779.

[14] Simons, D. B., *River and Canal Morphology*, in Shen, H. W. (editor) River Mechanics, Vol. II, Water Resources Publications, 1971.

[15] Woodward, S. M. and Posey, C. J., *Hydraulics of Steady Flow in Open Channels*, John Wiley, New York, 1941.

[16] Peterka, A. J., *Hydraulic Design of Stilling Basins and Energy Dissipators*. US Department of the Interior, Bureau of Reclamation, Eng. Mon. No. 25, Washington, 1964.

[17] Lemos, F. O., et al., *Dissipação de Energia em Obras Hidráulicas*. Seminário No. 223, LNEC, Nov. 1978.

[18] Rhone, T. J., *Baffled Apron as Spillway Energy Dissipator*, Proc. ASCE, **103**, HY12, Dec. 1977, pp. 1391–1401.

[19] Belchey, G. L., *Hydraulic Design of Stilling Basin for Pipe or Channel Outlets*, US Department of the Interior, Bureau of Reclamations, Res. Rep. No. 24, Washington, 1971.

[20] Escande, *Barrages*. Hermann, Paris, 1937.

[21] Escande, Barrages: *Déversoirs à Seuil Creager Déprimé*, Le Génie Civil, March 1941.

[22] US Corps of Engineers, *Hydraulic Desing Criteria*.

[23] USDI Bureau of Reclamation, *Studies of Crests for Overfall Dams*. Bulletin, 3, Denver, Colorado, 1948.

[24] Wagner, W. E., *Morning-Glory Shaft Spillways*, Trans. A.S.C.E., **121**, 1956, p. 345.

[25] Lemos, F. O. and Abecassis, F., *Descarregadores e Obras de Desvio*. Seminário No. 130, LNEC, Lisbon, 1972.

[26] Quintela, A. C.; Ramos, C. M., *Protecção Contra a Erosão de Cavitação em Obras Hidráulicas*, Tech. Memo. No. 539, LNEC, Lisbon, 1980.

[27] Lemos, F. O., *Critérios para o Dimensionamento Hidráulico de Barragens Descarregadoras*, Internal Report, LNEC, Lisbon, Sep. 1978.

[28] USDI — Bureau of Reclamation, *Design of Small Dams*, US Government Printing Office, 2nd edition, Washington. 1973.

7

Flows in porous media

A. DEFINITIONS AND GENERAL LAWS

7.1 Characteristics of porous media

Natural porous media are mainly alluvia consisting of granular material or cracked compact rocks. *Artificial porous media* are earthworks, the most important of which are earth dams. This chapter will deal with flows of water through porous media.

A porous medium is said to be *homogeneous* when, at any point, resistance to the flow is the same in relation to a given direction. Owing to the irregularity of natural porous media it is, however, necessary to define the *scale of homogeneity*; for instance, an alluvium with grains of about 1 mm in diameter will be homogeneous at the scale of dm^3; a rock mass can only be considered homogeneous for dimensions of about 100 times the greatest dimension of the blocks.

A porous medium is said to be *isotropic* if, whatever the direction considered, resistance to the flow or another property is the same. Most natural porous media are *anisotropic*, i.e. they are not isotropic. In fact, in the case of cracked rocks, the cracks of tectonic origin are usually oriented in directions that are parallel and perpendicular to the compressions that gave rise to them; the rock has the appearance of parallelepipeds cut by parallel fissures that constitute a favoured direction for the flow; the regime of the flow will depend on the geometry of the fissures and their infill material. Also in sedimentary formations, the intercalations of layers of different characteristics provide greater facility for flow in a horizontal sense and they are therefore anisotropic (see Section 7.7(b)).

These media can, however, be considered homogeneous provided that a sufficiently large scale of homogeneity is established.

A medium consisting of granular material which we shall call soil† is

---

† For agronomists the term 'soil' is more restricted in its sense, since it is usually confined to the upper part of alluvia or decomposed rock in which there is organic matter, air and water.

characterized, from the geometrical point of view, by various parameters that are indicated below.

Grain size — Defined by the grain-size curve corresponding to the percentage, in weight, of the grains whose diameter is less than a given diameter: for example, d_{10} will represent the diameter which is such that 10% of the material, in weight, consists of particles with a smaller diameter. The *minimum diameter* will be presented by d_{m}; the *maximum diameter* by d_{M}; the *equivalent diameter* is usually taken as being $d_{\mathrm{e}} = d_{10}$. From the point of view of grain size, soils can be classified as shown in Table 125.

Void ratio, e — ratio of the volume of the pores (or voids), v_{p}, to the volume v_{g} occupied by the grains: $e = v_{\mathrm{p}}/v_{\mathrm{g}}$.

Relative porosity, n — ratio of the volume of the pores, v_{p}, to the total volume, $v_{\mathrm{t}} = v_{\mathrm{p}} + v_{\mathrm{s}}$, in which v_{s} is the volume of the solid skeleton; i.e. $n = v_{\mathrm{p}}/v_{\mathrm{t}}$.

Coefficient of saturation, σ — ratio of the volume occupied by the water, v_{a}, to the volume v_{p} of the pores: $\sigma = v_{\mathrm{a}}/v_{\mathrm{p}}$.

Effective porosity, or specific yield, n_{e} — ratio of the volume that can be occupied by water capable of draining by gravity, v_{pa}, to the total volume: $n_{\mathrm{e}} = v_{\mathrm{pa}}/v_{\mathrm{t}}$.

Field capacity or specific retention, n_{r} — ratio of the volume of voids of the soil occupied by water that is retained against the action of gravity (by capillarity), v_{r}, to the total volume $n_{\mathrm{r}} = v_{\mathrm{r}}/v_{\mathrm{t}}$.

The following ratios result from these definitions:

$$n = n_{\mathrm{e}} + n_{\mathrm{r}} \tag{7.1}$$

$$e = \frac{n}{1-n}; \quad n = \frac{e}{1+e} \tag{7.1a}$$

Table 127 gives the value of relative porosity, n, for some soils; Table 126 gives the value of n for certain soils and rocks.

7.2 Water in soils. Classification of aquifers

Water in soils is found in two fundamental zones: the saturated zone and the non-saturated or aerated zone. Fig. 7.1 gives a scheme that makes it possible to classify the various types of aquifer.

Accordingly, taking aquifer B as an example, it is initially an *unconfined* aquifer (zone (a)), because as the water level in it coincides with the level reached in an observation borehole, i.e. the water table relates directly to atmospheric pressure.

Continuing downstream in aquifer B, in the following zone (b) the flow is between two impermeable surfaces and the aquifer is described as a *confined or artesian*† aquifer. The same occurs in zone (d).

When the upper layer that limits the aquifer is permeable such that the

---

† So called since they were frequent in the region of Artois (France).

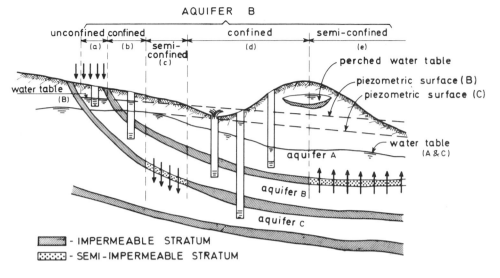

Fig. 7.1.

aquifer can lose or receive water — the phenomenon of leakage — the aquifer is known as *semi-confined*. Aquifer B in Fig. 7.1 also has two zones, (c) and (e), in which it behaves in this way.

There is also a special case in unconfined aquifers, the *perched water table*, which occurs when an impermeable formation appears between the saturated zone and the surface of the soil, giving rise to retention of infiltration water above that formation.

Geological formations can also be classified according to their water content and their capacity to transmit it not only as *aquifers* but also as *aquitards, aquicludes* and *aquifuges*.

An *aquitard* is a geological formation that contains water, but transmission of the water is extraordinarily slow. Aquitards are therefore of no use for withdrawing water under economically profitable conditions, but they may play an important part in recharging adjacent aquifers. Examples of aquitards are formations of silty or sandy clays.

An *aquiclude* is a geological formation that contains water in quantities that may go as far as saturation, but does not transmit it because the formation has very strong retention characteristics and thus prevents water from being withdrawn. The most characteristic formations covered by this definition are those with a high clay content.

Lastly, an *aquifuge* is a formation that contains no water, nor can it transmit water, examples being a granitic rock mass or a metamorphic formation that is neither weathered nor cracked.

In unconfined aquifers, owing to the effect of capillarity, the water rises above the water table, by a value λ (see Fig. 7.2). This water is known as *capillary water* and the upper limit constitutes the limit of the *saturation*

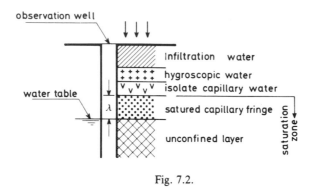

Fig. 7.2.

zone. The value of λ can be calculated by formula 1.3 given in Chapter 1; however, it is hard to determine the diameter of the interstices, so that, when dealing with porous media, it is more practical to use:

$$\lambda = \frac{c}{e\,d_e} \qquad\qquad (7.2)$$

in which: c = constant, varying between 0.1 and 0.5 cm^2; e = void ratio; d_e = equivalent diameter.

λ may have values from 3.0 m to 0.6 m in clays, or of only one millimetre in coarse sand.

Above the saturation zone there is also *isolated capillary water*, associated with the finer interstices of the soil, in which capillarity is more accentuated.

At a higher level there may also be *hygroscopic water*, fixed by adsorption on the surface of the particles of the soil.

Even further towards the surface, after rainfall, is *infiltration water* that penetrates to the aquifer by the action of gravity.

7.3 Specific storage. Storage coefficient. Specific yield

There are several parameters that characterize the quantity of water that can be extracted from an aquifer.

In the case of confined aquifers, *the specific storage of the aquifer*, S_s, of dimension L^{-1}, is defined as the volume of water that can be released per unit volume of the aquifer, associated with a unit drawdown in the piezometric surface. Jacob proposed the following expression:

$$S_s = \rho g\,(\alpha + n\beta) \qquad\qquad (7.3)$$

in which: n = total porosity; ρ = density of the water; g = acceleration of gravity; α = vertical compressibility of the solid skeleton of the porous medium; β = volumetric compressibility of the water, $\beta = 1/\varepsilon$ (see Section 1.11).

The *Storage coefficient*, S, dimensionless, is defined as the volume of water released by a vertical column of the aquifer of unit cross-section, when the mean piezometric head of the column diminishes by one unit.

In the case of confined aquifers, the storage coefficient is given by $s = b.S_s$, in which b is the thickness of the aquifer, and normally S varies between 10^{-3} and 10^{-5}. In the case of unconfined aquifers, the storage coefficient is given by $S = b.S_s + n_e$, in which n_e is the specific yield. The value of n_e normally varies between 0.01 and 0.3. Since $n_e \gg b.S_s$, current practice is to consider $S = n_e$, in unconfined aquifers.

7.4 Darcy's law

Consider a tube of length L, filled with porous material and connecting two reservoirs of constant level, Z_1 upstream and Z_2 downstream (Fig. 7.3).

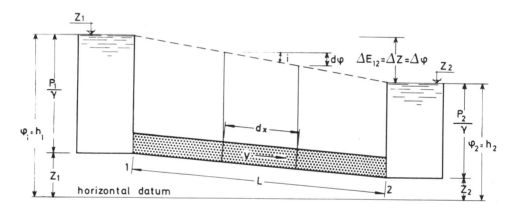

Fig. 7.3.

The head or piezometric potential, $\varphi = z + p/\gamma$, which is also usually represented by h, will be φ_1 in the initial cross-section 1 and φ_2 in the final cross-section 2. Since the velocities are very low, the energy line practically coincides with the piezometric surface.

The total energy loss between 1 and 2 will be $\Delta E_{12} = Z_1 - Z_2 = \Delta Z$.

As in the case of flows under pressure, the unit head loss (per unit length and per unit weight of flow) is defined as the ratio:

$$i = \frac{\Delta E}{L} = \frac{\Delta Z}{L} = -\frac{\partial \varphi}{\partial x} = -\operatorname{grad} \varphi \qquad (7.4)$$

The velocity of filtration, V, also known as specific discharge, is defined as the ratio of discharge (Q) to the total area of the tube. The effective velocity is V/n_e.

The Reynolds number of the flow is $\mathbf{R}_e = Vd/v$, in which d is a characteristic diameter related to the shape of the grains and their arrangement, but of imprecise definition and therefore giving a certain dispersion of the results, while v, of course, is the coefficient of kinematic viscosity of the fluid discharged, in this case water. In general, a value of d is taken that is equal to d_{50} of the grain-size curve. In fissured terrain $d = 2e$ is assumed, e being the width of the fissure.

Generally speaking, as for flows under pressure, we can take $i = fV^2/2gd$.

For low velocities under laminar flow, $f = a/\mathbf{R}_e = av/Vd$ (equation (4.1)):

$$i = \frac{av}{Vd} \frac{1}{d} \frac{V^2}{2g} = \frac{av}{2gd^2} \times V = \frac{1}{K}V \tag{7.5}$$

in which:

$$K = \frac{2gd^2}{av} \tag{7.6}$$

is known as *the hydraulic conductivity* and in which a is a constant (see Section 7.5). K is a characteristic of the permeability of the medium and has the dimensions $L\,T^{-1}$ of a velocity.

We can therefore write:

$$V = K\ i\ = -K \text{ grad } \varphi \tag{7.7}$$

This expression is known as *Darcy's law*, valid for $\mathbf{R}_e < 1$[†] and in fact this is generally the case when dealing with groundwater flow.

When the Reynolds number increases, there is a transition from laminar flow to turbulent flow. This transition is unlike that experienced in conduits under pressure and its theory is extremely complex (see, for example, [1]).

7.5 Hydraulic conductivity (permeability). Intrinsic permeability

(a) Homogeneous soils

From the form of equation (7.6) it is seen that *the hydraulic conductivity,*[‡] K, depends on the one hand on the liquid, in particular its viscosity, v, and on the other hand on the characteristics of the porous medium given by:

[†] Exceptionally, the law will remain valid for values of $\mathbf{R}_e = 5$ or even 10. The value of $\mathbf{R}_e$, may also be related to the effective velocity, which is more correct theoretically, but less practical.
[‡] Also sometimes called *permeability*.

$$K_0 = \frac{d^2}{a} \tag{7.8}$$

which is known as *intrinsic permeability*, with dimensions L^2.

The value of K may be affected by the chemical composition of the water if clays are present, the state of flocculation of which can vary. The same may happen as a result of chemical precipitation, or the expulsion or introduction of fine particles transported in the flow.

K is expressed in m/s or m/day. K_0 is usually expressed in m^2, although the Darcy unit $= 10^{-12}$ m^2 can also be used.†

Table 128 gives some values of K for certain porous media. The value of K can be determined in laboratory, as explained in Section 7.4 for soil samples. Generally speaking, however, permeability is determined by means of pumping tests in the field using methods developed in the study of well hydraulics (Section 7.9).

There are also standardized tests, among which mention may be made of *Lugeon* tests, which consist of injecting water into a borehole and measuring the discharge that is absorbed for various pressures. A Lugeon unit (UL) corresponds to the absorption of one litre per minute, per metre of borehole, under a pressure of 10 bars. In the case of a homogeneous soil that is not very permeable, UL is approximately 10^{-7} m/s.

Lefranc tests are also used: in these, a measurement is made of the discharge required to maintain a given level in a tube driven into the soil, which then permits the calculation of K.

(b) Stratified porous media

Consider a stratified system consisting of n layers of different material. b_i is taken as the thickness of each layer, i (Fig. 7.4), and K_i its hydraulic conductivity, in the previously defined sense of the term.

In the case of *vertical* movement (Fig. 7.4a), the discharge q (per unit surface area) crosses the various layers 'in series'. Taking Δh_i as the head loss in layer i, then, according to Darcy's law, $q = K_i \Delta h_i / b_i$ or $\Delta h_i = q b_i / K_i$, so that we have $\Delta h = q \Sigma b_i / K_i$.

For the whole system it is possible to define an *equivalent coefficient of vertical hydraulic conductivity*, K_v, so that $q = K_v \Delta h / L$, and we have:

$$\Delta h = q \frac{L}{K_v} \tag{7.9}$$

By making the two values of Δh equal, and bearing in mind that $L = \Sigma b_i = b$ (thickness of the system), we shall have:

$$\frac{1}{K_v} = \frac{1}{b} \Sigma \frac{b_i}{K_i} \quad \text{or} \quad K_v = \frac{b}{\Sigma b_i / K_i} \tag{7.9a}$$

† 1 Darcy $= \dfrac{1 \text{ cm/s} \times 1 \text{ centipoise}}{1 \text{ atm/cm}} = 10^{-12} \, m^2$.

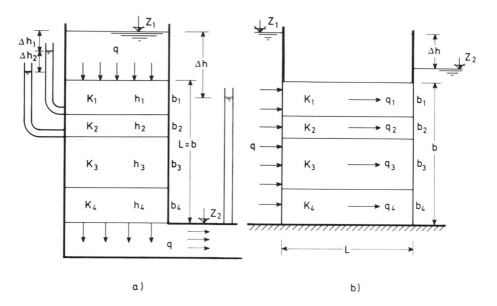

Fig. 7.4.

Horizontally (Fig. 7.4(b)), the flow takes places 'in parallel', and each layer has a discharge of $q_i = b_i K_i \Delta h/L$. The total discharge will be:

$$q = \Sigma q_i = \Sigma b_i . K_i . \Delta h/L \qquad (7.10)$$

By defining for the whole an *equivalent coefficient of hydraulic conductivity* so that, we have: $q = K_H b \Delta h/L$.

$$K_H = \frac{1}{b} \Sigma b_i K_i \qquad (7.10a)$$

In all cases $K_H > K_v$, except in fractured rocks.

Example. Determine the equivalent coefficient of vertical and horizontal hydraulic conductivity in an aquifer consisting of three layers, defined as follows: an upper layer of silty sand, of thickness 6 m and $K_1 = 10^{-5}$ m/s; an intermediate layer of coarse sand, of thickness 3 m and $K_2 = 10^{-3}$ m/s; a lower layer of clay, of thickness 8 m and $K_3 = 10^{-9}$ m/s. According to equations (7.9a) and (7.10a), we shall have: $b = \Sigma b_i = 22$ m.

$$K_v = b/\Sigma \frac{b_i}{K_i} = 22 \div \left(\frac{10}{10^{-3}} + \frac{2}{10^{-2}} + \frac{10}{5 \times 10^{-3}} \right) = 1.05 \times 10^{-4} \text{ m/s}$$

$$K_H = \frac{1}{b} \Sigma b_i K_i = \frac{1}{22} (10 \times 10^{-3} + 2 \times 10^{-2} + 10 \times 5 \times 10^{-5}) = 1.38 \times 10^{-3} \text{ m/s}.$$

The vertical hydraulic conductivity is mainly conditioned by the most impermeable layer, while the horizontal hydraulic conductivity is mainly conditioned by the most permeable layer.

7.6   Transmissivity

In an aquifer of thickness b, and width l, the discharge, according to Darcy's law, will be $Q = Kbli$. *Transmissivity*, T, is the name used for the product Kb, whose dimensions are $L^2 T^{-1}$:

$$T = Kb \qquad\qquad (7.11)$$

Transmissivity, T, is expressed in m²/s or in m²/day.

If the hydraulic conductivity, K, varies along the thickness, b, of the aquifer, the transmissivity, T, will be:

$$T = \int_0^b K\,dz \qquad\qquad (7.11a)$$

7.7   Unconfined flow. Seepage face

The unconfined flow, represented in Fig. 7.5 is to be considered.

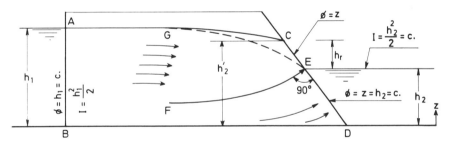

Fig. 7.5.

AB is taken to be the surface that demarcates the porous medium upstream and CD the line that demarcates it downstream. The streamline that ends at E, that is to say, the line FE, must be normal to CD, since ED constitutes an equipotential of the flow defined by $\varphi = h_2 = $ const. If the free surface coincided with GE, all of the water flowing between the two streamlines, GE and FE, would appear at a point E, and this would mean a theoretically infinite velocity. In reality, the free surface does not coincide with GE but with GC. The flow surface, EC, is that which gives outlet to the water passing between GE and FE. This surface is known as the *seepage face* and it appears whenever a phreatic layer flows outside the porous medium: earth dams, trenches, wells, etc.

The difference between the vertical permeability K_v and the horizontal permeability K_H $(K_v < K_H)$ accentuates the existence of the seepage face.

The distance measured vertically between the points C and E, which define the seepage face, is known as *the seepage depth, h_r*.

B. STEADY FLOW

7.8 Flow into trenches

(a) Confined aquifer

Consider a horizontal, confined aquifer, which is homogeneous and isotropic, of thickness, b, and is cut by a trench that is also horizontal. Assume that the flow is steady, i.e. that the layer is fed by a discharge that is equal to that which is withdrawn from it: fed by a river, lake, etc., at a distance x_0 from the trench (Fig. 7.6). This flow corresponds to the complex potential

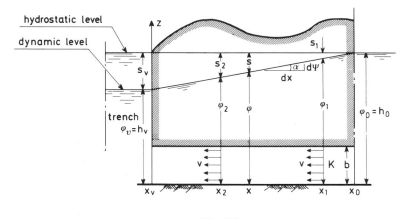

Fig. 7.6.

$w = az = a(x + iy)$ (see Section 2.32, example (a)). Hence the velocity potential is $\phi = ax$, and $V = \partial\phi/\partial x$. By Darcy's law is $V = -K\partial\phi/\partial x$. Hence the velocity potential ϕ is related with piezometric potential or head, $\varphi = (z + p/\gamma)$, by:

$$\varphi = -\frac{\phi}{K} \qquad (7.12)$$

Take b the thickness and l the width of the aquifer, the discharge, Q is $Q = Vbl = -Kbl.\,\partial\varphi/\partial x$. Hence $\varphi = (Qx)/(Klb) + \text{constant}$,

Take x_0 to be the distance from the recharge zone (or from the zone that is not disturbed by the flow), and φ_0 as the head corresponding to that distance. The *drawdown, $s = \varphi - \varphi_0$*, of the head at point x is given by

$$s = \varphi_0 - \varphi = \frac{Q}{Klb}(x_0 - x) = \frac{Q}{KT}(x_0 - x) \qquad (7.13)$$

that is to say, the drawdown at any point is proportional to the discharge. In the trench it will be:

$$s_v = \frac{Q}{KT} \cdot x_0 \tag{7.14}$$

(b) Unconfined aquifer

Consider the unconfined aquifer in Fig. 7.7, which is homogeneous and isotropic and based on a horizontal impermeable layer. This aquifer is cut by a horizontal trench that occupies its entire thickness and the aquifer receives recharge at the same rate at which water drains to the trench, i.e. steady state.

Analysis of this type of flow was developed by Dupuit[†], whose fundamental hypothesis, known as *Dupuit's hypothesis*, states that in any cross-section, at a distance x from the wall of the trench, the distribution of velocities is uniform and at all elevations in that cross-section their value is given by the equation:

$$V = -K(dh/dx) = K \tan \theta \tag{7.15}$$

whereas, according to Darcy's law, the surface streamline is given by: $V = Ki = K \sin \theta$.

Dupuit's hypothesis is thus valid when θ is small, which does not occur in the immediate vicinity of the trench where the steep water table gradient results in a seepage face. In other words, Dupuit's hypothesis is equivalent to considering the equipotential surfaces as vertical.

However, taking $Q = Vhl = Vhl$, integration of equation (7.15), $hdh = (Q/lK)dx$, enables us to determine the *form of the free surface*, given by:

$$h^2 - h_0^2 = \frac{2Q}{lK}(x - x_0) \tag{7.15a}$$

This equation is only valid for cross-sections away from the trench.

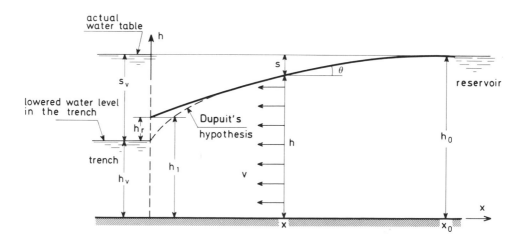

Fig. 7.7.

† Dupuit (1804–1866).

Near the trench, it is necessary to take into account the steep drawdown curve and the *seepage depth*, h_r, given by empirical expressions, such as Vibert's quoted by [11]:

$$h_r = -\frac{1}{2}x_0 + \frac{1}{2}\sqrt{x_0^2 + 4(h_0 - h_v)^2} \tag{7.16}$$

In practice, in order to determine the drawdown curve, the initial seepage point is joined to the curve given by equation (7.15a).
The discharge will be given by:

$$Q = Kl\frac{(h_0^2 - h_v^2)}{2x_0} \tag{7.17}$$

Dupuit's hypothesis gives correct values for the discharge, but not for the free surface where it is necessary to introduce the seepage depth near the trench.

If the trench does not completely cut all the layers it will be necessary to take account of the flux through the bottom and study the flow lines by means of flow nets (see Section 2.35). This point will be taken up again in 7.21, when the drainage of soils is studied.

7.9   Fully-penetrating wells in confined aquifers. Thiem's equation

A *well* is taken here to include a dug well and a borehole. A well is said to be *fully-penetrating* when it traverses the whole aquifer as far as a lower impermeable layer.

Consider a fully-penetrating well, opened in a horizontal confined aquifer. The flow is assumed to be steady, i.e. that the aquifer is fed along a cylindrical surface, at a distance r_0: a well located in the centre of a circular island in a lake, for example (Fig. 7.8).

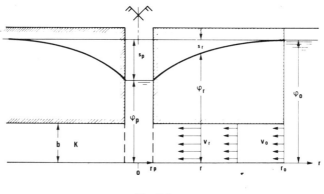

Fig. 7.8.

The flow corresponds to the radial flow defined by $W = a\ln z$ (see Section 2.32 example (b)). The potential function ϕ, and the flow function Ψ are, in cylindrical coordinates (equation 2.59a): $\phi = -K\varphi = a\ln r$; $\Psi = a\theta$ in which (equation 2.59c) $a = q/2\pi$, with q the discharge per thickness of the layer.

According to Darcy's law, $V_r = -K\partial\varphi/\partial r$. Taking b the thickness of the layer, the discharge is: $Q = 2\pi rb V_r = -2\pi Kbr . \partial\varphi/\partial r$; hence $d\varphi = -(Q/2\pi Kb).(\partial r/r)$. By integration we thus have:

$$s = \varphi_0 - \varphi = \frac{Q}{2\pi Kb} . \ln\left(\frac{r_0}{r}\right) = 0.366\frac{Q}{T}\log\left(\frac{r_0}{r}\right) \qquad (7.18)$$

and

$$s_p = 0.366\frac{Q}{T}\log\left(\frac{r_0}{r_p}\right) \qquad (7.19)$$

This equation that relates the drawdown, s_p, in the well to Q, in steady state flow, is the *characteristic curve of the well*. This equation represents a straight line in a system of Cartesian coordinates (Q,s_p).

In practice, wells are pumped at varying discharges, Q, which will produce their respective drawdowns, s_p, once steady state has been reached for each stage.

The expression of the horizontal variation of the piezometric surface will be given by the drawdown s, at a distance r from the well, by equation (7.18) which is known as *Thiem's equation*.

Using semilogarithmic coordinates $(s, \log r)$, this equation yields a straight line: it is a *characteristic of the aquifer* (Fig. 7.9).

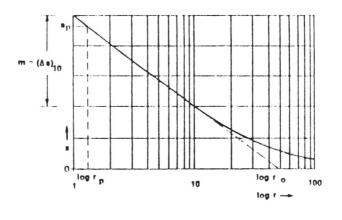

Fig. 7.9.

In practice, when studying an aquifer, observation boreholes or piezometers are installed at various distances from the pumping well and, for a given constant discharge, the drawdowns, s_i are measured in the piezometers. The points $(s, \log r)$ are plotted and the line of best-fit drawn. For

$s = 0$, $\log(r_0/r) = 0$, which corresponds to the distance r_0, and the radius of influence can thus be determined. For $r = r_p$, the drawdown s_p, in the well is obtained.†

The gradient of the straight line is:

$$m = 0.366\frac{Q}{T} \qquad (7.20)$$

so that the value of the transmissivity of the aquifer is given by T.

$$T = \frac{0.366\,Q}{m} \qquad (7.20a)$$

The value of m can be easily calculated by choosing two drawdown values (Δ_s) which coincide with a single log-cycle on the r-axis. After obtaining m and given Q, it is possible to determine the transmissivity T (see Fig. 7.9).

The general equation (7.18) is valid for two points (s_1, r_1) and (s_2, r_2), whence;

$$\Delta s_{1.2} = s_1 - s_2 = \frac{Q}{2\pi T}\ln\left(\frac{r_2}{r_1}\right) = 0.366\log\left(\frac{r_2}{r_1}\right) \qquad (7.21)$$

Example. A well is being pumped in a confined aquifer and has achieved steady state at a constant discharge of 0.05 m³/s. There occurs a drawdown of 6.5 and 4.5 m in two observation wells situated 10 m and 30 m away from the pumping well, respectively. Knowing that the radius of the well is 0.15 m, calculate the transmissivity of the aquifer, the theoretical drawdown in the pumping well in operation and its radius of influence.

By resorting to equation (7.20a) we have, successively:

$$T = 0.366\frac{0.05}{(6.5-4.5)}\log\left(\frac{30}{10}\right) = 0.00437 \text{ m}^2/\text{s}$$

$$s_p = 6.5 + 0.366\frac{0.05}{0.00437}\log\left(\frac{10}{0.15}\right) = 14.14 \text{ m}$$

$$r_0 = 10 \times 10^{(6.5 \times 0.00437/0.366 \times 0.05)} = 355 \text{ m}$$

7.10 Fully-penetrating wells in confined aquifers with overlapping parallel flow

When there is flow in the aquifer before pumping of the well, the total flow is the sum of the two equipotential flows: the initial flow, $w_1 = k i_0 z$, in which i_0 is the gradient of undisturbed

† Owing to the head losses in the well, there may be a significant divergence between the theoretical value thus determined and the real value $(s_p)_r$ obtained directly from the equation of the well. *The equivalent radius of the well, r_e,* is the term given to the value of r that in this equation corresponds to the drawdown $(s_p)_r$.

parallel flow, and the flow to the well, $w_2 = q/(2\pi k)\ln(z)$, both equations being complex. In these equations the thickness of the layer is $b = 1$.

Superposition of the two flows gives the following equations† for the equipotentials and for the streamlines:

$$\Phi = K\varphi = Ki_0 + \frac{q}{2\pi}\ln r = Ki_0 x + \frac{q}{2\pi}\ln \sqrt{x^2 + y^2} \tag{7.22}$$

$$\Psi = Ki_0 y + \frac{q}{2\pi}\theta = Ki_0 y + \frac{q}{2\pi}\arctan\left(\frac{x}{y}\right) \tag{7.22a}$$

The streamlines appear as shown in Fig. 7.10. The stagnation point, E, corresponds to the

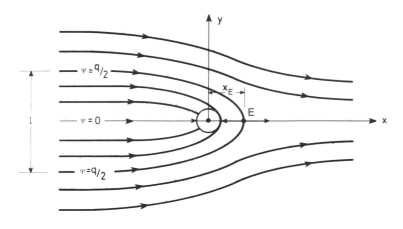

Fig. 7.10.

point at which the velocity of the uniform flow is equal and contrary to the inrush velocity of the well. The coordinates of E are as follows:

$$x_E = \frac{q}{2\pi Ki_0}; \quad y_E = 0 \tag{7.22b}$$

The streamlines that limit the flow to the well join at this point, and correspond to $\varphi = \pm q/2$.

The width of the strip limited by these streamlines tends to the value $l = q/ki_0$, which is obvious as the discharge of the well for $b = 1$ will be $q = lki_0$, with i_0 being the gradient of the uniform flow before the well.

7.11 Fully-penetrating wells in unconfined aquifers

In this case the water-bearing layer is limited by the water table and by the underlying impermeable layer that is considered to be horizontal. When pumping is started, a drawdown occurs in the free surface near the well, and a *cone of depression* is created (Fig. 7.11).

† See also the solution in Section 2.32 example e, for the case of a spring.

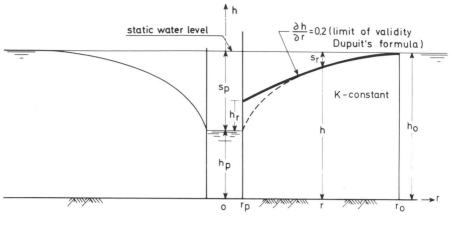

Fig. 7.11.

The flow can no longer be considered plane, as happened in the case of confined aquifers. The discharge is given by:

$$Q = \pi K \frac{h_0^2 - h_p^2}{\ln{(r_0/r_p)}} = 1.36 K \frac{2h_0^2 - h_p^2}{\log{(r_0/r_p)}} \tag{7.23}$$

This expression is the same as that deduced by Dupuit, based on direct application of Darcy's law.

Since $h_0^2 - h_0^2 = (h_0 + h_p)(h_0 - h_p) = s_p(h_0 + h_p)$, we have:

$$s_p = 0.73 \frac{Q}{K} \frac{\log{(r_0/r_p)}}{(h_0 + h_p)} \tag{7.24}$$

This is the *characteristic curve of the well* and it is not linear, since the term h_p, a function of s_p, appears on the right-hand side of the equation.

In practical applications it is easier to use a specific discharge defined by:

$$q_e = \frac{Q}{s_p} \tag{7.25}$$

Since, however, $s_p = h_0 - h_p$, we have:

$$q_e = \frac{Q}{s_p} = \frac{1.36 K}{\log{(r_0/r_p)}}(2h_0 - s_p) \tag{7.25a}$$

which is another way of presenting the characteristic curve of the well, such that q_e is a function of s_p.

For $s_p = 0$ we have $Q = 0$; for $s_p = h_0$ we should obtain the maximum discharge value, Q_M, corresponding to an infinite velocity.

The relationship between Q and Q_M is easily deduced, and we have:

$$Q = Q_M\left[1 - \left(\frac{h_p}{h_0}\right)^2\right]$$ (7.26)

whence:

$$h_p = h_0 \sqrt{1 - \frac{Q}{Q_M}}$$ (7.26a)

By calculating dh_p/dQ, we find that this derivative tends to infinity when Q tends to Q_M, that is to say, when h_p tends to zero. This means that the flow is unstable for very small values of h_p.

In practice the well must be operated for drawdown values between $0.5\,h_0$ and $0.75\,h_0$, as well as for reasons of economy, since the discharge is not proportional to the drawdown. The corresponding values of Q are $0.75\,Q_m$ and $0.94\,Q_m$, respectively.

Example. In an unconfined aquifer a well is being pumped in steady state flow, at a constant discharge of 0.1 m³/s, resulting in a drawdown in the well of 2.5 m. Since it is known that the aquifer has a thickness of 15 m, indicate the maximum value of the discharge that can be abstracted from it, and the most advisable values.

From equation (7.26), we shall have:

$$Q_M = \frac{0.1}{1 - \left|\dfrac{15 - 2.5}{15}\right|^2} = 0.33 \text{ m}^3\text{/s}$$

The recommended pumping values should fall in the range $0.75\,Q_M = 0.25$ m³/s and $0.94\,Q_M = 0.31$ m³/s.

Considering the *water table gradient*, Dupuit's formula is only valid for low gradients: $\partial h/\partial r < 0.2$, for which values we then have:

$$h_0^2 - h^2 = \frac{Q}{\pi K}\ln\left(\frac{r_0}{r}\right) = 0.73\frac{Q}{K}\log\left(\frac{r_0}{r}\right)$$ (7.27)

or

$$s = \frac{0.73}{h_0 + h}\frac{Q}{K}\log\left(\frac{r_0}{r}\right)$$ (7.28)

In the neighbourhood of the well there is a sudden lowering of the water table, and a seepage face appears. There are various formulae for calculating the seepage depth. Vibert's formula, quoted by [11], is given below:

$$h_r = r_p \ln\frac{x_0}{r_p} + \sqrt{\left(r_p\ln\frac{x_0}{r_p}\right)^2 + (h_0 - h_p)^2} \qquad (7.29)$$

The water table between the point at which $\partial h/\partial r = 0.2$ and the well can be plotted freehand, connecting the last point given by Darcy's equation and the start of the seepage face.

At a point far away from the well, and with very small drawdown, s, compared to h_0, that is, for $s = h_0 - h \ll h_0$, it can be taken that:

$$h_0^2 - h^2 = (h_0 + h)(h_0 - h) \simeq 2h_0 . s \qquad (7.30)$$

whence:

$$s = 0.366\frac{Q}{Kh_0}\log\left(\frac{r_0}{r}\right) \qquad (7.31)$$

According to [2] in this case an unconfined aquifer can be regarded as confined, and from pumping tests the various values obtained for s, at different distances, r, with Q constant, can be plotted on semilogarithmic paper $(s, \log r)$. For $s = 0$ the value of r_0 is obtained; for $r = r_p$ the theoretical value of s_p is obtained. The straight line gradient is:

$$m = 0.366\frac{Q}{Kh_0} \qquad (7.32)$$

m can be obtained in the same way as for confined aquifers, so that (Fig. 7.12) $m = (\Delta s)_{10}$, and thus:

$$K = 0.366\frac{Q}{h_0(\Delta s)_{10}}. \qquad (7.33)$$

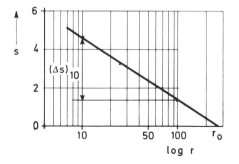

Fig. 7.12.

If the drawdown is large in relation to h_0, the observed data can be corrected (Jacob's correction):

$$s' = s - \frac{s^2}{2h_0} \qquad (7.34)$$

in which s' is the corrected drawdown.

Example. In an unconfined aquifer whose initial saturated thickness was 20 m, a pumping test was carried out with $Q = 60$ m³/hour, and the following drawdowns observed:

r(m)	$r_p = 0.25$	10	25	60	100
s(m)	$s_p = 6.0$	4.6	3.4	2.0	1.3

Calculate the permeability of the aquifer and the radius of influence.

Solution. From Fig. 7.12 we have $(\Delta s)_{10} = 3.3$ m, so that the permeability of the aquifer will be:

$$K = 0.366 \frac{Q}{h_0 (\Delta s)_{10}} = 0.366 \frac{60 \times 24}{20 \times 3.3} = 8 \text{ m/day} \ .$$

7.12   Partially penetrating wells

The study of partially penetrating wells, or more simply partial wells, is best carried out by trying to establish a flow net based on the characteristics of the aquifer (see Section 2.35), bearing in mind that the flow will take place both through the walls of the well and through the bottom. Nevertheless, as a general indication the following expressions are given, although many others are to be found in specialized texts.

(a) Confined aquifer

According to Custódio [2] the following formula, which gives the difference between the drawdown in a partial well $(s_p)_p$, and the drawdown, s_p, that would be obtained in the same aquifer and for the same discharge, if the well was fully penetrating, may be used:

$$(s_p)_p - s_p = \frac{Q}{2\pi T} \frac{1-\delta}{\delta} \left[\ln \frac{4b}{r_p} - F(\delta, \varepsilon) \right] \qquad (7.35)$$

in which $\delta = l/b$ is the ratio of the length of the well screen, l, to the aquifer thickness, b, and ε is the relative eccentricity of the screen zone: $\varepsilon = (|a_1 - a_2|)/2b$ (Fig. 7.13). $F(\delta, \varepsilon)$ is shown in Table 132.

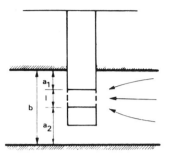

Fig. 7.13.

For the same length of the well screen, l, and the same discharge, the drawdown is minimal when the screen is at the centre of the aquifer ($a_1 = a_2$); the drawdown is greatest when it is at one of the extremes ($a_1 = 0$ or $a_2 = 0$).

Example. Calculate the drawdown expected in a partially penetrating well driven at the site of the well mentioned in Section 7.9 and subject to the same pumping conditions. Assume that the thickness of the aquifer is 10 m and that the well screen is located between 2 m below the confining layer and 3 m above the base of the aquifer.
In accordance with the respective definitions, we have:

$$\delta = \frac{10 - (2 + 3)}{10} = 0.50 \quad \text{and} \quad \varepsilon = \frac{|2 - 3|}{2 \times 10} = 0.05$$

so that $F(\delta, \varepsilon) = 3.433$.
Equation 7.35 thus indicates that;

$$(s_p)_p = 14.14 + \frac{0.05}{2\pi \times 0.00437} \times \frac{1 - 0.50}{0.50} \left(\ln \frac{4 \times 10}{0.15} - 3.433 \right) = 18.06 \text{ m}$$

According to Schneebeli [1], if the thickness b of the aquifer is considerable in comparison with the length of well screen l, that is $b \gg l$, the equipotential lines will be hemispherical about the well, such that:

$$Q = 2\pi K l \frac{(s_p)_p}{\ln (2r_d/r_p)} \tag{7.36}$$

(b) Wells in unconfined aquifers

No satisfactory solution is known from the theoretical point of view. Schneebeli recommends adoption of the previous formulae, substituting b ($h_0 - h_p$) by $1/2$ ($h_0^2 - h_p^2$).
In anisotropic formations, the discharge depends solely on the horizontal hydraulic conductivity, K_H. The water table, on the other hand, depends also on the vertical hydraulic

conductivity K_V. In order to determine the free piezometric surface, a shortening of the distance r by the value of K_V/K_H, is all that is required.

If $K_H/K_V > 1$, which results from stratified anisotropy, the free piezometric surface will not be lowered to such an extent.

7.13 Influence of the position related to the aquifers

If the well does not occupy the centre of the circular cone of depression, the assumption which has been made so far, it will be necessary to modify the previous formulae, according to [1].

(a) Well inside a circular recharge boundary

Take d as the distance from the well to the centre of the circular drawdown cone of radius r_0 (Fig. 7.14(a)). Define a fictitious radius of influence, r_a, to substitute r_0 in the previous formulae:

$$r_a = \frac{r_0^2 - d^2}{r_0} \tag{7.37}$$

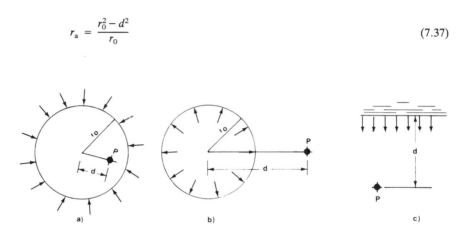

Fig. 7.14.

For $d/r_0 = 0.5$, the ratio is $Q_a/Q_0 = 1.05$, with Q_0 being the discharge of a well at the centre and Q_a the discharge in an eccentric well. If the well is very close to the extremity of the layer, the ratio is greater. For $d/r = 0.9$, we have $Q_a/Q_0 = 1.3$.

(b) Well outside a circular recharge basin

This case would occur near a circular recharge basin (Fig. 7.14(b)).

The fictitious radius will be:

$$r_a = \frac{d^2 - r_0^2}{r_0} \tag{7.38}$$

(c) Well close to a linear recharge boundary

This case may occur near a river (Fig. 7.14(c)). Taking d as the distance to the source, the fictitious radius of influence will be $r_a = 2d$.

7.14   Group of fully penetrating wells. Reciprocal influences

(a) *Confined aquifers*
Consider n wells, $P_1, P_2 \ldots P_i \ldots P_j \ldots P_n$ which traverse an aquifer under pressure, of thickness b (Fig. 7.15).

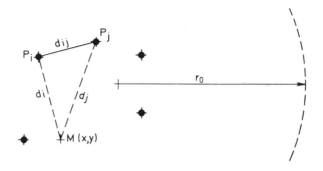

Fig. 7.15.

Taking:

d_i — distance of a point M (x, y) in the aquifer, from the pumping well P_i
d_{ij} — distance of well P_i from well P_j
r_{pi} — radius of well P_i
r_0 — common radius of influence of the n wells
i, j — vary from 1 to n
b — thickness of the aquifer.

The drawdowns are linear. Thus the effects overlap and at any point the drawdowns will be the sum of the drawdowns that each well would cause at that point. At a point M we therefore have:

$$s(M) = \frac{1}{2\pi Kb} \sum_{i=1}^{n} Q_i \ln\left(\frac{r_0}{d_i}\right) \tag{7.39}$$

When the discharges are known, the drawdown in well P_i will be:

$$s(P_i) = \frac{1}{2\pi Kb} \left[Q_i \ln\left(\frac{r_0}{d_i}\right) + \sum_{\substack{j=1 \\ (j \neq i)}}^{n} Q_i \ln\left(\frac{r_0}{d_{ij}}\right) \right] \tag{7.40}$$

When the drawdowns in the n wells are known, application of expression 7.40 gives n linear equations in n unknowns, which are the discharges Q_i. All that is necessary, therefore, is to solve the system of linear equations obtained.

(b) Unconfined aquifers
In unconfined aquifers, and if the drawdown is small, the previous equations can be applied. Otherwise the following expression can be used:

$$h_0^2 - h^2 = \sum_{i=1}^{n} Q_i \ln \frac{r_0}{d_i} \qquad (7.41)$$

7.15   Wells in aquifers with defined boundaries: image well method
In this chapter it has so far been considerd that the wells are in extensive aquifers or in circular formations, with a constant piezometric head at the periphery. Geological and morphological features of various kinds, however, limit real aquifers and cause distortions in the depression cones around the pumping wells. The method used in this case is the so-called *image well method* for calculating the influence of such aquifer boundaries on flow to the well. For more complicated cases the reader should consult [2] and [4].

In the case of an impermeable linear barrier the drawdown in the pumping well, calculated by the images method, is equal to the superposition of two drawdowns. One is the drawdown referred to a well in an infinite aquifer and the other is the drawdown of a well subject to the same pumping rate and located on the other side of the boundary and the same distance from it, in a line perpendicular to the barrier, passing through the real well (see Section 2.32, example *d*, Fig. 2.24).

In the case of a recharge boundary the superposition is not done with a pumping well but with a recharge well equidistant from the boundary and with the a rate of recharge equal to the rate of pumping in the real well (see Section 2.32, example (c), Fig. 2.24).

C.   UNSTEADY FLOW

7.16   General considerations
In the cases mentioned previously, steady flow has always been considered, i.e. the rate of recharge is equal to the rate at which the aquifer is pumped. In most real cases as water is withdrawn by the well a cone of depression forms about the well which increases in extent as pumping proceeds. The flow is therefore unsteady but may eventually reach a state in which the variations in level are so small that the flow can be considered steady. The basic equations of unsteady flow are given below.

7.17   The Theis equation
Consider a well in a confined aquifer of constant thickness, or a shallow unconfined aquifer with a lower horizontal impermeable layer; h_0 is taken as the initial head on the aquifer base. At a distance r from the well the piezometric surface drops by s, as a volume v is withdrawn. The drawdown, s, is a function of time, t, and distance, r:

$$s = f(r, t)$$

In the case of an unconfined aquifer, three basic assumptions should be satisfied: the saturated thickness (h_0) is small in comparison with its lateral extent, the water table variation is slow, and the water table gradient is weak (far from the pumping well).

Theis† deduced the following expression:

$$s = \frac{Q}{4\pi T} W(u) = 0.08 \frac{Q}{T} W(u) \tag{7.42}$$

in which:

$$u = \frac{Sr^2}{4Tt} \tag{7.43}$$

S is taken to be the storage coefficient and T the transmissivity of the aquifer.

The function $W(u) = \int_u^\infty \frac{e^{-u}}{u} du$ is known as the *well function* and is shown in Table 130. Values of the function outside the limit of the table may be found in other mathematical tables.

T and S are usually determined from pumping test data using the following procedure (Fig. 7.16):

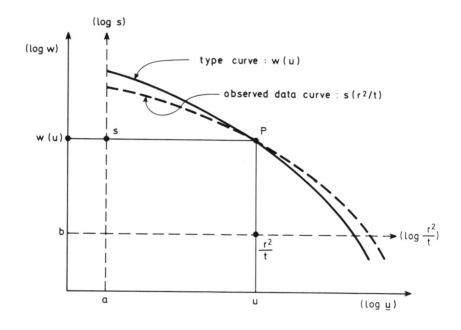

Fig. 7.16.

---

† For more details see [3].

(1) A plot on logarithmic paper of $W(u)$ versus u is drawn, arbitrarily choosing values of u, and reading the value of $W(u)$ from the tables; the resulting curve is known as the *Theis type curve*.

(2) On identical cycle transparent log-paper, the observed values of drawdown (s) are plotted against values of r^2/t for various piezometers installed at a distance r from the well, at time t. Thus the resulting curve is the *observed data curve*.

(3) The data curve is superimposed on the type curve, keeping the axes of the two curves parallel, until trial and error adjustment gives a position where the plotted points coincide as far as possible with the type curve.

(4) An arbitrary point, the match-point, is selected as the overlapping position of the two sheets of graph paper †. This point is defined by four coordinates: $(W(u), u)$ on the base graph and $(s, r^2/t)$ on the transparent graph.

The match-point usually corresponds to a pumping time of more than 12 hours.

Equations (7.42) and (7.43), define the relationship between the match-point coordinates, and in log-form are given by:

$$\log s = \log\left(\frac{Q}{4\pi T}\right) + \log W(u) = \log b + \log W(u) \qquad (7.44)$$

$$\log u = \log\left(\frac{S}{4T}\right) + \log\left(\frac{r^2}{t}\right) = \log a + \log\left(\frac{r^2}{t}\right) \qquad (7.45)$$

This means that equation (7.44) can be solved for T by using the coordinates of the match-point, s and $W(u)$. In the same way, equation (7.43) can be solved for S by using u and r^2/t of the match-point and the value of T that has already been calculated, i.e.:

$$T = \frac{Q}{4\pi s} W(u); \quad S = \frac{4uTt}{r^2} \qquad (7.46)$$

Example. In a well located in a confined aquifer, a pumping test is carried out with unsteady flow. By superimposing the resulting data curve on the Theis curve, and selecting a suitable match-point, the following pairs of coordinates have been derived:

$$u = 6 \times 10^2 \;\; ; \;\; W(u) = 2.3$$
$$s = 1.05 \text{ m}; r^2/t = 1.04 \text{ m}^2/\text{s}$$

Given that constant discharge from the well is 0.05 m³/s, calculate:

(a) the transmissivity and storage coefficient of the aquifer;
(b) the drawdown in an observation well located 100 m from the pumping well, after 24 hour's abstraction.

For (a), using equation (7.46) we have:

$$T = \frac{0.05}{4\pi} \times \frac{2.3}{1.05} = 0.0087 \text{ m}^2/\text{s} \; ; s = 4 \times 0.0087 \times \frac{6 \times 10^{-2}}{1.04} = 2 \times 10^{-3}$$

For (b), since for 24 hours $r^2/t = 0.116$, we have:

† To clarify this procedure, in Fig. 7.16 the point P of maximum coincidence was chosen.

$$u = \frac{2 \times 10^{-3}}{4\pi \times 0.0087} \times 0.116 = 6.67 \times 10^{-3} \rightarrow W(u) = 4.44 \; ;$$

$$s = \frac{0.05 \times 4.44}{4\pi \times 0.0087} = 2 \text{ m}$$

This method can be applied to unconfined aquifers provided that the remaining basic assumptions for the Theis equation are satisfied. Namely, the drawdowns occurring, s, are insignificant when compared with the saturated thickness ($b = h_0$) of the aquifer ($s < 0.1b$), the variations of the piezometric surface must be slow and the water-table gradient is small a long way from the well. When applying the Theis equation to unconfined aquifers, the value of the storage coefficient, S, is equal to the specific yield, n_e.

7.18 Simplified Jacob's formula

The function $W(u)$ can be expanded in series, as follows:

$$W(u) = \int_u^\infty \frac{e^{-u}}{u} du = 0.5772 - \ln u - \sum_1^\infty (-u)^n \frac{1}{n \cdot n!} \qquad (7.47)$$

For very small values of u ($u < 0.03$), the series terms become negligible after the first two terms, and expressed in decimal logarithmic form, we have the simplified Jacob's formula:

$$s = 0.183 \frac{Q}{T} \log 2.25 \frac{T}{S} \frac{t}{r^2} \qquad (7.48)$$

The formula can also be written as follows:

$$\frac{s}{Q} = \frac{0.183}{T} \log 2.25 \frac{T}{S} + \frac{0.183}{T} \log \frac{t}{r^2} \qquad (7.49)$$

The plot of specific drawdown (s/Q) against $\log(t/r^2)$ for pumping test data forms a straight line on semi-logarithmic paper.

Its gradient is then given by (Fig. 7.17):

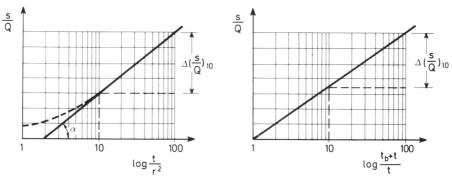

Fig. 7.17.   Fig. 7.18

$$\tan \alpha = \frac{0.183}{T} \tag{7.50}$$

which enables the formation transmissivity to be obtained.

In order to obtain $\tan \alpha$, the quickest way is to find the drawdown difference, $\Delta (s/Q)_{10}$, for a single log cycle of t/r^2. We then have:

$$T = \frac{0.183}{\Delta \left(\dfrac{s}{Q}\right)_{10}} \tag{7.51}$$

The straight line cuts the axis of the abscissae at a point t_0/r^2, defined by $s/Q = 0$, that is:

$$\frac{2.25 \, T t_0}{S r^2} = 1 \tag{7.52}$$

from which is obtained the storage coefficient of the formation, S:

$$S = 2.25 \, T \frac{t_0}{r^2} \tag{7.53}$$

7.19 Recovery curve

It is assumed that after a certain time of pumping, t_b, at constant discharge, Q, pumping is shut down. Analysis of the resulting recovery curve can be very useful for studying the aquifer. During the recovery period, it can be considered that based on the principle of superimposed flows, the drawdown resulting from zero discharge is equivalent to the sum of the drawdown, s_1, corresponding to the previously pumped discharge, Q, and the rise in level, s_2, that would result from the fictitious recharge, $-Q$, of the well, equal but negative.

$$s = s_1 + s_2 \tag{7.54}$$

By applying Jacob's formula to s_1 and s_2, it is at once deduced that:

$$s = 0.183 \frac{Q}{T} \log \frac{t_b + t}{t} \tag{7.55}$$

in which s represents the drawdown observed after time t, measured from the shutdown in pumping.

As in the previous case, by plotting on semilogarithmic paper, s/Q and $\log (t_b + t)t$, the transmissivity will be given by (Fig. 7.18):

$$T = \frac{0.183}{\Delta \left(\dfrac{s}{Q}\right)_{10}} \tag{7.56}$$

in which $\Delta (s/Q)_{10}$ has the same significance as was mentioned previously.

7.20 Radius of influence of a well

In the case of steady state flow it was assumed that the water withdrawn from the aquifer was replenished through a concentric cylinder, the radius of which was made equal to the distance from the well beyond which the aquifer would remain practically unaltered; that is, the distance at which the effect of the well was negligible.

This value, i.e. the radius of influence of the well, r_0, can be determined from a study of flow to the well in an unsteady state.

If in Jacob's formula it is taken that:

$$r_0^2 = 2.25\frac{Tt}{S} \text{ or } r_0 = 1.5\sqrt{\frac{Tt}{S}} \tag{7.57}$$

we shall have:

$$s = 0.183\frac{Q}{T}\log\frac{r_0^2}{r^2} = 0.366\frac{Q}{T}\log\frac{r_0}{r} \tag{7.58}$$

For $r = r_0$, $s = 0$, that is, r_0 is the radius of influence of the well.

In the real case r_0 depends on the formation characteristics, T and S, and also depends on the pumping time, t.

The variation of r_0 with t is obtained by deriving:

$$\frac{dr_0}{dt} = 0.75\sqrt{\frac{Tt}{S}} \tag{7.59}$$

that is, as the pumping time, t, increases, the variation of r_0 diminishes.

D. SOIL DRAINAGE

7.21 Drainage of soils under steady state

Drainage of soil, of course, is essential to the life of plants. Drainage can be carried out by means of ditches or drains, and it is necessary to determine their horizontal spacing.

(a) Dupuit's hypothesis — Ellipse or Donnan's equation

In Section 7.8 flow to trenches was studied. In the case of an unconfined soil layer, and based on Dupuit's hypothesis, according to equation (7.15): $V = -K(dh/dx)$.

Assume that a soil experiencing rainfall of intensity j, has ditches evenly spaced by the distance L (Fig. 7.19(a)).

In order for the water table to remain constant, it is necessary that the precipitation jx, in a strip with a width equal to unity, shall be equal to the discharge to the ditch.

We thus have the equation:

$$jx = -K\frac{dh}{dx}h \tag{7.60}$$

that is to say:

$$jx\,dx = -Kh\,dh \tag{7.60a}$$

By integrating this equation between $x = 0$ (highest point of the water table at the mid-point between ditches and equal to a water depth h_0), and $x = L/2$ (in the ditch), corresponding to a depth of h_v, we shall have;

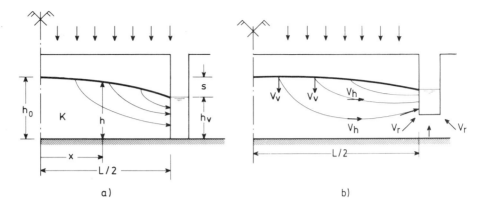

Fig. 7.19.

$$\int_0^{L/2} jx\,dx = \int_{h_0}^{h_v} -Kh\,dh \tag{7.60b}$$

and

$$j\left[\frac{x^2}{2}\right]_0^{L/2} = K\left[\frac{h^2}{2}\right]_{h_v}^{h_0} \tag{7.60c}$$

which leads to the following value for spacing of the drains:

$$L^2 = \frac{4K}{j}(h_0^2 - h_v^2) = \frac{4K}{j}(h_0 + h_v)(h_0 - h_v) \tag{7.60d}$$

By making $s = h_0 - h_v$, we have:

$$L^2 = \frac{1}{j}(8Kh_v s + 4Ks^2) \tag{7.61}$$

In general, K and j are expressed in metres per day and h and s in metres, so that L is given in metres. This equation is known as *the equation of an ellipse or Donnan's equation*.

(b) Hooghoudt's formula

As a rule drainage is effected by ditches or drains that do not reach the impermeable layer; the streamlines will no longer have the almost parallel course assumed in Fig. 7.6, in which the ditch cuts the whole layer. The situation is shown in Fig. 7.19(b). The velocities have vertical, V_v, horizontal, V_h and radial, V_r, components.

When the depth of water in the ditch is roughly equal to its width, flows to pipes or ditches are similar.

For such cases, many formulae have been deduced, the best being Hooghoudt's formula, which is essentially identical to the previous one:

$$L^2 = \frac{1}{j}(8Kes + 4Ks^2) \tag{7.62}$$

in which the thickness of the layer, h_v, is substituted by a smaller equivalent thickness, e. The value of the equivalent thickness is a function of the diameter of the drain d, the thickness b_0 of the permeable layer under the drain, and of the distance L between the drains. The value of e is shown in tables; see for example [6].

(c) Ernst's equation

Consider (Fig. 7.20) a soil consisting of one upper layer of mean thickness b_1

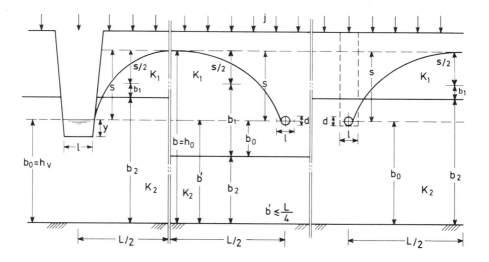

Fig. 7.20.

and hydraulic conductivity K_1, and a second layer of thickness b_2 and hydraulic conductivity K_2, resting on an impermeable stratum. A drainage system is established by means of ditches (left part of the figure) or by pipes (right part of the figure). These drains can be placed either in the upper layer (central part of the figure) or in the lower layer (outer parts of the figures).

Bearing in mind the symbols used in Fig. 7.20, where the discharge per unit area, j is equal to the rainfall intensity or irrigation, under steady flow, the total head loss, s, corresponds to the sum of the head losses resulting from the vertical, horizontal and radial flows.

$$s = s_v + s_h + s_r \qquad (7.63)$$

According to [6] the vertical flow, with velocity V_v, takes place mainly in the upper part of the two layers, and it is considered that the relevant saturated layer is equal to s in the case of the pipes and to $s + y$ in the case of ditches.†

We thus have (cf. equation (7.13)):

$$s_v = j\frac{s+y}{K_v} \tag{7.64}$$

The horizontal flow, with velocity V_h, is assumed to take place across the thickness $b = b_1 + b_2$, unless the thickness is very great, in which case the total thickness of the layer below the drains will be taken as equal to $L/4$.

The horizontal flow will be approximately (cf. equation (7.64)):

$$s_h = j\frac{L^2}{8\Sigma(K_i b_i)} = j\frac{L^2}{8(K_1 b_1 + K_2 b_2)} \tag{7.65}$$

The radial flow, with velocity V_r, takes place mainly in the layer b_0 below the drains, and is given by:

$$s_r = j\frac{L}{\pi k_r}\ln\frac{ab_0}{u} \tag{7.66}$$

in which u takes on the following values:

— in the case of ditches it is equal to the wetted perimeter;
— in the case of pipes, it is more difficult to select, but as a rule it is taken as:

$$u = l + 2d \tag{7.67}$$

in which l is the width of the trench excavated for placing the pipe and d is the diameter of the pipe. If the trench is filled with filtering material, d can be substituted by the depth of that material.

According to [6] *the method of applying the procedure is as follows.*

(1) In the case of a homogeneous soil ($b_2 = 0$), where the thickness, b_1, is less than $L/4$, it is assumed that $a = 1$ and $K_r = K_1$, so that:

$$s = j\left(\frac{s+y}{K_1} + \frac{L^2}{8K_1 b_1} + \frac{L}{\pi K_1}\ln\frac{b_0}{u}\right) \tag{7.68}$$

If the thickness is greater than $L/4$, it is assumed that $b_1 = L/4$.

Since the first term is independent of L, to facilitate the calculation, we

---

† Strictly speaking, half of these values should be taken, i.e. about half the saturated layer above the drain. For more illustration see [6].

may pass it to the left-hand side of the equation. In most cases its value is so small that it can be disregarded.

(2) In the case of soil consisting of two layers, b_1 and b_2, of permeability K_1 and K_2, such that the thickness of the layer in which the drain is placed is less than $L/4$ (previous case), the following conditions may be examined:

(a) the drain is placed in the lower layer, Layer 2 (left and right extremities of Fig. 7.20): if $K_1 < K_2$, the vertical resistance of the second layer can be disregarded, assuming that the vertical flow is concentrated in a layer of thickness b_1; for horizontal flow, since $K_1 < K_2$ and, as a rule in this case $b_1 < b_2$, it will suffice to use the term $K_2 b_2$; for radial flow it is considered that it takes place in thickness b_0, but limited to $L/4$; it is also assumed that $a = 1$.
We shall thus have:

$$s = j\frac{b_1}{K_1} + \frac{L^2}{8K_2b_2} + \frac{L}{\pi K_2}\ln\frac{b_0}{u} \qquad (7.69)$$

(b) if the drain is entirely in the upper layer (central part of Fig. 7.20), of thickness b_1, the complete equation is applied:

$$s = j\left(\frac{y+s}{K_1} + \frac{L^2}{8(K_1b_1 + K_2b_2)} + \frac{L}{\pi K_1}\ln\frac{ab_0}{u}\right) \qquad (7.70)$$

in which a can assume the following values:

$K_2 > 20K_1$ $- a = 4$
$0.1K_1 < K_2 < 20K_1$ $- a$ is given in Graph 135
$K_2 < 0.1K_1$ $- a = 1$; in this case the lower layer can be regarded as impermeable and the problem can be treated as if it was homogeneous soil.

Example. A soil consists of two layers, an upper one 2 m thick of permeability $K_1 = 0.5$ m/day, and a lower layer 3 m thick of permeability $K_2 = 5$ m/day. The second layer lies on an impermeable stratum.
    It is wished to drain a flow of $j = 0.02$ m/day and maintain the water table at a minimum depth of 0.5 m. Calculate the spacing of drains, if they are placed at a depth of 1 m and have a diameter of 0.1 m.

$b_1 = 2 - 0.5 = 1.5$ m; $b_2 = 3$ m; $b_0 = 2$ m $- 1$ m; $K_1 = 0.5$ m/d; $K_2 = 5$ m/d; $j = 0.02$ m/d; $s = 0.5$ m; $u = 1 + 2d = 0.5 + 2 \times 0.1 = 0.7$; $K_2/K_1 = 10$; $b_2/b_0 = 3$; $a = 3.8$ (Graph 131).

We thus have:

$$s = j\frac{s}{K_1} + j\frac{L^2}{8(K_1 b_1 + K_2 b_2)} + j\frac{L}{\pi K_1}\ln\frac{ab_0}{u}$$

$$0.5 = 0.02\frac{0.5}{0.5} + 0.02\frac{L^2}{8(0.5 \times 1.25 + 5 \times 3)} + 0.02\frac{L}{\pi \times 0.5}\ln\frac{3.8\times1}{0.7}$$

$$0.5 = 0.02 + 0.00016 L^2 + 0.021539 L$$

$$1.6 L^2 + 215.39 L - 4800 = 0$$

$$L = \frac{-215.39 \pm \sqrt{215.39^2 + 4 \times 1.6 \times 4800}}{2 \times 1.6}$$

$$L = \frac{-215.39 \pm 277.69}{3.174} = 19.67\,m \approx 20\,m$$

7.22   Soil drainage in conditions of unsteady flow

In unsteady flow the water level declines progressively after reaching a maximum depth, usually a few hours after the rain stops. The calculations attempt to obtain the drawdown which is required at the end of a given time.

Many formulae are found in the literature, including : Van der Leur, Sine and others. For more illustration see [6].

BIBLIOGRAPHY

[1] Schneebeli, G., *Hydraulique Souterraine*, Collection du Centre de Recherches et d'Essais de Chatou, Eyrolles, Paris, 1966.

[2] Custódio, E., *Hidrologia Subterrânea*, Ediciones Omega, Barcelona, 1976.

[3] Ferris, J. G., *Groundwater*, Chapter 7, in C. O. Wisler, C. E. F. Brater (eds.), Hydrology, John Wiley, New York, 1959, pp. 198–272.

[4] Bear, J., *Hydraulics of Groundwater*. McGraw-Hill, Israel, 1979.

[5] Todd, D. K., *Groundwater Hydrology*, John Wiley, New York, 1980.

[6] ILRI (International Institute for Land Reclamation and Improvement), *Drainage. Principles and Applications*, Publication 16, Wageningen, The Netherlands, 1973.

[7] Guyon, G., *Calcul de la Distance entre les Drains dans un Systeme de Drainage*. Terres et Eaux, No. 50, 1957.

[8] Carlier, M., *L'Hydraulique des Nappes de Drainage pour Canalisations Souterraines,* Annales de l'Institute Technique du Batiment et des Travaux Publics, Mai, 1963.

[9] Horn, J. W., *Principes Fondamentaux du Drainage des Terres*, Annual Bulletin of the International Commission on Irrigation and Drainage, 1964.

[10] Sine, L., *Le Dimensionement Rationel d'un Reseau de Drainage Agricole,* Annales de Gembloux, 1965.

[11] Carlier, M., *Hydraulique Génerale et Apliquée*, Collection du Centre de Recherches et d'Essais de Chatou, Eyrolles, Paris, 1972.

8

Flow measurements. Orifices and weirs

A. MEASUREMENTS OF WATER LEVELS AND PRESSURES

8.1 Instruments to measure the water level directly

This section is concerned with the simplest types of instruments used to measure water level. Among others there are the following:

(a) Staff gauges

These consist of a graduated linear scale, with the zero duly marked, reading of which gives the position of the water level directly.

Precision of the measurements varies greatly according to how the scale is placed, and whenever possible it is advisable to place it in a stilling well in communication with the watercourse, in order to eliminate the effects of agitation of the free surface caused by the wind or the actual turbulence of the flow. In any case, it is difficult to obtain absolute reading errors of less than 1 cm.

(b) Sounding rods

The sounding rod is a portable scale that gives the depth of the water level in relation to the bottom. Reading errors depend on the state of agitation of the free surface, and again it is hard to obtain readings with absolute errors of less than 1 cm.

(c) Sounding lines

These substitute sounding rods for greater depths or faster currents. The absolute reading errors are usually greater, and they may easily reach more than 15 cm.

(d) Point gauges

These consist of a thin vertical rod, pointed at its lower end, connected to a scale graduated usually in millimetres, associated with a vernier scale (Fig. 8.1(a)).

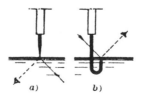

Fig. 8.1

The construction of the scale, its movement system and the way in which the reading is carried out have a great influence on the error. Under good conditions, however, the error may be less than 0.2 mm or even 0.1 mm.

In this type of gauge the pointed end is adjusted to coincide with the water surface; the contact with the water surface may be detected by movement of the surface meniscus or electrically by using an electric torch whose circuit is closed when the point touches the water on the way down. The ascendent operation can be influenced by the effect of capillarity on the point.

There is no advantage in the point being extremely sharp, and it is preferable for it to be rounded with a radius of 0.2 to 0.3 mm.

(e) Hook gauges

Somewhat greater accuracy is obtained with a *hook gauge* in which the end of the vertical rod is bent round to form a U with the point facing upwards (Fig. 8.1(b)). The principle of operation is the same as for the preceding instrument. This means that whereas a point gauge must touch the liquid surface, on entering, in a descending movement, that of the hook gauge must touch the liquid surface on leaving, in an ascending movement, which is when the reading is taken.

The hook gauge is more advantageous than the point gauge whenever the observer has to look downwards.

(f) Water level recorders

These not only measure but also record variations in the water level. The simplest system consists of a float connected to a line that in turn is connected to a recording system (balanced float gauge). There are other electrically-based systems, but description of them lies outside the scope of this book†.

8.2 Piezometers

A piezometer is a tube whose lower part is connected to the receptacle that contains the liquid and whose upper part is freely open to the atmosphere. The level in the piezometer gives directly the position of the piezometric line, i.e. the elevation corresponding to the term $z + p/\gamma$. As shown in Fig. 8.2, $p = \gamma h$.

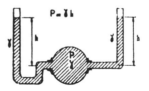

Fig. 8.2

Example. A pipe carries oil, with relative density $\delta = 0.80$. It is wished to find the pressure in the pipe, knowing that in a piezometer the oil rises 5 m above the pipe. Since $\delta = 0.80$ is $\gamma = \rho g = 800 \times 9.8 = 7840$ N/m³. Since $p/\gamma = 5$ m, we have $p = 5$. $\gamma = 5 \times 7840 = 3.9$ N/cm².

8.3 U-tube manometers

In the case of *very high pressures* the piezometer is advantageously substituted by a U-tube (Fig. 8.3(a)), in which is placed a liquid of specific weight γ' greater than the specific weight γ of the fluid in question.

† For example, consult [1] and [6].

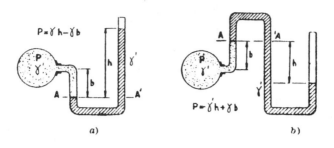

Fig. 8.3

The surface AA' is an equipotential, and the pressures at A and A' must then be equal.

We then have: $p + \gamma b = \gamma'h$, so that:

$$p = \gamma'h - \gamma b \qquad (8.1)$$

In the case of *very low pressures*, the arrangement shown in Fig. 8.3(b) is adopted, using a manometric liquid of specific weight that is less than that of the working fluid.

Since the pressure at A and A' are equal: $p - \gamma b = -\gamma'h$, and thus

$$p = -\gamma'h + \gamma b \qquad (8.2)$$

this expression being formally identical to that of (8.1).

8.4   Differential manometers

These are used for measuring the pressure difference between two points of a system in which a liquid flows. Two piezometers placed side by side can function as differential manometers. As a rule the two tubes are connected to form a U and are filled with a liquid that is different from the flowing fluid. If the specific weight, γ', of the manometric liquid is greater than the specific weight γ of the flowing fluid, the arrangement used is that shown in Fig. 8.4a. The pressure difference between the two points, 1 and 2, is therefore:

$$\frac{p_1 - p_2}{\gamma} = \Delta h = \left(\frac{\gamma'}{\gamma} - 1\right)\Delta h' = K\Delta h' \qquad (8.3)$$

If the specific weight of the manometric liquid is less than that of the flowing fluid, the reverse arrangement must be used, i.e. the U of the manometer must be made to face upwards (Fig. 8.4(b)).

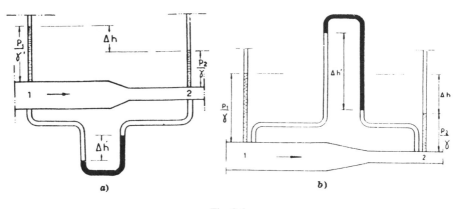

Fig. 8.4

In an analogous manner we thus have:

$$\frac{p_1 - p_2}{\gamma} = \Delta h = \left(1 - \frac{\gamma'}{\gamma}\right)\Delta h' = K\Delta h' \qquad (8.3a)$$

The constant $K = \left|1 - \dfrac{\gamma'}{\gamma}\right|$ is known as *manometric constant*.

If $K > 1$, the manometer is reducing, i.e. $\Delta h'$ will be less than Δh. If $K < 1$, the manometer will be amplifying, i.e. $\Delta h'$ will be greater than Δh.

Table 133 indicates manometric liquids in general use.

Example. In order to measure the difference in pressure between two points of a conduit in which water flows ($\gamma \simeq 10\,000$ N/m³), a differential manometer has been used in which the manometric liquid is benzol ($\gamma' = 8740$ N/m³).
The difference in level of the two arms of the manometer is $\Delta h' = 1.0$ m.
We thus have;

$$\Delta h = \Delta h' \ (1 - \gamma'/\gamma) = 1.0 \ (1 - 8\,740/10\,000) = 0.216 \text{ m of water column.}$$

8.5   Standards for installing piezometers and manometers

When installing a piezometer or manometer, the following general standards must be borne in mind:

(1) The pressure tappings must, whenever possible, be on a straight length of pipe that is long enough upstream and downstream.

(2) If a large number of pressure tappings are drilled in the pipe wall at the same cross-section and led into a piezometer ring, the pressure taken is more reliable than that obtained from a single pressure tapping. The piezometer ring automatically approximates such an average.

(3) The axis of the pressure tapping must be exactly perpendicular to the wall of the pipe.

(4) It is necessary to avoid all roughness inside the pipe, at the pressure tappings, in order to prevent local pressure variations.

(5) The diameter of the pressure tappings must be small, as a rule 2 to 3 mm for pipes of up to 30 cm diameter and 3 to 4 mm for larger diameters.

(6) Transmission of pressure to the measuring apparatus can be done by the liquid itself, by means of air or electrically. Generally speaking, for a given distance electrical conduction is

cheaper than pneumatic conduction. Hydraulic conduction for great distances (about 30 m) is very hard to achieve, owing to inertia, air bubbles, etc.; the diameter of the connection generally varies between 5 and 25 mm, according to the distance, and it is essential that all joints are completely sealed.

(7) If the apparatus indicates the pressure by itself (as is the case with *Bourdon* manometers), it is necessary to pay attention to the difference in elevation, h_0, between the point at which it is wished to measure the pressure and the actual measuring apparatus. If the apparatus is h_0 higher than the measurement point, it is necessary to add γh_0 to the reading; if the apparatus is lower, that value must be subtracted from it. In the case of differential manometers, the two tubes must be placed side by side and parallel, thus eliminating the errors due to variations in specific weight caused by temperature variations.

(8) In the case of hydraulic transmission, air bubbles must be completely eliminated, but this is difficult when depressions are being measured, and so the transmission must be as short as possible.

(9) When it is wished to record very rapid pressure variations, it is necessary to avoid transmission tubes that are deformable or very long. It is also necessary for the measuring apparatus itself to have sufficiently low inertia for it to be sensitive to such pressure variations[†].

(10) The manometric fluid must be chosen according to the measuring requirements. The manometric fluids shown in Table 133 are those in current use.

B. MEASUREMENT OF VELOCITIES

8.6 Floats

The most obvious technique of measuring velocities consists of using floats. This technique is valid mainly for surface velocities.

The principle of measurement offers no difficulty: measurement of the distance covered during a certain time.

Realization of this principle is, however, sometimes difficult because as a rule the floats do not follow previously chosen trajectories; they generally tend to move with the flow. Floats are also subject to the vagaries of winds and local surface currents which may drive them far off their courses.

As order of magnitude it can be supposed that in currents with a certain regularity the mean velocity, U, is about 0.7 to 0.8 of the maximum measured velocity of the flow near the surface.

It must be remembered, however, that this ratio may be higher than 1, especially in deep channels with almost vertical sides, in which the maximum velocities no longer occur near the surface.

8.7 Pitot-static tubes

A Pitot-static tube consists basically of two tubes which record *static pressure* (point A) *and stagnation pressure* (point B) (Fig. 8.5). The second tube records the *total head* which is the sum of the *pressure head* plus the *velocity head*: $p/\gamma + V^2/2g$. The difference between the two measurements, Δh, gives the value of the velocity head. Theoretically, we would then have $V = \sqrt{2g\Delta h}$; in practice, however, not all the velocity head is transformed into pressure head, and it is necessary to introduce a correction coefficient, a function of the shape of the orifices and of their location (instrument coefficient). We thus have:

---

† Common piezometers and manometers are incapable of detecting very rapid variations in pressure. It is advisable to use pressure transducers, of which there is a considerable range, according to need.

$$V = c\sqrt{2g\Delta h} \qquad (8.4)$$

In the case of liquids that flow with low turbulence, and if the Reynolds number, in which the diameter of the orifice is taken for the linear parameter, does not fall below 100, it can be considered that $c \approx 1$. Values of the Reynolds number lower than 100 only occur, as a rule, in very viscous liquids.

In the case of very turbulent flows, the value of c diminishes, and 0.98 can be taken as mean value. Only by calibrating the apparatus for the same conditions is it possible correctly to fix the value of c.

In order to achieve greater precision of measurement, the difference in pressure, Δh, can be measured by means of a differential manometer whose arms connect at A' and B' (Fig. 8.5).

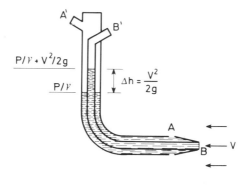

Fig. 8.5

As far as possible, the axis of the dynamic pressure tap orifice, B, must be placed parallel to the flow. It has been found experimentally, however, that deviations of up to 10° hardly alter the indication of the velocity, so that excessive care is not required in keeping parallelism†.

Whenever the tube is yawed from parallel, the value of the velocity shown must be multiplied by $\cos \theta$, θ being the angle of deviation.

8.8 Pitot cylinder

The Pitot cylinder is a variant of the Pitot tube that makes it possible to eliminate errors of direction, provide that the plane π is known, containing the velocity vector whose intensity it is wished to measure. The apparatus

† Prandtl's pitot-static tube, designed to be insensitive to small angles of yaw, gives a variation of only 1% in its coefficient at an angle of yaw of 19°.

consists of a cylinder in which, in a straight cross-section, there are two pressure taps making an angle of $\alpha = 78°\ 30'$ to one another† (Fig. 8.6).

The procedure is as follows:

(1) the cross-section of the cylinder in which the pressure tappings are situated is aligned with the plane π (known by hypothesis) of the velocity vector, and the two extremities of the cylinder are connected to a differential manometer;

(2) the cylinder is turned until the difference in pressure shown by the manometer is zero. In this position the velocity is directed according to the bisector of the angle α of the tappings, and the direction of the velocity in plane π is thus known (Fig. 8.6, position 1);

(3) the cylinder is turned half of angle α, that is to say 39° 15′ (Fig. 8.6, position 2). The reading of the manometer in this position gives the magnitude of the velocity by a process analogous to that of the Pitot tube.

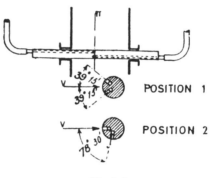

Fig. 8.6

8.9   Pitot spheres

The Pitot sphere is another variant that gives, at a point, the direction and magnitude of the velocity, without the need to know *a priori* the plane of the velocity vector. It consists basically of a sphere, with five pressure tappings (Fig. 8.7).

The sphere is placed at the end of a shaft that can rotate on its axis, the angle of rotation, ψ, being recorded on a horizontal scale.

During the calibration operation, the angle of inclination, δ, of the shaft can be varied in conditions where the static pressure, h, and the velocity, V, are known. After a given δ has been fixed, the shaft is turned until the readings 4 and 5 on the manometer coincide: $h_4 = h_5$.

The readings h_1, h_2 and h_3 are also taken and the following coefficients are established, these being, within certain limits, only characteristics of the apparatus, i.e. they are independent of the velocity:

---

† This angle was chosen since it was found that it corresponded best to the fundamental objectives of the apparatus.

$$\text{Coefficient of direction } K_d = \frac{h_3 - h_1}{h_2 - h_4} \tag{8.5}$$

$$\text{Coefficient of velocity } K_v = \frac{h_2 - h_4}{V^2/2g} \tag{8.6}$$

$$\text{Coefficient of pressure } K_h = \frac{h_2 - h}{V^2/2g} \tag{8.7}$$

The operation is repeated for other values of δ and the curves of K_d, K_v and K_h are obtained as functions of δ. These curves are characteristic of each apparatus.

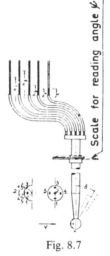

Fig. 8.7

In order to measure the velocity and its direction at a point, the sphere is placed at that point with the shaft perpendicular to the axis of the channel. The zero of the horizontal scale is adjusted with the axis of the channel and the shaft is turned until piezometers 4 and 5 give equal readings, $h_4 = h_5$, and then the horizontal angle, ψ, is read; the values of h_1, h_2 and h_3 are also recorded and the value of K_d is determined. Knowing K_d, the value of δ is determined from the calibration curve and this, together with the value of ψ, completely defines the direction of the velocity. Knowing δ, the values of K_v and K_h are also obtained from the calibration curves and these give the magnitude of the velocity, V, and of the pressure, h, at the point being studied.

8.10 Current meters

The current meter consist essentially of a rotating element whose speed of rotation varies with the local velocity of flow. The relationship between these variables is found by calibration in laboratory tests, with the current meter moving at a given velocity, and the water stationary. The method of counting the number of rotations is specific to each apparatus.

Current meters fall into two main classes, depending upon the design of the rotating elements; these are the cup type and vane (propeller) type.

The calibration equation of these devices is known as the characteristics curve of the current meter, and is of the following type:

$$V = a + bn \qquad (8.8)$$

in which V is the velocity, n the number of rotations per unit time, a and b two constants characteristic of each apparatus.

The calibration tests must be repeated periodically because the conditions of the current meter change with its operation.

The effect of obliquity of the current is harder to evaluate in current meters than in Pitot tubes, and it is difficult to give simple principles.

It is also necessary to take account of the turbulence of the flow, and so measurement at each point has to last for a certain time (5 to 10 minutes, or even more).

The measuring time must be controlled, and readings taken, at some characteristic points of the flow, at the end of 5, 10, 15 minutes, until the value of the velocity given by the apparatus is sensibly constant.

8.11 Current meters for sea currents
There is a form of current meter that is specially adapted to the recording of maritime currents. There are various types of this apparatus on the market which when placed at a point for several days record the direction and instantaneous magnitude of the current at that point.

The technique of installation is specific to each type of apparatus.

8.12 Measurements of the mean velocity in a cross-section
Given a cross-section A, normal to the direction of the flow, and the distribution of various velocities in that cross-section, the mean velocity in that cross-section is given by:

$$U = \frac{1}{A} \int_A V \, dA \qquad (8.9)$$

Practical realization of the integral can be carried out from plots of the isotachs, by multiplying the area between two isotachs by the mean of the velocities corresponding to those isotachs.

In some cases empirical rules are followed for determining the measurement points and preparing the results, these methods being confirmed by other measurement methods.

For rectangular channels, the best procedure is that indicated in Fig. 8.8 and used by the S.I.A.S. (Société des Ingénieurs et Architects Suisses).

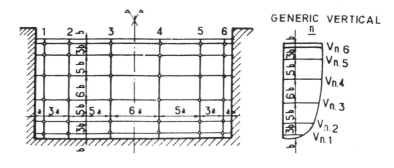

Fig. 8.8

The mean velocity, on a generic vertical, n, is given by

$$V_n = \frac{1}{12}(V_{n.1} + 2V_{n.2} + 3V_{n.3} + 3V_{n.4} + 2V_{n.5} + V_{n.6})$$ (8.10)

The mean velocity in the cross-section is:

$$U = \frac{1}{12}(V_1 + 2V_2 + 3V_3 + 3V_4 + 2V_5 + V_6)$$ (8.11)

The velocities at the various points can be determined by means of current meters or Pitot-static tubes.

C. DISCHARGE MEASUREMENT IN PRESSURE CONDUITS

8.13 Volumetric methods

The most precise method of measuring discharges, whether in pressure conduits or open-channel steady flows, is from their own definition: volume flowing in unit time. Thus by measuring the total quantity of fluid collected in a measured time, the mean discharge during that time is obtained. This method is only feasible for small discharges.

For measuring volumes, suitably calibrated tanks are used; for measuring the time chronometers are used. The techniques vary a great deal according to the nature of the problem and the desired precision. The simplest method consists of having a stop-watch with manual operation, and a calibrated tank.

In these conditions, if the tank has a capacity for at least one minute's flow, the error may be as low as 1%.

8.14 Pressure drop apparatus

Pressure drop apparatus are those that measure the discharge in closed-conduits, by measuring a pressure drop caused by a contraction (Fig. 8.9).

The device inserted in the conduit is known as *primary element*, a term that covers the pressure tappings and the part of the tube in which the apparatus is fitted. The apparatus required for measuring a pressure drop is the *secondary element*.

The principal primary elements are orifice meters, flow nozzles and Venturi meters, for each of which there are several standardized types.

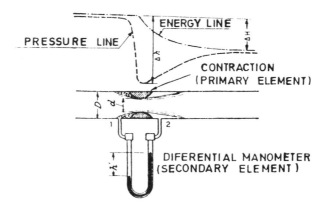

Fig. 8.9

The fluid measured can be compressible (gas) or incompressible (liquid). It is essential to know the specific weight of the fluid, γ, the coefficient of viscosity, v, and also, in the case of a gas, the adiabatic constant, $K\dagger$.

The diameter of the conduit will be represented by D and its area by A; the diameter of the constricted throat will be represented by d and its area by a. The relationship between these two areas is:

$$\sigma = \frac{a}{A} = \left(\frac{d}{D}\right)^2 \tag{8.12}$$

The Reynolds number is usually defined in relation to the diameter of the conduit or of the *throat*. In relation to the conduit it will thus be $\mathbf{R}_e = UD/v$, in which U is the mean velocity in the conduit.

If the fluid is gaseous, the pressure variation in the passage through the

---

† The *adiabatic constant*, K, of a gas is the quotient of the specific heat at constant pressure and the specific heat at constant volume. It varies with temperature and with pressure. As a first approximation, $K = 1.31$ can be taken for heated water vapour and $K = 1.40$ for diatomic perfect gases (O_2, H_2, N_2 CO).

primary element must be less than 0.8 of the *critical expansion*.† This value is given by Table 134 as a function of σ^2 and K.

The flow rate, in mass per unit time (kg/s), is given by:

$$Q_m = \alpha \varepsilon \frac{\pi d^2}{4} \sqrt{2\rho_1 \, \Delta p} \tag{8.13}$$

in which: d = diameter of the *throat* in m; $\Delta p = p_1 - p_2$ = pressure drop in N/m^2; ρ_1 = density in cross-section 1, in kg/m^3 (constant in the case of liquids); α = coefficient of discharge; ε = coefficient of compressibility (α and ε are nondimensional coefficients, determined experimentally). In the case of incompressible fluids (liquids) $\varepsilon = 1$.

Upstream of the primary element the conduit must be straight for a length that is 20 to 30 times its diameter, D; downstream of the primary element, the conduit must be straight for a length that is 10 to 15 times its diameter, D.

For a length of $2D$ upstream, the diameter of the conduit must not differ by more than 0.5% from the mean diameter of that length.

(1) Orifice meter ISA 1932

This type of orifice meter consists of an orifice with a concentric square-edge, in a thin plate which is clamped between the flanges of a pipe (Fig. 8.10 (a)). The upstream face of the plate must be flat and smooth. The downstream face of the plate must also be smooth, although the finish need not be to the same standard as the upstream face. The diameter of the orifice must be calibrated to an accuracy of $\pm 0.001 \, D$.

The downstream pressure tappings are located at the *vena contracta* to obtain a large pressure differential across the orifice.

These tappings usually consist of annular slots that open into piezometric chambers that are also annular (upper part of Fig. 8.10): type 'A' pressure tappings.

The slot giving communication from the annular chambers to the tube must be not more than $0.02D$. It must also be between 5 mm and 1 mm.

If $D > 400$ mm, it is permissible, though not recommended, to use individual pressure tappings (lower part of Fig. 8.10(b)): type 'B' pressure tappings.

The values of α are given by Table 140, the values of ε by Graph 144.

Orifice meters are usually constructed in steel, with diameters of 100 mm and upwards.

(2) Flow nozzle ISA 1932

This is represented in Fig. 8.10(b); it may have any absolute dimensions provided that the lesser diameter is more than 20 mm and that $\sigma \leqq 0.45$. For $\sigma = 0.45$, the detail shown in Fig. 8.11 must be adopted: the upstream surface of the nozzle must be bored out until the entrance diameter is equal to that of the pipe.

The nozzle must be carefully prepared: the diameter, d, of the cylindrical section, must

---

† When a fluid flows through an orifice, there is a value of the differential pressure above which the discharge does not increase when there is diminution of pressure downstream, but only when the pressure increases upstream. This value of the relative differential pressure is known as *critical expansion*:

$$x_c = \frac{p_1 - p_2}{p_1}$$

The flow is said to be subcritical, critical or hypercritical according to whether the relative differential pressure is lower, equal to or higher than the *critical expansion*.

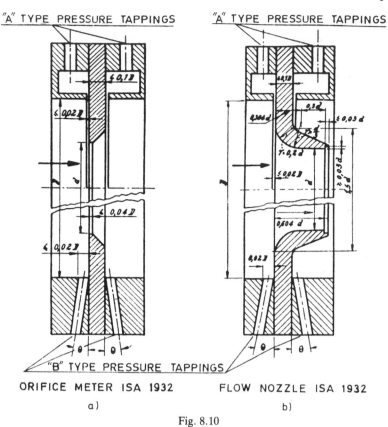

Fig. 8.10

have dimensions with an error of less than $0.001\,D$. Furthermore, in a length equal to $2\,D$, the diameter of the pipe must be within a tolerance of 1%.

The coefficient α is independent of $\mathbf{R}_e$ for sufficiently high values of this parameter, and is then only a function of σ.

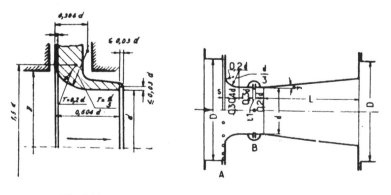

Fig. 8.11.   Fig. 8.12

As regards the pressure tappings, the considerations concerning the orifice meters apply. The values of α and ε are given in Table 136 and Graph 139.

They are usually constructed with a minimum of $D = 200$ mm.

(3) Venturi meters

There are several shapes of Venturi meters. For the one shown in Fig. 8.12, the coefficients of discharge, α, are given in Table 137, these being valid for a Reynolds number in the conduit that is higher than 10^5.

These nozzles are usually constructed with a minimum of $D = 150$ mm.

8.15 Mechanical meters

Mechanical meters are divided into two major groups:

(a) *Velocity meters* — based on measurement of the rotation velocity of a propeller or analogous device that is set in motion by action of the fluid. They consist basically of the following components: *driving system*; *transmission*, normally of the reducer type, which ensures transmission of the movement of the driving system to the measuring and recording mechanism; *recorder*, usually of an integrating type; *housing*, the structure containing the above mentioned components, provided with an entrance tube and an outlet tube, through which passes the fluid to be measured.

(b) *Volumetric meters* — based on the measurement of volumes, in which the driving system consists of a disc, piston or wheel which under the action of the water moves inside a measuring chamber and for each oscillation, translation or rotation propels a certain volume of liquid.

Variants of these two types are *proportional meters* and *compound meters*.

In *proportional meters*, the driving system is installed in a by-pass of the conduit in such a way that only part of the discharge measured passes through it, the total discharge being proportional to that part.

Compound meters consist of two meters, a main one (usually velocity type) fitted in the main conduit and a secondary one (usually volumetric type) fitted in a by-pass. An automatic cut-off valve directs the liquid to be measured, according to the value of the discharge, to one of these two meters; large discharges are measured in the main conduit and small discharges in the by-pass.

On the market there are many makes and models that differ as regards calibre, nominal discharge, service pressure, head loss that they introduce, fitting position, measuring capacity, sensitivity and precision, type of integrator and transmission, etc. Special reference may be made to screw meters provided with a *fitting saddle*, intended for installation in existing conduits with a diameter of at least 100 mm.

Measurement capacity may vary from 1 l/s for meters with a calibre of 15 mm to 1 m³/s for calibre of 800 mm.

Although the mechanical meters referred to are normally only integrating, they can be transformed into meters or recorders of discharges by using suitable accessories.

8.16 Magnetic and ultrasonic flowmeters

Magnetic flowmeters are based on the following measuring principles. When an electrical conductor passes through an electromagnetic field, an electromotive force or voltage is induced in the conductor that is proportional to the velocity of the conductor.

In the case of magnetic flowmeters, the conductor is the flowing liquid being measured.

Their main components are a sensing unit and the metering and recording apparatus where voltage output values are converted into discharge or volume values.

The *sensing unit* is made up of the following principal parts: a *measuring conduit*, which is an insulated, non-magnetic length of the conduit in which the water flows; an *inducer*, consisting of two magnetic coils in diametrically opposite positions in relation to the conduit, in such a way as to create an electromagnetic field normal to the axis of the conduit; *two measuring electrodes*, placed on the walls of the conduit, between the magnetic coils; and a *signal converter*, which is a suitable electrical system that usually amplifies of the signal and converts it into a rate-of-flow indication on a meter.

A source of electrical power is needed to activate the magnetic field and a transmitter is used to record or to transmit the rate-of-flow signals to desired locations. The models available on the market differ in their manufacturing details, particularly in metering characteristics, dimensions, ways in which they are connected to the conduit, material of the tubing, electrodes and conduit linings.

Ultrasonic flow meters are based on the emission of a sound signal that is collected by a receiver at opposite side of the conduit and reflected to the point of emission. The time between emission and reception provides a measure of the velocity of the flow. These meters are very precise and do not introduce head losses.

8.17   Measurement of discharges by means of elbows (elbow meters)

The elbow meter utilizes the pressure difference between the outside and inside of a pipe bend to measure the discharge (Fig. 8.13).

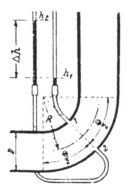

Fig. 8.13

We thus have:

$$Q = \mu K A \sqrt{2g\Delta h} \tag{8.14}$$

in which μ is the coefficient of discharge given by Table 138, K is a coefficient of shape given in the same table, based on the radius of the curve, R, and diameter of the pipe, D; $A = \pi D^2/4$ is the cross-section of the conduit; $\Delta h = h_2 - h_1$ is the difference in pressure between the outside and inside of the pipe bend, both measured on the bisector of the angle.

8.18   Factors to be taken into account in choosing a discharge meter in pressure conduits

Because of their physical arrangement and the delicate functioning of the driving system, mechanical meters have their field of application practically confined to clean water systems.

It would only be possible to apply propeller meters to the measurement of wastewater discharge in conduits of reasonable dimension.

If there is a possibility of the water being measured transporting slimy matter or causing calcareous deposits the use of volumetric meters, especially disc meters, is to be avoided.

Pressure drop apparatus can be used in measuring clean or wastewater provided that in the latter case the sediment load is not too great and the liquid can be considered homogeneous, as usually occurs with wastewater pumped through conduits.

In such a case any of the pressure drop apparatus mentioned can be used, from the classic Venturi meter, which on account of its profile prevents the formation of zones of dead water likely to give rise to deposits of solid particles transported, to orifice meters.†

In any case, in measuring wastewater discharge by means of pressure drop apparatus, it is always necessary to use models with pressure tappings without a piezometric chamber, since these chambers are easily clogged.

Discharges measured by means of pressure drop apparatus must, as far as possible, be constant or subject to slow and insignificant variations; with discharges that are relatively constant, use can be made of horizontal propeller meters, these often being used in pumping conduits; in situations of appreciable fluctuations in the discharge but within the range of small discharges, volumetric methods can be used to advantage; important fluctuations and significant discharges make it advisable to use vertical-axis propeller meters (if head losses are not of great importance) or compound or proportional meters, in those cases in which it is important to restrict head losses.

Magnetic and ultrasonic meters have a general application and tend to substitute the previous kinds since they do not disturb flows.

D ORIFICES

8.19 Definitions
An *orifice*, in the hydraulic sense, is a regular shaped opening in the wall or bottom of a vessel, through which the liquid contained in that vessel leaves it, the contour of the opening remaining completely submerged, i.e. below the free surface. An orifice is assumed to be made with a *thin wall* or *sharp edge* whenever the liquid is in contact only with the inner edge of the orifice.

— The *jet* is the liquid stream that leaves the orifice.
— The *head* is the depth of water that governs the flow.
— An *additional pipe* is an orifice with the walls prolonged to a length of 2 or 3 diameters, or an opening made in a container with sufficiently thick walls.
— *Approach velocity* is the velocity at which the liquid reaches the vessel. Orifice are very accurate arrangements for measuring discharges.

8.20 Small dimension orifices: Torricelli's formula; coefficients of contraction, velocity and discharge
In an orifice of small dimensions in relation to the head, at the bottom of a thin walled vessel the jet has the form shown in Fig. 8.14.

† The most suitable orifice meters for measuring wastewater discharge are eccentric orifices, although they have the drawback of not being standardized.

The jet velocity, in a cross-section in which the streamlines are parallel (*vena contracta* cross-section), is given theoretically by *Torricelli's formula*.

$$V = \sqrt{2g\,(h + \delta)} \tag{8.15}$$

in which $h + \delta$ is the total head over the *vena contracta* cross-section, h being the difference in elevation between the free surface and the plane of the orifice and δ the vertical distance between the plane of the orifice and the *vena contracta* cross-section.

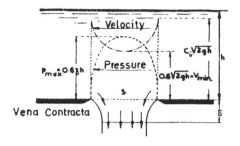

Fig. 8.14

For circular orifices, δ is approximately equal to the radius of the orifice. For high heads, δ may in effect be disregarded. Should the orifice be in a lateral wall, in all cases $\delta = 0$.

At the *vena contracta* cross-section, the streamlines are parallel and the pressure is practically equal to atmospheric pressure. In the cross-section of the orifice the pressure varies approximately between atmospheric pressure† at the edges and a maximum of $0.6\,\gamma\,h$ in the centre; the velocity varies between a minimum in the centre, equal to $0.6\sqrt{2gh}$ and a maximum at the edges, equal to $C_v\sqrt{2gh}$ (Fig. 8.14).

C_v is a coefficient that represents the influence of friction and viscosity and is known as *coefficient of velocity*; its values varies between 0.96 and 0.99. In the *vena contracta* cross-section, the mean velocity will thus be:

$$U = C_v\sqrt{2g\,(h + \delta)} \tag{8.15a}$$

Coefficient of contraction, C_c is the name given to the ratio between the cross-section at the *vena contracta*, A_c, and the cross-section of the orifice A:

$$C_c = \frac{A_c}{A} \tag{8.16}$$

---

† Equivalent to saying that the gauge pressure is zero.

The value of the coefficient of contraction is generally greater than 0.5.

From a practical point of view, the influence of the coefficient of velocity, the coefficient of contraction and distance δ is represented by a single coefficient known as *coefficient of discharge*, also termed the *orifice coefficient*, μ, whose expression is thus:

$$\mu = C_v C_c \sqrt{\frac{h + \delta}{h}} \tag{8.16a}$$

As has been said, for high heads δ is practically negligible.

The discharge that takes place through the orifice in the bottom or walls of reservoirs is always, from a practical point of view, given by the following formula:

$$Q = \mu A \sqrt{2gh} \tag{8.17}$$

in which μ is the cocfficient of discharge, A is the area of the orifice and h is the head over the centre of the orifice.

As an approximate value it can be assumed that for all liquids $\mu = 0.6$, whatever the shape of the orifice; 0.63 and 0.59 can be taken as extreme values and for very low heads the value may even be 0.7.

If it is intended to determine the discharge with great precision, the values used for μ must be those shown in Tables 140 and 141, obtained by various experimenters for given cases.

If there is an approach velocity, U_0, in the direction of the axis of the orifice, the formula for discharge can be written:

$$Q = \mu A \sqrt{2g\left(h + \frac{U_0^2}{2g}\right)} = \mu A \sqrt{2g\,H} \tag{8.17a}$$

8.21 Large dimension orifices

In the case of an orifice of large dimensions being located at the bottom of a reservoir, the preceding formulae are still valid. If, however, an orifice of large dimensions occurs in the wall of a reservoir, it is no longer valid to consider the head, h, as constant in the whole cross-section of the orifice and it will be necessary, from a formal point of view, to carry out an integration in order to obtain the discharge.

In the case of rectangular orifices of width l and height $h_2 - h_1$ (Fig. 8.15), the discharge is given by:

$$Q = \frac{2}{3}\mu' l\sqrt{2g}(h_2^{3/2} - h_1^{3/2}) \tag{8.18}$$

If there is an approach velocity, U_0, we have:

$$Q = \frac{2}{3}\mu' l \sqrt{2g}\left[\left(h_2 + \frac{U_0^2}{2g}\right)^{3/2} - \left(h_1 + \frac{U_0^2}{2g}\right)^{3/2}\right] \tag{8.18a}$$

It is hard to determine the value of the coefficient μ', but it can be taken as $\mu' = 0.60$.

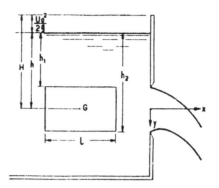

Fig. 8.15

In general, however, in practical cases only equation (8.17) is used, considering that h is the head over the centre of gravity of the orifice, with the corrections being introduced in the value of the coefficient of discharge, which is determined experimentally (see Tables 140 to 146).

8.22 Orifice with incomplete or suppressed contraction
Total contraction does not always occur: if not all the contour of the orifice has a sharp edge, the contraction is said to be *partial*; if the edges of the orifice are very close to the walls or the bottom of the vessel, the contraction is said to be *incomplete* or *partially suppressed*; if the orifice is supported on a wall, it is said that on that side the contraction is *suppressed*.

Tables 141 to 143 give indications on the coefficients of discharge in various cases of partial, incomplete and suppressed contraction.

8.23 Gates
In general, flow through gates is similar to flow through orifices.

The value of μ is given by Graph 144 for sluice gates, by Graph 146 for inclined flat gates and by Graph 145 for sector gates.

The value of h considered is the head above the bottom of the channel and not above the centre of gravity of the orifice.

Some indications of the distribution of pressure are given in Table 146(b).

The coefficients of discharge that are given presuppose that the nappe downstream is suitably aerated.

In the case of a bottom outlet, the discharge with aeration is given by:[†]

$$Q_{ar} = K(\mathbf{F}_r - 1)^n \qquad (8.18b)$$

with $\mathbf{F}_r = V/\sqrt{gh_c} = \sqrt{2H/h_c}$, in which H is the head and h_c the water depth in the *vena contracta* cross-section.

The exponent n takes on values of 0.85 to 1.4, and it can be accepted that $n = 1$. The coefficient K depends on the geometry, but the following values can be used: flat gate in circular tunnel, with well designed grooves, $K = 0.0255 - 0.04$; ditto, with progressive passage from *circular* cross-section to *rectangular* cross-section and again to *circular* cross-section, with well designed transitions and grooves, $K = 0.04–0.06$; poor transitions with detachment of the flow downstream, $K = 0.08$ to 0.12.

8.24   Partially or completely submerged orifices

An orifice is said to be *completely submerged* when the water downstream is at a level that is above the upper edge of the orifice (Fig. 8.16(a)).

The orifice is said to be *partly submerged* when the water downstream is at a level between the upper edge and lower edge of the orifice (Fig. 8.16(b)).

In the case of completely submerged orifices, the discharge is given by:[‡]

$$Q = \mu'A[U_2 + \sqrt{2gh + U_1^2 - U_2^2}] \qquad (8.19)$$

in which μ is the coefficient of discharge, h is the difference in level between upstream and downstream; U_1 is the mean velocity upstream and U_2 the mean velocity downstream.

If velocities U_1 and U_2 are negligible, we have:

$$Q = \mu'A\sqrt{2gh} \qquad (8.20)$$

in which h is the difference in level between upstream and downstream.

Experiments show that the above mentioned values of μ, for non-submerged orifices, are little affected by submersion. According to Weisbach, between the two coefficients there is the relationship $\mu' = 0.986\,\mu$. Graph 144 and 145 enable submerged flow to be studied in the case of sluice or sector gates.

Partially submerged orifices (Fig. 8.16(b)) are considered to be decom-

[†] Kalinske and Robertson, *Entrainment of Air in Flowing Water — Closed·Conduit Flow*, Transactions ASCE **108**. Quoted by [13].
[‡] Application of Bernoulli's theorem and of the expression for Borda head losses.

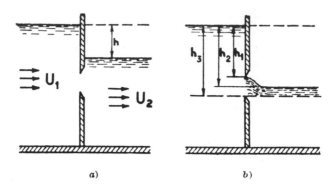

Fig. 8.16

posed into two parts, one of which is free and the other submerged. The discharge will then be:

$$Q = \mu_1 l (h_3 - h_2)\sqrt{2gh_2} + \frac{2}{3}\mu_2 l \sqrt{2g}(h_2^{3/2} - h_1^{3/2})$$ (8.21)

The values of μ_1 and μ_2 are uncertain but can be taken as equal to 0.60.

8.25 Additional pipes

As has been mentioned, the term *additional pipe* is used for a short pipe with a length approximately equal to the distance between the plane of the orifice and the plane of the *vena contracta* cross-section, for a sharp-edged orifice.

   If the shape of the additional pipe is such that it does not touch the form of the flow corresponding to the sharp-edged orifice, then it has no effect on the flow.

   If the shape of the additional pipe exactly fits the form of the jet between the plane of the orifice and the plane of the *vena contracta* section, and if in formula (8.17) the outlet cross-section is considered as equivalent to the *vena contracta* cross-section, then the coefficient of discharge, μ, is equal to the coefficient of velocity, C_v.

   As a rule the shape of the additional pipe alters the form of the flow and the contraction and thus the coefficients of discharge; Table 147 gives the coefficients of discharge for non-submerged additional pipes and Table 148 gives those for submerged additional pipes.

   In the case of internal cylindrical pipes, the coefficient of discharge is about 0.5.

   In the case of external cylindrical pipes, if the flow is established very rapidly (rapid opening of the valve connecting the pipe to the reservoir), the flow may discharge without touching the wall of the pipe and the influence of the pipe in the flow is therefore zero. If, however, the opening takes place slowly, so that the water touches the walls of the pipe or if by some other means it is obliged to adhere to the pipe, then the cross-section of the pipe is completely filled; the flow is highly turbulent and loses its clear appearance. By applying Bernoulli's theorem we find that in the *vena contracta* reductions in pressure are created with a value of about 0.75 h and the coefficient of discharge takes on a value of about 0.8. The pressure reduction can obviously not have values higher than atmospheric pressure, i.e. $0.74\, h < 10$ m water column, so that: $h < 14$ m water column.

For values of h that are higher than, or close to that value, the flow separates from the walls of the pipe and it no longer has any effect.

In the case of divergent external pipes, the pressure reductions generated in the *vena contracta* also increase the discharge. This increase, however, is confined to the maximum admissible reduction, on the one hand, and on the other hand to the maximum angle of divergence, which cannot be greater than 8°, in order to prevent the flow from separating from the wall and flowing freely.

With the nozzles of hoses, the value of the coefficient of discharge varies from 0.95 to 0.98 according to the type of nozzle. Generally it can be taken that $\mu = 0.97$ (consult [5]).

8.26 Culverts

From the point of view of hydraulics, the term *culvert* in this paragraph signifies a short pipe intended for carrying water, normally with a low head.

The discharge can be determined by considering the closed-conduit flow and taking into account the intake, friction and outlet head losses. From a practical point of view, however, it is easier to use equation (8.17) and to represent all such losses by the coefficient of discharge, μ, which is given for some cases in Table 150.

The coefficients are valid provided that at the intake the head over the upper edge is at least equal to the sum of the velocity head and the intake loss. As a rule it can be taken that this sum is equal to $1.5\, U^2/2g$.

8.27 Shape of the jet

In the case of vertical orifices of large dimensions, some phenomena of deformation of the jet can easily be seen.

If the initial cross-section is circular, the jet will tend towards an elliptical cross-section with a horizontal major axis.

In the case of square initial cross-sections, the vein only remains square as far as the *vena contracta*, after which a cross-section of octagonal appearance occurs, with the sides at 45° dominating, so that at a certain distance from the orifice the cross-section is again square, but with vertices displaced by 45° in relation to the orifice.

In the case of rectangular cross-sections much longer than wide, the shape takes on the aspect of a double T: the stem of the double T is parallel to the length of the orifice in the initial part; it is then inverted and the greater dimension becomes normal to the length of the orifice.

A triangular orifice gives a jet that is inverted, taking the shape of a star with three radii perpendicular to the sides of the orifice.

If the axis of the orifice is vertical, directed upward, the maximum water elevation attained will only be equal to the velocity head for low velocities in the *vena contracta*. For higher velocities, the maximum elevation z of the jet is less than the head h over the centre of the orifice, due to the resistance of the air and to the liquid elements that are falling. According to Cappa,[†] the head z is given by the expression:

$$z = \frac{h}{\alpha + \beta h + \gamma h^2}$$

(8.22)

in which the coefficients α, β and γ were determined for circular orifices, convergent conical pipes and conoidal pipes, i.e. pipes whose axial cross-section is given by curves with inward convexity (nozzles of hoses).

The values of α, β and γ are given by Table 149.

In the case of inclined jets, in which the axis makes an angle θ with the horizontal, the theoretical equation of the trajectory, supposing the air resistance to be zero, will be:

$$z = x \tan \theta - \frac{gx^2}{2V^2}(1 + \tan^2 \theta)$$

(8.23)

V being the velocity in the *vena contracta* and x and z the horizontal and vertical axes contained in the plane of the flow, originating in the centre of the *vena contracta*.

The maximum horizontal range, determined experimentally by Freeman[†], is obtained for $\theta = 45°$ when the head varies from 3.5 m to 7 m; for heads of about 10 m, the angle that gives the maximum ranges varies between 34° and 40°; for heads of about 30 m, it varies between 30° and 34°.

8.28 Approximate measurement of discharges by means of the path of the jet

(a) Vertical conduit

In the case of a conduit placed vertically (Fig. 8.17), in which the jet reached an elevation h, the discharge can be determined by *Dawrence and Braunworth's* formula[‡] (1906), which in SI units is written:

$$Q = 5.47 \, d^{1.25} \, h^{1.35} \quad \text{for} \quad h \leqslant 0.37 \, d$$

(8.24)

$$Q = 3.15 \, d^{1.99} \, h^{0.53} \quad \text{for} \quad h \geqslant 1.4 \, d$$

(8.25)

For $0.37 \, d < h < 1.4d$ the discharge is slightly less than that given by either of the formulae.

The error is about 15% to 20%.

(b) Horizontal conduit

In the case of a completely full horizontal conduit (Fig. 8.18(a)), *Greeve*[‡] (1928) established the following expression:

$$Q = c \frac{\pi d^2}{4} \sqrt{g \frac{x^2}{2y}}$$

(8.26)

† Quoted by [6].
‡ Quoted by [12].

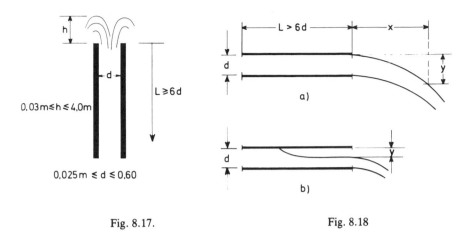

Fig. 8.17.   Fig. 8.18

Arising from the difficulty in measuring x and y, the error is about 10% to 15%.

Graph 151 gives the discharge when y is measured at a distance of $x = 0.305$ m.

For partially full conduits (Fig. 8.18b) with a depth $y_e = d - y$ that is less than $0.56\,d$, the measurement must be made at the outlet from the conduit, the discharge being given by Graph 151.

E. SHARP-CRESTED WEIRS

8.29 Definitions

A weir can be regarded as an incomplete orifice. Weirs are said to be *sharp-crested* weirs, also termed as thin-plate weirs, when the upper edge over which the water flows, i.e. the *crest*, is very narrow (1 to 2 mm — see Fig. 8.19).

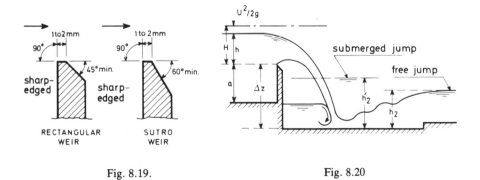

Fig. 8.19.   Fig. 8.20

The term *head* is given to the difference in level between the energy line upstream and the crest of the weir. As a rule, the energy line practically coincides with the free surface far from the fall zone. Bearing in mind Fig. 8.20, we then have $h \approx H$.

Errors resulting from this approximation are corrected by means of the coefficient of discharge.

Sharp-crested weirs are very accurate devices for flow measurement.

8.30   Rectangular weirs without lateral contraction: Bazin weirs

There have been a great many observations on *Bazin weirs*, the name usually given to a rectangular weir without lateral contraction effects, and this therefore allows high accuracy in measuring discharges. Since these end effects are suppressed by the channel side walls, this type of weir is sometimes termed the 'suppressed weir'.

The formula for the discharge, in metric units, is:

$$Q = \frac{2}{3}\mu' l \sqrt{2g}\; h^{3/2} = \mu l \sqrt{2g}\; h^{3/2} \tag{8.27}$$

in which l is the crest length and h is the head.

In the following expressions for μ, a represents the difference in level between the crest of the weir and the bottom of the channel (see Fig. 8.20).

(a) *Bazin* (1898):

$$\mu = \left(0.405 + \frac{0.003}{h}\right)\left[1 + 0.55\left(\frac{h}{h+a}\right)^2\right] \tag{8.28}$$

The limits of application, with errors of 1% to 2%, are 0.08 m $< h <$ 0.20 m; 0.20 m $< a <$ 2.0 m.

(b) *Rehbock*

$$\mu = \frac{2}{3}\left(0.605 + \frac{1}{1050\,h - 3} + 0.08\frac{h}{a}\right) \tag{8.28a}$$

$h > 0.05$ m is recommended.

(c) In 1929, Rehbock presented a simplification of his formula, giving the following, in metric units:

$$Q = \left(1.782 + 0.24\frac{h}{a}\right)l h_e^{3/2} \tag{8.28b}$$

in which $h_e = h + 0.0011$. The values obtained agree with those obtained by formula 8.28a.

(d) *S.I.A.S.* (Societé des Ingénieurs et Architects Suisses — 1947):

$$\mu = \frac{2}{3}0.615\left(1 + \frac{1}{1000\,h + 1.6}\right)\left[1 + 0.5\left(\frac{h}{h+a}\right)^2\right] \tag{8.28c}$$

The limits of application are: 0.025 m $< h <$ 0.8 m; $a > 0.3$ m; $h \le a$.
The Rehbock and S.I.A.S. formulae give practically identical values.

In order to obtain high precision in a rectangular weir some care has to be taken in its construction:

(a) Lateral contraction must be completely eliminated; the channel in which it is installed must therefore have very smooth, perfectly vertical walls and the length of the crest must be exactly equal to the width of the channel.

(b) The weir height must not be very low and the upper edge must be sharpened, as shown in the Fig. 8.19.

(c) The channel upstream must have a length of at least 20 h; the necessary precautions must be taken to ensure that the approach velocities to the weir are uniformly distributed and that the flow is essentially two-dimensional.

(d) Reading of the head must be taken at a distance from the crest that is never less than 5 h or even 10 h.

(e) Aeration must be complete in order to provide atmospheric pressure beneath the nappe. For this purpose, when necessary, air must be provided by vents and the pressure under the nappe must be controlled by means of a manometer. Otherwise the air beneath the nappe will be exhausted, causing a reduction of pressure, Fig. 8.21(a), with a corresponding increase in discharge for a given head; this is known as *adherent nappe*.

When the head increases, the nappe tends to separate from the vertical plate; if, however, the aeration is insufficient, a rarefied, unstable zone is formed under the nappe, which in those conditions is said to be *depressed* (Fig. 8.21(b)).

The nappe will be *free* when the air can circulate easily under it and there is permanent replacement of air brought down by the flow (Fig. 8.21(c)).

Should there be a flow in which, for any reason, it is impossible completely to replace the air removed by the flow, there occurs a rise in the water level below the weir nappe, which in this case is said to be *flooded from below* (8.21(d)).

When the downstream level rises, the weir becomes *drowned* (Fig. 8.21(e)).

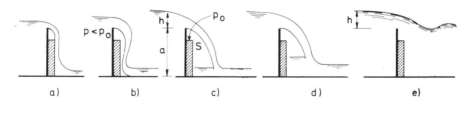

Fig. 8.21

8.31   Ventilation of the nappe

Concerning the air demand and ventilation requirements of sharp-crested weirs, Howe (1955)[†] proposed the following formula, valid per unit length:

$$q_{air} = 0.1 \ q \ (h/h_p)^{3/2} \tag{8.29}$$

in which h is the head over the crest and h_p is the water depth in the pool beneath the nappe, which is a function of the downstream level, of the discharge and of the drop, Δz. If a free jump is formed downstream of the weir, i.e. the flow is not drowned, h_p can be calculated by:

$$h_p = \Delta z \left(\frac{q^2}{g \Delta z^3} \right)^{0.22} \tag{8.30}$$

† Howe, J. W. *et al.*, *Aeration Demand of a Weir Calculated*, Civil Engineering, **25**, 5 (1955). Quoted by [12].

In the case of a submerged jump, $h_p \approx h_2$ (Fig. 8.20).

If p/γ is the pressure reduction admissible under the nappe, expressed in water column, the aeration pipe must have a hydraulic diameter D that satisfies:

$$\left(\frac{p}{\gamma}\right)^2 = \frac{\rho_{ar}}{\rho_{ag}} \left[\frac{fL}{D} + K_e + K_c + K_a\right] \frac{V_{ar}^2}{2g} \tag{8.31}$$

in which: p/γ = admissible pressure reduction, related to the measurement error (see Table 152(b)); f = friction factor given by Moody's diagram (Section 4.5 — in a first approximation it can be taken that $f = 0.02$); L = length of the aeration pipe; D = diameter of the aeration pipe; K_e = coefficient of loss, per intake ($K_e \approx 0.5$); K_c = coefficient of head loss at the bend (it can be taken that, approximately, $K_c = 1.1$); K_a = coefficient of loss at the expansion of the outlet (it can be taken that, approximately, $K_a = 1.0$); V_{air} = velocity of the air.

The error ε_q in the discharge with pressure reduction can be calculated approximately by[†]:

$$\varepsilon_q = 20 \, (p/\gamma h)^{0.92} \tag{8.31a}$$

If the 'ventilation error' is to be limited to 0.2% and one vent is provided at each side, the diameter of ventilation duct being, D_v, the following empirical equation can be adopted [16].

$$D_v = 0.11 \, Hl^{0.5}$$

where all dimensions are in metres.

Example

Calculate the quantity of air required for complete aeration of a weir in the following conditions: length $l = 6.00$ m; depth of water over the crest $h = 0.70$ m; discharge $Q = 6$ m³/s; $h_p = 0.90$ m. The aeration is by means of a steel pipe with a length of 2.5 m.

Equation (8.29) gives the air discharge required for complete aeration under these conditions:

$$Q_{air} = 0.1 \times 6 \times (0.7/0.9)^{3/2} = 0.412 \text{ m}^3/\text{s}$$

Assuming a diameter of 0.30 m for the pipe, from the air discharge of 0.412 m³/s the air velocity is:

$$V_{air} = 0.412/(3.14 \times 0.15^2) = 5.83 \text{ m/s}$$

Table 11 gives $\rho_{air} = 1.22$ kg/m³, so that (formula 8.31):

$$\left(\frac{p}{\gamma}\right)^2 = 1.22 \times 10^{-3} \left[\frac{0.02 \times 2.50}{0.30} + 2.6\right] \frac{5.83^2}{2 \times 9.8} = 0.006 \text{ m}^2$$

and hence:

$$p/\gamma = 0.08 \text{ m}$$

which according to Table 152(b) gives an error of over 2%.

† Johnson (1935), Hickox (1942), quoted by [12].

8.32   Rectangular weir with side contractions

Side contractions takes place when the width L of the channel is greater than the width l of the weir crest (Fig. 8.22).

Among the many formulae proposed for this type of weir, the one chosen is that of *Kindsvarer and Carter* (1957)†, adopted by ISO‡. For application of this formula see Graph 153.

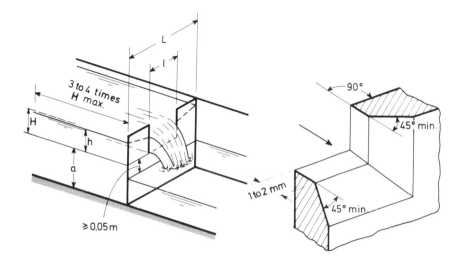

Fig. 8.22

As a practical formula for approximate calculations, Francis's formula is indicated, in S.I. units:

$$Q = 1.83(l - 0.2\text{H})\text{H}^{3/2} \tag{8.32}$$

The head must be measured at a distance of at least 2.0 m upstream of the weir. The extra width, $L—l$, of the channel must be at least equal to $6\,h$.

8.33   V-notch weir (triangular weir)

In a triangular, or V-notch weir, the profile along the crest is a triangle, and as a rule the bisector of the vertex is vertical (Fig. 8.23).

Among many formulae that of Kindsvater (1957) is recommended by ISO. See Table 154.

As a practical formula for approximate calculations, that of *Gourley* and *Grimp* is suggested, in S.I. units:

$$Q = 1.32\ tan\ \frac{\alpha}{2} h^{2.47} \tag{8.33}$$

† Kindsvater, C. E. and Carter, R. W., *Discharge Characteristics of Rectanular Thin-plate Weirs*, ASCE, Journal of the Hydraulics Division, **83**, HY6 (1957). Quoted by [12].
‡ International Standards Organisation.

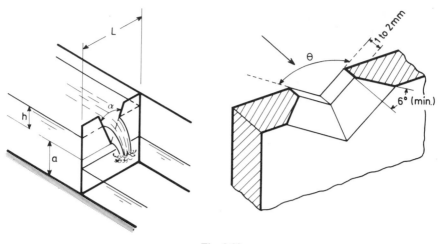

Fig. 8.23

in which Q is the discharge in m³/s; h is the head over the vertex in m and α is the angle of the vertex.

The extra width, $L-l$, must be at least equal to $3l/2$.

For $\alpha = \pi/2$, Thompson's formula is also indicated, in S.I. units:

$$Q = 1.42\ h^{5/2}$$

(8.33a)

8.34 Trapezoidal weir. Cipolletti weir

This is a trapezoidal notched sharp-crested weir with the sides of the notch, not vertical, having sharp edges. That most frequently used is known as the *Cipolletti weir*[†] (Fig. 8.24), in which the sides slope with inclination of 1H:4V. The inclination was to provide additional area at the flanks to counter-balance the reduction due to side-contraction effects, thus in theory, producing a very simple discharge equation.

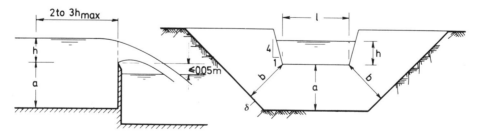

Fig. 8.24

The basic equation for the discharge of a Cipolletti weir is:

$$Q = \frac{2}{3}\mu l\sqrt{2g}\ h^{3/2}$$

(8.34)

† Cipolletti, C. (1886).

in which $\mu = C_d C_v$ and C_d, within the limits of application, is equal to 0.63 and C_v equal to 1.0 in the situation in which the approach velocity is negligible.

Accordingly, we have:

$$Q = 0.42 \, l\sqrt{2g} \, h^{3/2} = 1.860 \, lh^{3/2} \qquad (8.34a)$$

with Q in m³/s, l and h in m.

The limits of application, apart from those indicated in Fig. 8.24, are: $0.06 \, \text{m} < h < 0.60 \, \text{m}$; $h/l \leqq 0.5$; $a > 2h$ with a minimum of 0.30 m; $b > 2h$ with a minimum of 0.30 m.

8.35 Circular weir

The discharge is given by:

$$Q = \mu\phi d^{5/2} \qquad (8.35)$$

with d in dm and Q dm³; ϕ is a function of the degree of filling, h/d, given by Table 155, and μ is given by:

$$\mu = 0.555 + \frac{d}{110 \, h} + 0.041 \, \frac{h}{d} \qquad (8.35a)$$

Hégly's experiments on a weir with a diameter of 1 m and the distance from the lowest point of the weir to the bottom of the channel varying from 0.40 m to 0.80 m, approximately, allowed him to establish the formula:

$$Q = \mu A\sqrt{2gh} \qquad (8.35b)$$

in which A is the wetted area on the weir crest corresponding to the head h; the coefficient of discharge, μ, is given by:

$$\mu = \left(0.350 + \frac{0.002}{h}\right) \left[1 + \left(\frac{A}{A'}\right)^2\right] \qquad (8.35c)$$

in which A' is the wetted area of the approach channel (h is expressed in m).

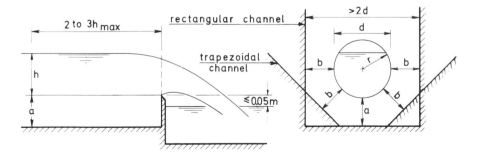

Fig. 8.25

Ramponi proposed a generalization of the preceding formula to a generic diameter d, making:

$$\mu = \left(0.350 + 0.002\frac{d}{h}\right)\left[1 + \left(\frac{A}{A'}\right)^2\right]$$ (8.35d)

Ramponi's experiments were carried out over weirs with a diameter of 100, 200, 300 and 400 millimetres. Calibration of various weirs at LNEC† confirmed the validity of this formula up to diameter of 400 mm. The formula does not seem to be applicable for greater diameters.

8.36 Proportional notch. Sutro weir
In this type of weir, the most common being known as a *Sutro weir* (1908), there is a linear relationship between head and discharge. The profile of the cross-section is defined by a rectangular zone associated with a curved zone (Fig. 8.26). The profile of the curved zone is defined by the equation:

$$\frac{x}{l} = 1 - \frac{2}{\pi}\tan^{-1}\sqrt{\frac{y}{b}}$$ (8.36)

where l and b define the length and height of base rectangle respectively.

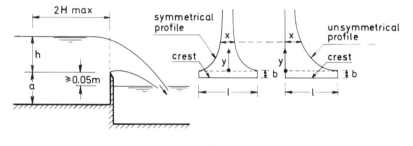

Fig. 8.26

Table 156 gives the values x/l and y/b that correspond to the equation (8.36).

The most current forms of this type of weir are those corresponding to the symmetrical and asymmetrical profiles (Fig. 8.26).

The equation for the discharge is:

$$Q = \mu l\sqrt{2gb}\left(h - \frac{b}{3}\right)$$ (8.36a)

in which μ is the coefficient of discharge (see Table 164).

† Laboratório Nacional de Engenharia Civil, Lisbon.

The application limits are: $h \geqq 2b$ with a minimum of 0.03 m; $x \geqslant$ 0.005 m; $b \geqslant 0.005$ m; $l \geqslant 0.15$ m; $l/a > 1$; $L/l > 3$.

8.37 Choice of a sharp-crested weir

In the description of the various types of weir, their limits of application have been defined. Sharp-crested weirs generally give high precision.

The rectangular weir is more precise ($\varepsilon = 1$ to 2%). For low discharges, however, the precision is attenuated and the sensivity is accentuated, so that below 30 l/s a triangular weir must be used.

Table 157 shows examples of the absolute errors in some cases of rectangular and triangular weirs, related to the error in measuring the head.

The precision of the values measured with a Cipolletti weir ($\varepsilon = 5\%$) is far less than that obtained in equal conditions with rectangular and triangular weirs.

Circular and parabolic weirs have errors due to the formulae ($\varepsilon = 2\%$).

To the errors of formula must be added those due to reading of the head and other errors, in accordance with the theory of errors[†].

8.38 Inclined weir

An inclined weir has a crest perpendicular to the axis of the channel, but the plane that contains it is sloping in relation to the vertical.

Consider the angle of inclination, α (Fig. 8.27); we find that in accordance with Boussinesq's theory the coefficient of discharge of this type of weir is equal to the product of the coefficient of discharge of a vertical weir and a coefficient $K = 1 - 0.3902\dfrac{\alpha}{180}$.

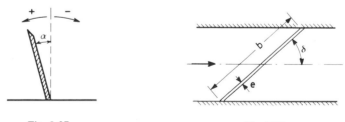

Fig. 8.27.   Fig. 8.28

† It must be remembered that if y is a variable that is a function of several variables x_i, the absolute error ε_a based on y in the absolute errors ε_{ai} of the different variables x_i is given by the expression:

$$\varepsilon_a = \Sigma \frac{\partial y}{\partial x} \, \varepsilon_{ai}$$

The relative error will be:

$$\varepsilon = \frac{\varepsilon_a}{y}$$

8.39 Oblique weir

An oblique weir is a weir in a vertical plane whose crest is oblique in relation to the longitudinal axis of the channel in which it is situated (Fig. 8.28).

The discharge is given by *Aichel's* formula (1953)†:

$$q = \left(1 - \frac{h}{a}\beta\right)q_n \tag{8.37}$$

in which q_n is the discharge by a weir of the same type placed perpendicular to the axis; h is the head; a is the distance from the crest to the bottom of the channel and β a non-dimensional function of the angle δ (Table 166).

8.40 Lateral weir

A lateral weir is placed in the wall of a channel, parallel to its axis.

The shape of the free surface along a lateral weir and the head over the crest are determined by considering that the specific energy relative to the bed remains constant along the channel and that the flow rate is reduced as the discharge takes place.

Accordingly, bearing in mind equation 6.2 and Fig. 6.2, which show the relationship of the water depth h with the discharge Q, for energy H, constant, the water depth decreases in supercritical flow and increases in subcritical flow (Fig. 8.29).

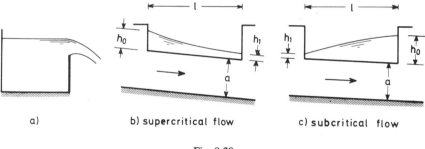

Fig. 8.29

In *supercritical flow* (Fig. 8.29(b)) the flow is controlled from upstream. Thus from depth h_0 upstream (which may be the uniform depth in the case of $I > I_c$ or another depth imposed on the flow), the discharge is calculated by finite differences and from the curve $h(Q)$, with H constant, the water depths in the channel and corresponding discharges are determined.

In *subcritical flow* (Fig. 8.29(c)), the flow is controlled from downstream and the procedure is similar from depth h_0 downstream, which may be the uniform depth if $I < I_c$ or another depth imposed on the flow.

† Quoted by [12].

The discharge can in either case be determined by Dominguez's formula[†]:

$$Q = \phi \mu l \sqrt{2g} \ h_0^{3/2} \tag{8.38}$$

The values of ϕ and μ are given by Table 159 for sharp-crested weirs and for broad-crested weirs with rounded edges and sharp edges.

F.   BROAD-CRESTED WEIRS

8.41   General considerations

The term *broad-crested weir* is applied to a structure over which the streamlines can run parallel (or very nearly parallel) at least over a short distance. Thus it can be assumed that there is in the so-called control cross-section a hydrostatic pressure distribution.

In order to establish this situation it is necessary that $0.08 \leqq H/b\text{-}0.50$ (Fig. 8.30). For lower values of H/b, the energy losses over the weir are no longer negligible; for higher values there will be a curvature of the streamlines and the pressure distribution will not be hydrostatic.

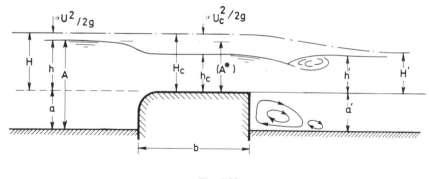

Fig. 8.30

The water level downstream will be such that the flow is not submerged, i.e. critical flow exists on crest of the weir.

If H is the head over the crest, h_c the water depth and U_c the mean velocity in the control cross-section, we shall have $H_c = H$, and:

$$H = h_c + \alpha \frac{U_c^2}{2g} = h_c + \alpha \frac{Q^2}{2gA_c^2} \tag{8.39}$$

By taking $\alpha = 1$:

† Dominguez, J., *Déversoirs Latéraux*, Hidraulica, 1945. Quoted by [19].

$$Q = A_c\sqrt{2g(H - h_c)} \tag{8.39a}$$

In order to take account of secondary effects such as: incomplete parallelism of the streamlines, non-uniform distribution of velocities, etc., a *coefficient of discharge*, C_d, is introduced, this being a function of the shape of the weir crest profile, approach conditions, etc. The value of C_d is given for various shapes of crest by the tables and graphs indicated below.

In practice, however, it is easier to measure the water depth, h, upstream of the weir, rather than the energy, H, which amounts to disregarding the kinetic energy of the flow upstream and thus the approach velocity. A *coefficient of velocity*, C_v, is therefore introduced, its value being given by Graph 160 for various cross-sections depending on the ratio C_dA^*/A in which A is the wetted area in the measurement cross-section corresponding to a depth equal to $(h + a)$; A^* is the wetted area in the control cross-section with a local water depth equal to h_c.

In order to make equation (8.39(a)) easier to use, it is also possible to present the critical depth h_c in the control cross-section, as a function of the energy H and consequently a function of the water depth h upstream, by an equation of the type $h_c = Kh$.

The final formula will thus be as follows:

$$Q = C_d\, C_vA_c\sqrt{2g}\ h\ (1 - K) \tag{8.40}$$

the coefficient of discharge having been taken as $\mu = C_dC_v$.

This formula and its derivatives, which are presented in the next sections, is valid if we have a non-submerged hydraulic jump (see sections 6.16 to 6.21 and Fig. 6.21). If a submerged jump occurs, Graph 113 gives data about the discharge reduction.

Expressions of this general equation for particular cases are given below.

8.42 Rectangular profile weir without lateral contractions
In the case of the rectangular profile weir with a crest width, l, we shall find $A_c = lh_c$ and $h_c = 2H_c/3 = 2H/3 \approx 2h/3$.

Formula 8.40 thus becomes:

$$Q = C_d\, C_v\frac{2}{3}lh\sqrt{2g\ (1 - 2/3)h} \tag{8.41}$$

The value of C_v is given by Graph 168. The value of C_d depends on constructional arrangements as indicated below and are given in Table 161. The width of the channel is represented by l.

(a) *Square-edged broad-crested weir* (Fig. 8.31(a))
This type of weir is standardized (British Standards Institution, BS 3680: 1969). The weir block must be placed in a rectangular channel and the

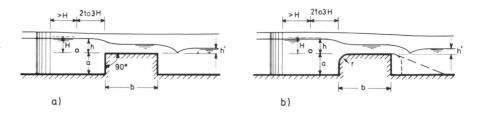

Fig. 8.31

surfaces must be smooth. Special attention must be paid to the upstream edge being well defined and the angle 90°.

(b) *Round-nosed horizontal-crested weir* (Fig. 8.31(b))
Again there is a standardized crest (BS 3680: 1969). The edge that defines the intersection of the upstream face with the horizontal crest is replaced by a rounded nose of radius r, so as to prevent separation of the flow. The downstream face may be vertical, sloping downstream or with the downstream vertex rounded.

The values of C_d are given in Table 161.

8.43   Triangular-throated weir†
Sometimes it is very advantageous to use a triangular control cross-section when it is necessary to measure a wide range of discharges. The discharges will be given with reasonable accuracy even when they are fairly small.

The geometrical data for this type of weir are defined in Fig. 8.32.

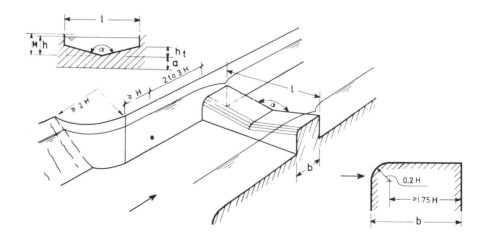

Fig. 8.32

† According to [12].

Returning to equation (8.40) and bearing in mind the geometry of the cross-section, there are two cases:

(a) The water level in the control cross-section does not exceed the depth h_t, at which the triangular cross-section ends and the walls become vertical. By simple algebric operations, equation (8.40) becomes:

$$Q = C_d C_v \frac{16}{25} \left[\frac{2}{5} g \right]^{0.5} \tan\frac{\alpha}{2} h^{2.5} \tag{8.42}$$

valid for $h \leqq 1.25\, h_t$, with $h_t = \frac{1}{2} l \tan^{-1}\alpha/2$.

The value of C_d and C_v are given in Table 162.

(b) If the water level in the control cross-section rises above the triangular cross-section, then, from geometrical considerations, equation (8.40) becomes valid for $h > 1.25\, h_t$:

$$Q = C_d C_v l \frac{2}{3} \left[\frac{2}{3} g \right]^{0.5} \left(h - \frac{1}{2} h_t \right)^{1.5} \tag{8.43}$$

8.44   Straight drop structures

Whenever a free drop, often termed a free-overfall, occurs in a channel, this can be used to measure discharge (Fig. 8.33).

(a)   *Rectangular channel without lateral contraction*

Upstream of the drop cross-section, in a zone where the streamlines are still parallel, the flow is critical, so that $h_c = 2/3\, H_c$; by taking the Coriolis coefficient as $\alpha = 1$, we have $Q = l\sqrt{2g}h_c^{3/2}$ (see critical flow formula, Section 6.6, in which l represents the width of the channel).

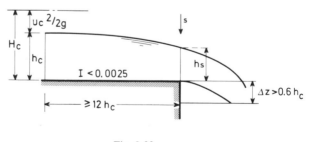

Fig. 8.33

In practice, it is difficult to determine exactly in which cross-section the flow is critical; it is easier to use dividers or a ruler to measure the water depth exactly in the cross-section s at the edge of the free-overfall.

Experiments by Rouse (1936), confirmed by other researchers, showed that $h_s = 0.715\, h_c$. According to [12] the discharge will thus be given by:

$$Q = l\sqrt{g}\left[\frac{h_s}{0.715}\right]^{3/2} \tag{8.44}$$

The error is about 2% to 3% if the value of h_s is measured in the middle of the channel. The sides of the channel must be parallel and the edge of the overfall must be well defined and perpendicular to the flow as must the length of the approach channel. This length must be greater than 12 h_c and its slope, in this reach, should preferably be zero and in anycase not exceed 0.0025. The water depth must be such that $h_s \geqq 0.03$ cm; the drop Δz must be greater than 0.6 h_c. The width of the channel must be $l > 3\ h_s$ and not less than 0.30 m.

(b) *Free drop in trapezoidal channel with rectangular control cross-section*
According to [12] the discharge is given by:

$$Q = C_d\, C_v\, \frac{2}{3}\left[\frac{2}{3}g\right]^{0.5} lh^{3/2} = klh^{3/2} \tag{8.44a}$$

The value of C_d is given by Graph 163.
It is recommended that $0.09 \text{ m} \leqslant h < 0.90$ m; $b = 1.25\ l$; $h'/h < 0.20$, Fig. 8.34.

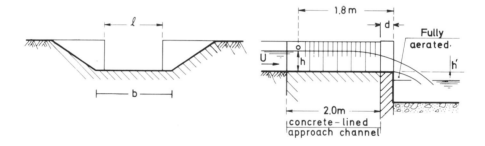

Fig. 8.34

G.   SHORT BROAD CRESTED WEIRS

8.45   General considerations
For the purpose of categorization, the term *short broad-crest* will be applied to the crests of weirs which, although not having sharp edges, are not sufficiently broad for parallel streamlines to be established on them and, consequently, the pressure distribution is not hydrostatic.

For this reason the coefficient C_d is higher than that of a broad-crested weir and the flow may give rise to pressures over the crest that are lower than atmospheric pressure.

In drowned flow conditions the same as above may be said. Attention is likewise called to Sections 6.16 to 6.21, Fig. 6.21 and Graph 144.

8.46   Short broad-crest weirs in rectangular channels without lateral contraction

(a)   *Trapezoidal and triangular profile*

For trapezoidal and triangular profiles, with l the width of the channel, the following formula is valid:

$$Q = \mu l \sqrt{2g}\ h^{3/2} \tag{8.45}$$

for which the values of the coefficients of discharge are given by Tables 164 and 165.

(b)   *Crump weir*

The Crump weir is a particular case of the triangular profile without lateral contraction, with the upstream and downstream slopes of 2/1 and 5/1 respectively (horizontal/vertical), on which Crump[†] carried out systematic tests.
      The discharge is given by the expression:

$$Q = C_d\ C_v \frac{2}{3} [2g]^{0.5}\ l h_e^{5/2} \tag{8.46}$$

The values of C_d and C_v are given by Graph 167; h_e is an equivalent depth that takes into account secondary effects due to viscosity and surface tension. This strict accuracy is only justified in high-precision laboratory measurements and has no application in field measurements.

(c)   *Sundry short profiles*

In general, only a scale model study will enable the coefficient of discharge to be fixed, but an indication is given by Table 174.

(d)   *WES crest*

The WES crest profile studied in 6.3 can also be used to measure discharges.

(e)   *Profile defined by beams*

A profile defined by beams (Fig. 8.35) is classified as follows:

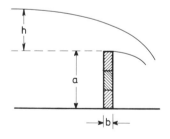

Fig. 8.35

if $h > 2h$ — sharp-edged weir

if $h \leqq 1.5\ b$ — broad-edged weir

if $1.5b < h < 2b$ — weir in which the two types of flow can be observed

---

† Crump, E. S., *A New Method of Gauging Stream Flow with Little Afflux by Means of a Submerged Weir of Triangular Profile*, Proc. Inst. Civil Engineers, March 1952.

According to [19] the formula of the Belgian Mechanics' Society enables the discharge to be calculated in this type of weir:

$$Q = \mu l \sqrt{2g}\, h^{3/2} \tag{8.47}$$

$$\mu = 0.41067 \left(1 + \frac{1.8}{1000\,h}\right) \left[1 + 0.55\left(\frac{h}{h+a}\right)^2\right]\left(0.70 + 0.185\,\frac{h}{b}\right) \tag{8.48}$$

Limits of application are: $0.10\ \text{m} \leqslant h \leqslant 0.80\ \text{m}$; $0.30\ \text{m} \leqslant a \leqslant 1.50\ \text{m}$; $0.030\ \text{m} \leqslant b \leqslant 0.23\ \text{m}$; $h \leqslant a$.

H.   VENTURI FLUMES. PARSHALL FLUMES

8.47   Long Venturi flumes
cause of a similarity to Venturi tubes (see Section 8.14), the term *Venturi flume* is given to a device for measuring discharges in channels by means of a reduction in the cross-section that can be achieved by contraction, raising the bottom or by both means.

Theoretical concepts presented in Chapter 6 related to the discharge equation and performance of spillways (Sections 6.16 to 6.21) are also valid for these flumes.

The coefficients of discharge and other informations concerning their dimensions and methods to minimize the flow measurement error may be obtained by consulting national and international standards (e.g. BS 3680, Part 4C: 1981).

8.48   Parshall flumes
Parshall flumes are of the Venturi flume type and were standardized and calibrated by Parshall (1922).

Table 168 gives the geometric characteristics of such flumes of 25.4 mm (1″) to 15.240 m (50′) [12].

The discharge is given by the expression:

$$Q = Kh^u \tag{8.49}$$

Table 169 shows the limits of application and the values of K and u in the preceding formula [12].

When the ratio h'/h is greater than that given by Table 169, the discharge is reduced, due to submersion, by a value ΔQ as shown in Graph 170 [12].

The probable error is $= 3\%$ for non-submerged flow. In the case of submerged flow the precision diminishes and above a submersion of 95% the values are completely unreliable.

The head loss caused by a Parshall flume is greater than the difference $(h - h')$. Graph 171 gives the head losses for Parshall flumes of 300 mm width and above; for lower values there are no data available [12].

BIBLIOGRAPHY

[1] Addison, H., *Hydraulic Measurements*. Chapman & Hall, London, 1949.

[2] A.S.M.E., *Fluid Meters. Their Theory and Application*, 4th edn., American Society of Mechanical Engineers, New York, 1937.

[3] I.S.A., *Rules for Measuring the Flow of Fluids by Means of Nozzles and Orifice Plates*, Bulletin, 9.

[4] I.S.O., *Proposition de Rédaction d'une Norme Internationale de Mesure de Débit des Fluids au Moyen de Diaphragmes, Tuyères ou Tubes de Venturi*, International Standards Organization, 1954.

[5] Lansford, W., *The Use of an Elbow in a Pipe for Determinining the Rate of Flow in the Pipe*, University of Illinois, Bulletin No. 289, December, 1936.

[6] Linford, A., *Flow Measurements and Meters*, 2nd edn., E. and F. N. Spon, London, 1961.

[7] Marangoni, C., *Prove su Venturimetri Unificati I.S.A.*, La Ricerca Scientifica, Leglio, 1940.

[8] Marchetti, M., *Considerazioni sulle Perditte di Carico Dovute à Bocchelli e Diagrammi di Misura*, Istituto di Idraulica e Costruzioni Idrauliche, Milano, 1953.

[9] Gentilini, B., *Efflusso dalle Luci Soggiacenti alle Paratoie Piane Inclinate e a Settore*, Memorie e Studi dell'Istituto di Idraulica e Costruzioni Idrauliche, Milano, 1941.

[10] Marchetti, M., *Efflusso da Lancie e Bocchelli Anticendi*, Memorie e Studi dell'Istituto de Idraulica e Costruzioni Idrauliche, Milano, 1947.

[11] Rouse, H., *Engineering Hydraulics*, Chapman & Hall Limited, London, 1950.

[12] Bos, M. G. (Ed.), *Discharge Measurement Structures*, IRLI (International Institution for Land Reclamation and Improvement), Wageningen, The Netherlands, 1976.

[13] Levin, L., *Formulaire des Conduites Forcées Oléoducts et Conduits d'Aeration*, Dunod, Paris, 1968.

[14] U.S.B.R., *Water Measurement Manual*, 2nd edn, US Department of Interior. US Government P.O., Washington DC, 1967.

[15] Troskolansky, A., *Theorie et Pratique des Mesures Hidrauliques*, Dunod, Paris (no date).

[16] Ackers, P., White, W., Perkins, J., and Harrison, A., *Weirs and Flumes for Flow Measurement*, John Wiley, New York, 1978.

[17] Schlag, A., *La Normalisation Internationale des Methodes de Mesure de Débit*, La Tribune du Cebedeau, **24**, 327, 1971.

[18] Lefebvre, J., *Mesure des Débits et des Vitesses des Fluides*, Masson, Paris, 1986.

[19] Bos, G. M., Replogle, J. A., and Clemens, A. J., *Flow Measuring Flumes for Open Channel Systems*, John Wiley, New York, 1984.

9

Centrifugal pumps

9.1 Definitions and classification

Pump is the term given to a hydraulic machine capable of raising the pressure of a fluid, that is to say of communicating energy to it.

In *centrifugal pumps*† the increase in pressure is due to centrifugal force provided to the fluid by a *wheel or impeller* that moves inside a *body or casing* which guides the fluid from its entry until its exit.

Regarding the form of the impeller, there are three basic classes:

(1) *Centrifugal pumps, properly speaking, or radial flow pumps* — in these pumps the pressure is mainly developed by action of centrifugal force. The liquid enters axially through the centre and leaves radially by the periphery. If the inlet is on one side only, they are said to be *single pump volute* (Fig. 9.1(a)). If the inlet is through two sides, they are known as

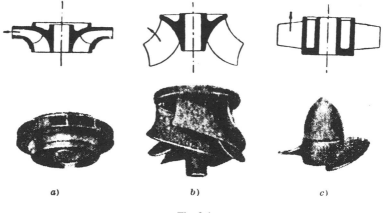

a)   *b)*   *c)*

Fig. 9.1

† Apart from *centrifugal pumps*, there are also *rotary pumps* and *reciprocating pumps*.

double (or twin) pump volute. This type of pump is used mainly for large pumping heads.

(2) *Mixed-flow pumps* — the head is partly developed by centrifugal force, and partly by the lift of the vanes on the liquid. The liquid enters axially and leaves in a direction mid-way between axial and radial (Fig. 9.1(b)). This type of pump is mainly adapted to medium pumping heads.

(3) *Axial-flow pumps* — the head is developed mainly by the propelling or lifting action of the vanes on the liquid. The liquid enters axially and leaves almost axially. This type of pump is well adapted to low pumping heads.

In radial or mixed-flow pumps, the impeller may be *open or closed:* that in Fig. 9.1(b) is open; and that in Fig. 9.1(a) is closed.

Regarding the form of the casing, there are the following basic types:

(1) *Volute or spiral type pump* — designed so that equal velocities will be maintained around the impeller and the velocity of the water will be reduced when it passes to the outlet cross-section (Fig. 9.2(a)).

Fig. 9.2

(2) *Turbine-type pump* — with a cross-section that is constant and concentric with the impeller, which in this case has around it fixed diffuser vanes that guide the flow and reduce the velocity of the water, transforming the kinetic energy into potential pressure energy. Energy is added to the liquid in a number of impulses. (Fig. 9.2(b)).

Regarding the number of impellers, pumps are classified into:

(1) *Single stage,* when there is only one impeller.
(2) *Multistage,* when there is more than one impeller. In the case of multistage pumps, they may be connected in *series* or in *parallel* (see Sections 9.12 and 9.11). There may also be a series of stages linked in parallel, in which case the connection is mixed.

In terms of the sense of rotation, pumps are classified as *direct-sense or inverse-sense*.

In order to determine the sense of rotation of a horizontal-axis pump, it is necessary to look at the pump from the side of the motor that drives it: if the axis turns from left to right (clockwise) the pump is an inverse-sense pump, otherwise it is a direct-sense pump. In order to determine the direction of rotation of a vertical-axis, it must be looked at from above.

Regarding the position of the axis, pumps are classified into *horizontal-axis, vertical-axis or inclined-axis* pumps.

Vertical-axis pumps may have: a *suspended casing,* which allows submerged installation, the pump being suspended from a vertical pipe that acts as discharge piping and inside which passes the shaft of the pump, *or a supported casing,* in which case it is installed in a dry chamber.

Suspended-casing pumps may have only the impeller submerged and the motor connected to the shaft, outside the water, or the motor may also be submerged.

Centrifugal pumps are characterized by the *discharge,* by the *pumping heads*, by the *net positive suction head,* by the *power* and *efficiency*, by the *rotation speed* and by the *specific speed.*

9.2 Discharge

Discharge is the volume of liquid pumped per unit time. It is usually measured in cubic metres per second or in litres per second. It will be represented by Q.

9.3 Pumping head

Pumping head is the increase in pressure that the pump can communicate to the fluid. It is usually expressed in metres of column of the liquid or in Newtons per square centimetre. It will be represented by H.

With reference to Fig. 9.3, the following symbols will be used:

Z_1	— Elevation of the free surface in the supply tank.
Z_2	— Elevation of the free surface in the delivery tank. (Should the free surfaces of the delivery or supply tanks not be at atmospheric pressure, Z_1 and Z_2 would be the elevations reached by the liquid in piezometric tubes placed in the tanks.)
Z_b	— Elevation of the axis of the pump, in the case of horizontal-axis pumps: elevation of the intake cross-section, in the case of vertical-axis pumps.
Z_a	— Elevation of the energy line at the suction flange of the pump.
Z_c	— Elevation of the energy line at the discharge flange of the pump.

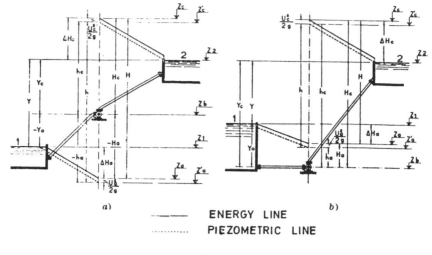

———— ENERGY LINE
············ PIEZOMETRIC LINE

Fig. 9.3

Z'_a	— Elevation of the hydraulic grade line at the suction flange.
Z'_c	— Elevation of the hydraulic grade line at the discharge flange.
$Y_a = Z_1 - Z_b$	— *Static suction head*. This will be negative in Scheme (a), positive in Scheme (b). If negative it is also termed the *static suction lift*.
$Y_c = Z_2 - Z_b$	— *Static discharge head*.
$Y = Z_2 - Z_1$	— *Total static head*.
$h_a = Z'_a - Z_b$	— *Total suction head:* reading off a manometer fitted on the intake flange of the pump and relating to the datum of elevation Z_b, expresed in liquid column. This will be negative (lift) in Scheme (a) and positive in Scheme (b).
$h_c = Z'_c - Z'_b$	— *Discharge pressure head:* reading off a manometer fitted on the discharge flange and relating to datum Z_b expressed in liquid column.
$h = Z'_c - Z'_a$	— *Total discharge head:* reading off a differential manometer between the inlet and discharge flanges, expressed in liquid column.
U_a	— Velocity in the inlet cross-section of the pump.
U_c	— Velocity in the discharge cross-section of the pump.
$\dfrac{U_a^2}{2g}$	— Velocity head in the inlet cross-section.
$\dfrac{U_c^2}{2g}$	— Velocity head in the discharge cross-section.

$$H_a = Z_a - Z_b$$
$$= h_a + \frac{U_a^2}{2g}$$

— *Total suction head*.† This is negative in Scheme (a), positive in Scheme (b).

$$H_c = Z_c - Z_b$$
$$= h_c + \frac{U_c^2}{2g}$$

— *Total discharge head.*

$$H = Z_c - Z_a$$ — *Total head.*
$$\triangle H_a = Z_1 - Z_a$$ — Local and friction energy losses in the suction piping.
$$\triangle H_c = Z_c - Z_2$$ — Local and friction energy losses in the discharge piping.

The following algebraic relationships occur:

$$Y = Y_c - Y_a \tag{9.1}$$

$$h = h_c - h_a \tag{9.2}$$

$$H = H_c - H_a \tag{9.3}$$

If $\quad U_a = U_c \tag{9.3a}$

then we also have $h = H$.

9.4   Net positive suction head

The net positive suction head (npsh), is the difference between the total suction head, *related to absolute pressure*, and the vapour pressure of the liquid and is represented by H_0.

We thus have, in metres:

$$H_0 = p_0 + H_a - h_v \tag{9.4}$$

in which:

p_0 — Atmospheric pressure, in m.
H_a — Total suction head in m; negative in Scheme (a), positive in Scheme (b) (Fig. 9.3).‡
h_v — Vapour pressure in m.

† The term *total suction lift* is used in the case of Scheme (a); *total suction head* in Scheme (b).

‡ Note that the suction friction head includes the friction in the pipe and in all fittings in the suction line.

If p_0 and h_v are expressed in N/m^2, we then have, taking γ to represent the specific weight in N/m^3:

$$H_0 = \frac{p_0 - h_v}{\gamma} + H_a \tag{9.4a}$$

It is always necessary that $H_0 > 0$, whence in Scheme (b) $|H_a| < p_0 - h_v$: in this way cavitation phenomena are avoided in pumps and they are prevented from becoming de-primed, for excessive values of H_a. In fact H_a is theoretically limited to atmospheric pressure and in practice can never exceed about 8 m water column.

Examples
(1) A pump delivers water at a temperature of 4°C and at an atmospheric pressure 760 mm of mercury ($p_0 = 10.13 \ N/cm^2$; $\gamma = 9810 \ N/m^3$). The reading on a manometer in the suction piping, reduced to the axis of the pump, is $h_a = -2.00 \ N/cm^2$†; the velocity in the suction cross-section is $U_a = 1.5$ m/s. Determine the net positive suction head, H_0.
    We shall have:

$$h_a = -2.00 \ N/cm^2 = -1.96 \text{ m of water column}$$

$$\frac{U_a^2}{2g} = \frac{1.5^2}{2g} = 0.15 \text{ m } H_a = h_a + \frac{U_a^2}{2g} = -2 + 0.115 \approx -1.89 \text{ m}$$

$$p_0 = 10.13 \ N/cm^2 = \frac{101\ 300}{9\ 810} \text{ m} = 10.33 \text{ m of water column}$$

so that: $h_v = 0.08$ m (Table 9)

$$H_0 = 10.33 - 1.89 - 0.08 = 8.36 \text{ m}$$

(2) A pump delivers hot water at 80°C ($\gamma = 9533 \ N/m^3$ and $h_v = 4.83$ m, Table 9) at an elevation of 1000 m ($p_0 = 8.93 \ N/cm^2$, Table 11). The reading on a manometer, in the suction piping, reduced to the axis of the pump, is $h_a = 1.00 \ N/cm^2$‡. The velocity in the suction cross-section is $U_a = 1.5$ m/s.
Determine the net positive suction head H_0.
We shall have the following:

$$h_a = 1.00 \ N/cm^2 = \frac{10\ 000}{9533} = 1.05 \text{ m of water column at 80°C}$$

$$\frac{U_a^2}{2g} = 0.11 \text{ m}$$

$$H_a = h_a + \frac{U_a^2}{2g} = 1.05 + 0.11 = 1.16 \text{ m of water column at 80°}$$

$$p_0 = 8.93 \ N/cm^2 = \frac{89\ 300}{9\ 533} = 9.4 \text{ m of water column at 80°C}$$

$$h_v = 4.83 \text{ m}$$

† Gauge pressure.
‡ Gauge pressure.

so that:

$$H_0 = 9.4 + 1.16 - 4.83 = 5.73 \text{ m}$$

It should be noted that although in the suction piping the relative pressures are higher in Example (2) than Example (1), the net positive suction head is less in Example (2) than in Example (1). Accordingly, in hot water installations ($T > 70°C$) or installations with very volatile liquids, it is always necessary to use arrangements of Type (b) (Fig. 9.3) in order to prevent cavitation or to avoid de-priming of the pump.

9.5 Power and efficiency
The definitions are as follows:
— *Effective power of the pump*, P_u: power corresponding to the work done by the pump. Taking Q as the discharge in m³/s and H as the total head in metres, we have, with γ expressed in N/m³:

$$P_u = \gamma Q H \text{ Watt} = \frac{\gamma Q H}{1000} \text{ kW} \tag{9.5}$$

— *Power absorbed by the pump*, P_a: power delivered to the shaft of the pump.
— *Efficiency of the pump*, $\eta = \dfrac{P_u}{P_a}$: ratio between the effective power and the absorbed power.

We thus have:

$$P_a = \frac{\gamma Q H}{1000 \times \eta} \text{ kW} \tag{9.5a}$$

In an analogous manner there will be definitions of the effective power of the motor, P_u', power absorbed by the motor, P_a' and efficiency of the motor, η'.

In the case of rigid transmission, $P_a = P_u'$. In the case of belt transmission, taking η'' as the efficiency of the transmission, $P_a = \eta'' P_u'$.

Given below as examples are the values for efficiency of centrifugal pumps, such efficiency being considered as good.

Characteristics	High pressure		Low pressures			High discharge		
Q(l/s)	3	25	2	25	100	150	1000	2000
η	0.56	0.78	0.53	0.81	0.84	0.86	0.90	0.91

Efficiency is, however, far lower for liquids of high or even moderate viscosity. In these cases there is an increase in absorbed power, a head reduction and some reduction in the discharge. Whenever it is necessary to know these quantities in greater detail, tests must be carried out beforehand with the particular liquid to be pumped.

9.6 Rotation speed

The rotation speed, n, is the number of rotations made by the impeller per unit time. Its value has an appreciable influence on the functioning of the pump. The following general rules apply:

(1) The discharges, Q, in a first approximation, can be considered as proportional to the rotation speed, n:

$$\frac{Q}{Q_1} \approx \frac{n}{n_1} \tag{9.6}$$

(2) The heads, H, in a first approximation, vary proportionally to the square of the speed:

$$\frac{H}{H_1} \approx \left(\frac{n}{n_1}\right)^2 \tag{9.7}$$

(3) The absorbed power varies, in a first approximation, proportionally to the cube of the speed:

$$\frac{P}{P_1} \approx \left(\frac{n}{n_1}\right)^3 \tag{9.8}$$

(4) The efficiency under operating conditions is practically independent of the rotation speed.

We thus have, approximately:

$$\frac{P}{P_1} \approx \left(\frac{H}{H_1}\right)^{3/2} \approx \left(\frac{Q}{Q_1}\right)^3 \approx \left(\frac{n}{n_1}\right)^3 \tag{9.9}$$

9.7 The 'hill diagram'

The surface that relates the efficiency, η, to the discharge, Q, the pumping head, H, and the rotation speed has the shape of a *hill*. The point of maximum efficiency corresponds to the nominal characteristics of the pump: Q_0, H_0, P_0, n_0. Representation of the variation of η with Q and H takes on the appearance of Fig. 9.4, which is shown in nondimensional form. The same graph shows a family of the curves $H(Q)$ in which the defined parameter is the rotation speed n.

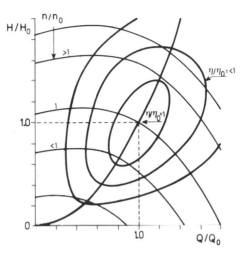

Fig. 9.4

We can write:

$$H = a n^2 + b \, n \, Q + c \, Q^2 \tag{9.10}$$

in which a, b and c can be determined from three known points of the curve $H(Q)$.

This combination is commonly referred to as '*hill diagram*'.

In order to resolve a specific case, it may not be possible to find standard manufactured pumps whose nominal characteristics coincide with requirements. The actual operating conditions may vary, either due to variations of manometric head (the most usual case) or because of variations in the discharge.

For each pump, operating conditions are therefore defined, these being also indicated by the manufacturers, so as to maximise efficiency, prevent cavitation and ensure stability.

9.8 Specific speed

Specific speed, n_s is the speed in revolutions per unit time at which a geometrically similar pump would run to deliver unit discharge when a total head is equal to unity.†

---

† It is also possible to define as the specific speed the revolution per unit time made by a geometrically similar pump that develops unit power when the total head is equal to unity.

The expression for n_s will be:

$$n_s = n \, \frac{\sqrt{Q}}{H^{3/4}} \tag{9.11}$$

The specific speed is the same for all similar pumps, and for any given pump does not change with the speed.

When the specific speed is used to characterize a 'type' of pump, it must be calculated for the point of optimum efficiency.

For a multistage pump, the pumping head of each stage is taken for calculating the specific speed. Also for the specific speed of a double-suction pump, the discharge of the former must be divided by 2 or its specific speed divided by $\sqrt{2}$.

Examination of the expression (9.10) shows that for the same conditions of head and discharge, pumps of higher specific speed are more rapid and therefore smaller; also for the same rotation speed and the same discharge, pumps of higher specific speed will function with a smaller pumping head, or for the same speed and same head they will function with higher discharges.

The numerical value of n_s varies with the units chosen for n, Q and H.

In S.I. units, Q is expressed in m³/s and H in metres; n should be given in revolutions per second. It is, however, current practice to express n in revolutions per minute and that is the way in which the following values are presented, considering that:†

$$n = \frac{n(\text{r.p.m.}) \, \sqrt{Q(\text{m}^3/\text{s})}}{H^{3/4}(\text{m}^{3/4})} \tag{9.11a}$$

$$H_0 = p_0 - h_v + H_a' = 9.9 + H_a' \tag{9.12}$$

In general for n_s less than 70 or 80, pumps are *single-suction radial-flow pumps*; for n_s between 70 and 80 up to about 120, they are *double-suction radial-flow or mixed-flow pumps*; for values of n_s between 120 and 150 up to 170, they are *mixed-flow pumps*. For higher values of n_s they are *axial-flow pumps* (Fig. 9.1(c)). As a rule the latter are used in conditions with high discharge and low pumping head. This classification, however, is not a strict one.

† In American specialized literature it is taken that:

$$n = \frac{n(\text{r.p.m.}) \, \sqrt{Q(\text{U.S.gls/min})}}{H^{3/4}(\text{ft})^{3/4}} \tag{9.11b}$$

In order to convert n_s, expressed thus, to n_s as expressed in formula (9.11(a)), divide by 51.648.

9.9   Suction limits

The specific speed is the parameter which, together with the net positive suction head is the best guide to check the suitability of a pump chosen from the point of view of suction.

A pump of low specific speed will operate more effectively with large pumping capacities than another pump of higher speed.

If the suction head is very great, it is sometimes necessary to use slower and larger pumps; on the other hand, with low suction heads or with positive suction heads, speed must be increased and a cheaper pump can be used.

An increase in speed, without proper suction conditions, causes disturbances in the functioning of the pump. Account must be taken of the references and information given by the manufacturers.

9.10   Characteristic curves. Operating point

Characteristic curves show the interrelation of pump head, H, capacity, Q, power, P, and efficiency, η, for a specific impeller diameter and casing size.

It is usual to plot total head, power, and efficiency against capacity at constant speed (Fig. 9.5).

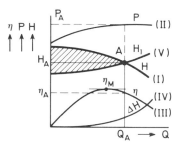

Fig. 9.5

Curve (I)   — Total head, H, versus discharge
Curve (II)   — Absorbed power, P, versus discharge
Curve (III) — Efficiency, η, versus discharge

Apart from the characteristic curves of the pump there are also *characteristic system-head curves*. Still bearing in mind Fig. 9.5, we have:

Curve (IV) — Total head loss $\triangle H$ (friction plus local head losses, in the system), versus the discharge. These head losses are determined according to the indications given in Chapter 4, referring to closed conduit-flows

Curve (V)   — Curve $H_1 = Y + \triangle H$, versus the discharge. This curve is obtained from the previous one, by adding to $\triangle H$ the term Y, corresponding to the total geometric head

When the characteristic curves have been plotted, the *operating point of the pump* is obtained from intersection of curve (V) with curve (I); to this point, A, defined by (H_A, Q_A), corresponds a given absorbed power, P_A and a given efficiency, η_A.

If curve (I) is situated entirely under curve (V), the pump is unsuitable for the installation defined by (V), and will therefore give no discharge.

The following general observations are also made (see [2]).

(1) The point corresponding to $Q = 0$ on curve (I) characteristic of the pump (shutoff head), must be higher than the point corresponding to $Q = 0$ of curve (V) characteristic of the system.

(2) The area between the vertical axis and curves (I) and (V) (shaded area of Fig. 9.5) must be as large as possible, compatible with a good efficiency.

(3) The greater the radius of curvature of curves (I) and (V), the greater will be the functioning stability of the pump.

(4) The operating point must be situated slightly beyond the point corresponding to the point of maximum efficiency, η_M, in order to take account of an eventual diminution of discharge due to ageing of the installation.

According to [2], Fig. 9.6 shows, qualitatively and as an example, the typical forms of

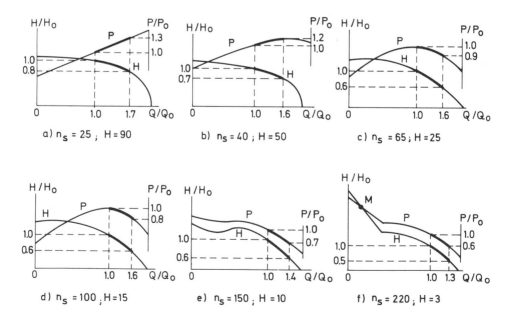

Fig. 9.6

curves $Q(H)$ and $P(H)$, taking as parameter the specific speeds. Q_0, H_0, P_0 are taken to represent the values corresponding to the optimum functioning point, that is, maximum efficiency. On each curve a thick line is used to mark a functioning zone in which efficiency does not drop more than 5% below the maximum value.

The possible shapes of the characteristics of the pumps suggest the following deductions (see Fig. 9.7).

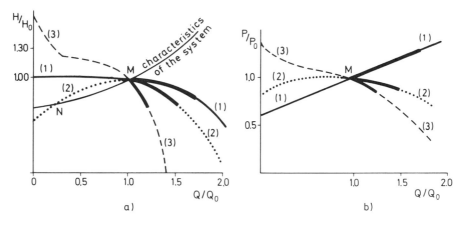

Fig. 9.7

(1) In terms of operating *stability*, if the characteristic curve of the system cuts curve $H(Q)$ at only *one point*, M, operation *will be stable* (curves 1 and 3). If the characteristic curve of the system cuts curve $H(Q)$ at *two points*, N, and M — (curve 2) — only the operating condition corresponding to point M is stable: in effect, at point N, if the discharge increases, H also increases, which causes an increase in discharge of the system and so on, successively, until the operating point moves to M; on the other hand, if the discharge of the pump falls below that corresponding to point N, the pumping head becomes lower than that required by the system; this further reduces the discharge, and the pump falls to point $Q = 0$, from which it can only depart by means of the special arrangement referred to below. Performance is also unstable if the two points M and N are very close together, or if the characteristic of the system is tangent or nearly tangent to curve $H(Q)$.

(2) Regarding *starting up*, this will take place without any difficulty in (case 1), with the discharge valve closed ($Q = 0$), corresponding to the minimum power value; on the other hand, in case 3, start-up must take place with the valve open since the power grows when the discharge decreases.

In (case 2), start-up is impossible, since the pumping head, for $Q = 0$, is higher than that which the pump achieves for the same discharge. In this case, start-up is only possible by creating a by-pass between the discharge (before the valve) and the intake of the pump; this starts with the by-pass open and, therefore, with a lesser static head: then the discharge valve is opened and the by-pass valve closed.

(3) With respect to *efficiency*, curve $\eta(Q)$ is relatively flat for radial pumps; for axial pumps the acceptable efficiency zone is very small and this calls for special care in choosing the operating point of the pump. The maximum possible efficiency value usually decreases with n_s.

9.11   Operation of pumps in parallel

Suppose that several pumps, for example three, are operated in parallel (Fig. 9.8) and their curves, $H = f(Q)$ are (1), (2) and (3), respectively. Curve

H, corresponding to the combination (curves $(1) + (2) + (3)$) is obtained by adding the individual abscissae of the curves, H. The operating point, A, is the intersection of the sum curve, thus obtained, with the curve H_1 characteristic of the system.

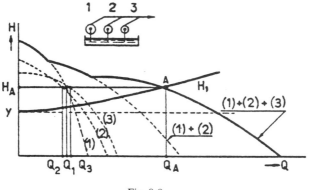

Fig. 9.8

The discharge delivered by each pump is obtained by plotting the horizontal line that passes through A and finding its intersection with curves (1), (2) and (3). The total discharge will be the sum of the discharges of the various pumps; that is, when pumps are operated in parallel the performance is obtained by adding the capacities at the same head:

$$Q = Q_1 + Q_2 + Q_3 \qquad (9.13)$$

The total head is equal to H_A for each pump and for the whole.

When connected in parallel, the total discharge is always lower than the sum of the discharges of each pump operating separately.

When the curve $Q(H)$ of the pumps is of type (b) shown in Fig. 9.7(a), problems may arise in parallel operating. In fact, bearing in mind Fig. 9.9(a), it may happen that the characteristic $Q(H)$ of the whole is still less

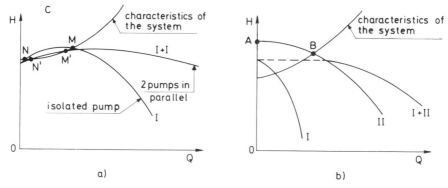

Fig. 9.9

favourable than the characteristic of that pump when isolated. The operating point MN of a single pump may in fact give rise in parallel operating to the nearest $M'N'$, and this increases instability even further.

As an extreme case, it may happen that the resulting characteristic no longer cuts the characteristic of the system, i.e. operation of the second pump causes a failure of the system as a whole. For these reasons when pumps have characteristic curves of this type they must only be used whenever there is full knowledge of the conditions in which they are going to operate.

Fig. 9.9(b) illustrates another parallel operation under poor conditions. Between A and B only one pump is operating; the other is idling (without discharge) and this may harm the pump by its overheating.

9.12   Operation of pumps in series
In this case, curve H of the combination is obtained by adding up the ordinates of the curves H corresponding to each pump (Fig. 9.10).

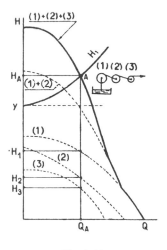

Fig. 9.10

The operating point, A, is obtained by the intersection of this with the curve H_1 characteristic of the pumping system.

The discharge of all the pumps is, of course, the same and is equal to Q_A. The heads relating to each pump are H_1, H_2 and H_3. For pumps in series the performance is obtained by adding the heads at the same capacity. It must be noted, however, that the operation of pumps in series may cause serious difficulties; in effect, pump (3) will naturally be constructed to take a pressure that will be lower than the pressure to which it is in fact subjected.

Instead of pumps operating in series it is therefore preferable to use multistage pumps.

In fact, a multistage pump is a connection in series of various impellers fitted in the same casing; the characteristic curves are obtained in an analogous way to that previously described, from the characteristic curve of each impeller. The efficiency of the pump usually increases with the number of stages, up to a certain limit, which is explained by the proportional reduction of the friction in the bearings that is approximately the same for one or more stages. Above a certain number of stages this reduction is no longer significant.

It may, however, happen that the pumps have to be operated in series, *separated from one another* as shown in Fig. 9.11 . This may occur in order to reduce pressure in the pipeline, which would be higher if a single pump were used (Fig. 9.11(a)) or in order to increase the carrying capacity of an existing pipeline (Fig. 9.11(b)).

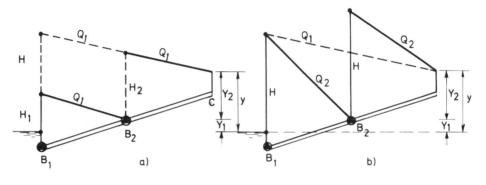

Fig. 9.11

In the first case, maximum pressures, H_1 and H_2, are obtained in the conduit that are lower than the pressure H corresponding to the use of a single pump to carry the same discharge Q_1.

In the second case, by inserting pump 2 it is possible to increase the discharge from its initial value, Q_1, to a value $Q_2 > Q_1$.

However, the operation in series of pumps that are situated apart requires special attention to the geometrical heads, y_i, of each pump.

Consider, for example, the case of two pumps with characteristic curves B_1 and B_2 that are different and two lengths of piping upstream and downstream of B_2, with characteristic curves I_1 and I_2 that are also different. The total hydrostatic head, y, is known.

As has been explained, the value of the discharge then comes from the intersection of the characteristic curve of the series $(B_1 + B_2)$ with the characteristic curve of the pumping system $(y + I_1 + I_2)$.

Once the discharge is known, we know the pressure head of pump H_1, corresponding to an elevation head $y_1 = H_1 - I_1$.

In practice, the curve I_1 moves according to axis H until it intersects curve B_1 at the point corresponding to discharge Q and the value of y_1 is read off axis H.

The procedure is the same for y_2 or y_2 is calculated by simple difference: $y_2 = y - y_1$.

9.13 Mixed connection

The combination of pumps in parallel and series is analysed by plotting the curve $H = f(Q)$ for each series, as shown in Section 9.12, and adding the abscissae of those curves, as shown in Section 9.11. A common case is the connection in parallel of several multistage pumps.

The combination in series of pumps operating in parallel is analysed by first obtaining the curves corresponding to the various connections in parallel, as shown in Section 9.11, and then making the series, as explained in Section 9.12.

9.14 Starting-up conditions

The power P of a rotary machine is given, as we know, by $P = nB$, with n being the number of revolutions and B the torque. In SI the units are: n, revolutions per second (rev/s); and B, Newton metre (Nm). We shall thus have P in Nm/s.

Let us see how n and B vary during the start-up until the normal operating speed of the pump is reached, in principle the nominal speed, n_0.

During the start-up period, the start-up torque must be sufficient to overcome friction in the bearings, stuffing, etc., and the inertia of the pump in order to increase the power of the pump during acceleration, until it reaches the normal value that is considered to coincide with the nominal value P_0.

As we have seen previously, in radial pumps the power is minimal for zero discharge; in axial pumps the power is a maximum for zero discharge. It is therefore convenient for the former to start-up with the valve closed ($Q = 0$) and the latter with the valve open.

Fig. 9.12 shows two typical curves of the relationship between (B/B_0) and (n/n_0) during the start-up of radial-flow pumps.

Curve (I) corresponds to starting-up with the valve closed: the start-up torque, B, grows to about 0.55 of its nominal value, B_0, until the pump reaches speed $n = n_0$; the maximum power at start-up corresponds to the value of the power of the characteristic curve $P(H)$ that occurs for $Q = 0$; the valve is then opened, the flow starts, the discharge rises to the normal discharge, Q_0, and the same occurs with the power, which reaches its value P_0.

Curve (II) corresponds to start-up of the same pump with the valve open. The velocity increases to to a value n_i, when H reaches the value y of the elevation head at which pressure is built up and allows the discharge to begin (Fig. 9.12(b)); the speed then increases to n_0, up to the operating point (H_0,

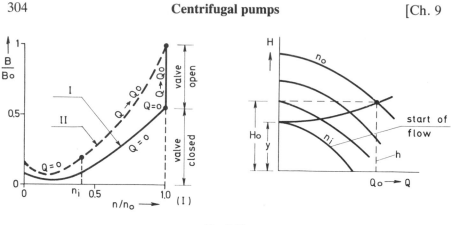

Fig. 9.12

Q_0). The start-up torque is greater than that which would occur if the valve was closed.

In an axial pump, on the other hand, the starting torque with a closed valve would be far greater than with the valve open and would be even greater than the torque corresponding to nominal operating conditions by as much as 30 to 40%. Start-up must therefore take place with the valve open, this phenomenon being qualitatively similar to that described previously for the case of the radial pump with open valve.

The recommendation of the manufacturers should always be respected.

9.15 Motor conditions

The motor driving a centrifugal pump must not only comply with the nominal operating conditions, but also satisfy the starting conditions. This means that the starting torque of the motor must always be greater than the starting torque of the pump. For stationary applications, alternating-current motors are the most common choice for pump drives.

A three-phase motor, as we know, can start-up with star (Y) or delta ($\triangle$) connections. The star-delta starting is begun with the motor connected in star up to a point at which the torque of the motor approaches the torque of the pump, although always greater.

At this point it is changed over to the delta connection and the pump reaches its nominal speed.

Start-up in delta (on-line start-up) absorbs a current that is far higher than the nominal value, although its lasts only a short time. The recommendations of the manufacturers should always be respected.

For high-power pumps, specialists must always be consulted.

9.16 Installation, operation and maintenance

If a centrifugal pump is installed in proper conditions and care is taken with its operation and maintenance, it lasts a fairly long time and as a rule is free of breakdowns. Given below are the main points to be borne in mind, for the most common pump.

(1) The pump must be installed as close as possible to the liquid to be pumped, in order to

avoid large suction heads. In the suction piping it is advantageous to avoid bends and any pipe fittings that increase head losses.

(2) The pump must be protected against flooding. The electric motor must be installed in a dry location (except in the case of submersible pumps).

(3) The foundation must be strong enough to ensure proper alignment of the electric pump unit.

(4) The pump unit must be positioned and levelled before it is connected to the suction and discharge piping. These must be made to fit into the respective flanges without any force being exerted on the pump, in such a way that the bolts are used solely for tightening the sealing. This means that the suction and discharge piping must have supports that are very close to the pump.

(5) In the discharge pipe there must be an operating/control valve that is used for starting and stopping. Particularly in radial-flow pumps, it is advisable to close the valve before stopping the motor, especially when the pump is subject to a high discharge head.

(6) Between the control valve and the pump there should be a convenient check valve or similar device for protecting the pump in case the motor stops suddenly (see Chapter 10).

(7) The suction pipe must not allow any ingress of air especially when there is a large suction head. It must never be completely horizontal but always have a slope, however slight, up toward the pump. High points must be avoided completely.

(8) When it is necessary to insert a reducer cone, this must be eccentric (Fig. 9.13), in order to prevent the formation of air pockets in its upper section.

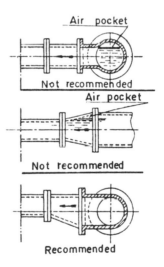

Fig. 9.13

(9) If a bend is necessary on the suction side, a long-radius elbow must be used. All kinds of throttling must also be avoided in the suction pipe.

(10) If the suction head is not very large it is advisable to use a foot-valve that must ensure a free cross-section at least equal to the cross-section of the suction pipe.

(11) It is necessary to prevent the entry of foreign bodies into the suction piping (except in the case of pumps specially designed for the purpose). For this a rack or grating system must be provided. These components must have an effective cross-section that is never less than that of the suction pipe and, furthermore, must be such that the velocity through the effective cross-section does not exceed about 0.60 m/s.

(12) Before the pump is started up it must be lubricated according to the manufacturer's instructions.

(13) Care must be taken to ensure that the pump is primed (suction pipe and pump full of water) before it is started up, otherwise damage may be caused to components that depend on

water lubrication for their proper functioning. Under no circumstances must a pump be started up unless it is full.

(14) As has been explained, a radial-flow pump, for high or medium heads, requires less power in start-up if the discharge valve is closed. This valve must therefore be closed before the start-up. This effect is, however, less marked for pumps with a high specific speed. For mixed-flow pumps the power required, with the valve remaining closed, may be equal to or even greater than the power required if the valve is open. For axial-flow pumps, the pump must be started up with the discharge valve open.

(15) It is necessary to avoid any extreme tightening on the stuffing box of the pump. At the optimum tightening point a drop of water must come out of the stuffing box.

(16) To prevent transmission of vibration to the pump casing, both the suction and discharge piping should be independently supported at a point close to the inlet and outlet flanges.

9.17 Breakdowns

Given below are the main breakdowns in centrifugal pumps, and their most frequent causes.

(1) *No discharge* — pump not primed; very low speed; very large suction or discharge head; impeller completely blocked; wrong sense of rotation.

(2) *Little discharge* — air pockets in the suction or in the casing; low speed; large discharge head; too large a suction head or net positive suction head insufficient; impeller partly clogged; mechanical defects or impeller damaged; foot-valve too small or not sufficiently submerged; wrong sense of rotation.

(3) *Low pressure* — velocity very low; air or gases in the liquid; mechanical defects or impeller damaged; diameter of the impeller too small; wrong direction of rotation.

(4) *Loss in suction following a period of satisfactory functioning* — entry of air into the suction pipe; clogging of the suction pipe; suction head too high or net positive suction head too low; air or gases in the liquid; stuffing box worn.

(5) *Excessive energy consumption* — head lower than envisaged, causing the pumping of a high discharge; specific weight or viscosity of the liquid too high; mechanical defects such as misaligned axis; rotating components' bearing wear or over tight.

9.18 Information to be given to the manufacturer when choosing a centrifugal pump

It is of general advantage to provide the following data:

(1) Scheme of the installation, with elevations, as far as possible in accordance with Fig. 9.3.
(2) Liquid to be pumped (common name). Discharge desired, including variation in suction or discharge level.
(3) Main corrosive agents: sulphuric acid, hydrochloric acid, etc. (in case of mixtures, state percentages).
(4) pH, in the case of aqueous solutions.
(5) Impurities or other elements not specified in (3) (metal salts or organic matter, even in small percentages).
(6) Specific weight of the solution at a given temperature.
(7) Maximum, minimum and normal temperatures.
(8) Vapour pressure at those temperatures.
(9) Viscosity.
(10) Entrained or dissolved air in the liquid (free of air, some air, or saturated).
(11) Other dissolved gases (in p.p.m. or cm^3 per litre).
(12) Solids in suspension (specific weight, quantity, fall velocity, diameter of the particle, hardness characteristics).
(13) Kind of service (continuous, intermittent, other types).
(14) Indicate whether attack of the metal is undesirable or allowable to some degree.
(15) Indicate any previous experience.
(16) Indicate what can be regarded as economic duration (sometimes frequent substitution of the pumps may be more economical than installing a very expensive pump).
(17) Motor sizes for the power required over the head range on which the pumps will operate.

BIBLIOGRAPHY

[1] Addison, H., *A Treatise on Applied Hydraulics*. Chapman & Hall, London, 1948.

[2] Foulquier, A., *Exploitations des Pompes Centrifuges*, Révue Générale de l'Hydraulique, No. 58, July–August 1950.

[3] Hydraulic Institute USA, *Standards of Hydraulic Institute*, New York, 1955.

[4] Matthiessen, *Bombas*, Editorial Labor, Madrid, 1954.

[5] Anderson, H. H., *Centrifugal Pumps*, The Trade and Technical Press, Morden, Surrey, England.

[6] Macintyre, A. J., *Bombas e Instalações de Bombeamento*, Editora Guanabara Dois, Rio de Janeiro, 1980.

[7] Karassik, I. J. *et al.*, *Pump Handbook*, McGraw-Hill, New York, 1976.

[8] Troskolañski, A. T., *Les Turbopompes*, Editions Eyrolles, Paris, 1977.

10

Transient flow in pressure conduits†
protection of pipelines

A. ELASTIC WAVES. WATER HAMMER

10.1 Qualitative aspect

Consider a pump which feeds a tank through a pipe AB, Fig. 10.1.

If the flow through the pump is suddenly cut off at time $t = 0$, and the non-return valve immediately downstream is closed, the fluid immediately next to the valve is stopped (Fig. 10.1(a)); for the time being, however, the fluid further downstream continues to move as though nothing had happened. Consequently the pressure in the fluid next to the valve is reduced and the pipe (no longer assumed perfectly rigid) contracts slightly as a result of the reduction in pressure. The next element of fluid now finds a pressure reduction in front of it; therefore it too comes to rest and contracts the pipe slightly. This phenomenon creates a temporary availability of a mass of fluid to enable the adjacent downstream element of fluid to remain in motion for an instant longer, and when it stops it is decompressed and provides a volume that allows motion of the next element, and so on successively.

In this way a wave of reduced pressure, sometimes called a wave of rarefaction, is generated that travels along the pipe with the velocity of an elastic wave, until the whole pipe is subject to the reduced pressure thus caused. If c is the velocity of propagation of this wave (*celerity*), it will reach the tank at a time $t = l/c$ after the valve closure, in which l is the length of the pipe between the pump and the tank.

This means that at time t, the pressure in cross-section B, in the passage from the pipe to the tank, is less than the pressure in the tank, and this causes a reflected wave, i.e. another wave is produced which returns along the pipe

---

† *Transient flow* means the unsteady flow that occurs in the passage from one steady flow to another steady flow. It does not cover the unsteady flows caused by the permanent action of a source of disturbance, such as hydraulic resonance.

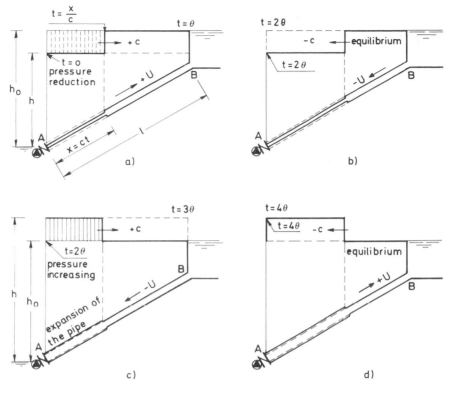

Fig. 10.1.

to the starting section of the first wave. This wave re-establishes the original values of the pressure in the successive elements of fluid that had previously been decompressed and the fluid density and the pipe diameter return to their original values in the successive cross-sections of the pipe (Fig. 10.1(b)). This reflected wave travels from the tank to the valve with velocity $-c$ (celerity) reaching the valve in time $2t$, after the beginning of the phenomenon. Since the movement of the fluid was taking place from the tank to the valve, the fluid immediately next to the valve is obliged to stop. This reduction in kinetic energy causes a local pressure increase, which leads to compression of the fluid and expansion of the pipe. This combined process of compression of the fluid and expansion of the pipe is transmitted along the pipe to the tank, where it arrives at the end of time interval $3t$ (Fig. 10.1(c)).

When this pressure wave reaches the tank, since the pressure in it is now lower than the pressure in the conduit, the flow is reversed and this allows a return to the initial pressure conditions and velocity until, at time $4t$, this wave reaches the cross-section of the valve and conditions are created for starting the whole process again with a new wave of rarefaction (Fig. 10.1(d)).

If there were no head losses, this pulsatory phenomenon would be maintained indefinitely. In practice, energy is gradually dissipated by friction and imperfect elasticity, and so the waves diminish in intensity and die away.

This situation of a sudden cut-off of the pump is equivalent to that of the sudden stopping of a turbine or the sudden closing of a valve upstream of a long pipe.

In the case of the sudden closing of a valve at the downstream end of a pipe (the case usually presented in specialized books) or the sudden cut-off of a turbine (or pump) downstream of a pipe, the first wave would be one of overpressure; at the end of time interval t, another wave is produced in the reservoir which returns along the pipe to the starting point of the first wave where it would arrive in time interval $2t$; then a wave of reduced pressure would follow and this would reach the reservoir in time interval $3t$, followed by its reflection that would reach the valve in time interval $4t$, when the phenomenon would restart.

In the following we shall refer specifically to the various possibilities of establishing transient flow with valve closure and opening upstream or downstream of a pipe: the cut-off of a pump (or turbine) upstream corresponds qualitatively to a closure upstream, just as its start-up corresponds to an opening upstream; the cut-off of a turbine (or pump) downstream corresponds qualitatively to a closure downstream; its start-up corresponds qualitatively to an opening downstream.

Since transient phenomena are extremely complex, the aim of this chapter is to help physical perception of the phenomenon and enable simple cases to be resolved, as well as preliminary design for the purpose of comparing solutions.

For calculating water hammer references [1–5] may be consulted, for example.

10.2   Velocity of an elastic wave. Influence of the pipe

(a) The velocity of propagation of the elastic wave (celerity) is considered, in a first approximation, as a characteristic of the flowing fluid and of the pipe wall.

Representing by ε the bulk modulus of the fluid, defined in Section 1.11, and by E the elastic modulus† of the material of the pipe, for circular pipes free of constraints we have:

$$c = \left[\rho \left(\frac{1}{\varepsilon} + \frac{1}{E} \cdot \frac{d}{e} \right) \right]^{-1/2} \tag{10.1}$$

in which d represents the pipe internal diameter and e the thickness of the pipe walls. In the case of water, the values of ρ and ε are given by Table 9.

† The elastic (or Young's) modulus E is the ratio between the stress and strain, in the elastic phase. Its dimensions are $ML^{-1}T^{-2}$.

In the theoretical case of an undeformable pipe, that is to say, $E = \infty$, we should have $c = \sqrt{\varepsilon/\rho}$, which corresponds to the velocity of propagation of sound in water (≈ 1400 m/s).

Table 172 gives the value of E for various materials.

Example
Calculate the velocity of propagation of the pressure wave in a normal steel pipe ($E = 2.1 \times 10^{11}$ N/m²) in which water flows at 20°C ($\rho = 998$ kg/m³; $\varepsilon = 21.39 \times 10^8$ N/m²); the pipe diameter is 0.5 m and the thickness of the wall is 6 mm.
We thus have:

$$c = \left[998 \left(\frac{1}{21.39 \times 10^8} + \frac{1}{2.1 \times 10^{11}} \cdot \frac{500}{6} \right) \right]^{-1/2} =$$

$$= [998 (0.0468 \times 10^{-8} + 0.0397 \times 10^{-8})]^{-1/2} = 1076 \text{ m/s}$$

(b) The following are formulae for the value of c in other types of pipe (according to [1]).

(1) *Reinforced concrete* — The elastic modulus of steel is taken and an equivalent thickness $e = e_m [1 + (E_b/E_m) (e_b/e_m)]$ is used, in which e_b is the thickness of the concrete; e_m is the thickness equivalent to the total cross-section of the metal in a cross-section normal to the axis of the pipe.
(2) *Steel pipe in a tunnel in rock, with grouting of concrete between the pipe and the tunnel*

$$c = \left\{ \rho \left[\frac{1}{\varepsilon} + \frac{d_a}{E_a e} (1 - \lambda) \right] \right\}^{-1/2} \tag{10.2}$$

and

$$\lambda = \left[\frac{d^2}{4eE_a} \right] \times \left[\frac{d_a^2}{4e\,E_a} + \frac{d_b^2 - d_a^2}{4d_b\,E_b} + \frac{m+1}{2m\,E_r} d_a \right]^{-1} \tag{10.2a}$$

in which $1/m$ is the Poisson's ratio[†] of the rock. The subscripts a, b and r refer to the steel, concrete and rock, respectively, d_b being the value of the external diameter of the concrete.
(3) *Tunnel excavated in rock*

$$c = \left[\rho \left(\frac{1}{\varepsilon} + \frac{2}{E_r} \right) \right]^{-1/2} \tag{10.3}$$

(4) *Pipe with thick walls* — Expression (10.1) is used with d/e multiplied by the factor ψ, defined by:

$$\psi = 2 \frac{e}{d} \left(1 + \frac{1}{m} \right) + \left(1 + \frac{e}{d} \right)^{-1} \tag{10.4}$$

† Poisson's ratio is the ratio between the lateral and axial unit strains. The lateral deformations are measured in a plane normal to the applied force.

(5) *Pipeline made up of lengths with different characteristics* — in simplified calculations a weighted value can be used in the following form:

$$\bar{c} = l / \sum_i (l_i / c_i)$$

(10.5)

in which: $\bar{c}$ = equivalent celerity; l = total length of the pipeline; l_i, c_i = length and corresponding celerity of each section.

For a more precise analysis, specialized books should be consulted or analytical methods used that take account of the different characteristics of the various parts of the pipeline.

(c) The previous values refer to circular pipes. Graph 184 gives indications of the influence of the shape of the transverse cross-section on the value of c. Information on the influence of solid suspended material and dissolved air can be obtained in [5] and [6].

10.3 Theoretical analysis of water hammer. Allievi's equations

Application of the momentum equation (see Section 2.21) to a length of horizontal pipe, considering the head losses to be zero, enables us quickly to obtain the first equation of motion†
or *equation of dynamics*:

$$\frac{\partial U}{\partial t} = g \frac{\partial h}{\partial x}$$

(10.6)

in which U is the flow velocity, h is the pressure head in the pipe ($h = p/\gamma + z$) and x is measured along the axis of the pipe.

The *equation of continuity* will be written:

$$\frac{\partial U}{\partial x} = \frac{g}{c^2} \frac{\partial h}{\partial t}$$

(10.7)

By differentiating equations (10.6) and (10.7) with respect to x and with respect to t, we obtain the following system of differential equations:

$$\frac{\partial^2 h}{\partial t^2} = c^2 \frac{\partial^2 h}{\partial x^2}$$

(10.8)

$$\frac{\partial^2 U}{\partial t^2} = c^2 \frac{\partial^2 U}{\partial x^2}$$

(10.8a)

known as the *vibrating cord* system of equations.

Integration of the previous system leads to the following equations:

$$h - h_0 = F\left(t - \frac{x}{c}\right) + f\left(t + \frac{x}{c}\right)$$

(10.9)

$$U - U_0 = \frac{g}{c}\left[F\left(t - \frac{x}{c}\right) - f\left(t + \frac{x}{c}\right)\right]$$

(10.9a)

F and f are two functions which depend on the basic principles governing the discharge and on the boundary conditions.

† Consult [7], for example.

Interpretation of these equations is simplified if one assumes that an observer moves in the conduit, in the direction of the flow, at velocity c; the distance covered by the observer will be $x = ct + b$, so that $t - x/c = -b/c = $ constant; for the observer, therefore, if $(t - x/c)$ is constant, $F(t - x/c)$ will also be constant. Consider another observer moving in a direction contrary to the flow, at velocity $-c$; for this observer, by analogous reasoning f will be constant. F and f therefore represent two waves, the first proceeding in the direction of the flow and the second in the opposite direction.

In most practical cases, the pipe ends in a reservoir, R, whose dimensions are large enough in relation to the pipe for the elevation of water in the reservoir not to change during the phenomenon of water hammer. The *characteristic curve* of the reservoir, i.e. the curve that relates the head, h, in the reservoir to the discharge in the pipe, is then $h = h_0$ (constant). Thus, at point $B(x = l)$ we always have $h = h_0$ and, from the first equation, $F(t - l/c) = f(t + l/c)$. This equation proves that the wave F is totally reflected in the reservoir and gives rise to a wave f that is equal and moving in the opposite direction, as in fact has been affirmed in the qualitative analysis of the phenomenon given above.

10.4 Rapid closure or opening. Joukowsky's formula

Our discussion so far has been based on the concept of *instantaneous closure or opening*, i.e. one of duration $T = 0$, which is physically impossible. In reality, the variation in discharge (caused by the stopping of a pump, for example) takes place in time $T \neq 0$, and elementary waves are generated as the stopping operation takes place.

There will be a *rapid closure* whenever the time of annulment of the discharge, T, is not more than $2l/c = 2t$, that is to say, not more than the time corresponding to the transmission and return of an elastic wave. Accordingly, at the cross-section where the closure is carried out, there is no noticeable change in pressure due to the appearance of the reflected waves.

In the case of $T > 2t$, pressure waves are still being generated when the first reflected waves arrive, attenuating the effect of those waves. The manoeuvre in this case is called a *slow closure*.

Similar considerations apply to rapid and slow opening of valves, in which case the pressure waves are ones of rarefaction.

At the cross-section of the valve, at any time less than $2t$, $f = 0$, since f does not yet exist at the cross-section of the valve, where it only arrives at time $2t$; moreover, as the valve is closed, $U = 0$. By substituting these values, $f = 0$ and $U = 0$, in equations (10.9) and (10.9a), we obtain *Joukowsky's formula*:†

$$\Delta h_J = h - h_0 = \pm \left(\frac{cU_0}{g} \right) \tag{10.10}$$

† Also known as Allievi's formula, sometimes using the symbol $\Delta h_A = cU_0/g$.

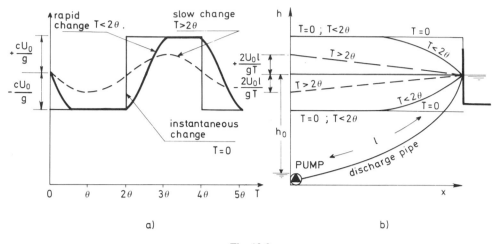

Fig. 10.2.

This equation is valid at the valve when the duration of closure or opening is $T \lesssim 2t$ (rapid).

In more general terms it can be written:

$$\Delta h = \pm \frac{c}{g} \Delta U \qquad (10.10a)$$

The minus sign corresponds to a sudden closure upstream (as shown in Fig. 10.1); the plus sign corresponds to a sudden closure downstream.

If we make $c = 1000$ m/s and $g = 10$ m/s², we shall have $h - h_0 = 100 \, U_0$, which shows that under these conditions the depression and subsequent overpressure may be very high.

Fig. 10.2(a) shows qualitatively what takes place at the cross-section of the pump, according to the description in Section 10.1 (closure upstream of the pipe) for instantaneous, rapid and slow changes.

Fig. 10.2(b) shows the pressure variation along the pipe, for the three cases considered: in the case of instantaneous change, $T = 0$, the maximum pressure reduction and overpressure occur in the whole pipe; in the case of a rapid change, $T < 2t$, they only occur in the initial part of the pipe, the shorter T is, the longer will be that part: in the case of slow changes, $T > 2t$, it is assumed, as a first approximation, that the reductions and increases in pressure vary linearly along the pipe, but this does not correspond to reality, as will be seen in due course.

To determine the maximum pressure variations in slow changes see Section 10.5.

The amplitude of the elastic wave is modified when the value of the celerity changes and at bifurcations.

If Δh is the amplitude of the wave that encounters a junction, and h' the amplitude that proceeds to one of the branches, i, of the junction, with cross-section area a_i, we have the following relationship (Stephenson [8]):

$$\frac{\Delta h'}{\Delta h} = \frac{2a_i/c_i}{a_1/c_1 + a_2/c_2 + ..a_i/c_i} \tag{10.11}$$

10.5 Slow closure or opening. Parameters of the pipe

Joukowsky's formula is only applicable in the case of rapid changes.

Various formulae have been deduced for the case of slow changes, in which the variation of the cross-section area is linear with time.

In the characterization of slow changes it is current practice to use a *pipeline parameter*, defined thus:

$$A^* = \frac{c\,U_0}{g\,h_0} = \frac{\Delta h_J}{h_0} \tag{10.12}$$

which represents the influence of the elastic effects on the transient flow in question. In the particular case in which variation in the discharge follows a linear law, it can be proved (see [5]) that the maximum pressure variation, Δh, complies with the following expression (*Michaud's formula*):

$$\frac{\Delta h}{h_0} = \frac{2A^*\theta}{T} = \frac{2lU_0}{gh_0T} \tag{10.13}$$

In practice, however, a linear opening or closure does not, as a rule, mean a linear variation of the cross-section area and it is necessary to take into account the temporal variation of the cross-section for each case. A common type of opening and closure, however, is one in which the shaft, driven by a motor of constant velocity, moves uniformly. A linear variation of the shaft displacement results in variations of the areas and consequently of the discharges that are appropriate to each type of valve.

If Z is the position of the valve shaft in relation to a reference position (closed position $Z = 0$), and Z_0 the elevation corresponding to fully open, we shall have:

$$\frac{Z}{Z_0} = \left(1 - \frac{t}{T}\right) \tag{10.14}$$

Based on equation (10.14) Graph 175 gives the variation in open cross-section and the values of the overpressures, for the following types of valves:

— *globe valve*, in which the linear variation of the valve shaft corresponds to a linear variation of the cross-section area, and consequently of the discharge, thus falling within the previous cases, with the graph shown being equivalent to the well known *Allievi's graph*;
— *circular gate valve*;
— *rectangular gate valve*;
— *needle valve*;
— *butterfly valve*;
— *plug valve*.

Graph 176 [9] also considers the case of a gate valve in which the shaft moves, not at uniform velocity, but with uniform acceleration.

In order to limit the values of the overpressures due to opening and closure of valves, special control of the valve motion can be applied (see [2] and [10]).

10.6   Minimum and maximum pressure after sudden stoppage of a pump

Graph 188 makes it possible to determine the maximum and minimum pressures lines along a pipeline in the case of sudden stoppage of a pump, taking account of the head losses and assuming instantaneous annulment of the discharge. From this graph it can be verified that the value of the maximum pressure reduction immediately next to the pump is less than Δh_j (the greater the value of the initial head loss, the less it will be), the maximum overpressure being less than that corresponding to the case in which the head losses are zero. For head losses higher than $0.7 \, \Delta h_j$, the maximum pressure immediately next to the pumps does not exceed the initial value of the pressure.

If the annulment of the discharge is not sudden (namely if the inertia of the pump is not negligible), the effects are qualitatively similar to those of head loss.

10.7   Breakdown of the flow

Since water hardly supports any tensile forces, whenever pressure in the pipe reaches very low values (-8 m water column), there will be a release of dissolved gases, and there may be a *breakdown of the flow* owing to the creation of large gaseous bubbles within the fluid (vapour cavities). If the cavity formation takes place downstream of a valve or in an elevated part of the pipeline, this phenomenon may result in the separation of columns of liquid: *separation of the flow*, sometimes termed 'slug' flow.

For a simplified analysis of flow separation downstream of a valve or in an elevated part of the pipeline [11] and [12] may be consulted. Very precise analysis of the phenomenon can only be done with the aid of a computer.

10.8   Influence of air bubbles on water hammer

The existence of air bubbles in conduits may be due to the following: the entry of air through vortices in the intakes; the entry of air either through air valves or surge chambers; the gradual release of dissolved air; defective filling of the conduit when insufficient care is taken to eliminate the air completely.

In this case it is necessary to take it into account when analysing water hammer.

B.   MASS OSCILLATION. SURGE SHAFTS

10.9   Water hammer when there is a surge shaft

As was explained in the study of water hammer, the length, l, of the pipeline is an important factor in the intensity of water hammer, for a given valve operating time. Whenever topographical conditions allow, it will be convenient to locate a *surge shaft or surge tank* in the pipeline, consisting of a tank with the water surface exposed to atmospheric pressure. This device is normally used for protecting pipelines feeding turbines.

It can also be used to protect discharge pipelines although the topographical conditions are generally less favourable. In either case the shaft may be upstream of the pipeline to be protected, in which case it is known as an *upstream shaft* (Fig. 10.3(a) and (c)) or downstream of the pipeline to be protected, when it is called a *downstream shaft* (Fig. 10.3(b) and (d)).† In order to confirm this point, see Fig. 10.4, corresponding to an upstream shaft in a pumping system.

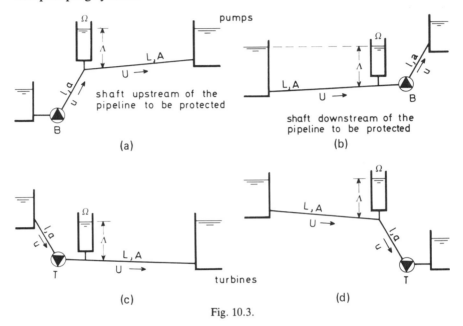

Fig. 10.3.

Both the reduced and excess pressure elastic waves, Δh, that are generated in the pump (point B), divide into two when they reach the shaft (point C), one of them, Δh_r, proceeds to the shaft and is reflected; the other, Δh_t, is transmitted, through the pipeline that it is intended to protect, towards the reservoir.

The part of the wave transmitted, h_t, depends on the ratio of the areas of the cross-sections of the pipes and the surge shaft, the type and length of the shaft and the functional variation of the discharge.

Bearing in mind Figs. 10.3 and 10.4, the following symbols will be used:

l —   length of the pipeline directly connected to the pumps (or turbines);
a —   cross-section of that pipe;
u —   velocity in the pipe; u_0, initial velocity;
L —   length of pipeline to be protected;
A —   cross-section of the pipe to be protected;
U —   velocity in the pipe to be protected; U_0, initial velocity. U is positive in the shaft-reservoir direction, negative in the opposite direction;

† For a detailed analysis of the behaviour of surge shafts, consult [14–16].

Λ — distance from the water surface in the shaft to the connection cross-section;
Ω — cross-section of the surge shaft.

The case in which the cross-sections of the pipes and shaft are equal, $a = A = \Omega$, is easier to deal with and, moreover, it occurs most frequently in pumping systems in which the shaft consists of a long pipe situated on a slope.

Bernhart [17] studied the phenomenon of the wave not being fully reflected, in the case of a valve being closed downstream (cases 10.3(b) and (d)). From the results, Graph 178 was prepared representing the effect of reflection in the case of the closure of a valve upstream of the pipeline (Case 10.3(a) — stopping the pump).† It was assumed that the partially reflected wave is equal in both cases, even though the first corresponds to an overpressure wave and the second to a depression wave.

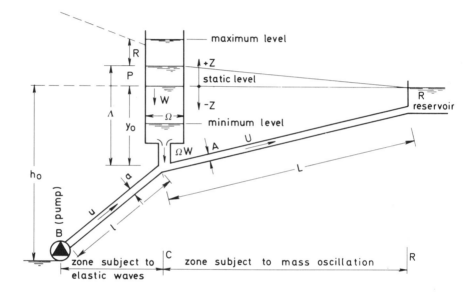

Fig. 10.4.

If $a = A = \Omega$ (usual in pumping systems with shafts consisting of pipelines along a slope), the percentage of wave transmitted may be considerable, up to about $0.65\,\Delta h_j$ in the case of a sudden closure (see Graph 178).

In the case of shafts with a throttled orifice there is no significant effect on the value of transmission of the elastic wave, unless the ratio between the area of the orifice, σ, and the area of the pipe, a, is such that $\sigma < 0.25\,a$ (Graph 179).

The influence of the distance, Λ, from the water surface to the connection of the surge shaft, on the wave transmitted, Δh_t, is given in Graph 180.

In simple cases, the pipeline — L, A — downstream can, as a first approximation, be studied in relation to the part Δh_t of the wave transmitted, in the same way that the pipeline — l, a — was studied in relation to the wave Δh.

In complex cases it is necessary to consult specialized references or make use of a calculation method that takes account of the reflection and transmission of elastic waves, at the base of the surge shaft.

† Also valid for 10.3(c).

10.10   Mass oscillation in the shaft-reservoir zone. Fundamental equations
Although the effect of water hammer (elastic waves) is eliminated or at least
reduced by the surge shaft, another completely different phenomenon arises
— that of mass oscillation between the shaft and the reservoir.

In the case of a stoppage of the pumps, for example in the *SR* zone,
(shaft-reservoir), due to the effect of inertia there is continuation of the flow
that is fed by the water in the surge shaft. When the flow stops, since the
effect of inertia also causes the level in the surge shaft to fall below the level
in the reservoir, there is reversal of the flow from *R* to *S*. Also due to inertia
the water in the shaft rises above the level in the reservoir, thus creating a
pendular, oscillatory flow, the energy of which will be absorbed by head
losses in the pipe.

This phenomenon of *mass oscillation* is distinct from that of the propagation of elastic waves; whereas the former, of a pendular type, is due to the
action of gravity, the latter is due to the compressibility of water. In some
specialized literature the two phenomena are called *rigid water column* and
elastic water column, respectively.

Fig. 10.5(a) shows the case of a *downstream surge shaft*, which is more
usual in turbines; Fig. 10.5(b) shows the case of an *upstream surge shaft*,
which is more usual in pumps.

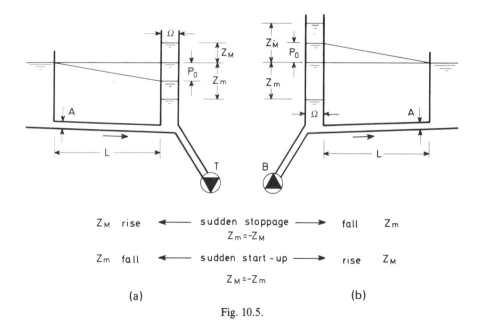

Fig. 10.5.

In the case of a *sudden stoppage*, the rise in case (a) corresponds to a fall
in case (b); in the case of a *sudden start-up*, the fall in case (a) corresponds to
a rise in case (b).

Bearing in mind Figs. 10.4 and 10.5, we have, apart from the symbols already defined, the following:

Z — level in the surge shaft, measured from the static level (positive above the static level, negative below it);

Z_M — maximum value of Z;

Z_m — minimum value of Z;

$W = dZ/dt$ — velocity of the oscillation of the water surface level in the surge shaft (positive upwards): $W_0 = Q_0/\Omega$;

Q — pump discharge as a function of time: Q_0 initial discharge;

P — total head loss (friction and local) in the pipe protected by the surge shaft. If P_0 is the loss for the initial discharge Q_0, then: $P = \pm P_0 (UA/Q_0)^2$; P will take the same sign as Q;

R — head loss in the passage from pipe to shaft if there is a throttled orifice (throttling loss). If R_0 is the value that would correspond to the passage of discharge Q_0 to the surge shaft, then $R = R_0 (\Omega W/Q_0)^2$.

By applying the fundamental equation of dynamics, $f = ma$, to the water mass in the pipe, which is equivalent to disregarding the water mass in the shaft, we have:

$$\gamma(Z + P + R)A = - \rho L A \frac{dU}{dt} \qquad (10.15)$$

Dividing by A and bearing in mind that $\gamma = \rho g$, the following differential equation is obtained:

$$\frac{L}{g} \left(\frac{dU}{dt} \right) + Z + P + R = 0 \qquad (10.15a)$$

From the equation of continuity, with Q the discharge from the pumps, we have:

$$AU = \Omega W + Q \qquad (10.15b)$$

10.11   Surge shafts with constant cross-section ($P = 0$; $R = 0$). Non-dimensional parameters

Consider the simplest case of a shaft of constant cross-section, Ω, without a throttled orifice at the base, in an instantaneous sudden stoppage ($Q = 0$), disregarding the head losses in the upstream pipe ($P = 0$).

Bearing in mind that $W = dZ/dt$ and that for $Q = 0$ (pumps stopped) the only movement, either in the shaft or in the pipe, is due to mass oscillation, then the equation of continuity (10.15(b)) gives

$$U = \frac{\Omega}{A} W = \frac{\Omega}{A} \cdot \frac{dZ}{dt}$$

By introducing this value into (10.15(a)) we have:

$$\frac{L\Omega}{gA} \cdot \frac{d^2Z}{dt^2} + Z = 0 \qquad (10.16)$$

The integral of this equation is the sinusoid:

$$Z = Z_* \sin \frac{2\pi t}{T_*}$$

(10.16a)

of which the period is:

$$T_* = 2\pi \sqrt{\frac{L\Omega}{gA}}$$

(10.16b)

These equations characterise a pendular-type movement.

Let us see how the maximum amplitude, Z_*, is determined. After the pumps stop ($Q = 0$), movement continues in the conduit CR downstream of the shaft, changing from velocity U_0 to the final velocity $U_1 = 0$.

During this time the water has fallen from level $Z = 0$ to the minimum level $-Z_*$ equivalent to a volume $\forall = \Omega Z_*$ whose centre of gravity is at level $Z_{*G} = -Z_*/2$ (Fig. 10.6).

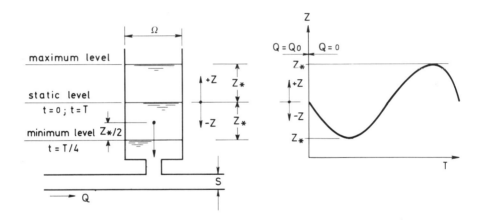

Fig. 10.6.

Disregarding head losses, the total energy of the system will remain constant and there will be a simple transformation of the potential energy, E_p, into kinetic energy, E_c. We then have:

$$E_{p1} + E_{c1} = E_{p0} + E_{c0}$$

(10.17)

Taking as datum the plane $Z = 0$, we have $E_{p0} = 0$ and:

$$E_{p1} = \gamma\Omega Z_* \frac{Z_*}{2} = \frac{\gamma\Omega Z_*^2}{2} \; ;$$

(10.17a)

also $E_{c1} = 0$ and:

$$E_{c0} = \frac{1}{2} mV^2 = \frac{1}{2} \rho LA U_0^2 = \frac{1}{2} \rho LA \frac{Q_0^2}{A^2} = \frac{1}{2} \rho \frac{L}{A} Q_0^2;$$

(10.17b)

so that:

$$\frac{\gamma \Omega}{2} Z^2 = \frac{1}{2} \rho \frac{L}{A} Q_0^2 \tag{10.17c}$$

Since $\gamma = \rho g$, we thus have the expression of the two parameters:

$$Z_* = \pm Q_0 \sqrt{\frac{L}{gA\Omega}} \tag{10.18}$$

$$T_* = 2\pi \sqrt{\frac{L\Omega}{gA}} \tag{10.18a}$$

These parameters Z and T are fundamental in studying more complex cases.

Similar reasoning demonstrates that in the case of an instantaneous change from discharge Q_0 to discharge Q_1, the initial drawdown will be:

$$Z = (Q_0 - Q_1) \sqrt{\frac{L}{gA\Omega}} \tag{10.19}$$

In studying surge shafts with constant cross-section, it is of help to make use of the following *non-dimensional parameters* [15]:

$$z = Z/Z_* ; \; p = P/Z_* ; \; q = Q/Q_0 ; \; u = U/U_0 ; \; t' = T/T_* ; \; w = W/W_0 \tag{10.20}$$

where:

$$W_0 = Q_0/\Omega \tag{10.21}$$

10.12   Surge shaft with constant cross-section. Effect of head loss ($P \neq 0$; $R = 0$)

(a) Instantaneous total stoppage
In the case of a sudden stoppage, taking into account the head losses, Graph 193(a) gives the first rise and the first and second drawdown.

(b) Instantaneous total start-up
In the case of instantaneous total start-up of the pumps there will be a rise in the surge shaft. If the start-up is total, i.e. if there is instantaneous passage from zero discharge to a discharge, Q_0, the rise is given by Graph 193(b), which also gives the value of the first drawdown; this case is only likely to occur in a pumping system that has only one pump, starting up with the valve open.

(c) Partial instantaneous start-up

If the discharge passes suddenly from $n\%$ to 100%, the first rise is given by Graph 193(c). The zone to the right of $a\ b$ correspond to an aperiodic motion.

(d) Linear start-up

If the discharge passes linearly from $n\%$ to 100% in time T, the value of the rise of water in the shaft is given by Graph 181(d).

Examples
   (a) A discharge line with length $L = 5000$ m and cross-section $A = 1$ m^2, in which there is a discharge $Q_0 = 1$ m^3/s, is protected in its first part by a surge shaft of constant cross-section, $\Omega = 5$ m^2. Calculate the oscillation in the shaft, in the case of a sudden stoppage of the pumps and considering the head loss in the pipeline, $P_0 = 4.5$ m.
   We thus have:

$$Z_* = Q_0 \sqrt{L/gA\Omega} = 1 \times \sqrt{5000/(10 \times 1 \times 5)} = 10 \text{ m}$$

$$T_* = 2\pi \sqrt{L\Omega/gA} = 2 \times 3.14 \sqrt{5000 \times 5/(10 \times 1)} = 314 \text{ s}$$

$$p_0 = P_0/Z_* = 4.5/10 = 0.45$$

From Graph 181(a) we obtain $z_m = -0.73$, so that:

$$Z_m = z_m \times Z_* = -0.73 \times 10 = -7.3 \text{ m}$$

   From Graph 181 we can also calculate the first rise, which reaches the value $0.5 \times 10 = 5.0$ m and the second drawdown, which reaches the value $-0.38 \times 10 = -3.8$ m, that is to say, about half the first.
   The effect of head loss on absorption of the oscillations is very evident.
   (b) Calculate the rise in the surge shaft of the previous example, when the discharge passes suddenly from 0 to 1 m^3/s (sudden start-up of the pumps).
   We have:

$$z_M = 1 + 0.125\ p_0 = 1.06;$$

hence:

$$Z_M = Z_* \times z_M = 10.6 \text{ m}.$$

   Graph 181(b) gives the first drawdown equal to $0.35\ Z = 3.5$ m (above static level, since it has a positive sign).
   (c) Calculate the rise in the surge shaft when the discharge passes instantaneously from 0.35 m^3/s to a total discharge of 1 m^3/s.
   The initial discharge is 35% of the total. The increase is $100 - n = 65\%$. Graph 181(c) gives $z_M = 0.78$, so that $Z_M = 7.8$ m.
   (d) Calculate the rise in the surge shaft when the discharge passes linearly from 0 to 1 m^3/s in time $T = 40$ s.
   We shall have $t' = T/T_* = 40/314 = 0.3$; Graph 181(d), for $p_0 = 0.45$ and $t' = 0.13$ gives $z_m = 1.02$ and $Z_M = 10.2$ m.

10.13   Surge shafts of constant cross-section with throttled orifice ($R \neq 0$)
In order to reduce the amplitude of the oscillations and thus the dimensions
of the shaft, a head loss, R, can be introduced into the connection of the shaft
to the pipe, by means of a throttled orifice (Fig. 10.7). R_0 will be taken to
represent the throttling head loss corresponding to the passage of a discharge Q_0 through the throttled orifice.†

In the drawdown phase, the water flows from the shaft to the pipe. The
existence of the head loss on one hand reduces the discharge from the shaft
and therefore limits the downsurges; on the other hand it may cause, at the
pipe, a piezometric head, Y_m, lower than that corresponding to the water
level, Z_m, in the shaft.

In the rising phase, the flow is from the pipe to the shaft; the pressure,
Y_M, at the start of the pipe may be higher than that corresponding to the
water level, Z_m, in the shaft.

We shall always find:

$$Y = |Z| + |R| \tag{10.22}$$

In practice, the reduction in pressure due to the throttled orifice may
have more drawbacks than the increase in pressure. Accordingly, in order to
prevent an exaggerated low pressure, an asymmetrical throttling can be
used, causing an outlet loss, R', that is less than the inlet loss, R'' (Fig.
10.7(b)). This enables the height of the shaft to be reduced without creating
inconvenient depressions. For analysis of the devices that can create R' and
R'', see Section 10.18 in this chapter.

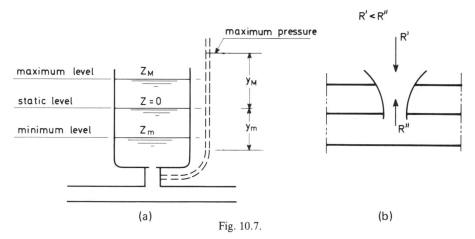

Fig. 10.7.

The non-dimensional parameter $r = R/Z_*$ is also defined; to which
corresponds the parameter $r_0 = R_0/Z_*$. In the same way r' and r'_0, are

---

† The throttled orifice, of course, also tends to favour the transmission of elastic waves, due to
water hammer, to the pipe for which protection is desired downstream of the shaft (see Section
10.9).

defined corresponding to the outlet losses and r'' and r''_0, corresponding to the inlet losses.

Graph 199(a) gives values of the maximum drawdown, z_m, corresponding to sudden stoppage of the pumps. The line SS' divides the graph into two zones: an upper zone in which the value of y_m, corresponding to the maximum lowering of pressure in the pipe, has an absolute value higher than z_m, which must be taken into account; in the lower zone $y_m < z_m$. In the case of partial stoppage, the effect of the orifice is reduced and the drawdown may be higher than that corresponding to total stoppage, which occurs to the right of the line AB of the graph.

Graph 182(b) gives values of the maximum rise, z_M and of the maximum overpressure, y_M (in the case of $y_M > z_M$). In the case of instantaneous total start-up, the zone on the right of the line in the graph corresponds to an aperiodic motion, i.e. the water surface level rises to the dynamic level without passing it, so that there will be no oscillation.

Examples
(a) Solve the previous example ($p_0 = 0.45$; $Z_* = 10$), assuming that at the base of the shaft there is an orifice which causes a head loss.

Solution
In order to reduce the transmission of water hammer, the area of the orifice will be taken as 0.25 of the cross-section of the pipe (see Graph 178), so that:

$$\sigma = 0.25 \qquad a = 0.25 \text{ m}^2$$

$$d = 0.56 \text{ m}$$

The velocity through the orifice will be:

$$U_0 = Q_0/\sigma = 1/0.25 = 4 \text{ m/s}$$

The throttling loss, making $\mu = 0.6$ in formula (8.17), will thus be:

$$R_0 = (U^2/2g)\,(1/\mu^2) = 1.4 \text{ m}$$

to which corresponds $r_0 = R_0/10 = 1.4/10 = 0.14$ and Graph 182(a) gives $z_m = 0.68$ or $Z_M = 6.8$ m.[†]
The rise following a total sudden start-up would be (Graph 182(b)):

$$z_M = 0.97 \text{ and } Z_M = 9.7 \text{ m.}$$

(b) Solve the previous example, trying to reduce the drawdown to 5 m.
We have:

$$p_0 = 4.5 \qquad Z_* = 10$$

† Equation 10.30 gives the value of R_0 with greater accuracy.

Therefore we must have:

$$Z_m = \frac{5}{10} = 0.5$$

The graph gives $r_0' = 0.6$, so that $R_0' = 6$ m.
Making $U_0 = \mu \sqrt{2gh} = 0.6 \sqrt{2g \times 6} = 6.6$ m/s, the area of the orifice must be

$$\sigma = \frac{Q_0}{U_0} = \frac{1}{6.6} = 0.15 \text{ m}^2$$

corresponding to a diameter of 0.44 m.

C.   PROTECTIVE DEVICES

10.14   General considerations

Protection of a pipeline in which the flow takes place by gravity can generally be satisfactorily achieved, provided that the closing or opening of the valve takes place slowly enough, so that any overpressures that are generated will be confined to reasonable values.

In discharge lines, however, it is not possible to control the amplitude of the phenomenon so easily, since stoppage of the pumps may occur suddenly owing to a power failure.

If the profile of the pipeline is close to the initial piezometric line, the sudden stoppage may cause pressures that are lower than atmospheric pressure; the pressure must not drop below vapour pressure of the liquid in pipe (≈ -8 m water column) in order to prevent vapour cavities or even rupture of the flow. Furthermore, as has been explained, the lower pressure in the first phase, the higher will be the following overpressure phase. The most common method of limiting the pressure reduction is therefore to continue feeding the water even after power failures. A profile of the pipeline shown in Fig. 10.8(a) will be more suitable, from the point of view of low pressures, than the one shown in Fig. 10.8(b).

In general, it is accepted that the characteristics of the pump, for a given rotation speed, correspond to the steady flow curves.

This may, however, not happen. In effect, it may occur that at a certain lower rotation speed the compression head, generated by the pump, no longer reaches the head required for generating the flow, but is, however, higher than the suction head. In such a case there will be a reversal of the flow, which brings about a more sudden stoppage of the pump, which may still be rotating in the correct sense. This more sudden stoppage increases the effects of water hammer.

After stoppage of the pumping unit, if the flow continues in a reverse direction the pump will also reverse its rotation and work like a turbine.

If there are check valves, they tend to close when the flow is reversed, so

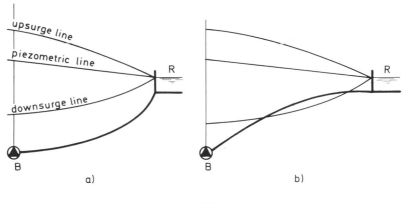

Fig. 10.8.

that flow reversal is prevented and stoppage of the pump occurs solely according to its inertia.

Given below are the principal devices available for reducing the effects of water hammer.

Structural dimensioning of the pipe must take account of the excess and reduced pressures.

In a *buried pipe*, with well compacted fill, the latter may be beneficial in resisting overpressures but may aggravate the effects of reduced pressures.

From a practical point of view, the pipe should be designed to resist the eventual maximum overpressures and reductions that may occur, even though, owing to the short duration of variations in pressure, it is possible to accept a reduction in the coefficients of safety.

10.15 Inertia of the pump units and inertia flywheel

The fitting of an inertia flywheel on the shaft of the electric pump unit will enhance the inertia of the unit and the stopping time can be increased, thus reducing the effects of water hammer.

The use of flywheels is very limited, since if the length of the pipeline is more than a few hundred metres, the weight of the flywheel quickly becomes excessive and this system ceases to be economical. Moreover, the heavier the flywheel, the greater must be the power of the motor in order to overcome the inertia of the flywheel at start-up. This situation may lead to impracticable starting currents that may preclude the satisfactory start-up of the motor.

The inertia of the pump unit itself, increased by the inertia of a flywheel, may however, in some cases, be a simple means of solving the problem.

For the purpose of quick calculation or preliminary design, Graph 195 shows the effect of the inertia of pump units, enhanced by a flywheel. In order to make use of that graph, the following parameters are defined [18]:

$$\text{Inertia parameter:} \quad J = \frac{\eta \, I \, n^2 \, c}{180 \gamma A l U_0 h_0} \tag{10.23}$$

in which: η — efficiency of the pump units; I — moment of inertia of the rotating parts of the pump units, plus that of the flywheel in kg . m^2; n— pump rotation speed in rev/min; c — celerity; A and l — cross-section area and length of the pipe; U_0 and h_0 — velocity and head under normal operating conditions.

$$\text{Pipeline parameter:} \quad A^* = \frac{cU_0}{gh_0} \tag{10.24}$$

Graph 183 gives the values of maximum pressure, h_M/h_0, and of minimum pressure, h_m/h_0, if flow is allowed in an inverse direction through the pump. If there is a check valve and the head losses are unimportant, the maximum overpressure value, above the static level, may be taken as equal to the value of the maximum pressure reduction, below that same level. The occurrence of sub-atmospheric pressures or rupture of the flow may aggravate the overpressure values.

Stephenson [8] gives the following practical rule: if the parameter $In^2/(\gamma A l h_0)$ exceeds 0.01 (rev/min)^2, pump inertia may reduce the downsurge by at least 10%.

The effect of head loss can be evaluated from Kinno and Kennedy [18].

Example
An electric pump unit in normal operation discharges $Q_0 = 21.1$ l/s to a discharge head $h_0 = 125.5$ m, with the motor turning at a speed of 1450 rev/min. The length of the pipeline is $l = 890$ m and its diameter is $d = 225$ mm ($A = 0.04$ m^2). The celerity is $c = 1200$ m/s. Efficiency is $\eta = 0.75$.

The moment of inertia of the pump unit is $I_g = 17$ kg . m^2. In order to increase the inertia a flywheel has been coupled to the unit, the flywheel having a moment of inertia of $I_v = 21$ kg . m^2.

We then have:

$$J = \frac{\eta \, I \, n^2 \, c}{180 \, \gamma \, A \, l \, U_0 h_0} = \frac{0.75 \times 38 \times 1450^2 \times 1200}{180 \times 9800 \times 0.04 \times 890 \times 0.5 \times 125.5} = 17.88$$

$$A^* = \frac{c \, U_0}{g \, h_0} = \frac{1200 \times 0.5}{9.8 \times 125.5} = 0.487 \approx 0.5$$

The graph gives $h_M/h_0 = 1.0$ and $h_m/h_0 = 0.87$.

Accordingly $h_m = 0.87 \times 125.5 = 109.2$ m and $h_M = 125.5$ m if flow can take place in the reverse direction. If there is a check valve, the maximum pressure will be equal to $h_m = 1.13 \times 125.5 = 141.8$ m.

If there is no flywheel, we have $J = 8$ with $h_M/h_0 = 1.02$ and $h_m/h_0 = 0.77$, so that $h_m = 96.6$ m and $h_M = 128$ m without a check valve and $h_M = 154.4$ m if there is a check valve.

10.16 Surge shafts

A surge shaft protects the downstream zone and reduces the intensity of water hammer in the zone upstream of the shaft. Its use is greatly limited by topographical conditions. Regarding its effect, see Section 10.9; for its dimensioning, see Sections 10.10 to 10.11.

10.17 Air vessels
(a) General considerations
When topographical conditions are unsuitable for using a surge shaft, use may be made of a vessel whose upper part contains compressed air, the lower part containing a certain volume of water. Its effects are similar to those of a surge shaft with a variable cross-section. Immediately after stoppage of the pumps, water is forced out of the air vessel by the compressed air in the upper portion to the pipeline thus reducing the amplitude of the negative surge due to water hammer. During the second period, the direction of flow is reversed and the water flows back into the air vessel, compressing the air. This cycle of events is repeated until the amplitude of the oscillations is reduced by friction. Fig. 10.9 illustrates such a scheme.

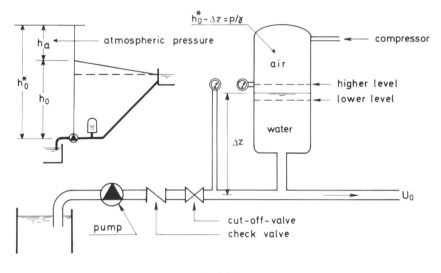

Fig. 10.9.

In order to improve the effect of the air vessel, it is possible to place — at the interconnection between the air vessel and the pipe — a device which is usually designed to provide a greater resistance to flow into the air vessel than to outflow, as in the case of a surge shaft.

Generally a short pipe, Borda type, is used (Fig. 10.10(a)).

It is also possible to use a flap with an orifice in the centre (Fig. 10.10(b)) so that in the vessel-pipe direction the passage cross-section corresponds to the whole of the flap area, while in the pipe-vessel direction the flap closes and the cross-section is reduced to that of the orifice, thus causing an asymmetrical head loss. When the flap closes, however, new water hammer

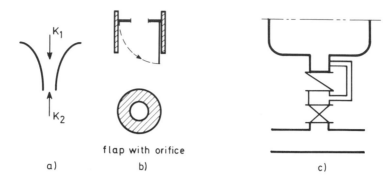

flap with orifice

a)   b)   c)

Fig. 10.10.

is caused, and the smaller the cross-section of the orifice, the greater will be the water hammer.

Use can also be made of a device with check valve and a small orifice open by-pass (Fig. 10.10(c)), which is hydraulically equivalent to the flap valve.

In calculations associated with air vessels, it is necessary to take account of the laws of air expansion and compression in the thermodynamic processes:

$$h\forall^n = \text{constant} \tag{10.25}$$

in which $\forall$ is the volume of air, $n = 1.4$ for adiabatic expansion and $n = 1$ for isothermal expansion. In practice it can be taken that $n = 1.3$.

In general it is also assumed that the air vessel is installed immediately downstream of the pumps, that the pumps stop instantaneously and there will be no reverse flow through them, so that it becomes essential for there to be a check valve downstream of the pumps.

(b) Design criteria

The following symbols will be used:

h_0^* — initial absolute pressure head
h_c^* — absolute pressure head in the pipe
h_R^* — absolute pressure head in the vessel, a function of t
U_0 — steady state velocity
U — flow velocity at instant t
l — length of the pipe
A — cross-section of the pipe
$\forall_c = LA$ — volume of the pipe if it has a constant cross-section or $\forall_c = \Sigma L_i A_i$ in the case of a telescopic pipeline
$\forall_0$ — volume of air in the air vessel at pressure headh_0^*
$\forall$ — volume of air in the air vessel at pressure head h_R^*
αW^2 — head loss caused by the flap valve in the pipe-air vessel direction, W being the velocity in the orifice

The following parameters are considered [19]:

$$\text{Pipeline parameter:} \quad A^* = \frac{c\,U_0}{gh_0^*} \tag{10.26}$$

$$\text{Air vessel parameter:} \quad B^* = \frac{U_0^2}{gh_0^*} \cdot \frac{\mathcal{V}_c}{\mathcal{V}_0} \tag{10.27}$$

$$\text{Throttling parameter:} \quad C^* = \alpha W^2/h_0^* \tag{10.28}$$

A simple procedure for quick determination of the volume of an air vessel with a flap valve is as follows.

(a) Parameter A^* is calculated. From Graph 184(a), bearing in mind the longitudinal profile of the conduit, the value of parameter B^* is chosen which originates a downsurge curve along the pipe, without causing any inconvenience; from B^* the volume of air, $\mathcal{V}_0$, in the air vessel is determined (head losses in the pipe are disregarded). If none of the curves satisfies the requirement, this implies a large volume in the air vessel and another solution would have to be adopted, for example feed tanks at high points (see Section 10.19);

(b) from Graph 184(b), knowing A^* and B^*, the optimum head loss is determined for the orifice of the flap, i.e. that which produces a maximum pressure in the pipe, with a value identical to the maximum pressure in the air vessel. It is to be noted that from the value of h_M^*/h_0^*, since A^* is known, B^* and Δh can also be determined;

(c) with the help of Graph 184, it is possible to calculate again the maximum and minimum pressures in the pipe, next to the vessel connection. Another orifice can be chosen, or other dimensions for the air vessel.

To calculate the orifice size and its head loss, the procedures for shaft chambers with throttling are valid.

The volume of air is calculated from B^*. The volume of the vessel must be such that for minimum pressure there is no passage of air to the pipe. This volume will be given by:

$$\mathcal{V} = \mathcal{V}_0 \, (h_0^*/h_m^*)^{1/1.3} \tag{10.29}$$

In addition it is prudent to allow a safety margin.

Maunier and Puech [20] advise an air vessel volume of 1.2 times the maximum volume of air calculated.

As a first approximation the outlet diameter can be considered to be about one-half the main pipe diameter [8].

A compressor must be provided for replacing the air that is dissolved in the water. Instead of air it is also possible to use a gas that is less soluble, nitrogen, or place a film on the water–air interface to prevent or restrict passage of the air into solution.

Example
A discharge pipe of diameter $d = 800$ mm ($A = 0.503$ m^2) and length $l = 4550$ m, carries a discharge $Q_0 = 80\,000$ m^3/day (0.926 m^3/s). It is necessary to determine the volume of the air vessel, so that the overpressure due to the maximum upsurge is not more than $h_M = 220$ m water column and the pressure reduction due to the minimum downsurge does not fall below $h_m = 120$ m water column.

The celerity is $c = 1000$ m/s, the elevation head is $\Delta z = 196$ m and the head loss in the conduit is $P_0 = 7.4$ m. The pipe connecting the conduit to the air vessel has a diameter of 500 mm and the throttle a diameter of 200 mm.

Solution
The flow velocity is $U_0 = Q/A = 0.926/0.503 = 1.84$ m/s; also $cU_0/g = 1000 \times 1.84/9.8 = 188$ m.
The absolute manometric discharge head will be $h_0^* = (\Delta Z + P_0 + \text{atmospheric pressure}) = 196 + 7.4 + 10.3 = 203.7$ m. Hence:

$$A^* = (cU_0)/(gh_0^*) = 188/203.7 = 0.92 \approx 1$$

Since it is wished that $h_m^* = h_m + \text{atmospheric pressure} = 120 + 10.3 = 130.3$ m, we shall have $h_m^*/h_0^* = 130.3/203.7 = 0.63$. Graph 184(a), for $A^* = 1$ and $h_m^*/h_0^* = 0.63$ indicates $B^* = 0.3$.†
Since the volume of the conduit is $V_c = LA = 4550 \times 0.503 = 2289$ m³, we have:

$$V_0 = \frac{U_0^2 \, V_c}{B^* g \, h_0^*} = \frac{1.84^2 \times 2289}{0.3 \times 9.8 \times 203.7} = 13 \ m^3$$

For maximum pressure, we shall have $h_M^*/h_0^* = (220 + 10.3)/203.7 = 1.13$.
From Graph 184(b) we have $C^* = \alpha W^2/h_0^* = 2.5$, since $\alpha = K/2g$, we thus have $\alpha W^2 = KW^2/2g = 2.5 \times 203.7 = 509.25$ m.
By making $K = 1/\mu^2 = 1/0.6^2 = 3$, we have $W^2/2g = 509.25/3 = 170$ m and $W = 58$ m/s, which gives a cross-section of $0.926/58 = 0.016$ m².
At minimum pressure, h_m^*, the air occupies the maximum volume $V_u = V_0 \, (h_0^*/h_m^*)^{1/1.3} = (203.7/130.3)^{1/1.3} = 18.3$ m³. For safety it is taken that $V = 20$ m³.

10.18   Factors related to head loss orifices in surge shafts and air vessels

In the case of a symmetrical orifice, if d is the diameter of the orifice and D the diameter of the piping connecting the vessel to the discharge pipe, the head loss through the orifice will be of the form $KU_0^2/2g$ and for the purpose of preliminary design the following expression proposed by Idel'cik, for sharp-edged orifices, can be used

$$K = \left(1 + 0.707 \ \sqrt{1 - \frac{d^2}{D^2} - \frac{d^2}{D^2}}\right)^2 \left(\frac{D^2}{d^2}\right)^2 \tag{10.30}$$

The increase in velocity in the orifice causes a lowering of pressure downstream of it; if the velocity is high and consequently the pressure is very low, gas bubbles may be released and cause *cavitation* problems.

The *incipient cavitation*, corresponding to the velocity, U_i, occurs intermittently over a restricted area.

The design criterion usually adopted is the *critical cavitation*, corresponding to the velocity U_c; this is a continuous light cavitation with acceptable noise and vibration. No major damage can be expected even after long operation.

As soon as the mean pressure of the jet is equal to vapour pressure, the velocity tends to be constant for that value, and a *discharge blockage* occurs; the corresponding flow velocity is termed *choking velocity*, U_{ch}.

Criteria of *cavitation velocities* may also be used. Graph 185 taken from

† A reasonable profile is assumed for the pipe. In the case of a less favourable profile, $B = 0.1$ might be assumed; in the case of a very favourable profile, $B = 0.8$ might be taken.

[21] enables the cavitation conditions to be determined in sharp-edged orifices, at the inlet of the vessel (e.g. an orifice in a flap valve).

Dubin and Guéneau [19] proposed the following expression for dimensioning the diameter of the orifice in a flap valve:

$$d = \frac{D}{1.2\sqrt[4]{\dfrac{cA}{Q_0}}} \tag{10.31}$$

in which: d = diameter of the flap valve orifice; D = diameter of the discharge pipe; A = cross-section of the discharge pipe; c = velocity of the elastic wave in the discharge pipe; Q_0 = initial discharge.

Ruus [22] recommends that the local head loss should not exceed 0.6 of the initial absolute static head.

Dupont [23] considers that the ratio between the cross-section of the discharge pipe and the throttling cross-section should be between 13 and 17.

10.19   Feed tanks

A feed tank, also termed a discharge tank, differs from a surge shaft in that, during normal functioning, it is isolated from the pipe by a check valve (Fig. 10.11).† When a pressure occurs in the pipe below that imposed by the water level in the tank, the valve opens and the consequent flow prevents sub-atmospheric pressure. By preventing column separation, this flow offers protection against high pressures associated with cavity collapse. The tank is supplied via a by-pass, with a float valve.

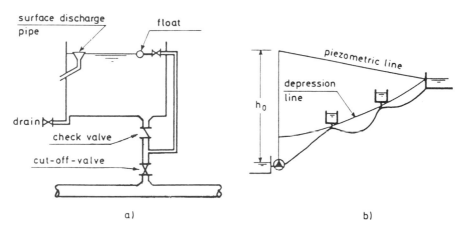

Fig. 10.11.

Feed tanks are particularly appropriate for protecting high points of a conduit. As a rule, one tank is installed on the first point, and others on the

† The tank water surface would be subjected to atmospheric pressure but would be below the hydraulic grade line, as opposed to that of a surge shaft.

following ones, if necessary. The tank can of course only operate if the water surface is above the lowest level to which the head in the pipeline would otherwise drop following pump stoppage.

Graph 186(a) provides data for dimensioning a single feed tank on the condition that the tank is close to the pump or that there is a check valve in the pipe, immediately upstream of the tank, in order to prevent positive waves from being reflected in the direction of the pump and increasing the overpressure that would occur without a tank. The graph gives the value of the maximum pressure, Δh_M, measured above the static level; it also gives the volume discharged and the discharge time. If there is no check valve, Graph 186(b) must be used.

In the case of several tanks installed at successive high points of a pipeline, the minimum pressure that can occur in a zone is that correspond-ing to the water level in the preceding tank. Graph 186(c) gives data for the calculation of two tanks: the first close to the pump or with a check valve in the pipe, upstream; the second with check valve in the pipe, upstream. The distance between the tanks is equal to the distance between the second tank and the reservoir.

Either of the graphs mentioned can serve as a guide but more complex cases must be analysed by computer or graphically.

10.20 Pump by-pass check valve
When the downsurge due to water hammer leads to lower pressures than the level of the pump sump, i.e. when the discharge head is far less than cU_0/g, the low pressure can be alleviated by installing a check valve in a by-pass in parallel with the pump between the suction and the discharge lines (Fig. 10.12).

When there is a by-pass functioning after the pump stops, this causes a

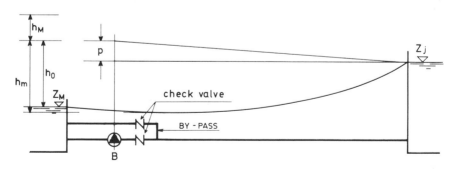

Fig. 10.12.

local head loss, ΔH_0 at discharge Q_0.

This head loss conditions the maximum lowering of pressure, Δh_m, which in accordance with Almeida [11], is given by:

$$\Delta h_{\mathrm{m}} = A^* \psi h_0 \tag{10.32}$$

in which:

$$\psi = 1 - 0.5 \left[\sqrt{(A^*/\Delta h_0)^2 - 4\,(1 - A^*)/\Delta h_0} - A^*/\Delta h_0 \right] \tag{10.32a}$$

A^* being the pipeline parameter (equation (10.12)) and $\Delta h_0 = \Delta H_0/h_0$.
The maximum overpressure the pump is given by:

$$\Delta h_{\mathrm{M}} = \Delta h_{\mathrm{m}} - P \tag{10.33}$$

where P is the head loss in the conduit.
   The mean time for annulment of the discharge is given by [11]:

$$\overline{T_{\mathrm{A}}} = \theta(A^* + 1) \tag{10.34}$$

   The minimum volume, $\forall$, of water in the pump sump is given empirically by [11]:

$$\frac{\forall c}{L Q_0} = K\,(0.52\,A^* + 1.8) \tag{10.35}$$

in which K is a safety coefficient ($5 < K < 10$).
   Since the pump does not stop instantaneously, a discharge will pass through it and join that which flows through the by-pass.
   The check valves of the by-pass may be special, quick-opening and slow-closing valves thus acting rapidly on the depression and reducing the effect of the counter-shock respectively.
   In the light of the characteristics of the valves, the phenomenon must be analysed using appropriate methods for calculating the water hammer.

10.21   Relief valves

There are a number of sophisticated water hammer relief valves on the market, their characteristics being given by the manufacturers. These valves are usually installed in the discharge pipe.
   Whenever safety conditions require it, installation of two or more valves in parallel, in order to ensure that at least one is operational, is recommended.
   Relief valves only provide protection from overpressures, so that it is advisable to determine whether the pressure reduction is acceptable. Even for the overpressure, since the opening time is about five seconds or more, relief valves will only be efficient in pipes whose length is such that the valve functions before arrival of the overpressure wave ($t > 2l/c$).

In short, relief valves are particularly indicated in the case of large pumping heads and pipes that are not too short.

Special attention must be paid to their design, quality of manufacture and maintenance.

10.22   Combined protective arrangements

The use of more than one of the protective devices that have been described above needs to be checked by means of mathematical models. Specialized literature should be consulted for formulating such models.

10.23   Unconventional devices

References are sometimes made to unconventional devices, such as:

— the use of two-way air valves;
— the use of intermediate check valves, usually consisting of flap valves with an orifice;
— the use of electronic control systems which, by measuring pressures, operate various valves and ensure the desired pressure;
— the installation of very flexible pipelines in order to reduce the velocity of the elastic waves;
— the use of air vessels along the pipeline, with elastic membranes, of the type used in small hydropneumatic discharge systems in buildings;
— the use of fuse-joints which break under high overpressures;
— the use of air vessels with automatic air admission, in such a way that the internal pressure is not less than atmospheric pressure, and which work partly as a surge shaft.

Use of these devices calls for the greatest care, and mathematical modelling of the phenomenon should be attempted, though this is not easy and always requires experience.

BIBLIOGRAPHY

[1] Fox, J. A., *Hydraulic Analysis of Unsteady Flow in Pipe Networks*, Macmillan, London, 1977.
[2] Wylie, E. B., Streeter, V. L., *Fluid Transients*, MacGraw-Hill, 1978.
[3] Chaudhry, M. H., *Applied Hydraulic Transients*, Van Nostrand, 1979.
[4] Watters, G. Z., *Modern Analysis and Control of Unsteady Flow in Pipelines*, Ann Arbor Science, 1979.
[5] Almeida, A. B., *Regime Hidráulico Transitório em Condutas Elevatórias*, Ph. D. thesis, Lisbon, 1981.
[6] Choan, R. K., *Wave Propagation in Two-phase Mixtures*, The City University, Res. Mem. No. ML78, 1978.
[7] Parmakian, J., *Waterhammer Analysis*, Prentice-Hall, New York, 1955.

[8] Stephenson, D., *Pipeline Design for Water Engineers*, Elsevier, Amsterdam, 1976.

[9] Wood, D. S., *Water Hammer Charts for Various Types of Valves*, Proc. ASCE, **99**, HY1, 1973.

[10] Driels, M., *Predicting Optimum Two-stage Valve Closure*, ASME, Paper 75–WA/FE–2, 1975.

[11] Almeida, A. B., *Manual de Protecção Contra o Golpe de Ariete em Condutas Elevatórias*. LNEC, Lisbon, 1981.

[12] Thorley, A. R. D., Enever, K. J., *Control and Suppression of Pressure Surges in Pipelines and Tunnels*, CIRIA, 1979.

[13] Bergeron, P., *Du Coup de Bélier en Hydraulique au Coup de Foudre en Electricité*, Dunod, Paris, 1950.

[14] Stucky, A., *Cours d'Aménagement des Chutes d'Eau. Chambres d'Equilibre*, Lausanne, 1951.

[15] Calame, J., Gaden, D., *Theorie des Chambres d'Equilibre*, Ed. de la Concorde, Lausanne, 1926.

[16] Jaeger, C., *Fluids Transients in Hydroelectric Engineering Practice*, Blackie, London, 1971.

[17] Bernhart, H. M., *The Dependence of Pressure Wave Transmission through Surge Tanks on the Valve Closure Time*, Proc. 2nd. International Conference on Pressure Surges, London, BHRA, 1976.

[18] Kinno, H., Kennedy, J. F., *Water Hammer Charts for Centrifugal Pump Systems*, Proc. ASCE, **91**, HY3, 247–270, 1965.

[19] Dubin, C., Guéneau, A., *Détermination des Dimensions Caractéristiques d'un Réservoir d'Air sur une Installation Elévatoire*, La Houille Blanche, No. 6, 1955.

[20] Meunier, M., Puech, C. H., *Étude du Fonctionnement et du Dimensionnement des Ballons d'Air Anti-Bélier*, Bulletin Technique du Génie Rural, No. 124, C.T.G.R.E.F., Paris, 1978.

[21] Miller, D. S., *Internal Flows Systems*, BHRA, Fluid Engineering, London, 1977.

[22] Ruus, E., *Charts for Water Hammer in Pipelines with Air Chambers*, Canadian Journal of Civil Engineering, **4**, 3, 1977.

[23] Dupont, A., *Hydraulique Urbaine*, Vol. II., Paris, Eyrolles, 1974.

Tables and graphs

Guide to the arrangement of the tables and graphs

Tables and graphs are inter-mixed, but numbered sequentially

Please note that, only in this Tables and Graphs section, the decimal comma rather than the decimal point has been chiefly used.

Tables and graphs

0. General tables — Index

1. Conversion table: measures of length

Name \ Symbol	m	cm	in	ft	yd	mi	na mi
metre	1	0,01	0,0254	0,3048	0,9144	1609,3	1853,2
centimetre	100	1	2,54	30,48	91,44	160 935	185 325
inch	39,37	0,3937	1	12	36	63 360	72 963
foot	3,2808	0,0328	0,08333	1	3	5280	6080,2
yard	1,0936	0,01093	0,0278	0,3333	1	1760	2026,8
mile	$6,21 \times 10^{-4}$	$6,21 \times 10^{-6}$	$1,58 \times 10^{-5}$	$1,89 \times 10^{-4}$	$5,68 \times 10^{-4}$	1	1,151
nautical mile	$5,39 \times 10^{-4}$	$5,39 \times 10^{-6}$	$1,37 \times 10^{-5}$	$1,64 \times 10^{-4}$	$4,92 \times 10^{-4}$	0,8684	1

2. Conversion table: measures of area

Name \ Symbol	m²	sq in	sq ft	sq yd	ac	ha	sq mi
square metre	1	$6,452 \times 10^{-4}$	0,0929	0,8361	4047	10^4	2 589 998
square inch	1550	1	144	1296	6272.640	155×10^5	4 014 489 600
square foot	10,76	0,00694	1	9	43.560	107 639	27 878 400
square yard	1,196	$7,716 \times 10^{-4}$	0,1111	1	4840	11 960	3 097 600
acre	$2,471 \times 10^{-4}$	$1,594 \times 10^{-7}$	$2,296 \times 10^{-5}$	$2,066 \times 10^{-4}$	1	2,471	640
hectare	10^{-4}	$6,452 \times 10^{-8}$	$9,29 \times 10^{-6}$	$8,361 \times 10^{-5}$	0,4047	1	259
square mile	$3,861 \times 10^{-7}$	$2,491 \times 10^{-10}$	$3,587 \times 10^{-8}$	$3,228 \times 10^{-7}$	$1,563 \times 10^{-3}$	$3,861 \times 10^{-3}$	1

Are (a) = 100 m². Hectare (ha) = 100 a. 1 km² = 100ha.

3.   Conversion table: measures of volume

Name \ Symbol	m^3	cu in	U.S. gal	imp gal	cu ft
cubic metre	1	$1,639\times10^{-5}$	$3,785\times10^{-3}$	$4,542\times10^{-3}$	$28,317\times10^{-3}$
cubic inch	61023,4	1	231	277,274	1728
U.S. gallon	264,17	0,004329	1	1,200	7,4805
Imperial gallon	220,08	0,003607	0,83311	1	6,2321
cubic foot	35,31	$5,787\times10^{-4}$	0,13368	0,16046	1

4.   Conversion table: measures of discharge

Name \ Symbol	m^3/s	U.S. gps	imp gps	ac ft pd	sec ft or cu sec
cubic metre per second	1	$3,785\times10^{-3}$	$4,542\times10^{-3}$	$14,276\times10^{-3}$	$28,317\times10^{-3}$
U.S. gallon per second	264,17	1	1,2	3,771	7,480
Imperial gallon per second	220,08	0,8333	1	3,142	6,232
acre feet per day	70,0	0,2652	0,3183	1	1,9835
cubic foot per second	35,31	0,1337	0,1605	0,5042	1

5. Conversion table for sundry units

In order to reduce unit A to unit B, multiply A by K_1
In order to reduce unit B to unit A, multiply B by K_2

Unit A	Factor K_1	Factor K_2	Unit B
Velocity			metre per second
foot per second (ft/s)	0.3048	3.2808	(m/s)
metre per minute			centimetre per second
(m/min)	1.677	0.6	(cm/s)
kilometre per hour			
(km/hr)	0.278	3.6	m/s
mile per hour (mi/hr)	1.609	0.621	km/hr
knot (nautical miles			
per hour)	1.853	0.539	km/hr
foot per second	0.6818	1.4667	mi/hr
Force			
pound (lb)	0.45359	2.2046	kilogram force (kgf)
libra	32.174	3.11×10^{-2}	*poundal* (pdl)
dyne	1.02×10^{-3}	981	gram force (gf)
poundal	14.098	7.09×10^{-2}	gf
Newton (10^5 dyne)	0.102	9.81	kgf
metric ton(t)	1000	10^{-3}	kgf
short ton	0.907	1.102	metric ton
long ton	1.016	0.984	metric ton
long ton	1.120	0.89286	short ton
ounce (oz)	28.350	3.527×10^{-3}	gf
Mass			
slug	14590.0	6.854×10^{-5}	gram (mass) (g)
slug	32.174	3.11×10^{-2}	pound (mass) (lb)
kilogram	0.102	9.81	metric unit of mass (m.u.m.)
Specific weight			
pound per cubic foot	15.710	6.37×10^{-2}	dyne per cubic centi-
(lb/ft^3)			metre (dyne/cm^3)
pound per cubic foot	16.02	6.24×10^{-3}	kgf/m^3
Newton per cubic metre	0.102	9.81	kgf/m^3
(N/m^3)	0.102	9.81	kgf/m^3
Specific mass			
slug per cubic foot	0.5154	1.9402	g/cm^3
slug per cubic foot	52.54	1.90×10^{-2}	m.u.m./m^3 (kg m^{-4} s^2)
kilograms (mass) per m^3	0.102	9.81	m.u.m./m ^{3}
gram per cm^3	1000	10^{-3}	kg/m^3
Pressure			
bar, megabar or			dyne/cm^2 (bar,
hectopieze	10^6	10^{-6}	microbar)
millibar	10^{-3}	10^3	bar
bar	1.0197	0.9807	kilogram force per square
			centimetre (kgf/cm^2)
atmosphere	1.033	0.968	kgf/cm^2
atmosphere	10.33	9.68×10^{-2}	metre of water
atmosphere	760	1316×10^{-3}	millimetre of mercury
atmosphere	14.697	0.06804	pound per square
			inch (lb/in^2)
atmosphere	29.921	0.03342	inch of mercury
atmosphere	33.901	0.0295	foot of water
millimetre of mercury	13.6×10^{-1}	73.6	metre of water
metre of water	0.1	10	kgf/cm^2
millimetre of mercury	13.6×10^{-4}	735.6	kgf/cm^2
pound per square foot			
(lb/ft^2)	0.488×10^{-3}	2.049×10^3	kgf/cm^2
pound per square inch			
(lb/in^2)	0.0703	14.225	kgf/cm^2
inch of mercury	0.03453	28.964	kgf/cm^2

5. (contd.)

Unit A	Factor K_1	Factor K_2	Unit B
foot of water	0.0310	32.308	kgf/cm^2
kgf/cm^2	9.81×10^4	1.02×10^{-5}	Pascal
atmosphere	10.133×10^4	9.689×10^{-6}	Pascal
metre of water column	9.81×10^3	1.02×10^{-4}	Pascal
millimetre of mercury, column	133.416	0.0075	Pascal
bar	10^3	10^{-5}	Pascal
Energy			
kilogram-metre (kgm)	9.81	0.102	Joule
kilogram-metre	0.272×10^{-5}	3.67×10^5	kilowatt hour (kWh)
erg	10^{-7}	10^7	Joule
calorie	0.427	2.342	kilogram-metre
Power			
horsepower (Hp) (B.U.)†	0.7456	1.341	kilowatt (kW)
horsepower†	1.014	0.986	horsepower (C.U.)
horsepower (C.U.)‡	0.736	1.36	kW
megawatt (MW)	1000	0.001	kW
kilogram-metre per second (kgm/s)	0.0133	75	horsepower (C.U.)
erg per second (erg/s)	10^{-7}	10^7	Watt (W) (joule per second)
Watt	0.102	9.81	kgm/s
kilowatt (kW)	1.360	0.7353	horsepower (C.U.)
Dynamic viscosity			
pound.second per square foot (lb.s/ft^2)	478.78	2.09×10^{-3}	dyne.s/cm^2 (poise)
lb.s/ft^2	4.876	0.205	kgf.s/m^2
dyne.s/cm^2 (poise)	1.02×10^{-2}	98	kgf.s/m^2
Newton.s/m^2	0.102	9.81	kgf.s/m^2
Newton.s/m^2	10	0.1	dyne.s/cm^2
Kinematic viscosity			
square foot per second (ft^2/s)	929.03	1.08×10^{-3}	cm^2/s (stokes)
ft^2/s	0.0929	10.8	m^2/s
Surface tension			
pound per foot (lb/ft)	14.594	0.68×10^{-4}	dyne/cm
lb/ft	1.488	0.672	kgf/m
Newton/m	0.012	9.81	kgf/m
Concentration			
pound per cubic foot (lb/ft^3)	16.02		gram per litre (g/l)
ounce per Imperial gallon	6.24	0.16	g/l
pound per U.S. gallon	119.82	8.35×10^{-1}	g/l
Angular measures			
sexagesimal degree (°)	1.11111	0.9	centes.degree (gr)
sexagesimal minute (′)	0.01852	54	gr
sexagesimal second (″)	0.00031	3240	gr
sexagesimal degree	0.01745	57.296	radian (rad)
sexagesimal minute	291×10^{-6}	3437.75	rad
sexagesimal second	4848×10^{-9}	206264.8	rad
Angular velocity			
revolution per day	7.2722×10^{-5}	1.375×10^4	radian/s (rad/s)
revolution/min	1.0472×10^{-1}	9.5493	rad/s
revolution/s	6.2832	0.1592	rad/s
degree/s	1.7453×10^{-2}	57.2967	rad/s
Angular acceleration			
revolution/s^2	6.2832	0.1592	radian/s^2 (rad/s^2)
revolution/min^2	1.7453×10^{-1}	572.967	rad/s^2
revolution/min.s	0.10472	9.5493	rad/s^2

†British units.
‡Continental units.

6. Equivalence between degrees Baumé and densities at 15.6°C
for liquids denser than water
Compiled from [6]

°B	0	1	2	3	4	5	6	7	8	9
0	1,0000	1,0069	1,0140	1,0211	1,0284	1,0357	1,0432	1,0507	1,0584	1,0662
10	1,0741	1,0821	1,0902	1,0985	1,1069	1,1154	1,1240	1,1328	1,1417	1,1508
20	1,1600	1,1694	1,1789	1,1885	1,1983	1,2083	1,2185	1,2288	1,2393	1,2500
30	1,2609	1,2719	1,2832	1,2946	1,3063	1,3182	1,3303	1,3426	1,3551	1,3679
40	1,3810	1,3942	1,4078	1,4216	1,4356	1,4500	1,4646	1,4796	1,4948	1,5104
50	1,5263	1,5426	1,5591	1,5761	1,5934	1,6111	1,6292	1,6477	1,6667	1,6860
60	1,7059	1,7262	1,7470	1,7683	1,7901	1,8125	1,8354	1,8590	1,8831	1,9079

7. Equivalence between degrees Baumé and densities at 15.6°C
for liquids less dense than water
Compiled from [6]

°B	0	1	2	3	4	5	6	7	8	9
10	1,0000	0,9929	0,9859	0,9790	0,9722	0,9655	0,9589	0,9524	0,9459	0,9396
20	0,9333	0,9272	0,9211	0,9150	0,9091	0,9032	0,8974	0,8917	0,8861	0,8805
30	0,8750	0,8696	0,8642	0,8589	0,8537	0,8485	0,8434	0,8383	0,8333	0,8284
40	0,8235	0,8187	0,8140	0,8092	0,8046	0,8000	0,7955	0,7910	0,7865	0,7821
50	0,7778	0,7735	0,7692	0,7650	0,7609	0,7568	0,7527	0,7487	0,7447	0,7407
60	0,7368	0,7330	0,7292	0,7254	0,7216	0,7179	0,7143	0,7107	0,7071	0,7035
70	0,7000	0,6965	0,6931	0,6897	0,6863	0,6829	0,6796	0,6763	0,6731	0,6669

8. Physical properties of fresh water at atmospheric pressure

$(g = 9.81 \text{ m/s}^2)$

Compiled from [1], [4], [13] and [20]

Temperature T (°C)	Density ρ (kg/m^3)	Specific weight γ (N/m^3) or kg/m.s	Dynamic viscosity μ (Ns/m^2)	Kinematic viscosity $\nu = \mu/\rho$ (m^2/s)	(cSt)	Surface tension (water with air) (N/m)	Vapour pressure h_v (metre of water column at 4°C)	Modulus of elasticity ε (N/m^2) approx. values
0	999,9	9 809,02	1776×10^{-6}	$1,78 \times 10^{-6}$	1,78	0,07564	0,062	$19,52 \times 10^8$
4	1 000,0	9 810,00	1570×10^{-6}	$1,57 \times 10^{-6}$	1,57	0,07514	0,083	
10	999,7	9 807,06	1315×10^{-6}	$1,31 \times 10^{-6}$	1,31	0,07426	0,125	$20,50 \times 10^8$
20	998,2	9 792,34	1010×10^{-6}	$1,01 \times 10^{-6}$	1,01	0,07289	0,239	$21,39 \times 10^8$
30	995,7	9 767,82	824×10^{-6}	$0,83 \times 10^{-6}$	0,82	0,07122	0,433	$21,58 \times 10^8$
40	992,2	9 733,48	657×10^{-6}	$0,66 \times 10^{-6}$	0,66	0,06965	0,753	$21,68 \times 10^8$
50	988,1	9 693,26	549×10^{-6}	$0,56 \times 10^{-6}$	0,56	0,06769	1,258	$21,78 \times 10^8$
60	983,2	9 645,19	461×10^{-6}	$0,47 \times 10^{-6}$	0,47	0,06632	2,033	$21,88 \times 10^8$
80	971,8	9 533,39	363×10^{-6}	$0,37 \times 10^{-6}$	0,37	0,06259	4,831	
100	958,4	9 401,90	275×10^{-6}	$0,29 \times 10^{-6}$	0,29	0,05896	10,333	

In the usual calculations of hydraulics, it is taken that $\rho = 1000 \text{ kg/m}^3$; $\gamma = 10000 \text{ N/m}^3$; $\nu = 10^{-6} \text{ m}^2/\text{s}$.

Note. The following values are given for the density of ice: 0°C, $\rho = 916,7 \text{ kg/m}^3$; -10°C, $\rho = 918,6 \text{ kg/m}^3$; -20°C, $\rho = 920,3 \text{ kg/m}^3$.

9. Physical properties of salt water

Compiled from [1]

(a) *Density ρ and specific weight γ $(g = 9.81 \text{ m/s}^2)$*

Temperature °C	Salinity = 30‰		Salinity = 35‰		Salinity = 40‰	
	ρ kg/m^3	γ N/m^3	ρ kg/m^3	γ N/m^3	ρ kg/m^3	γ N/m^3
0	1 024,11	10 046,52	1 028,13	10 085,96	1 032,17	10 125,59
5	1 023,75	10 042,99	1 027,70	10 081,74	1 031,67	10 120,68
10	1 023,08	10 036,42	1 026,97	10 074,58	1 030,88	10 112,93
15	1 022,15	10 027,29	1 025,99	10 064,96	1 029,85	10 102,83
20	1 020,99	10 015,91	1 024,78	10 053,09	1 028,60	10 090,57
25	1 019,60	10 002,28	1 023,37	10 039,26	1 027,15	10 076,34
30	1 018,01	9 986,68	1 021,75	10 023,37	1 025,51	10 060,25

(b) *Kinematic viscosity*, ν (salinity = 35‰): at 10°C — $1.40 \times 10^{-6} \text{m}^2/\text{s}$; at 15°C — $1.22 \times 10^{-6} \text{m}^2/\text{s}$ and at 20°C — $1.08 \times 10^{-6} \text{m}^2/\text{s}$.

Note. The following average values are indicated to the salinity of sea water: Atlantic Ocean 35.4‰; Indian Ocean 34.8‰; Pacific Ocean 34.9‰; Baltic Sea 37.8‰, Mediterranean Sea 34.9‰; Red Sea 38.8‰.

10. Values of atmospheric pressure
Compiled from [3]

Altitude (m)	Boiling point (°C)	Atmospheric pressure			
		(Atmosphere)	(N/cm^2)	(Metre of water)	(mm of mercury)
— 325	101,1	1,04	10,536	10,74	790
0	100,0	1,00	10,134	10,33	760
340	98,9	0,96	9,732	9,92	730
690	97,8	0,92	9,339	9,52	700
1 045	96,6	0,88	8,937	9,11	670
1 420	95,4	0,84	8,535	8,70	640
1 820	94,1	0,80	8,132	8,29	610
2 240	92,8	0,76	7,730	7,88	580
2 680	91,5	0,72	7,338	7,48	550
3 140	90,1	0,68	6,936	7,07	520

11. Physical properties of air at normal atmospheric pressure
Compiled from [13] and [20]

Temperature		Density	Specific weight	Dynamic viscosity μ		Kinematic viscosity ν	
(°F)	(°C)	ρ (kg/m^3)	γ (N/m^3)	poise (dyne.s/cm^2)	(N.s/m^2)	stokes (cm^2/s)	(m^2/s)
0	— 17,8	1,381	13,548	$1,62 \times 10^{-4}$	$1,62 \times 10^{-5}$	0,117	$1,17 \times 10^{-5}$
20	— 6,67	1,325	12,998	$1,68 \times 10^{-4}$	$1,68 \times 10^{-5}$	0,127	$1,27 \times 10^{-5}$
40	4,44	1,272	12,478	$1,73 \times 10^{-4}$	$1,73 \times 10^{-5}$	0,136	$1,36 \times 10^{-5}$
60	15,6	1,222	11,988	$1,79 \times 10^{-4}$	$1,79 \times 10^{-5}$	0,147	$1,47 \times 10^{-5}$
80	26,7	1,777	11,546	$1,84 \times 10^{-4}$	$1,84 \times 10^{-5}$	0,157	$1,57 \times 10^{-5}$
100	38	1,136	11,144	$1,90 \times 10^{-4}$	$1,90 \times 10^{-5}$	0,166	$1,66 \times 10^{-5}$
120	49	1,096	10,752	$1,95 \times 10^{-4}$	$1,95 \times 10^{-5}$	0,175	$1,75 \times 10^{-5}$
150	66	1,043	10,232	$2,03 \times 10^{-4}$	$2,03 \times 10^{-5}$	0,193	$1,93 \times 10^{-5}$
200	93	0,963	9,447	$2,15 \times 10^{-4}$	$2,15 \times 10^{-5}$	0,223	$2,23 \times 10^{-5}$

12. Values of the kinematic coefficient α and of the coefficient of momentum β

(a) *Rectilinear channels* $\alpha = 1.01$ to 1.10
   Natural watercourses $\alpha = 1.20$ to 1.50
   (Favre measured $\alpha = 1,74$ in the tailrace of a Kaplan turbine.)

(b) *According to Bazin, the following values are obtained* (C is the coefficient of Chézy's formula; see Section 4.6)

   Cross-sections of indefinite width $\alpha = 1 + 150 \times \dfrac{1}{C^2}$

   Very wide rectangular cross-sections $\alpha = 1 + 210 \times \dfrac{1}{C^2}$

   Semicircular cross-sections $\alpha = 1 + 240 \times \dfrac{1}{C^2}$

(c) *For distributions of velocities of a parabolic type shown in the figure, the following values of α are obtained according to the ratio* $\dfrac{V_1}{V_0}$ *[22].*

V_1/V_0	1/1	1/1,5	1/2	1/5	$V_1 = O$ $V_0 = 2\,U$
α	1,00	1,04	1,09	1,31	2,00

(d) *Also according to the experiments of Darcy and Bazin, the following indications can be given for the values of α [22].*
   Rectangular channels with wooden walls $\alpha = 1.052$
   Trapezoidal channels with wooden walls $\alpha = 1.048$
   Trapezoidal channels with masonry walls $\alpha = 1.071$
   Semicircular channels lined with cement $\alpha = 1.025$
   Semicircular channels lined with coarse sand $\alpha = 1.089$
   Trapezoidal channels lined with earth $\alpha = 1.100$

(e) *As general indication, the following examples are also given:*
   Horseshoe channels $\alpha = 1.07$
   Rectangular channels with obstacles $\alpha = 1.41$

(f) *In laminar flow, in circular cross-section,* $\alpha = 2.00$

(g) *The value of β falls between* 1.00 *and* 1.20 — There exists the following ratio:

$$\beta = 1 + \frac{\alpha - 1}{3} = \frac{\alpha + 2}{3}$$

(h) *According to Chow's indications, [35], α and β take the following values:*

Channels	Value of α			Value of β		
	Min.	Mean	Max.	Min.	Mean	Max.
Regular channels, spillways	1,10	1,15	1,20	1,03	1,05	1,07
Natural watercourses	1,15	1,30	1,50	1,05	1,10	1,17
Frozen rivers	1,20	1,50	2,00	1,07	1,17	1,33
Rivers with marginal fields flooded	1,50	1,75	2,00	1,17	1,25	1,33

(i) *The maximum known value is* $\alpha = 7.4$, *determined in a laboratory in spiral flow.*

13. Drag coefficients C of immersed bodies
(Formula (2.40))

Scheme	Description	C	Extent of validity	Characteristic area (A)	Characteristic dimension
	Flat plate aligned with flow	$1{,}33\ (\mathbf{R_e})^{-1/2}$ $0{,}074\ (\mathbf{R_e})^{-1/5}$	Laminar $\mathbf{R_e} < 10^7$	Area of plate	L
	Flat plate perpendicular to flow	$\begin{array}{c\|c} L/d & C \\ 1 & 1{,}18 \\ 5 & 1{,}2 \\ 10 & 1{,}3 \\ 20 & 1{,}5 \\ 30 & 1{,}6 \\ \infty & 1{,}95 \end{array}$	$\mathbf{R_e} > 10^3$	Area of plate	d
	Circular disc perpendicular to flow	$1{,}12$	$\mathbf{R_e} > 10^3$	Area of disc	D
	Flat plate perpendicular to flow without lateral detachment	$\begin{array}{c\|c} L/c & C \\ 5 & 1{,}95 \end{array}$	$\mathbf{R_e} > 10^3$	Area of plate	L
	Sphere	$24\ (\mathbf{R_e})^{-1/2}$ $0{,}47$ $0{,}20$	$\mathbf{R_e} < 1$ $10^3 < \mathbf{R_e} < 3 \times 10^5$ $\mathbf{R_e} > 3 \times 10^5$	Projected area	D
	Empty hemisphere	$0{,}34$	$10^4 < \mathbf{R_e} < 10^6$	Projected area	D
	Solid hemisphere	$0{,}42$	$10^4 < \mathbf{R_e} < 10^6$	Projected area	D

13 (contd.).   Drag coefficients C of immersed bodies
(Formula (2.40))

Scheme	Description	C		Extent of validity	Characteristic area (A)	Characteristic dimension
	Empty hemisphere	1,42		$10^4 < R_e < 10^6$	Projected area	D
	Solid hemisphere	1,17		$10^4 < R_e < 10^6$	Projected area	D
	Straight circular cylinder with axis normal to flow	L/d 1 5 10 20 30 ∞	C 0,63 0,8 0,83 0,93 1,0 1,2	$10^3 < R_e < 10^5$	Projected area	D
	Ditto, with elimination of detachment at the two bases	L/d 5	C 1 a 1,2	$10^3 < R_e < 10^5$	Projected area	D
	Straight circular cylinder with axis parallel to flow	L/d 5	C 0,9	$R_e < 10^3$	Projected area	D
	Parallelepiped with major axis normal to the flow	2,0		$R_e = 3,5 \times 10^4$	Projected area	D
	Hydrodynamic profile	L/d 5	C 0,06 a 0,1	$R_e > 2 \times 10^5$	Projected area	D

14. Position of the centre of gravity G, area A and square of the radius of gyration, K^2, of plane figures
Compiled from [23] and [27]

Figure	Position of the centre of gravity G	Area A	Square of the radius of gyration K^2†
	$V = \dfrac{2}{3}h =$ $= 0.6667\,h$	$A = \dfrac{1}{2}bh = 0.5\,bh$	$K_x^2 = \dfrac{h^2}{18} = 0.0556\,h^2$
	$V = \dfrac{a}{\sqrt{2}} =$ $= 0.707\,a$	$A = a^2$	$K_x^2 = \dfrac{a^2}{12} = 0.0833\,a^2$
	$V = \dfrac{h}{2} = 0.5\ h$	$A = bh$	$K_x^2 = \dfrac{h^2}{12} = 0.0833\,h^2$
	$V = \dfrac{ab}{\sqrt{a^2 + b^2}}$	$A = ab$	$K_x^2 = \dfrac{a^2b^2}{6(a^2 + b^2)}$
	$V = \dfrac{a\cos\theta + b\sin\theta}{2}$	$A = ab$	$K_x^2 = \dfrac{a^2\cos^2\theta + b\sin^2\theta}{12}$
	$V_1 = \dfrac{h}{3} \cdot \dfrac{2b + a}{b + a}$ $V_2 = \dfrac{h}{3} \cdot \dfrac{b + 2a}{b + 2}$	$A - h \cdot \dfrac{b + a}{2}$	$K_x^2 =$ $= \dfrac{h^2}{18} \cdot \dfrac{a^2 + 4ab + b^2}{(a + b)^2}$
	$V = R$	$A = \pi R^2 = 3.1416\,R^2$	$K_x^2 = \dfrac{R^2}{4} = 0.25\,R^2$
	$V_1 = 0.5756\,R$ $V_2 = 0.4244\,R$	$A = \dfrac{\pi R^2}{2} = 1.5708\,R^2$	$K_x^2 = 0.0699\,R^2$ $K_y^2 = \dfrac{R^2}{4} = 0.25\,R^2$
	$V = 0.5756\,R$	$A = \dfrac{\pi R^2}{4} = 0.7854\,R^2$	$K_x^2 = 0.0700\,R^2$

† It is to be noted that the value of K_x^2 is not changed by the figure being turned 180° in relation to the axis xx. This is what happens, for example, in the case of a triangle, if the vertex is turned downwards.

14. (contd.)   Position of the centre of gravity G, area A, and square of the radius of gyration K^2, of plane figures
Compiled from [23] and [27]

Figure	Position of the centre of gravity G	Area A	Square of the radius of gyration K^2
	$V = 0.2234\ R$	$A = R^2\left(1 - \dfrac{\pi}{4}\right)$ $= 0.2146\ R^2$	$K_x^2 = 0.0349\ R^2$
	$V_1 = R \times$ $\times\left(1 - \dfrac{2\sin\alpha}{3\alpha}\right)$ $V_2 = 2\,R\,\dfrac{\sin\alpha}{3\alpha}$	$A = \alpha R^2$ (α in radians)	$K_x^2 = \dfrac{R^2}{4}\left(1 + \dfrac{\sin 2\alpha}{2\alpha} - \dfrac{16\sin^2\alpha}{9\alpha^2}\right)$ $K_y^2 = \dfrac{R^2}{4}\left(1 - \dfrac{\sin 2\alpha}{2\alpha}\right)$
	$V_1 = R \times$ $\left(1 - \dfrac{4\sin 3\alpha}{6\alpha - 3\sin 2\alpha}\right)$ $V_2 = R\left(\dfrac{4\sin 3\alpha}{6\alpha - 3\sin 2\alpha} - \cos\alpha\right)$	$A = \dfrac{R^2}{2}$ $(2\alpha - \sin 2\alpha)$ (α in radians)	$K_x^2 = \dfrac{R}{4} \times$ $\left(1 + \dfrac{2\sin 2\alpha \cdot \sin^2\alpha}{2\alpha - \sin 2\alpha} - \dfrac{64}{9}\dfrac{\sin^6\alpha}{(2\alpha - \sin 2\alpha)^2}\right)$ $K_y^2 = \dfrac{R^2}{4}\left(1 - \dfrac{2}{3}\cdot \dfrac{\sin 2\alpha\,\sin^2\alpha}{2\alpha - \sin 2\alpha}\right)$
	$V = a$	$A = \pi ab$ $= 3.1416\,ab$	$K_x^2 = \dfrac{a^2}{4}$ $K_y^2 = \dfrac{b^2}{4}$
	$V_1 = \dfrac{3}{5}a = 0.6a$ $V_2 = \dfrac{2}{5}a = 0.4a$	$A = \dfrac{\pi}{2}ab =$ $= 1.5708\,ab$	$K_x^2 = 0.0582\,a^2$ $K_y^2 = 0.1698\,b^2$
	$V_1 = \dfrac{a}{2\tan\alpha}$ $V_2 = \dfrac{a}{2\sin\alpha}$	$A = \dfrac{1}{4}na^2\tan^{-1}\alpha$ $= 0.25\,n\tan^{-1}\alpha \cdot a^2$	$K_1^2 = \dfrac{12V_1^2 + a^2}{48}$ $K_2^2 = \dfrac{6V_2^2 - a^2}{24}$

Regular polygon with n sides	n	α	$1/2\ \sin\alpha$	$1/2\ \tan\alpha$	$0.25\ n\cot\alpha$
	5	$36°$	0,8506	0,6882	1,7205
	6	$30°$	1,0000	0,8660	2,5981
	7	$25°\ 42'\ 51''{,}4$	1,1537	1,0397	3,6339
	8	$22°\ 30'$	1,3066	1,2071	4,8284
	9	$20°$	1,4619	1,3737	6,1818
	10	$18°$	1,6180	1,5388	7,6942
	12	$15°$	1,9318	1,8660	11,1962

15. Flow in pipes — Index

16. Pipes, friction losses, summary

$$\Delta H = Li = Lf\frac{1}{D}\frac{U^2}{2g} = LbQ^2$$

Name	Field of application	f	b (circular pipes)	Remarks
Moody	This diagram is valid for any type of flow (laminar, smooth or rough turbulent flow) and for any fluid. *It must be used* whenever it is feared that the flow is not completely turbulent, namely in large industrial conduits or when the fluid is different from water at normal temperature.	Given by the diagram according to the Reynolds number, $\mathbf{R_e}$, and the relative roughness ε/D.	$0.0826\, f D^{-5}$	Moody's diagram is given in Graph 18. Auxilliary tables and graphs: 8, 9, 16.
Chézy	Formula initially deduced for channels, but can give good results in rough pipes, *rough turbulent flow*, provided that the coefficient of roughness is judiciously fixed.	$\dfrac{8g}{C^2}$	$\dfrac{6.48}{C^2} D^{-5}$	The value of C is given by Bazin's and Kutter's formulae. Auxilliary tables and graphs: 20, 21.
Darcy	Valid for cast piping and in *rough turbulent flow*.	$8\, gb$	$6.48\, b\, D^{-5}$	Auxiliary table: 23.
Manning	Valid for rough pipes or channels where there is a turbulent flow.	$\dfrac{124.6}{K_5^2} D^{-0.333}$	$\dfrac{10.3}{K_5^2} D^{-5.333}$	Auxilliary tables and graphs: 22.
Other monomial formulae	Valid for pipes and flow conditions for which they were deduced.	Varies according to the type of formula.	Varies according to the type of formula.	

Notes.   For studying the ageing of pipelines, see Tables 26, 27, 28.

17. Values of absolute roughness

Table prepared based on [17], [31], [32] and [64]

Characteristics	Roughness, ε, in mm		
	Lower	Upper	Normal
(1) Galleries			
Rock unlined in the whole perimeter	100	1000	—
Rock unlined but sill lined	10	100	—
(2) Concrete pipes			
Unusually rough — very rough shuttering wood, poor concrete with erosion wear, poor alignment at joints	0.6	3.0	1.5
Rough — attacked by angular matter transported, visible form marks, spalling	0.4	0.6	0.5
Granular — brushed surface, in good condition, smooth joints	0.18	0.4	0.3
Centrifugally cast	0.15	0.5	0.3
Smooth — steel forms smooth, average workmanship, smooth joints, new or nearly so	0.06	0.18	0.1
Very smooth — first-class workmanship, very smooth, steel forms, smooth joints	0.015	0.06	0.03
(3) Steel pipes — ends welded, interior continuous			
Severe tuberculation and encrustation	2.4	12.2	7.0
General tuberculation of 1 to 3 mm	0.9	2.4	1.5
Heavy brush-coated, enamels and tars, with thick coat	0.3	0.9	0.6
Light rust	0.15	0.30	0.2
Hot-asphalt-dipped	0.06	0.15	0.1
Centrifugally applied concrete lining	0.05	0.15	0.1
Centrifugally applied enamel lining	0.01	0.3	0.06
Natural bitmen (gelsonite), pistol-applied cold in thickness of 0.4 mm	—	—	0.042
Bituminous enamel (coal tar), brush-coated with 2 to 2.5 mm thickness	—	—	0.040
Ditto, applied with spatula	—	—	0.030
Ditto, hot application, flame-smoothed	—	—	0.012
(4) Steel piping: rivetted at transverse joints			
(Joints 5 to 10 metres apart, longitudinal seam welded)			
Severe tuberculation and encrustations	3.7	12.2	8.0
General tuberculation of 1 to 3 mm	1.4	3.7	2.5
Rusted	0.6	1.4	1.0
Heavy brush-coated asphalts and tars, in thick coat	0.9	1.8	1.5
Hot-asphalt-dipped, brush-coated graphite	0.3	0.9	0.6
New smooth pipe, centrifugally applied enamels	0.15	0.6	0.4
(5) Steel pipes: rivetted at transverse and longitudinal joints			
(Transverse joints 1.8 to 2.4 metres)			
Severe tuberculations and encrustations	6.0	12.2	9.0
General tuberculation of 1 to 3 mm			
3 rows of rivets at the transverse joints	4.6	6.0	5.0
2 rows of rivets at the transverse joints	3.0	4.6	3.5
1 row of rivets at the transverse joints	2.1	3.0	2.5
Fairly smooth			
3 rows longitudinal rivets 3 transverse rows	1.8	2.1	2.0
2 transverse rows	1.5	1.8	1.6
1 transverse row	1.1	1.5	1.3

17.   (Contd.) Values of absolute roughness

2 rows longitudinal rivets	3 transverse rows	1.2	1.5	1.3
	2 transverse rows	0.9	1.2	1.1
	1 transverse row	0.6	0.9	1.2
1 row longitudinal rivets	3 transverse rows	0.8	1.1	1.0
	2 transverse rows	0.5	0.8	0.6
	1 transverse row	0.3	0.5	0.4

(6) Wood stave

Excessive growth on walls, rough projecting staves with protruding joints	0.3	3.5	3.2
Used, in good condition	0.12	0.3	0.2
New, first-class construction	0.03	0.12	0.07

(7) Asbestos cement piping

		0.025	0.015

(8) Iron pipes

Wrought iron, rusty	0.15	3.00	0.6
Galvanized iron, lined cast iron	0.06	0.3	0.15
Unlined cast iron, new	0.25	1.0	0.5
Cast iron with corrosion	1.0	3.0	1.5
Cast iron, with deposit	1.0	4.0	2.0

(9) Stoneware pipes

Very well aligned joints		0.06		
In lengths of 1.0 m:	$D < 600$ mm	—	0.3	0.15
	$D > 600$ mm	—	0.6	0.3
In lengths of 0.6 m:	$D < 300$ mm	—	0.3	0.15
	$D > 300$ mm	—	0.6	0.3

(10) Wastewater pipes in use, when the new materials have roughness less than those indicated for pipes in use

With layers of slime less than 5 mm	0.6	3.0	1.5
With slimy or fatty encrustation les than 25 mm	6.0	30	15
With sandy solid material on the sill, deposited irregularly	60	300	150

(11) Smooth materials

Brass, copper, lead	0.04	0.010	0.007
Aluminium	0.0015	0.005	0.004

(12) Extra-smooth materials

Glass	0.001	0.002	—
Polyurethane + epoxy. Pistol applied without air and at ambient temperature, in a coat of 0.1 to 0.2 mm (without joints)	0.002	0.004	—
Vinyl (polyvinyl acetochloride or vinyl polychloride) Ditto	0.003	0.004	
Araldite (epoxy). Ditto	0.0025	0.003	—

18. Moody diagram

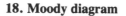

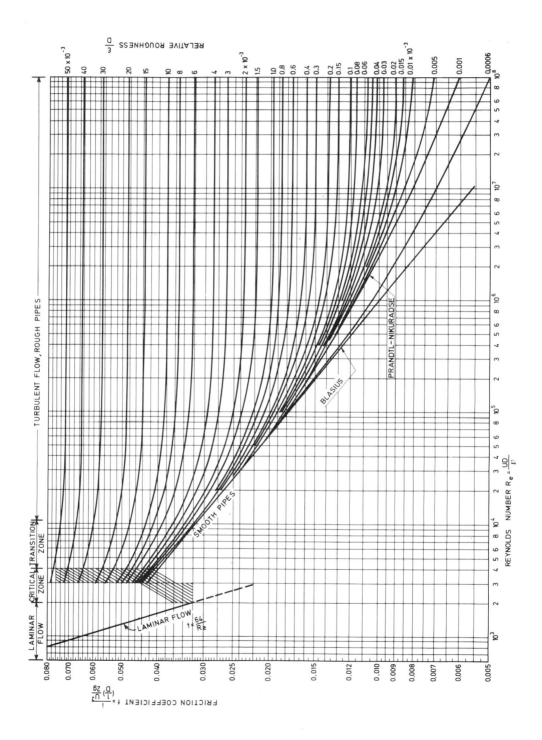

19.   Laminar flow in non-circular pipes

Values of a of formula $U = a\dfrac{gAI}{u}$

Extracted from [19]

Characteristics of the cross-section	a
Square	0.035
Rectangular — ratio between sides 1/2	0.029
Rectangular — ratio between sides 1/3	0.022
Rectangular — ratio between sides 1/4	0.017
Rectangular — ratio between sides 1/5	0.015
Rectangular — ratio between sides 1/10	0.008
Equilateral triangle	0.029
Elliptical with semi-axes a and b	$\dfrac{1}{4\pi}\dfrac{ab}{a^2+b^2}$
Circular (Poiseuille's formula)	$\dfrac{1}{8\pi} = 0.040$

20.   Values of K_B of Bazin's formula $C = \dfrac{87\sqrt{R}}{K_B + \sqrt{R}}$

(R in metres)

Characteristics of the piping	Values of K_B (m$^{1/2}$)
New asbestos cement pipes	0.06
New rolled steel pipes	0.10
New cast iron pipes	0.16
Concrete pipes in good condition	0.18
Used cast iron pipes	0.23
Cast iron pipes with encrustations	0.36

21.   Values of K_k of Kutter's formula $C = \dfrac{100\sqrt{R}}{K_K + \sqrt{R}}$

(R in metres)

Characteristics	Values of K_K (m$^{1/2}$)
New cast iron or concrete pipes	0.15 to 0.175
Cast iron pipes in current service, with water not causing encrustation	0.275
Cast iron pipes with many years service	0.35
Cast iron pipes with much encrustation and dirty water	0.45

22. Resolution of Manning formula for circular pipes

Values of a ($i = aQ^2$) given to D and K_s

(i in metres per kilometre; Q in cubic metre per second)

For i in metres per metre, divide the value of a given in the table, by 1000

D (m)	$K_s = 20m$	$K_s = 40m^{1/3}/s$	$K_s = 50m^{1/3}/s$	$K_s = 60m^{1/3}/s$	$K_s = 70m^{1/3}/s$	$K_s = 75m^{1/3}/s$
0,10	5543354,4	1385946,2	886936,7	615904,3	452509,9	394170,2
20	137524,0	34383,7	22003,8	15279,9	11226,2	9778,89
30	15822,8	3956,0	2531,65	1758,02	1291,63	1125,11
40	3411,86	853,031	545,898	379,081	278,514	242,606
0,50	1037,930	259,503	166,069	115,321	84,7273	73,8039
60	392,554	98,1462	62,8087	43,6155	32,0446	27,9133
70	172,529	43,1357	27,6047	19,1692	14,0838	12,2680
80	84,6455	21,1630	13,5433	9,4047	6,9097	6,0189
90	45,1651	11,2922	7,2264	5,0182	3,6869	3,2115
1,00	25,7500	6,4380	4,1200	2,8610	2,1020	1,8310
10	15,4897	3,8727	2,4783	1,7210	1,2644	1,1014
20	9,7387	2,4349	1,5582	1,0820	0,7950	0,6925
30	6,3550	1,5889	1,0168	0,7061	0,5188	0,4519
40	4,2803	1,0702	0,6849	0,4756	0,3494	0,3044
1,50	2,9627	0,7407	0,4740	0,3292	0,2418	0,2107
60	2,1000	0,5250	0,3360	0,2333	0,1714	0,1493
70	1,5198	0,3800	0,2432	0,1689	0,1241	0,1081
80	1,1205	0,2801	0,1793	0,1245	0,09147	0,07967
90	0,8398	0,2100	0,1344	0,09331	0,06856	0,05972
2,00	0,6388	0,1597	0,1022	0,07098	0,05215	0,04543
10	0,4925	0,1231	0,07880	0,05472	0,04020	0,03502
20	0,3843	0,09608	0,06148	0,04270	0,03137	0,02732
30	0,3032	0,07580	0,04851	0,03368	0,02475	0,02156
40	0,2416	0,06041	0,03866	0,02684	0,01972	0,01718
2,50	0,1943	0,04859	0,03109	0,02159	0,01586	0,01382
60	0,1577	0,03942	0,02523	0,01752	0,01287	0,01121
70	0,1289	0,03223	0,02063	0,01432	0,01052	0,009167
80	0,1062	0,02655	0,01699	0,01180	0,008668	0,007551
90	0,8807	0,02202	0,01409	0,009785	0,007189	0,006262
3,00	0,07350	0,01838	0,01176	0,008166	0,006000	0,005226
10	0,06171	0,01543	0,009873	0,006856	0,005037	0,004388
20	0,05210	0,01303	0,008336	0,005788	0,004253	0,003704
30	0,04421	0,01105	0,007074	0,004912	0,003609	0,003144
40	0,03771	0,009427	0,006033	0,004189	0,003078	0,002681
3,50	0,03230	0,008077	0,005169	0,003589	0,002637	0,002297
60	0,02780	0,006950	0,004448	0,003089	0,002269	0,001977
70	0,02402	0,006005	0,003843	0,002669	0,001961	0,001708
80	0,02084	0,005209	0,003334	0,002315	0,001701	0,001482
90	0,01814	0,004535	0,002902	0,002015	0,001481	0,001290
4,00	0,01585	0,003963	0,002536	0,001761	0,001294	0,001127

Unlined tunnels, very irregular and completed abandoned	$K_s = 20\,m^{1/3}/s$
Unlined tunnels with large protruding blocks	$K_s = 30$ to 40
Unlined, regular tunnels	$K_s = 50$
Riveted steel pipes or with a lot of welds. Tunnels of coarse or aged concrete and tunnels of masonry in a bad state	$K_s = 60$
Cast iron or concrete pipes with a lot of use, large encrustations. Tunnels of coarse masonry	$K_s = 70$
Concrete pipes with frequent joints; cast iron pipes in current service	$K_s = 75$

22.   (Contd.) Resolution of Manning formula for circular pipes
Values of a $(i = aQ^2)$ given to D and K_s
(i in metres per kilometre; Q in cubic metre per second)
For i in metres per metre, divide the value of a given in the table, by 1000

D (m)	$K_s = 80\mathrm{m}^{1/3}/\mathrm{s}$	$K_s = 85\mathrm{m}^{1/3}/\mathrm{s}$	$K_s = 90\mathrm{m}^{1/3}/\mathrm{s}$	$K_s = 100\mathrm{m}^{1/3}/\mathrm{s}$	$K_s = 110\mathrm{m}^{1/3}/\mathrm{s}$	$K_s = 120\mathrm{m}^{1/3}/\mathrm{s}$
0,10	346378,9	306983,4	273830,9	221734,2	183199,8	153922,3
20	8593,25	7615,89	6793,42	5500,96	4544,97	3818,63
30	988,693	876,244	781,615	632,911	522,920	439,351
40	213,192	188,944	168,539	136,474	112,625	94,7370
0,50	64,8555	57,4791	51,2717	41,5172	34,3021	28,8201
60	24,5289	21,7391	19,3914	15,7022	12,9734	10,9001
70	10,7806	9,5544	8,5226	6,9012	5,7018	4,7906
80	5,2891	4,6877	4,1813	3,3858	2,7974	2,3503
90	2,8222	2,5012	2,2311	1,8066	1,4926	1,2541
1,00	1,6090	1,4260	1,2720	1,0300	0,8510	0,7150
10	0,9679	0,8578	0,7652	0,6196	0,5119	0,4301
20	0,6085	0,5393	0,4811	0,3895	0,3218	0,2741
30	0,3971	0,3519	0,3139	0,2542	0,2100	0,1765
40	0,2675	0,2370	0,2114	0,1712	0,1415	0,1189
1,50	0,1851	0,1641	0,1463	0,1185	0,09791	0,08226
60	0,1312	0,1163	0,1037	0,08400	0,06940	0,05831
70	0,09497	0,08416	0,07508	0,06079	0,05023	0,04220
80	0,07001	0,06205	0,05535	0,04482	0,03703	0,03111
90	0,05248	0,04651	0,04149	0,03359	0,02776	0,02332
2,00	0,03992	0,03538	0,03156	0,02555	0,02111	0,01774
10	0,03077	0,02727	0,02433	0,01970	0,01628	0,01367
20	0,02401	0,02128	0,01898	0,01537	0,01270	0,01067
30	0,01894	0,01679	0,01498	0,01213	0,01002	0,008418
40	0,01510	0,01338	0,01193	0,009664	0,007985	0,006709
2,50	0,01214	0,01076	0,009600	0,007774	0,006423	0,005396
60	0,009852	0,008731	0,007788	0,006307	0,005211	0,004378
70	0,008055	0,007139	0,006368	0,005157	0,004261	0,003580
80	0,006635	0,005881	0,005246	0,004248	0,003509	0,002949
90	0,005503	0,004877	0,004350	0,003523	0,002910	0,002445
3,00	0,004593	0,004070	0,003631	0,002940	0,002429	0,002041
10	0,003856	0,003417	0,003048	0,002468	0,002039	0,001713
20	0,003255	0,002855	0,002573	0,002084	0,001722	0,001447
30	0,002763	0,002448	0,002184	0,001768	0,001461	0,001228
40	0,002356	0,002088	0,001863	0,001508	0,001246	0,001047
3,50	0,002019	0,001789	0,001596	0,001292	0,001068	0,0008970
60	0,001737	0,001539	0,001373	0,001112	0,0009187	0,0007719
70	0,001501	0,001330	0,001186	0,0009607	0,0007938	0,0006669
80	0,001302	0,001154	0,001029	0,0008334	0,0006886	0,000578
90	0,001133	0,001005	0,0008961	0,0007256	0,0005995	0,0005037
4,00	0,0009903	0,0008777	0,0007829	0,0006340	0,0005238	0,0004401

Pipes with coarse plastering; stoneware pipes; in thin plate and with
protruding welds, in fairly smooth masonry; in new cast iron — $K_s = 80\,\mathrm{m}^{1/3}/\mathrm{s}$
Concrete pipes or in steel embedded in concrete — $K_s = 85$
Concrete pipes very smooth; in finished wood; in metal plate without
protruding welds; in asbestos cement — $K_s = 90$ to 100
Galvanized iron pipes — $K_s = 100$ to 110
Brass and glass pipes, polyethylene or polyvinyl pipes — $K_s = 125$

23. Darcy's formula. Values of a of the formula $\Delta H = L a Q^2$

According to [24]

(a) *Cast iron pipes in service*

$$b = 0{,}000\,507 + 0{,}000\,012\,94 \times \frac{1}{D}$$

D (m)	$a = \dfrac{64b}{\pi^2 D^5}$	D (m)	$a = \dfrac{64\,b}{\pi^2 D^5}$	D (m)	$a = \dfrac{64\,b}{\pi^2 D^5}$	D (m)	$a = \dfrac{64\,b}{\pi^2 D^5}$
0,01	116 785 000	0,09	713,78	0,17	26,624	0,45	0,188 00
02	2 338 500	10	412,42	18	19,835	50	0,110 39
03	250 310	11	251,24	19	15,058	60	0,044 031
04	52 560	12	160,01	20	11,571	70	0,020 255
0,05	15 874	0,13	105,84	0,25	3,705 2	0,80	0,010 350
06	6 021	14	72,220	30	1,467 7	90	0,005 721 4
07	2 266	15	50,639	35	0,670 40	1,00	0,003 365 5
08	1 321,9	16	36,301	40	0,341 33		

(b) *Cast iron pipes, new*

Take half of the values given in (a)

(c) *Pipes in asphalted plate*

Take 1/3 of the values given in (a)

24. Flow in hoses. Values of C, $\dfrac{Di}{U^2}$, $\dfrac{\Delta D}{D}$ and $\dfrac{\Delta i}{i}$

Extracted from [9]

Characteristics	C $(m^{1/2}s^{-1})$	For a pressure increase of 10 N/cm²	
		$\dfrac{\Delta D}{D}$	$\dfrac{-\Delta i}{i}$
Very smooth rubber hose	68.2	0.0041	0.020
Coarse rubber hose	66.7	0.0034	0.017
Very smooth cotton and rubber hose	66.5	0	0
Coarse cotton and rubber hose	49.5	0.0018	0.009
Flax and hemp hose	43.3	0.0006	0.003
Top quality leather hose	53.9	0.0061	0.031

25. Comparison between the coefficients of the empirical formulae and absolute roughness
Extracted from [34]

(a) *Variation K_s and K_B with ε for various diameters.*

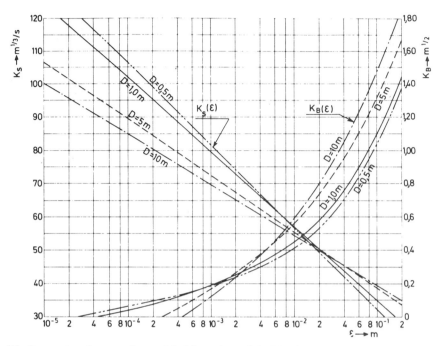

(b) *Comparison between the empirical formulae and the Moody diagram.*

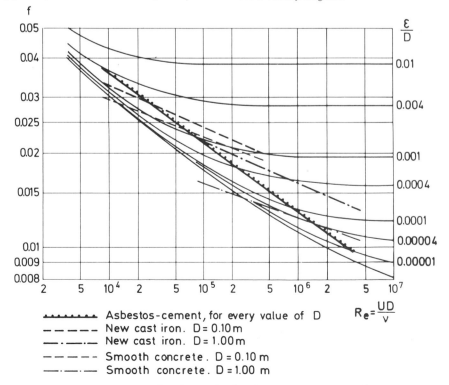

········· Asbestos-cement, for every value of D
— — — — New cast iron. D = 0.10 m
—·—·—· New cast iron. D = 1.00 m
— — — — Smooth concrete. D = 0.10 m
—·—·— Smooth concrete. D = 1.00 m

$$R_e = \frac{UD}{v}$$

26. Ageing of conduits

(a) *Roughness increase, in cast-iron and steel pipes that are unlined or lined internally with bituminous or equivalent products*†

Tendencies	Mean values (mm/year)	Degree of attack
1	0.025	Light
2	0.075	Moderate
3	0.25	Considerable
4	0.75	Severe
5	2.50	Very severe
6	7.50	Extreme

(b) *Reduction of carrying capacity, due to ageing of the pipes, for different types of water*‡

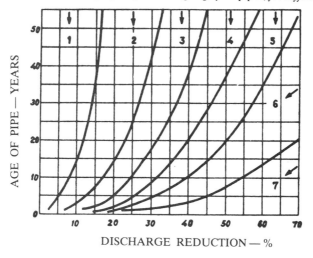

1 – Extreme cases of non-aggressive water. Small nodules.
2 – Non-aerated and practically non-corrosive filtered water. Slight general incrustation.
3 – Well water or hard water with little corrosive action. Larger incrustations, with nodules up to about 12 mm in height.
4 – Water from swampy regions with traces of iron and with organic matter, slightly acid. Large incrustations up to about 25 mm in height.
5 – Acid water of granitic rocks. Excessive incrustations and tuberculations.
6 – Extremely corrosive water; small pipes for fresh water, slightly acid.
7 – Extreme cases of highly aggresive water.

---

†Colebrook, C.F. and White, C.M., *The Reduction of Carrying Capacity of Pipes with Age*, Journal of the Institution of Civil Engineers. No. 1, 1937–38.
‡Price, A., *Kempe's Engineers' Year-Book*, Morgan Brothers, London, 1947. Quoted by [28].

27. Langelier's index

(a) *Values of K of expression* 4.26.†

Dry residue (p.p.m.)	K				
	0°C	10°C	15°C	20°C	30°C
—	2,45	2,23	2,12	2,02	1,86
20	2,54	2,32	2,21	2,11	1,95
40	2,58	2,36	2,25	2,15	1,99
80	2,62	2,40	2,29	2,19	2,03
120	2,66	2,44	2,34	2,23	2,07
160	2,68	2,46	2,36	2,25	2,09
200	2,71	2,49	2,38	2,28	2,12
240	2,74	2,52	2,42	2,31	2,15
280	2,76	2,54	2,44	2,33	2,17
320	2,78	2,56	2,46	2,35	2,19
360	2,79	2,57	2,47	2,36	2,20
400	2,81	2,59	2,48	2,38	2,22
440	2,83	2,61	2,51	2,40	2,24
480	2,84	2,62	2,52	2,41	2,25
520	2,86	2,64	2,54	2,43	2,27
560	2,87	2,65	2,55	2,44	2,28
600	2,88	2,66	2,56	2,45	2,29
640	2,90	2,68	2,58	2,47	2,31
680	2,91	2,69	2,59	2,48	2,32
720	2,92	2,70	2,60	2,49	2,33
760	2,92	2,70	2.60	2,49	2,33
800	2,93	2,71	2,61	2,50	2,34

28. Approximate values of pH_s according to the total alkalinity

Alkalinity (p.p.m.)	5	10	15	20	25	30	40	70	120	190
pH_s	10	9,7	9,3	8,9	8,7	8,5	8,1	7,8	7,5	7,2

†Lamont, P., *Formulae for Pipe-line Calculations*, International Water Supply Association, Third Congress, London, 1955.

29. Head losses in sudden enlargements
Compiled from [33]

$$\Delta H = K \frac{U_1^2}{2g} \qquad \eta = \frac{A_1}{A_2}$$

1. *Uniform distribution of velocities* ($\alpha = 1$; $\beta = 1$)

    1.1 — $\mathbf{R}_e < 10$ $K = 26/\,\mathbf{R}_e$ (Karev, 1953)

    1.2 — $10 < \mathbf{R}_e < 3.500$

Values of K

A_1/A_2	$\mathbf{R}_e$												
	10	15	20	30	40	50	10^2	2.10^2	5.10^2	10^3	2.10^3	3.10^3	$3,5.10^3$
0,1	3,10	3,20	3,00	2,40	2,15	1,95	1,70	1,65	1,70	2,00	1,60	1,00	0,81
0,2	3,10	3,20	2,80	2,20	1,85	1,65	1,40	1,30	1,30	1,60	1,25	0,70	0,64
0,3	3,10	3,10	2,60	2,00	1,60	1,40	1,20	1,10	1,10	1,30	0,95	0,60	0,50
0,4	3,10	3,00	2,40	1,80	1,50	1,30	1,10	1,00	0,85	1,05	0,80	0,40	0,36
0,5	3,10	2,50	2,30	1,65	1,35	1,15	0,90	0,75	0,65	0,90	0,65	0,30	0,25
0,6	3,10	2,70	2,15	1,55	1,25	1,05	0,80	0,60	0,40	0,60	0,50	0,20	0,16

    1:3 — $\mathbf{R}_e > 3.500$ $K = (1 - \eta)^2$

2. *Non-uniform distribution of velocities* (Idel'cik, 1984)

$$K = \eta^2 + \alpha - 2\,\eta\,\beta$$

α — *Coriolis coefficient*; β — *Boussinesq coefficient*   (Table 12).

30. Head losses in diffusers in pipes
Compiled from [33]

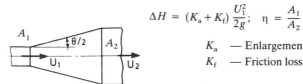

$$\Delta H = (K_a + K_f)\frac{U_1^2}{2g}; \quad \eta = \frac{A_1}{A_2}$$

K_a — Enlargement loss coefficient

K_f — Friction loss coefficient

1. *Loss from enlargement*: $K_a = \psi \cdot \phi (1-\eta)^2$

   1.1 *Values of ψ, corresponding to the distribution of velocities*

Values of ψ

V_M / U_1	θ^o						
	5	8	10	14	20	24	28
1,00	1,00	1,00	1,00	1,00	1,00	1,00	1,00
1,02	1,10	1,12	1,14	1,15	1,10	1,07	1,02
1,04	1,14	1,20	1,23	1,26	1,19	1,10	1,03
1,06	1,17	1,27	1,31	1,36	1,24	1,14	1,04
1,08	1,19	1,42	1,49	1,49	1,31	1,18	1,05
1,10	1,19	1,54	1,62	1,54	1,34	1,20	1,06
1,12	1,22	1,62	1,68	1,57	1,36	1,21	1,06
1,14	1,22	1,68	1,81	1,60	1,36	1,21	1,06
1,16	1,22	1,78	1,89	1,61	1,36	1,21	1,06

V_M — Maximum velocity in cross-section A_1

U_1 — Mean velocity in cross-section A_1

θ – Angle of aperture of the diffuser

   1.2   *Values of $\phi(1-\eta)^2$*

   1.2.1   *Truncated cone diffusers (conical diffusers)*

Values of $\phi(1-\eta)^2$

$\frac{A_1}{A_2}$	θ^o													
	3	6	8	10	12	14	16	20	24	30	40	60	90	180
0	0,03	0,08	0,11	0,15	0,19	0,23	0,27	0,36	0,47	0,65	0,92	1,15	1,10	1,02
0,05	0,03	0,07	0,10	0,14	0,16	0,20	0,24	0,32	0,42	0,58	0,83	1,04	0,99	0,92
0,075	0,03	0,07	0,09	0,13	0,16	0,19	0,23	0,30	0,40	0,55	0,79	0,99	0,95	0,88
0,10	0,03	0,07	0,09	0,12	0,15	0,18	0,22	0,29	0,38	0,52	0,75	0,93	0,89	0,83
0,15	0,02	0,06	0,08	0,11	0,14	0,17	0,20	0,26	0,34	0,46	0,67	0,84	0,79	0,74
0,20	0,02	0,05	0,07	0,10	0,12	0,15	0,17	0,23	0,30	0,41	0,59	0,74	0,70	0,65
0,25	0,02	0,05	0,06	0,08	0,10	0,13	0,15	0,20	0,26	0,35	0,47	0,65	0,62	0,58
0,30	0,02	0,04	0,05	0,07	0,09	0,11	0,13	0,18	0,23	0,31	0,40	0,57	0,54	0,50
0,40	0,01	0,03	0,04	0,06	0,07	0,08	0,10	0,13	0,17	0,23	0,33	0,41	0,39	0,37
0,50	0,01	0,02	0,03	0,04	0,05	0,06	0,07	0,09	0,12	0,16	0,23	0,29	0,28	0,26
0,60	0,01	0,01	0,02	0,03	0,03	0,04	0,05	0,06	0,08	0,10	0,15	0,18	0,17	0,16

In the interval $0 < \theta < 40°$ the following expression is valid.

$$\phi = 3.2 \tan\frac{\theta}{2}\sqrt[4]{\tan\frac{\theta}{2}}$$

30 (Contd.)   Head losses in diffusers in pipes

1.2.2   Truncated pyramid diffusers (rectangular diffusers)

If θ_1 and θ_2 are the angles of the aperture, a mean angle $\theta_m = \dfrac{\theta_1 + \theta_2}{2}$ is defined. If the cross-section is square, $\theta_1 = \theta_2 = \theta_m$

$\dfrac{A_1}{A_2}$	\multicolumn{12}{c}{$\theta°$ mean}
	2	4	6	8	10	12	16	20	28	40	60	180
0	0,03	0,06	0,10	0,14	0,19	0,23	0,34	0,45	0,73	1,05	1,10	1,10
0,05	0,03	0,05	0,09	0,13	0,17	0,21	0,31	0,40	0,66	0,94	0,99	0,99
0,075	0,03	0,05	0,08	0,12	0,16	0,20	0,29	0,38	0,62	0,90	0,94	0,94
0,10	0,02	0,05	0,08	0,11	0,15	0,19	0,28	0,36	0,59	0,85	0,89	0,89
0,15	0,02	0,04	0,07	0,10	0,14	0,17	0,24	0,32	0,52	0,76	0,79	0,79
0,20	0,02	0,04	0,06	0,09	0,12	0,15	0,22	0,29	0,47	0,67	0,70	0,70
0,25	0,02	0,03	0,06	0,08	0,11	0,13	0,19	0,25	0,41	0,59	0,62	0,62
0,30	0,01	0,03	0,05	0,07	0,09	0,11	0,17	0,22	0,36	0,51	0,54	0,54
0,40	0,01	0,02	0,04	0,05	0,07	0,08	0,12	0,16	0,26	0,38	0,40	0,40
0,50	0,01	0,01	0,02	0,03	0,05	0,06	0,08	0,11	0,18	0,26	0,27	0,27
0,60	0,01	0,01	0,02	0,02	0,03	0,04	0,05	0,07	0,12	0,17	0,18	0,18

In the interval $0 < \theta < 25°$ the expression $\phi \simeq 4.0 \tan\dfrac{\theta}{2}\sqrt[4]{\tan\dfrac{\theta}{2}}$ is valid.

1.2.3   Plane diffuser: two parallel faces; aperture according to the other two symmetrical faces with angle θ.

$\dfrac{A_1}{A_2}$	\multicolumn{13}{c}{$\theta°$}
	3	6	8	10	12	14	16	20	24	30	40	60	90	180
0	0,03	0,08	0,11	0,15	0,19	0,23	0,27	0,36	0,47	0,65	0,92	1,15	1,10	1,02
0,05	0,03	0,07	0,10	0,14	0,16	0,20	0,24	0,32	0,42	0,58	0,83	1,04	0,99	0,92
0,075	0,03	0,07	0,09	0,13	0,16	0,19	0,23	0,30	0,40	0,55	0,79	0,99	0,95	0,88
0,10	0,02	0,07	0,09	0,12	0,15	0,18	0,22	0,29	0,38	0,52	0,75	0,93	0,89	0,83
0,15	0,02	0,06	0,08	0,11	0,14	0,17	0,20	0,26	0,34	0,46	0,67	0,84	0,79	0,74
0,20	0,02	0,05	0,07	0,10	0,12	0,15	0,17	0,23	0,30	0,41	0,59	0,74	0,70	0,65
0,25	0,02	0,04	0,06	0,08	0,10	0,13	0,15	0,20	0,26	0,35	0,47	0,65	0,62	0,58
0,30	0,02	0,04	0,05	0,07	0,09	0,11	0,13	0,18	0,23	0,31	0,40	0,57	0,54	0,50
0,40	0,01	0,03	0,04	0,05	0,07	0,08	0,10	0,13	0,17	0,23	0,33	0,41	0,39	0,37
0,50	0,01	0,02	0,03	0,04	0,05	0,06	0,07	0,09	0,12	0,16	0,23	0,29	0,28	0,26
0,60	0,01	0,01	0,02	0,03	0,03	0,04	0,05	0,06	0,08	0,10	0,15	0,18	0,17	0,16

In the interval $0 < \theta < 40°$ the expression $\phi \simeq 3.2 \tan\dfrac{\theta}{2}\sqrt[4]{\tan\dfrac{\theta}{2}}$ is valid.

30 (Contd.). Head losses in diffusers in pipes

2. *Friction head losses*

The values of K_f given below corresponding to a friction coefficient $f = 0.02$ in the diffuser.

2.1 *Conical diffuser*

Values of K_f

$\dfrac{A_1}{A_2}$	θ°					
	2	3	6	10	14	20
0,05	0,14	0,10	0,05	0,03	0,02	0,01
0,075	0,14	0,10	0,05	0,03	0,02	0,01
0,01	0,14	0,10	0,05	0,03	0,02	0,01
0,15	0,14	0,10	0,05	0,03	0,02	0,01
0,20	0,14	0,10	0,05	0,03	0,02	0,01
0,25	0,14	0,10	0,05	0,03	0,02	0,01
0,30	0,13	0,09	0,04	0,03	0,02	0,01
0,40	0,12	0,08	0,04	0,02	0,02	0,01
0,50	0,11	0,07	0,04	0,02	0,02	0,01
0,60	0,09	0,06	0,03	0,02	0,02	0,01

2.2 *Rectangular diffusers*

$$K_f = \Delta_1 + \Delta_2$$

Δ_1 — corresponds to the angle θ_1 of two symmetrical faces
Δ_2 — corresponds to the angle θ_2 of the other two faces, also symmetrical.

Values of Δ_1 and Δ_2

$\dfrac{A_1}{A_2}$	θ°_1 , θ°_2					
	2	4	6	10	14	20
0,05	0,07	0,04	0,02	0,02	0,02	0,01
0,10	0,07	0,03	0,02	0,02	0,02	0,01
0,15	0,07	0,03	0,02	0,02	0,02	0,01
0.20	0,07	0,03	0,02	0,02	0,02	0,01
0,25	0,07	0,03	0,02	0,02	0,02	0,01
0,30	0,07	0,03	0,02	0,02	0,02	0,01
0,40	0,06	0,03	0,02	0,02	0,01	0,01
0,50	0,06	0,03	0,02	0,01	0,01	0,01
0,60	0,05	0,02	0,02	0,01	0,01	0,01

Example: $\theta_1 = 10°$; $\theta_2 = 4°$; $A_1/A_2 = 0.20$.
From the table: $\Delta_1 = 0,02$; $\Delta_2 = 0,03$; $K_f = \Delta_1 + \Delta_2 = 0.05$

30 (Contd.).   Head losses in diffusers in pipes

2.3 *Plane diffusers, two parallel faces* $(a_1 \neq a_2 ; b_1 = b_2)$; $a_1 b_1$, intake cross-section; $a_2 b_2$, outlet cross-section

2.3.1 — $a_1/b_1 = 0.5$

A_1	θ^o					
A_2	2°	4°	6°	10°	20°	40°
0,10	0,27	0,14	0,09	0,05	0,03	0,01
0,20	0,25	0,13	0,08	0,05	0,03	0,01
0,30	0,22	0,11	0,08	0,05	0,02	0,01
0,50	0,18	0,09	0,06	0,04	0,02	0,01

2.3.2 — $a_1/b_1 = 1.0$

A_1	θ^o					
A_2	2°	4°	6°	10°	20°	40°
0,10	0,40	0,20	0,13	0,08	0,04	0,02
0,20	0,37	0,18	0,13	0,07	0,04	0,02
0,30	0,33	0,17	0,11	0,07	0,03	0,02
0,50	0,25	0,13	0,08	0,05	0,03	0,01

2.3.3 — a_1/b_1 1.5

A_1	θ^o					
A_2	2°	4°	6°	10°	20°	40°
0,10	0,53	0,26	0,18	0,11	0,05	0,03
0,20	0,48	0,24	0,16	0,10	0,05	0,02
0,30	0,43	0,21	0,14	0,09	0,04	0,02
0,50	0,32	0,16	0,10	0,06	0,03	0,02

2.3.4 — $a_1/b_1 = 2.0$

A_1	θ^o					
A_2	2°	4°	6°	10°	20°	40°
0,10	0,65	0,33	0,22	0,13	0,06	0,03
0,20	0,60	0,30	0,28	0,12	0,06	0,03
0,30	0,53	0,26	0,18	0,11	0,05	0,03
0,50	0,39	0,19	0,13	0,08	0,04	0,02

3. *Head loss coefficient K in a curved diffuser* $\dfrac{dp}{dx}$ = constant (Formula No. 4.36)

$$\Delta H = K\frac{U_1^2}{2g} ; K = \phi_o(1.43 - 1.3\eta)(1 - \eta)^2$$

Values of ϕ_o

L/d_1 or L/a_1	0	0,5	1,0	1,5	2,0	2,5	3,0	3,5	4,0	4.5	5,0	6,0
ϕ_o in circular or rectangular cross-section	1,02	0,75	0,62	0,53	0,47	0,43	0,40	0,38	0,37	—	—	—
ϕ_o in plane cross-sections	1,02	0,83	0,72	0,64	0,57	0,52	0,48	0,45	0,43	0,41	0,39	0,37

L is the length of the diffuser; d_1 and a_1 are the diameter or one side of the inlet cross-section of the diffuser, respectively.

31. Head losses at the outlet of diffusers in an undefined medium

Compiled from [33]

$$K = (1 + \sigma)K_1; \Delta H = KU_1^2/2g$$

1. *Values of* σ

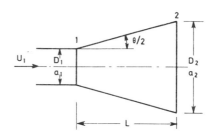

L/D_1 or L/a_1	1,0	2,0	4,0	6,0	10,0
σ	0,45	0,40	0,30	0,20	0,0

2. *Values of* K_1

2.1 Circular cross-section

$\dfrac{L}{D_1}$	$\theta°$										
	2	4	6	8	10	12	16	20	24	28	30
1,0	0,90	0,79	0,71	0,62	0,55	0,50	0,41	0,38	0,38	0,39	0,40
1,5	0,84	0,70	0,60	0,51	0,45	0,40	0,34	0,33	0,36	0,40	0,42
2,0	0,81	0,65	0,52	0,43	0,37	0,33	0,29	0,30	0,35	0,40	0,44
2,5	0,78	0,60	0,45	0,36	0,30	C.27	0,26	0,28	0,33	0,41	0,44
3,0	0,74	0,53	0,40	0,31	0,27	0,24	0,23	0,27	0,35	0,44	0,48
4,0	0,66	0,44	0,32	0,26	0,22	0,21	0,22	0,27	0,36	0,45	0,51
5,0	0,52	0,35	0,28	0,23	0,20	0,19	0,22	0,29	0,38	0,48	0,53
6,0	0,41	0,28	0,21	0,18	0,17	0,18	0,24	0,32	0,42	0,51	0,56
10,0	0,40	0,20	0,15	0,14	0,16	0,18	0,26	0,35	0,45	0,55	0,60

2.2 Square cross-section

$\dfrac{L}{a_1}$	$\theta°$								
	0	2	4	8	10	12	16	20	24
1,0	1,0	0,89	0,79	0,64	0,59	0,56	0,52	0,52	0,55
1,5	1,0	0,84	0,74	0,53	0,47	0,45	0,43	0,45	0,50
2,0	1,0	0,80	0,63	0,45	0,40	0,39	0,38	0,43	0,50
2,5	1,0	0,76	0,57	0,39	0,35	0,34	0,35	0,42	0,52
3,0	1,0	0,71	0,52	0,34	0,31	0,31	0,34	0,42	0,53
4,0	1,0	0,65	0,43	0,28	0,26	0,27	0,33	0,42	0,53
5,0	1,0	0,59	0,37	0,23	0,23	0,26	0,33	0,43	0,55
6,0	1,0	0,54	0,32	0,22	0,22	0,25	0,32	0,43	0,56
10	1,0	0,41	0,17	0,18	0,20	0,25	0,34	0,45	0,57

31 (contd.)   Head losses at the outlet of diffusers in an undefined medium

2.3 *Plane cross-section*: $a_1/b = 0.5$ to 2 (b constant width)

$\dfrac{L}{a_1}$	$\theta°$													
	0	2	4	6	8	10	12	16	20	24	28	32	36	40
1,0	1,00	0,95	0,89	0,84	0,79	0,75	0,70	0,64	0,58	0,55	0,52	0,51	0,50	0,51
1,5	1,00	0,93	0,86	0,78	0,71	0,66	0,61	0,53	0,49	0,46	0,45	0,45	0,46	0,48
2,0	1,00	0,90	0,80	0,72	0,65	0,59	0,54	0,47	0,42	0,41	0,41	0,42	0,45	0,50
2,5	1,00	0,88	0,76	0,66	0,59	0,53	0,48	0,42	0,38	0,38	0,39	0,42	0,46	0,51
3,0	1,00	0,86	0,72	0,62	0,54	0,48	0,43	0,37	0,36	0,36	0,38	0,42	0,47	0,54
4,0	1,00	0,83	0,66	0,55	0,46	0,41	0,37	0,33	0,32	0,34	0,38	0,42	0,49	0,58
6,0	1,00	0,76	0,56	0,45	0,37	0,32	0,30	0,28	0,30	0,34	0,40	0,47	0,56	0,65
10,0	1,00	0,67	0,43	0,33	0,27	0,25	0,24	0,25	0,30	0,37	0,45	0,53	0,63	0,73

a_1 and b are the dimensions of the inlet rectangular cross-section; a_2 and b are the dimensions of the enlarged cross-section.

32.   Head losses in sudden contractions

According to Karev (1953), quoted by [33]

$$\Delta H = K \frac{U_2^2}{2g} \qquad \eta = \frac{A_2}{A_1}$$

1. $-1 < \mathbf{R}_e < 8$ $K = 27/\mathbf{R}_e$

2. $-10 \leqslant \mathbf{R}_e \leqslant 10^4$ — Values of K

$\dfrac{\mathbf{R}_e}{A_2/A_1}$	10	20	30	40	50	10^2	2.10^2	5.10^2	10^3	2.10^3	4.10^3	5.10^3	10^4	$>10^4$
0,1	5,00	3,20	2,40	2,00	1,80	1,30	1,04	0,82	0,64	0,50	0,80	0,75	0,50	0,45
0,2	5,00	3,10	2,30	1,84	1,62	1,20	0,95	0,70	0,50	0,40	0,60	0,60	0,40	0,40
0,3	5,00	2,95	2,15	1,70	1,50	1,10	0,85	0,60	0,44	0,30	0,55	0,55	0,35	0,35
0,4	5,00	2,80	2,00	1,60	1,40	1,00	0,78	0,50	0,35	0,25	0,45	0,50	0,30	0,30
0,5	5,00	2,70	1,80	1,46	1,30	0,90	0,65	0,42	0,30	0,20	0,40	0,42	0,25	0,25
0,6	5,00	2,60	1,70	1,35	1,20	0,80	0,56	0,35	0,24	0,15	0,35	0,35	0,20	0,25

3 — $\mathbf{R}_e \geqslant 10^4$ $K = 0,5 \, (1 - \eta)$

33.   Dimensioning of contractions without caviatation†

The contraction can be drawn by means of two arcs of a cubic parabola. D_1/D_2 is calculated; C/L is fixed and from the graph a value of L/D_1 is to be found, to the right of the curve presenting C/L.

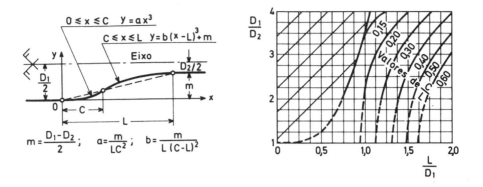

The head loss can be disregarded for a contraction dimensioned as above:

Example. $D_1 = 1.80$ m; $D_2 = 0.60$ m; $D_1/D_2 = \dfrac{1.80}{0.60} = 3$; $m = \dfrac{D_1 - D_2}{2} = 0.60$ m.

$C/L = 0.3$ is fixed. From the graph, for $D_1/D_2 = 3$ and $C/L = 0.3$, we have $L/D_1 > 1.4$. Take $L = 1.5 \times D_1 = 2.70$ m. We thus have $C = 0.3 \times L = 0.81$ m;

$$a = \frac{m}{C^2 L} = \frac{0.60}{0.81^2 \times 2.70} = 0.3387 \text{ m}^{-2};$$

$$b = \frac{m}{L(C - L)^2} = \frac{0.60}{2.7 \times (-1.89)^2} = 0.0622 \text{ m}^{-2}.$$

The coordinates of the transition curve in relation to the axes x, y indicated in the figure are then calculated from the expressions:

For $O \geqslant x \geqslant C$ $y = ax^3 = 0.3387 \, x^3$

For $C \geqslant x \geqslant L$ $y = b (x - L)^3 + \eta = 0.0622 (x - 2.7)^3 + 0.6$

---

† Rouse, H. and Hassen, M. M., *Cavitation Free Inlets and Contractions*, Mechanical Engineering, **71**, 3, 1949. Quoted by [21].

34. Head losses from a reservoir to a pipe

Compiled from [33] ($R_e > 10^4$)

$$\Delta H = K \frac{U^2}{2g}$$

(In non-circular conduits, take $D = 4R$, R being the hydraulic radius).

1. *Sharp-edge intake* — Values of K

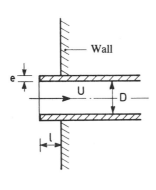

e/D	l/D						
	0	0,005	0,010	0,050	0,200	0,300	∞
0	0,50	0,63	0,68	0,80	0,92	0,97	1,00
0,004	0,50	0,58	0,63	0,74	0,86	0,90	0,94
0,008	0,50	0,55	0,58	0,68	0,81	0,85	0,88
0,012	0,50	0,53	0,55	0,63	0,75	0,79	0,83
0,016	0,50	0,51	0,53	0,58	0,70	0,74	0,77
0,020	0,50	0,51	0,52	0,55	0,66	0,69	0,72
0,024	0,50	0,50	0,51	0,53	0,62	0,65	0,68
0,030	0,50	0,50	0,51	0,52	0,57	0,59	0,61
0,040	0,50	0,50	0,51	0,51	0,52	0,52	0,54
0,050	0,50	0,50	0,50	0,50	0,50	0,50	0,50
∞	0,50	0,50	0,50	0,50	0,50	0,50	0,50

2. *Re-entrant conical intake* — values of K

l/D	θ°								
	0	10	20	30	40	60	100	140	180
0,025	1,0	0,96	0,93	0,90	0,86	0,80	0,69	0,59	0,50
0,050	1,0	0,93	0,86	0,80	0,75	0,67	0,58	0,53	0,50
0,075	1,0	0,87	0,75	0,65	0,58	0,50	0,48	0,49	0,50
0,10	1,0	0,80	0,67	0,55	0,48	0,41	0,41	0,44	0,50
0,15	1,0	0,76	0,58	0,43	0,33	0,25	0,27	0,38	0,50
0,25	1,0	0,68	0,45	0,30	0,22	0,17	0,22	0,34	0,50
0,60	1,0	0,46	0,27	0,18	0,14	0,13	0,21	0,33	0,50
1,0	1,0	0,32	0,20	0,14	0,11	0,10	0,18	0,30	0,50

3. *Flush conical intake* — Values of K

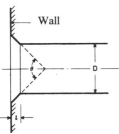

l/D	θ°								
	0	10	20	30	40	60	100	140	180
0,025	0,50	0,47	0,45	0,43	0,41	0,40	0,42	0,45	0,50
0,050	0,50	0,45	0,41	0,36	0,33	0,30	0,35	0,42	0,50
0,075	0,50	0,42	0,35	0,30	0,26	0,23	0,30	0,40	0,50
0,10	0,50	0,39	0,32	0,25	0,22	0,18	0,27	0,38	0,50
0,15	0,50	0,37	0,27	0,20	0,16	0,15	0,25	0,37	0,50
0,60	0,50	0,27	0,18	0,13	0,11	0,12	0,23	0,36	0,50

35. Dimensioning of intakes without cavitation†

According to [20]

The intake can be drawn by means of an elliptical arc. After calculating the semi-axis b, the ratio b/r is calculated. A value of a/r is fixed to the right of the line in the graph (unshaded zone of the graph).

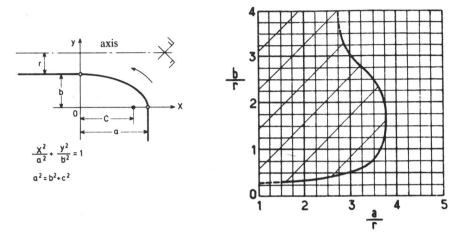

Example. $r = 0.20$ m. $b = 0.40$ m is fixed; we then have $\dfrac{b}{r} = \dfrac{0.40}{0.20} = 2$. In order for there to be no risk of caviation, we must have $\dfrac{a}{r} > 3.75$. *Take $a = 4\,r = 0.80$ m.* The equation of the ellipse will be: $\dfrac{x^2}{0.80^2} + \dfrac{y^2}{0.40^2} = 1$.

To draw the ellipse from knowledge of the focus, the focal semi-distance is $c = \sqrt{a^2 - b^2} = \sqrt{0.64 - 0.16} = 0.69\ m$.

36. Head losses in intakes near to a front wall

$$\Delta H = K\frac{U^2}{2g}$$

Nosova (1956) quoted by [33]

In the case of non-circular conduits take $D = 4R$, R being the hydraulic radius

Values of K for $\mathbf{R}_e > 10^4$

h/D	0,10	0,125	0,15	0,20	0,25	0,30	0,40	0,50	0,60	0,80
				$r/D = 0{,}2$						
K	—	0,80	0,45	0,19	0,12	0,09	0,07	0,06	0,05	0,05
				$r/D = 0{,}3$						
K	—	0,50	0,34	0,17	0,10	0,07	0,06	0,05	0,04	0,04
				$r/D = 0{,}5$						
K	0,65	0,36	0,25	0,10	0,07	0,05	0,04	0,04	0,03	0,03

†The same as for Table 33.

37. Head losses in partly open valves

$$\text{Values of } K \qquad \Delta H = K \frac{U^2}{2g}$$

(*U* is the velocity in the normal cross-section of the piping)

(a) *Sluice valves in circular pipe* (gate valve)[†]

$\frac{x}{D}$	K	$\frac{x}{D}$	K	$\frac{x}{D}$	K	$\frac{x}{D}$	K
0,181	41,21	0,250	22,68	0,417	6,33	0,583	1,55
0,194	35,36	0,333	11,89	0,458	4,57	0,667	0,77
0,208	31,35	0,375	8,63	0,500	3,27	1,000	

(b) *Sluice valves in rectangular pipe*[‡]

$\frac{S_o}{S}$	K	$\frac{S_o}{S}$	K	$\frac{S_o}{S}$	K	$\frac{S_o}{S}$	K
0,1	193, —	0,4	8,12	0,7	0,95	0,9	0,09
0,2	44,5	0,5	4,02	0,8	0,39	1,0	0,00
0,3	17,8	0,6	2,08				

(c) *Ball valves*[‡]

θ^o	K	θ^o	K	θ^o	K	θ^o	K
0		20	1,56	40	17,3	60	206, —
5	0,05	25	3,10	45	31,2	65	486, —
10	0,29	30	5,47	50	52,6	82	∞
15	0,75	35	9,68	55	106, —		

(d) *Butterfly valves*[‡][§]

θ^o	K	θ^o	K	θ^o	K	θ^o	K
0	≈0	20	1,54	40	10,8	60	118, —
5	0,24	25	2,51	45	18,7	65	256, —
10	0,52	30	3,91	50	32,6	70	750, —
15	0,90	35	6,22	55	58,8	90	∞

[†] *Kuichling's* experiments quoted by [19].
[‡] *Weisbach's* experiments quoted by [19].
[§] The head losses obtained by *Scimemi* are higher: $K = 0.53$ for the butterfly valve completely open and $K = 0.26$ for the bottom valve, also open.

37 (contd.)　　Head losses in partly open valves

For $\theta = 0$, K depends essentially on the ratio between the maximum thickness, e, of the body of the valve and the diameter of the pipe.

e/d	0,50	0,15	0,20	0,25
K ($\theta = 0$)	0,05 — 0,10	0,10 — 0,16	0,17 — 0,24	0,25 — 0,35

(e) *Check valves*† $\left(\dfrac{A_o}{A} = 0.535\right)$

α°	K	α°	K	α°	K	α°	K
15	90	30	30	45	9,5	60	3,2
20	62	35	20	50	6,6	65	2,3
25	42	40	14	55	4,6	70	1,7

†*Weisbach's* experiments quoted by [28].

38.　　Head losses in large gates‡

$$\Delta H = K \frac{U^2}{2g}$$

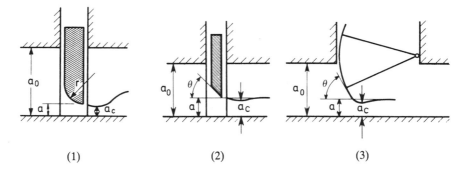

(1)　　　　　　　　(2)　　　　　　　　(3)

Relative aperture: $\quad a_r = \dfrac{a}{a_o}$;

Coefficient of contraction $\eta = \dfrac{a_c}{a} = \dfrac{1}{1 + \sqrt{c(1 - a_r^2)}}$

‡ According to Ckimitski, Coefficient of contraction of tainter gates, *Hydro-technical Construction* (in Russian), No. 11, 1964. Quoted by [30].

38 (Contd.) Head losses in large gates

In case (1): $c = 0.4/e^{1.6r/a}$

r/a	0	0,02	0,06	0,08	0,10	0,15	0,20	0,25	0,27	0,29	0,31
c	0,400	0,290	0,153	0,111	0,081	0,036	0,016	0,007	0,005	0,004	0,003

In cases (2) and (3): $c = 0.4 \sin^3 \theta$

θ	0	10	20	30	40	50	60	70	80
c	0	0,002	0,016	0,050	0,106	0,180	0,260	0,332	0,382

(b) *Head loss coefficient* $K = K_1 + K_2 = K_1 + \left[\dfrac{1}{\eta a_r} - 1 \right]^2$

K_1 is the head loss coefficient when the valve is completely open: 0.10 to 0.20 in cases (1) and (2); 0.20 to 0.35 in case (3).

39. Head losses in Howell–Bunger conical valves[†]

(valve completely open)

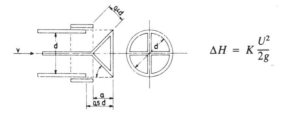

$$\Delta H = K \frac{U^2}{2g}$$

$2a/d$	0,10	0,20	0,30	0,40	0,50	0,60	0,70	0,80	0,90
K	30	9	4	2,2	1,3	0,92	0,69	0,67	0,67

[†]*According to N. Frenkel, Hydraulique*, Ed. Gossenergoikdat, 1956. Quoted by [31].

40.   Head losses in 90° bends of rectangular cross-section[†]

$$\Delta h = K \frac{U^2}{2g} - \text{Values of } K$$

r = Radius of the bend, based on the axis of the pipe
l = Width of the cross-section
h = Height of the cross-section

l/h ＼ r/h	5/3	1	2/3
6	0,09	0,16	0,38
3	0,15	0,22	0,55

[†]Wirt, L., quoted by [20].

41.   Head losses in 90° bends, of circular cross-section[‡]

$$\Delta H = K \frac{U^2}{2g} - \text{Values of } K$$

Only advisable for small diameters ($D \leqslant 0.50\,m$)
r = Radius of the bend based on the axis of the conduit.
U = Mean velocity

r (m) ＼ U m/s	0,60	0,90	1,20	1,50	1,80	2,10	2,40	3,00	3,65	4,60	6,10	9,15	12,20
0,00	1,03	1,14	1,23	1,30	1,36	1,42	1,46	1,54	1,62	1,71	1,84	2,03	2,18
0,08	0,46	0,51	0,55	0,58	0,60	0,63	0,65	0,69	0,72	0,76	0,82	0,90	0,97
0,15	0,31	0,34	0,36	0,38	0,40	0,42	0,43	0,46	0,49	0,51	0,54	0,60	0,65
0,30	0,21	0,23	0,25	0,26	0,28	0,29	0,30	0,31	0,33	0,35	0,37	0,41	0,44
0,60	0,19	0,21	0,22	0,23	0,24	0,25	0,26	0,28	0,29	0,31	0,33	0,36	0,39
0,90	0,18	0,20	0,22	0,23	0,24	0,25	0,26	0,27	0,29	0,30	0,33	0,36	0,39
1,20	0,18	0,20	0,21	0,23	0,23	0,25	0,26	0,27	0,28	0,30	0,32	0,35	0,38
1,50	0,18	0,20	0,21	0,22	0,23	0,24	0,25	0,27	0,28	0,29	0,32	0,35	0,38
1,80	0,18	0,19	0,21	0,22	0,23	0,24	0,25	0,26	0,28	0,29	0,31	0,35	0,37
2,10	0,19	0,21	0,22	0,23	0,24	0,25	0,26	0,28	0,29	0,31	0,33	0,36	0,39
2,40	0,21	0,23	0,25	0,26	0,27	0,28	0,29	0,31	0,32	0,34	0,37	0,41	0,44
3,00	0,26	0,29	0,31	0,32	0,34	0,35	0,36	0,38	0,40	0,42	0,46	0,50	0,54
4,60	0,37	0,41	0,43	0,46	0,48	0,50	0,52	0,55	0,57	0,61	0,65	0,72	0,77
6,10	0,45	0,51	0,54	0,57	0,60	0,62	0,64	0,68	0,72	0,75	0,81	0,90	0,97
7,60	0,50	0,56	0,59	0,63	0,65	0,69	0,71	0,75	0,79	0,83	0,89	0,99	1,06

[‡]Fuller W.E., *Loss of Head in Bends*, Jour. New Eng. Water Works Assoc., December, 1913. Quoted by [41].

42.   Corrective factors to be applied, by multiplication, to the factors of the preceding table, in the case of deflection angles other than 90°

Angle	0°	10°	20°	30°	40°	50°	60°	70°	80°	90°	100°	110°	120°
Corrective factor	0	0,20	0,38	0,50	0,62	0,73	0,81	0,89	0,95	1,00	1,04	1,09	1,12

43. Head losses in bends of circular cross-section with constant area

Compiled from [56]

$$\Delta H = K_b \frac{U^2}{2g}$$

$$K_b = K_1 . K_2 . K_3$$

(1) Values of K_1 in smooth pipes ($R_e = 10^6$)

r/d	θ (degrees)								
	10	20	30	40	50	60	70	80	90
1	0.02	0.03	0.06	0.08	0.12	0.15	0.18	0.22	0.24
2	0.02	0.04	0.06	0.08	0.10	0.11	0.13	0.15	0.16
4	0.03	0.06	0.08	0.09	0.11	0.12	0.14	0.15	0.17
6	0.04	0.06	0.09	0.11	0.13	0.14	0.16	0.18	0.20
8	0.04	0.07	0.10	0.12	0.14	0.16	0.18	0.20	0.22
10	0.05	0.08	0.11	0.13	0.16	0.18	0.20	0.22	0.24

(2) Values of K_2 ($R_e \neq 10^6$)

For $R_e = 10^6$; $K_2 = 1.0$

r/d	R_e					
	10^4	5×10^4	10^5	5×10^5	5×10^6	10^7
1.0	2.20	1.50	1.25	1.00	1.00	1.00
1.5	2.20	1.60	1.40	1.00	1.00	1.00
$\geqslant 2.0$	2.20	1.65	1.50	1.00	0.77	0.69

(3) Values of $K_3 = \dfrac{f_{rough}}{f_{smooth}} \leqslant 1.4$

To obtain f values consult Graph 18.

44. Head losses in single mitre bends

Compiled from [26]

$$\Delta H = K_b \frac{U^2}{2g}$$

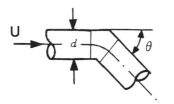

$$K_b = K_1 . K_2$$

(1) Values of K_1 in smooth pipes

θ	R_e			
(degrees)	2×10^4	6×10^4	10^5	2.5×10^5
10	0.06	0.05	0.04	0.03
20	0.12	0.10	0.08	0.06
30	0.19	0.16	0.14	0.12
40	0.28	0.23	0.22	0.19
50	0.38	0.34	0.32	0.29
60	0.50	0.48	0.48	0.47
70	0.68	0.67	0.66	0.65
80	0.88	0.88	0.88	0.88
90	1.13	1.13	1.13	1.13

(2) Values of $K_2 = \dfrac{f_{rough}}{f_{smooth}} \leqslant 1.4$

To obtain f values consult Graph 18.

45. Head losses in special bends†

$$\Delta H = K \frac{U^2}{2g} - \text{Values of } K$$

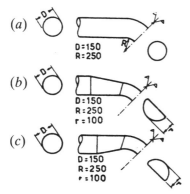

(a) D=150 R=250

(b) D=150 R=250 r=100

(c) D=150 R=250 r=100

R_e	Bend (a)	(b)	(c)
$0{,}5 \times 10^5$	0,195	0,140	0,080
$0{,}75 \times 10^5$	0,205	0,145	0,115
$1{,}0 \times 10^5$	0,215	0,165	0,130
$1{,}5 \times 10^5$	0,225	0,185	0,135
$2{,}0 \times 10^5$	0,230	0,190	0,140
$2{,}5 \times 10^5$	0,230	0,195	0,140

† According to [10].

46. Head losses in composite mitre bends

$$\Delta H = K \frac{U^2}{2g} - \text{Values of } K$$

K_s — Coefficient for smooth pipes

K_r — Coefficient for rough pipes (relative roughness 0.0022).

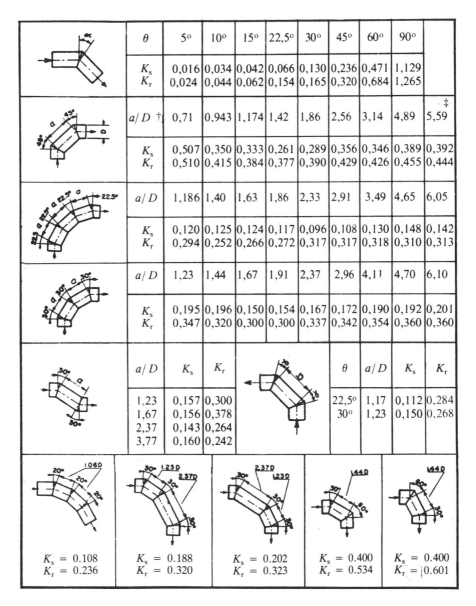

	θ	5°	10°	15°	22,5°	30°	45°	60°	90°	
	K_s	0,016	0,034	0,042	0,066	0,130	0,236	0,471	1,129	
	K_r	0,024	0,044	0,062	0,154	0,165	0,320	0,684	1,265	
	a/D †	0,71	0,943	1,174	1,42	1,86	2,56	3,14	4,89	5,59 ‡
	K_s	0,507	0,350	0,333	0,261	0,289	0,356	0,346	0,389	0,392
	K_r	0,510	0,415	0,384	0,377	0,390	0,429	0,426	0,455	0,444
	a/D	1,186	1,40	1,63	1,86	2,33	2,91	3,49	4,65	6,05
	K_s	0,120	0,125	0,124	0,117	0,096	0,108	0,130	0,148	0,142
	K_r	0,294	0,252	0,266	0,272	0,317	0,317	0,318	0,310	0,313
	a/D	1,23	1,44	1,67	1,91	2,37	2,96	4,11	4,70	6,10
	K_s	0,195	0,196	0,150	0,154	0,167	0,172	0,190	0,192	0,201
	K_r	0,347	0,320	0,300	0,300	0,337	0,342	0,354	0,360	0,360

	a/D	K_s	K_r		θ	a/D	K_s	K_r
	1,23	0,157	0,300		22,5°	1,17	0,112	0,284
	1,67	0,156	0,378		30°	1,23	0,150	0,268
	2,37	0,143	0,264					
	3,77	0,160	0,242					

$K_s = 0.108$ $K_r = 0.236$	$K_s = 0.188$ $K_r = 0.320$	$K_s = 0.202$ $K_r = 0.323$	$K_s = 0.400$ $K_r = 0.534$	$K_s = 0.400$ $K_r = 0.601$

† According to Schubert's experiments, quoted by [12].
‡ For higher value of a/D, K_s and K_r, are practically constants. For $a/D = 6,28$ is $K_s = 0,399$ and $K_r = 0,444$.

47. Head losses in combining junctions without transition[†]

$$\Delta H_{1.3} = K_{1.3}\,\frac{U_3^2}{2g} \quad ; \quad \Delta H_{2.3} = K_{2.3}\,\frac{U_3^2}{2g}$$

Values of $K_{1.3}$ and $K_{2.3}$,

1. *Angle of junction* $\theta = 30°$

   1.1 $A_3 = A_1$

Q_2/Q_3	0,0		0,1		0,2		0,4		0,8		1,0	
A_2/A_3	$K_{1.3}$	$K_{2.3}$	$K_{1.3}$	$K_{2.3}$	$K_{1.3}$	$K_{2.3}$	$K_{1.3}$	$K_{2.3}$	$K_{1.3}$	$K_{2.3}$	$K_{1.3}$	$K_{2.3}$
0,1	0	−1,00	0,02	0,21	−0,33	3,10	−2,15	13,5	−10,1	53,8	−16,3	83,7
0,2	0	−1,00	0,11	−0,46	+0,01	0,37	−0,75	2,95	−4,61	11,5	−7,70	17,3
0,3	0	−1,00	0,13	−0,57	0,13	−0,06	−0,30	1,15	−2,74	4,22	−4,75	6,33
0,4	0	−1,00	0,15	−0,60	0,19	−0,20	−0,05	0,59	−1,82	2,14	−3,35	2,92
0,6	0	−1,00	0,16	−0,62	0,24	−0,28	+0,17	0,26	−0,90	0,85	−1,90	0,89
0,8	0	−1,00	0,17	−0,63	0,27	−0,30	0,26	0,18	−0,43	0,53	−1,17	0,39
1,0	0	−1,00	0,17	−0,63	0,29	−0,35	0,36	0,16	−0,15	0,45	−0,75	0,27

   1.2 $A_3 = A_1 + A_2$

Q_2/Q_3	0		0,1		0,2		0,4		0,8		1,0	
A_2/A_3	$K_{1.3}$	$K_{2.3}$	$K_{1.3}$	$K_{2.3}$	$K_{1.3}$	$K_{2.3}$	$K_{1.3}$	$K_{2.3}$	$K_{1.3}$	$K_{2.3}$	$K_{1.3}$	$K_{2.3}$
0,06	0	−1,13	−0,10	+1,82	−0,81	10,1	−4,07	41,5	—	—	—	—
0,10	0,01	−1,22	+0,04	0,02	−0,33	2,88	−2,14	13,4	—	—	—	—
0,20	0,06	−1,50	0,16	−0,84	+0,06	0,05	−0,73	2,70	−3,59	11,4	−8,64	17,3
0,33	0,42	−2,00	0,51	−1,40	0,52	−0,72	+0,07	0,52	−2,19	3,30	−4,00	4,80
0,50	1,40	−3,00	1,36	−2,24	1,26	−1,44	0,86	−0,36	−0,82	1,18	−2,07	1,53

2. *Angle of junction* $\theta = 45°$

   2.1 $A_3 = A_1$

Q_2/Q_3	0,0		0,1		0,2		0,4		0,8		1,0	
A_2/A_3	$K_{1.3}$	$K_{2.3}$	$K_{1.3}$	$K_{2.3}$	$K_{1.3}$	$K_{2.3}$	$K_{1.3}$	$K_{2.3}$	$K_{1.3}$	$K_{2.3}$	$K_{1.3}$	$K_{2.3}$
0,1	0	−1,00	0,05	0,24	−0,20	3,15	−1,65	14,0	−8,10	55,9	−13,2	86,9
0,2	0	−1,00	0,12	−0,45	+0,17	0,54	−0,50	3,15	−3,56	12,4	−6,10	18,9
0,3	0	−1,00	0,14	−0,56	0,22	−0,02	−0,12	1,30	−2,10	4,90	−3,70	7,40
0,4	0	−1,00	0,16	−0,59	0,27	−0,17	+0,08	0,72	−1,30	2,66	−2,55	3,71
0,6	0	−1,00	0,17	−0,61	0,27	−0,26	0,26	0,35	−0,55	1,20	−1,35	1,42
0,8	0	−1,00	0,17	−0,62	0,29	−0,28	0,36	0,25	−0,17	0,79	−0,77	0,80
1,0	0	−1,00	0,17	−0,62	0,31	−0,29	0,41	0,21	+0,06	0,66	−0,42	0,59

[†] According to Levin's and Kaliev's formulae, quoted by [33].

47 (Contd.) Head losses in combining junctions without transition

2.2 $A_3 = A_1 + A_2$

Q_2/Q_3	0		0,1		0,2		0,4		0,8		1,0	
A_2/A_3	$K_{1.3}$	$K_{2.3}$	$K_{1.3}$	$K_{2.3}$	$K_{1.3}$	$K_{2.3}$	$K_{1.3}$	$K_{2.3}$	$K_{1.3}$	$K_{2.3}$	$K_{1.3}$	$K_{2.3}$
0,06	0,00	−1,12	−0,05	+1,82	−0,59	10,3	−3,21	42,4	—	—	—	—
0,10	0,06	−1,22	+0,11	0,06	−0,15	3,00	−1,55	13,9	—	—	—	—
0,20	0,20	−1,50	0,30	−0,85	+0,26	0,12	−0,33	3,00	−3,42	12,4	—	—
0,33	0,37	−2,00	0,48	−1,38	0,50	−0,66	+0,20	0,70	−1,60	3,95	−3,10	5,76
0,50	1,30	−3,00	1,27	−2,24	1,20	−1,50	0,90	−0,24	−0,68	1,60	−1,52	2,18

3. *Angle of junction* $\theta = 60°$

3.1 $A_3 = A_1$

Q_2/Q_3	0,0		0,1		0,2		0,4		0,8		1,0	
A_2/A_3	$K_{1.3}$	$K_{2.3}$	$K_{1.3}$	$K_{2.3}$	$K_{1.3}$	$K_{2.3}$	$K_{1.3}$	$K_{2.3}$	$K_{1.3}$	$K_{2.3}$	$K_{1.3}$	$K_{2.3}$
0,1	0	−1,00	0,09	0,26	0,00	3,35	−1,00	14,7	−5,44	58,5	−9,00	91,0
0,2	0	−1,00	0,14	−0,42	0,16	0,55	−0,16	3,50	−2,24	13,7	−4,00	21,0
0,3	0	−1,00	0,16	−0,54	0,23	−0,03	+0,11	1,55	−1,17	5,80	−2,30	9,70
0,4	0	−1,00	0,17	−0,58	0,26	−0,13	0,24	0,92	−0,64	3,32	−1,50	4,70
0,6	0	−1,00	0,17	−0,61	0,29	−0,23	0,37	0,45	−0,11	1,64	−0,68	2,11
0,8	0	−1,00	0,18	−0,62	0,31	−0,26	0,44	0,35	+0,16	1,12	−0,28	1,35
1,0	0	−1,00	0,18	−0,62	−0,32	−0,26	0,48	0,28	0,32	0,92	0,00	1,00

3.2 $A_3 = A_1 + A_2$

Q_2/Q_3	0		0,1		0,2		0,4		0,8		1,0	
A_2/A_3	$K_{1.3}$	$K_{2.3}$	$K_{1.3}$	$K_{2.3}$	$K_{1.3}$	$K_{2.3}$	$K_{1.3}$	$K_{2.3}$	$K_{1.3}$	$K_{2.3}$	$K_{1.3}$	$K_{2.3}$
0,06	0,00	−1,12	−0,03	+2,00	−0,32	10,6	−2,03	43,5	—	—	—	—
0,10	0,01	−1,22	+0,10	0,10	−0,03	3,18	−0,96	14,6	—	—	—	—
0,20	0,06	−1,50	0,19	−0,83	+0,20	0,20	−0,14	3,30	−2,20	13,7	—	—
0,33	0,33	−2,00	0,45	−1,37	0,49	−0,67	+0,34	0,91	−0,85	4,70	−1,90	6,60
0,50	1,25	−3,00	1,23	−2,13	1,17	−1,38	0,90	−0,02	−0,05	2,22	−0,78	3,10

47 (Contd.)   Head losses in combining junctions without transition

4.   *Angle of junction* $\theta = 90°$

4.1 $A_3 = A_1$

(a)   *Values of $K_{2.3}$*

A_2/A_3	Q_2/Q_3					
	0	0,1	0,2	0,4	0,8	1,0
↓	$K_{2.3}$	$K_{2.3}$	$K_{2.3}$	$K_{2.3}$	$K_{2.3}$	$K_{2.3}$
0,1	— 1,00	0,40	3,80	16,3	64,9	101
0,2	— 1,00	— 0,37	0,72	4,30	16,9	26,0
0,3	— 0,75	— 0,38	0,13	1,55	5,94	8,93
0,4	— 0,75	— 0,41	— 0,02	0,98	3,69	5,44
0,6	— 0,70	— 0,41	— 0,12	0,53	1,89	2,66
0,8	— 0,65	— 0,39	— 0,14	0,36	1,25	1,67
1,0	— 0,60	— 0,37	— 0,18	0,26	0,94	1,20

(b)   *Values of $K_{1.3}$*

Q_2/Q_3	0	0,1	0,2	0,4	0,8	1,0
$K_{1.3}$	0	0,16	0,27	0,46	0,60	0,55

4.2 $A_3 = A_1 + A_2$

Q_2/Q_3	0		0,1		0,2		0,4		0,8		1,0	
A_2/A_3	$K_{1.3}$	$K_{2.3}$	$K_{1.3}$	$K_{2.3}$	$K_{1.3}$	$K_{2.3}$	$K_{1.3}$	$K_{2.3}$	$K_{1.3}$	$K_{2.3}$	$K_{1.3}$	$K_{2.3}$
0,06	0,02	— 1,12	0,20	2,06	—	11,2	—	46,2	—	—	—	—
0,10	0,04	— 1,22	0,20	0,20	—	3,58	—	16,2	—	—	—	—
0,20	0,08	— 1,40	0,25	— 0,68	0,34	0,50	—	4,20	—	17,0	—	—
0,33	0,45	— 1,80	0,59	— 1,20	0,66	— 0,45	0,62	1,59	—	6,98	—	10,4
0,50	1,00	— 2,75	1,16	— 1,96	1,25	— 1,15	1,22	0,42	—	3,65	—	5,25

48. Head losses in combining junctions with transition and with $A_1 = A_3$

According to the formulae of Petermann, Kinne, Zusmanovic et al. quoted by [33].

$$\Delta H_{1.3} = K_{1.3} \frac{U_3^2}{2g}; \quad \Delta H_{2.3} = K_{2.3} \frac{U_3^2}{2g}$$

Values of $K_{1.3}$ and $K_{2.3}$

Form of transition	A_2/A_3	Q_2/Q_3											
$\theta = 45°$		0,1		0,2		0,3		0,4		0,6		1,0	
		$K_{1.3}$	$K_{2.3}$	$K_{1.3}$	$K_{2.3}$	$K_{1.3}$	$K_{2.3}$	$K_{1.3}$	$K_{2.3}$	$K_{1.3}$	$K_{2.3}$	$K_{1.3}$	$K_{2.3}$
$\dfrac{r}{D_2}=0,1$	0,122	0,10	0,00	−0,15	1,70	−0,50	4,30	0,90	−8,00	−3,20	19,5	−9,70	53,7
	0,34	0,10	−0,47	+0,07	−0,02	0,00	0,30	−0,14	0,80	−0,66	2,10	−2,90	5,40
	1,0	0,14	0,62	0,17	−0,04	0,19	−0,17	+0,16	0,00	+0,06	0,22	−0,58	0,38
$\dfrac{r}{D_2}=0,2$	1,0	0,14	−0,62	0,18	−0,04	0,18	−0,17	0,18	0,00	0,03	0,22	−0,61	0,38
$\delta = 8°$	0,122	0,10	−0,04	−0,10	+0,50	0,36	+1,80	−0,71	3,70	2,20	0,50	−7,10	22,5
	0,34	0,10	−0,58	+0,11	−0,03	0,09	0,00	+0,01	0,30	0,40	0,90	−1,95	2,10

48 (Contd.)   Head losses in combining junctions with transition

θ = 60°

Form of transition	A_2/A_3	Q_2/Q_3 0,1 $K_{1,3}$	0,1 $K_{2,3}$	0,2 $K_{1,3}$	0,2 $K_{2,3}$	0,3 $K_{1,3}$	0,3 $K_{2,3}$	0,4 $K_{1,3}$	0,4 $K_{2,3}$	0,6 $K_{1,3}$	0,6 $K_{2,3}$	1,0 $K_{1,3}$	1,0 $K_{2,3}$
$\frac{r}{D}=0,1$	0,122	0,10	0,00	0,04	2,17	−0,10	5,50	−0,44	10,2	−1,45	21,9	−6,14	60,0
	0,34	0,15	−0,43	0,20	0,00	+0,19	0,42	+0,11	1,00	−0,25	2,30	−1,65	6,18
	1,0	0,13	−0,60	0,19	−0,40	0,23	−0,14	0,23	0,01	+0,14	0,30	−0,30	0,53
$\frac{r}{D}=0,2$	1,0	0,13	−0,60	0,19	−0,40	0,23	−0,16	0,23	0,01	0,13	0,26	−0,35	0,50
$\delta = 8°$	0,122	0,15	−0,50	0,10	0,35	0,00	+1,40	−0,16	3,10	−0,78	7,50	−3,10	21,1
	0,34	0,15	−0,56	0,23	−0,30	0,25	0,00	+0,21	1,00	0,00	0,87	−0,75	2,00

θ = 90°

Form of transition	A_2/A_3	Q_2/Q_3 0,1 $K_{1,3}$	0,1 $K_{2,3}$	0,2 $K_{1,3}$	0,2 $K_{2,3}$	0,3 $K_{1,3}$	0,3 $K_{2,3}$	0,4 $K_{1,3}$	0,4 $K_{2,3}$	0,6 $K_{1,3}$	0,6 $K_{2,3}$	1,0 $K_{1,3}$	1,0 $K_{2,3}$
$\frac{r}{D}=0,1$	0,122	—	−0,50	—	1,35	—	4,60	—	8,40	—	23,6	—	—
	0,34	—	−0,36	—	0,10	—	0,54	—	1,10	—	2,62	—	7,11
	1,0	0,12	−0,60	0,23	−0,35	0,29	−0,10	0,32	0,10	0,36	0,43	0,35	0,87
$\frac{r}{D}+0,2$	1,0	0,08	−0,64	0,16	−0,45	0,21	−0,15	0,24	0,00	0,25	0,31	0,17	0,71
$\delta = 8°$	0,122	—	−0,50	—	1,00	—	+3,24	—	6,90	—	19,2	—	62,0
	0,34	—	−0,43	—	0,00	—	0,49	—	1,00	—	2,20	—	5,38

49. Head losses in dividing junctions without transitions

According to the formulae of Levin, Kinne *et al.*, quoted by [39].

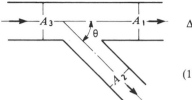

$$\Delta H_{3.1} = K_{3.1}\frac{U_3^2}{2g} \; ; \; \Delta H_{3.2} = K_{3.2}\frac{U_3^2}{2g}$$

Values of $K_{3.1}$ and $K_{3.2}$

(1) $A_1 = A_3$

U_2/U_3 / θ	0,0	0,2	0,4	0,6	0,8	1,0	1,5	2,0	3,0
				Values of $K_{3.2}$					
15°	1,0	0,65	0,38	0,20	0,09	0,07	0,22	1,10	7,20
30°	1,0	0,70	0,46	0,31	0,25	0,27	0,59	1,52	7,40
45°	1,0	0,75	0,60	0,50	0,51	0,58	1,00	2,16	7,80
60°	1,0	0,84	0,76	0,65	0,80	1,00	1,60	3,00	8,10
90°[1]	1,0	1,04	1,16	1,35	1,64	2,00	3,10	4,60	9,00
90°[2]	1,0	1,01	1,05	1,15	1,32	1,45	1,85	2,45	—
				Values of $K_{3.1}$					
θ 15 — 90°	0,40	0,26	0,15	0,06	0,02	0,00	—	—	—

[1] Values for $h_2/h_3 < 2/3$, in which h_2 and h_3 are the depths of sections A_2 and A_3.
[2] Values for $h_2/h_3 = 1$, in which h_2 and h_3 are the depths of sections A_2 and A_3.

(2) $A_3 = A_1 + A_2$ $B = 1$

U_2/U_3 / θ		0,1	0,4	0,6	0,8	1,0	1,2	1,4	1,6	2,0
					Values of $K_{3.2}$					
15°		0,81	0,38	0,19	0,06	0,03	0,06	0,13	0,35	0,98
30°		0,84	0,44	0,26	0,16	0,11	0,13	0,23	0,37	0,89
45°		0,87	0,54	0,38	0,28	0,23	0,22	0,28	0,38	0,73
60°		0,90	0,66	0,53	0,43	0,36	0,32	0,31	0,33	0,44
90°		1,00	1,00	1,00	1,00	1,00	1,00	1,00	1,00	1,00
θ					Values of $K_{3.1}$					
15°-60°		0,81	0,36	0,16	0,04	0,00	0,07	0,39	0,90	3,2
$\theta = 90°$	A_1/A_3 0-0,4	0,81	0,36	0,16	0,04	0,00	0,07	0,39	0,90	3,2
	0,5	0,81	0,40	0,23	0,16	0,20	0,36	0,78	1,36	4,0
	0,6	0,81	0,38	0,20	0,12	0,10	0,21	0,59	1,15	—
	$\geqslant 0,8$	0,81	0,36	0,16	0,04	0,00	0,07	—	—	—

50. Head losses in dividing junctions with transition and with $A_1 = A_3$

According to the formulae of Kaliev, Kinne *et al* quoted by [33]

(See figure of Table 48 reversing the directions of flow.)

$$\Delta H_{3.1} = K_{3.1} \frac{U_3^2}{2g} \; ; \; \Delta H_{3.2} = K_{3.2} \frac{U_3^2}{2g}$$

(1) *Values of $K_{3.2}$*

Form of transition	θ / Q_1/Q_3 / A_1/A_3	45°				60°				90°			
		0,1	0,3	0,6	1,0	0,1	0,3	0,6	1,0	0,1	0,3	0,6	1,0
$\frac{r}{D_2}=0,1$	0,122	0,40	1,90	9,60	30,6	0,90	2,70	12,0	36,7	1,20	4,00	17,8	—
	0,34	0,62	0,35	0,90	3,35	0,77	0,60	1,10	3,16	1,15	1,42	2,65	6,30
	1,0	0,77	0,56	0,32	0,32	0,84	0,67	0,53	0,62	0,85	0,77	0,78	1,00
$\frac{r}{D_2}=0,2$	1,0	0,77	0,56	0,32	0,32	0,84	0,67	0,53	0,62	0,85	0,74	0,69	0,91
$\delta = 8°$	0,122	0,40	0,90	5,40	17,4	0,70	1,30	5,40	16,6	0,90	3,40	17,3	—
	0,34	0,62	0,35	0,60	2,0	0,67	0,44	0,68	1,85	1,10	1,30	2,17	5,20

(2) *Values of $K_{3.1}$*

Take the values of $K_{1.3}$ of table 48

51. Head losses in three-way symmetrical combining junctions

According to [33]

Values of the coefficients of formula (4.39a)

θ	15	30	45
A	7,3	6,6	5,6
B	0,07	0,25	0,50
C	3,7	3,0	2
D	2,64	2,30	1,80

52. Head losses in slots

Compiled from [31]

See Fig. 4.15

$$\Delta H = K U^2/2g$$

(a) $\lambda < 6$

for $\lambda < 4$: $\eta = \dfrac{A + 0,25 . lb}{A}$; for $4 \leqslant \lambda \leqslant 6$: $\eta \; \dfrac{A + pb}{A}$

Values of K (formula (4.42))

η \ θ	7	10	15	20	30	45	60	90
1,1	0,0006	0,0008	0,0012	0,0016	0,0024	0,0034	0,0042	0,0048
1,2	0,002	0,003	0,004	0,006	0,009	0,012	0,015	0,017
1,4	0,006	0,009	0,014	0,018	0,026	0,037	0,046	0,053
1,6	0,011	0,016	0,024	0,032	0,047	0,067	0,092	0,094
1,8	0,016	0,023	0,035	0,046	0,067	0,095	0,117	0,135
2,0	0,021	0,030	0,045	0,059	0,086	0,122	0,149	0,172
2,5	0,030	0,043	0,064	0,085	0,124	0,175	0,215	0,248
3,0	0,037	0,052	0,078	0,103	0,151	0,213	0,261	0,302
4,0	0,045	0,064	0,095	0,126	0,184	0,260	0,319	0,368
6,0	0,052	0,074	0,110	0,146	0,213	0,302	0,370	0,427
8,0	0,055	0,078	0,117	0,154	0,226	0,319	0,391	0,451
10,0	0,056	0,080	0,120	0,158	0,232	0,328	0,401	0,463
20,0	0,059	0,084	0,125	0,165	0,241	0,341	0,417	0,482
40,0	0,059	0,085	0,126	0,167	0,243	0,344	0,422	0,487
100,0	0,060	0,085	0,126	0,167	0,244	0,345	0,423	0,489

(b) $\lambda > 6$: $K = K_1 + K_2$; K_1 *equal to the value of K in the preceding table*

K_2, *formula* (4.42c) with $\eta = \dfrac{A + pb}{A}$

Values of K_2

η \ λ	6,1	6,3	6,5	7	8	9	10	$\geqslant 15$
1,1	0,0004	0,001	0,002	0,003	0,005	0,006	0,007	0,008
1,2	0,001	0,004	0,006	0,011	0,018	0,022	0,024	0,027
1,4	0,004	0,011	0,018	0,032	0,052	0,063	0,071	0,081
1,6	0,007	0,020	0,032	0,056	0,089	0,109	0,122	0,139
1,8	0,010	0,028	0,044	0,078	0,125	0,154	0,171	0,195
2,0	0,013	0,036	0,056	0,099	0,158	0,194	0,216	0,247
2,5	0,019	0,051	0,081	0,143	0,228	0,280	0,311	0,356
3,0	0,023	0,064	0,100	0,176	0,282	0,346	0,385	0,440
4,0	0,029	0,080	0,126	0,223	0,356	0,438	0,487	0,556
6,0	0,037	0,099	0,156	0,275	0,440	0,540	0,601	0,687
8,0	0,040	0,109	0,172	0,303	0,485	0,596	0,662	0,757
10,0	0,043	0,116	0,182	0,321	0,513	0,630	0,701	0,801
20,0	0,048	0,129	0,203	0,357	0,572	0,702	0,781	0,893
40,0	0,050	0,136	0,213	0,376	0,602	0,739	0,823	0,940
100,0	0,052	0,140	0,220	0,388	0,621	0,762	0,848	0,969

53. Local head losses. Equivalent length of pipe

The lower values of K are recommended for accessories with flanges, namely for diameters of over 100 mm.

Accessory	Name	K extreme values	Equivalent length L of straight pipe in metres, for $K = 75\ m^{1/3}/s$ and for the following values of diameter D in mm. For $K_s \neq 75\ m^{1/3}/s$, multiply by $(K_s/75)^2$														
			12,5	25	50	75	100	125	150	175	200	250	300	350	400	450	500
	Globe valve	5,2	0,68	1,72	4,32	7,33	10,9	14,9	18,7	22,9	27,5	37,0	47,2	—	—	—	—
		10,3	1,34	3,3	8,3	14,1	21,0	28,7	36,0	44,1	52,8	71,1	90,7	—	—	—	—
	Sluice valve	0,05	0,006	0,02	0,04	0,07	0,11	0,14	0,18	0,22	0,26	0,35	0,45	0,56	0,67	0,78	0,90
		0,19	0,02	0,06	0,15	0,27	0,40	0,54	0,68	0,84	1,00	1,35	1,72	2,12	2,53	2,96	3,40
	Check valve	0,6	0,08	0,02	0,05	0,08	0,13	0,17	0,22	0,26	0,32	0,43	0,54	0,67	0,80	0,93	1,07
		2,3	0,30	0,76	1,90	3,24	4,83	6,60	8,30	10,1	12,1	16,4	20,9	25,6	30,4	35,8	41,2
	Horizontal-impulse non-return valve	8	1,04	2,64	6,60	11,3	16,8	23,0	28,8	35,3	42,2	56,9	72,9	89,1	106	124	143
		12	1,56	3,96	9,96	16,9	25,2	34,4	43,2	52,9	63,3	85,3	109	134	160	187	215
	Spherical non-return valve	65	8,45	21,4	53,9	91,7	136	187	234	287	343	462	590	724	866	1011	1164
		70	9,10	23,1	58,1	98,7	147	201	252	309	370	498	635	780	932	1089	1254

53. (Contd.)   Local head losses. Equivalent length of pipe

The lower values of K are recommended for accessories with flanges, namely for diameters of over 100 mm.

Accessory	Name	K extreme values	Equivalent length L of straight pipe in metres, for $K_s = 75$ m$^{1/3}$/s and for the following values of diameter D in mm. For $K_s \neq 75$ m$^{1/3}$/s, multiply by $(K_s/75)^2$														
			12,5	25	50	75	100	125	150	175	200	250	300	350	400	450	500
	Angle valve	2,1	0,27	0,66	1,66	2,82	4,20	5,74	7,20	8,82	10,68	14,2	18,1	22,2	26,6	31,1	35,8
		3,1	0,40	1,65	4,15	7,05	10,5	14,4	18.0	22,0	26,4	35,5	45,4	55,7	66,6	77,8	89,6
	"Y" valve	2,9	0,38	0,99	2,49	4,23	6,30	8,61	10,8	13,2	15,8	21,3	27,1	33,3	39,9	46,6	53,7
	Foot valve	$\simeq 15$	1,95	4,95	12,5	21,2	31,5	43,0	54,0	66,1	79,2	107	136	167	200	233	269
	Threaded union	0,02	0,00	0,01	0,02	0,03	0,04	0,06	0,07	0,09	0,10	0,14	0,18	0,22	0,27	0,31	0,36
		0,07	0,01	0,03	0,06	0,10	0,15	0,20	0,25	0,31	0,37	0,50	0,64	0,78	0,93	1,09	1,25
	Threaded reduction	0,05 †	0,01	0,02	0,04	0,07	0,11	0,14	0,18	0,22	0,26	0,35	0,45	0,56	0,67	0,78	0,90
		2,0	0,26	0,65	1,66	2,82	4,20	5,74	7,20	8,82	10,6	14,2	18,1	22,2	26,6	31,1	35,8

† These values are applicable when this accessory is used as a cross-section reducer. When it is used to increase the cross-section, the losses are about 1.4 times higher than those originating in a sudden enlargement.

53. (Contd.) Local head losses. Equivalent length of pipe

The lower values of K are recommended for diameters of over 100 mm.

Equivalent length L of straight pipe in metres, for $K_s = 75\ \mathrm{m^{1/3}/s}$ and for the following values of diameter D in mm. For $K_s \neq 75\ \mathrm{m^{1/3}/s}$, multiply by $(K_s/75)^2$

Accessory	Name	K extreme values	12,5	25	50	75	100	125	150	175	200	250	300	350	400	450	500
	45° bend, large radius, with flange	0,18	0,02	0,06	0,15	0,25	0,38	0,52	0,65	0,79	0,95	1,28	1,63	2,00	2,40	2,80	3,22
		0,20	0,03	0,07	0,17	0,28	0,42	0,57	0,72	0,88	1,07	1,42	1,81	2,22	2,66	3,11	3,58
	Annular bend standard	0,75	0,10	0,25	0,62	1,06	1,58	2,15	2,70	3,31	3,96	5,33	6,80	8,36	9,99	11,67	13,43
		2,2	0,29	0,73	1,83	3,10	4,60	6,31	7,92	9,70	11,6	15,6	20,0	24,5	29,3	34,2	39,5
	Annular bend consisting of two 90° curves	† 0,38	0,05	0,12	0,30	0,54	0,80	1,08	1,76	1,68	2,00	2,70	3,44	4,24	5,06	5,92	6,80
		‡ 0,25	0,03	0,08	0,21	0,35	0,53	0,72	0,90	1,10	1,32	1,78	2,27	2,79	3,33	3,89	4,48
	Rounded inlet	0,04	0,01	0,02	0,04	0,06	0,08	0,12	0,14	0,18	0,20	0,28	0,36	0,44	0,54	0,62	0,72
		0,05	0,01	0,02	0,04	0,07	0,11	0,14	0,18	0,22	0,26	0,35	0,45	0,56	0,67	0,78	0,90
	Sharp-edged inlet	0,47	0,06	0,16	0,39	0,66	0,99	1,35	1,69	2,07	2,48	3,34	4,26	5,24	6,26	7,31	8,42
		0,56	0,07	0,18	0,46	0,79	1,18	1,61	2,02	2,47	2,96	3,98	5,08	6,24	7,46	8,71	10,03

† 90° bends, normal.
‡ 90° bends, large radius.

53. (Contd.)   Local head losses. Equivalent length of pipe

The lower values of K are recommended for diameters of over 100 mm.

Accessory	Name	K extreme values	Equivalent length L of straight pipe in metres, for $K_s = 75\,m^{1/3}/s$ and for the following values of diameter D in mm. For $K_s \neq 75\,m^{1/3}/s$, multiply by $(K_s/75)^2$														
			12,5	25	50	75	100	125	150	175	200	250	300	350	400	450	500
	90° bend, normal, threaded	0,55	0,07	0,20	0,50	0,85	1,26	1,72	2,16	2,65	3,17	4,27	5,44	6,68	7,99	9,34	10,8
		0,9	0,12	0,30	0,75	1,27	1,89	2,58	3,24	3,97	4,75	6,40	8,16	10,0	12,0	14,0	16,1
	90° bend large radius, threaded	0,22	0,03	0,07	0,18	0,31	0,46	0,63	0,79	0,97	1,16	1,56	2,00	2,45	2,93	3,42	3,95
		0,60	0,08	0,20	0,50	0,85	1,26	1,72	2,16	2,65	3,17	4,27	5,44	6,68	7,99	9,34	10,8
	90° bend normal with flange	0,21	0,03	0,07	0,18	0,31	0,46	0,63	0,79	0,97	1,16	1,56	2,00	2,45	2,93	3,42	3,95
		0,30	0,04	0,10	0,25	0,48	0,63	0,86	1,08	1,33	1,59	2,14	2,72	3,34	4,00	4,67	5,40
	90° bend large radius, with flange	0,14	0,02	0,05	0,12	0,20	0,29	0,40	0,50	0,62	0,74	1,00	1,27	1,56	1,86	2,18	2,51
		0,23	0,03	0,08	0,19	0,32	0,48	0,66	0,83	1,01	1,21	1,64	2,09	2,56	3,06	3,58	4,12
	45° bend, normal, threaded	0,30	0,04	0,10	0,25	0,48	0,63	0,86	1,08	1,33	1,59	2,14	2,72	3,34	4,00	4,67	5,40
		0,42	0,05	0,14	0,35	0,59	0,88	1,20	1,51	1,85	2,22	2,99	3,81	4,68	5,59	6,54	7,52

53. (Contd.)   Local head losses. Equivalent length of pipe

The lower values of K are recommended for diameters of over 100 mm.

Accessory	Name	K extreme values	Equivalent length L of straight pipe in metres, for $K_s = 75$ m$^{1/3}$/s and for the following values of diameter D in mm. For $K_s \neq 75$ m$^{1/3}$/s, multiply by $(K_s/75)^2$														
			12,5	25	50	75	100	125	150	175	200	250	300	350	400	450	500
Re-entrant intake		0,62 †	0,08	0,20	0,51	0,87	1,30	1,78	2,23	2,73	3,27	4,08	5,6	6,91	8,26	9,65	11,1
		1,0	0,13	0,33	0,83	1,41	2,10	2,87	3,60	4,41	5,28	7,11	9,07	11,1	13,3	15,5	17,9
T, normal, threaded	From the line to the branch	0,85	0,11	0,28	0,71	1,20	1,78	2,44	3,06	3,75	4,49	6,04	7,71	9,47	11,3	13,2	15,2
		1,3	0,17	0,43	1,08	1,83	2,73	3,73	4,68	5,77	6,86	9,24	11,8	14,5	17,3	20,3	23,3
	From the branch to the line	0,92	0,12	0,30	0,76	1,29	1,93	2,64	3,31	4,06	4,86	6,54	7,34	10,2	12,3	14,3	16,5
		2,15	0,28	0,71	1,78	3,03	4,51	6,17	7,74	9,48	11,4	15,3	19,5	24,0	28,6	33,5	38,5
T, large radius, threaded	From the line to the branch	0,37	0,05	0,12	0,31	0,52	0,78	1,06	1,33	1,63	1,95	2,63	3,36	4,12	4,93	5,76	6,63
		0,80	0,10	0,26	0,66	1,13	1,68	2,30	2,88	3,53	4,22	5,69	7,26	8,91	10,6	12,4	14,3
	From the branch to the line	0,50	0,06	0,17	0,42	0,71	1,05	1,44	1,80	2,20	2,64	3,55	4,54	5,57	6,66	7,78	8,96
		0,52	0,07	0,17	0,43	0,73	1,09	1,49	1,87	2,29	2,75	3,70	4,72	5,79	6,93	8,09	9,31

† K diminishes when the thickness of the wall increases and the edges are rounded.

54. Head losses in trashracks normal to the flow

According to Berezinski (1958)[†] and L. Levin (1968),[‡] quoted by [31].

$$\Delta H = KU^2/2g.$$

U — mean velocity of the approach flow when the whole of the area is free.

(See Fig. 4.14.)

$$K = K_d \cdot K_f \cdot p^{1.6} \cdot f(b/a) \cdot \sin\varphi$$

K_d — Coefficient of deposits on trashrack considered 'dirty'
  1.1 to 1.2 for modern automatic cleaner
  1.5 for old automatic cleaner
  2 to 4 or more for hand cleaning, function of the characteristics of the watercourse

K_f — Shape coefficient of the bar cross-sections
  0.51 for elongated rectangular cross-sections
  0.35 for circular cross-section
  0.32 for elongated cross-section with semicircles at the ends

p — Ratio between the area obstructed by the trashrack and the total cross-section area. The e/a ratio varies in practice between 6% and 16%, but the value of p, which takes account of the whole structure of the trashrack, may reach values of 22% to 38%[§]

$f(b/a) = 8 + 2{,}3(b/a) + 2{,}4(a/b)$

b — dimension of the transverse cross-section of the bars in the direction of the flow
a — free space between bars
φ — angle of the trashrack to the horizontal

p	0,05	0,10	0,15	0,20	0,25	0,30	0,35	0,40	0,45	0,50
$p^{1,6}$	0,008	0,025	0,048	0,076	0,109	0,146	0,186	0,231	0,279	0,330

[†] Berezinsky, A., *Head losses through trashracks*, Hydrotechnical Construction No. 5, Moscow, 1958.
[‡] Levin, L., *Étude hydraulique des grilles des prises d'eau*. AIRH, 7th Congress. 1957.
[§] Novikov, A., *Head losses in trashracks*, Hydrotechnical Construction, No. 10, Moscow, 1957.

55.   Head losses in oblique trashracks

According to Spander (1928), quoted by [33]

$$\Delta H = KU^2/2g - \text{Values of } K.$$

$U = $ Velocity of the approach flow when the whole area is free.

The head loss coefficients in trashrack bars, under a given angle of attack, θ, are given by:

$$K = K_1 . K_2$$

in which K_1 is a function of both the shape of the bar and the angle of attack and K_2 is a function of both the ratio a/a_o and of the angle of attack.

Shape of bars

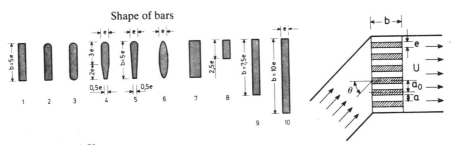

(1) *Value of* K_1

Shape number	θ									
	0°	5°	10°	15°	20°	25°	30°	40°	50°	60°
1	1,00	1,00	1,00	1,00	1,00	1,00	1,00	1,00	1,00	1,00
2	0,76	0,65	0,58	0,54	0,52	0,51	0,52	0,58	0,63	0,62
3	0,76	0,60	0,55	0,51	0,49	0,48	0,49	0,57	0,64	0,66
4	0,43	0,37	0,34	0,32	0,30	0,29	0,30	0,36	0,47	0,52
5	0,37	0,37	0,38	0.40	0,42	0,44	0,47	0,56	0,67	0,72
6	0,30	0,24	0,20	0,17	0,16	0,15	0,16	0,25	0,37	0,43
7	1,00	1,08	1,13	1,18	1,22	1,25	1,28	1,33	1,31	1,20
8	1,00	1,06	1,10	1,15	1,18	1,22	1,25	1,30	1,22	1,00
9	1,00	1,00	1,00	1,01	1,02	1,03	1,05	1,10	1,04	0,82
10	1,00	1,04	1,07	1,09	1,10	1,11	1,10	1,07	1,00	0,92

(2) *Value of* K_2

a/a_o	θ									
	0°	5°	10°	15°	20°	25°	30°	40°	50°	60°
0,50	2,34	2,40	2,48	2,57	2,68	2,80	2,95	3,65	4,00	4,70
0,55	1,75	1,80	1,85	1,90	2,00	2,10	2,25	2,68	3,55	4,50
0,60	1,35	1,38	1,42	1,48	1,55	1,65	1,79	2,19	3,00	4,35
0,65	1,00	1,05	1,08	1,12	1,20	1,30	1,40	1,77	2,56	4,25
0,70	0,78	0,80	0,85	0,89	0,95	1,05	1,17	1,52	2,30	4,10
0,75	0,60	0,62	0,65	0,70	0,75	0,85	0,95	1,30	2,05	3,90
0,80	0,37	0,40	0,45	0,50	0,55	0,64	0,75	1,06	1,75	3,70
0,85	0,24	0,25	0,30	0,36	0,42	0,50	0,60	0,88	1,40	3,50

56.   Stability of trashracks

Compiled from [31]

(a) *Strouhal's number*, S_t, *for different cross-sections of the bars, for flow around a single, isolate bar.*

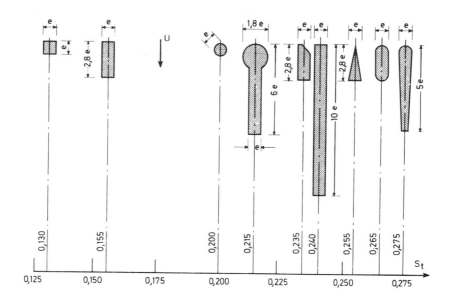

(b) *Coefficient, c, by which* S_t *is multiplied depending on the ratio* $(a + e)/e$.

$\frac{(a+e)}{e}$	1,5	2,0	2,5	3,0	3,5	4,0	4,5	5,0	6,0
c	2,15	1,7	1,4	1,2	1,1	1,05	1,03	1,01	1,0

57. Open channel flows: Uniform flow — Index

58.   Values of K_B — Bazin formula

	Characteristics	$K_B - m^{1/2}$
(1)	Channels of very smooth concrete; channels of planed boards, of major dimension aligned with the direction of the current; metal walls without rust and without protuberances at the joints. (The course of the channel must be in long reaches, connected by large-radius bends; the water must be clear).	0.06†
(2)	Concrete channels plastered, but not fully smoothed and with small protuberances at the joints. Channels of planed boards, with poorly adjusted joints. Metal channels with common welds, but without protuberances at the joints. Channels of regular masonry with square-set stones.	0.16
(3)	Concrete channels, partly plastered, with protruding joints, water not clear, mossy vegetation. Channels with common stone pitching	0.46
(4)	Fairly regular earth channels, eventually lined with stone pitching, without vegetation and wide bands. Channels of irregular masonry, with fairly smooth bottom due to slime deposits.	0.85
(5)	Earth channels, with regular cross-section and low vegetation on the bottom, little vegetation on the banks. Natural watercourses with regular routing, with little vegetation or sediments on the bottom.	1.30
(6)	Earth channels in poor maintenance condition, with vegetation on the bottom and banks. Earth channels made with mechanical excavators and poorly maintained.	1.75

59. Values of K_K — Kutter formula

	Characteristics	$K_K - m^{1/2}$
(1)	Very smooth concrete walls, semicircular cross-section	0.12†
(2)	Ditto, rectangular cross-section	0.15
(3)	Walls of planed boards, rectangular cross-section	0.20
(4)	Walls of unplaned boards, rectangular or trapezoidal cross-section; very regular mansory with square-set stones.	0.25
(5)	Walls of common masonry, carefull constructed.	0.35
(6)	Walls of repaired masonry	0.45
(7)	Walls of common stone pitching	0.55
(8)	Walls of common masonry, with muddy bottom	0.75
(9)	Walls of masonry, without maintenance	1.00
(10)	Small channels excavated in rock, of small dimensions; fairly regular earth channels, without vegetation	1.25 to 1.50
(11)	Earth channels in poor maintenance condition, with vegetation; natural watercourses, with earth bed	1.75 to 2.00
(12)	Earth channels completely abandoned; natural watercourses with gravel bed	2.50

† For extremely smooth channels, especially those of large dimensions, the coefficients K_B and K_K may take on negative values. In this case it is preferable to use the Manning formula.

60. Values of the roughness coefficient, K_s (m$^{1/3}$/s), in the Manning formula
Compiled from [35]

The figures printed in heavy type are the values generally recommended for design purposes.

	Type of channel and description	Maximum	Normal	Minimum
A.	*Partly full pipes*			
A.1	Metal			
(a)	polished brass	111	**100**	77
(b)	steel			
	1. lockbar and welded	100	83	71
	2. riveted and spiral	77	63	59
(c)	cast iron			
	1. coated	100	77	71
	2. uncoated	91	71	63
A.2	Nonmetal			
(a)	plastic material	125	111	100
(b)	glass	111	**100**	77
(c)	cement			
	1. next surface	100	91	77
	2. mortar	91	77	67
(d)	concrete			
	1. straight culvert, free of debris	100	91	77
	2. straight sewer with manholes, inlets, etc.	77	67	59
	3. unfinished, steel form work	83	77	71
	4. unfinished, smooth wood form work	83	**71**	63
	5. unfinished, rough wood form work	67	59	50
(e)	earthenware			
	1. vitrified sewer pipe	91	71	59
	2. vitrified sewer pipe, with manholes, inlets, etc.	77	67	59
	3. vitrified, subdrain with open joint	71	**63**	56
(f)	sewers with bends and connections	83	77	63
(g)	rubble masonry, cemented	56	40	33
B.	*Lined or precast channels*			
B.1	Metallic			
(a)	smooth steel surface			
	1. unpainted	91	**83**	71
	2. painted	83	77	59
(b)	corrugated	48	40	33
B.2	Non-metallic			
(a)	cement			
	1. smooth surface	100	91	77
	2. mortar	91	77	67
(b)	concrete			
	1. trowel-plastered	91	**77**	67
	2. plastered, with gravel on bottom	67	59	50
	3. non-plastered	71	59	50
	4. gunited, regular cross-section	63	53	43
	5. gunited, wavy cros-section	56	45	40
	6. over regular excavated rock	59	50	
	7. over irregular excavated rock	45	37	

60. (Contd.) Values of the roughness coefficient

	Type of channel and description	Maximum	Normal	Minimum
(c)	channel with concrete-lined bottom and sides of:			
	1. mortared stone	67	59	50
	2. mortared irregular stone	59	50	42
	3. dry rubble or riprap	50	33	29
(d)	gravel bottom with sides of			
	1. concrete	59	50	40
	2. mortared irregular stone	50	43	38
	3. dry rubble or riprap	43	30	28
(e)	masonry			
	1. cemented rubble	59	40	33
	2. dry rubble	43	31	29
(f)	dressed ashlar	77	67	59
(g)	asphalt			
	1. smooth	77	77	
	2. rough	63	63	
(h)	vegetal lining	33		2
C.	*Excavated or dredged channel*			
(a)	earth, straight and uniform			
	1. clean, recently completed	63	56	50
	2. with short grass, few weeds	45	37	30
(b)	earth, winding and sluggish			
	1. without vegetation	43	40	33
	2. with grass, some weeds	40	33	30
	3. dense weeds or aquatic plants in deep channels	33	29	25
	4. earth bottom and rubble banks	36	33	29
	5. stony bottom and weedy banks	40	29	25
	6. cobble bottom and clean banks	33	25	20
(c)	dragline — excavated or dredged			
	1. without vegetation	40	36	30
	2. little brush on banks	29	20	17
(d)	cuts rock			
	1. little rough	40	29	25
	2. very rough and uniform	35	25	20
(e)	neglected channels, tree trunks and bushes not cut			
	1. dense weeds, high as flow depth	20	13	8
	2. clean bottom, brushes on sides	25	20	13
	3. the same case of high vegetation	22	14	9
	4. dense brush, high stage	13	10	7

61. Values of the inverse roughness coefficient, $n = 1/K_s$, in natural watercourses

According to Cowan, quoted by [35] and [36].

$$n = \frac{1}{K_s} = a(n_0 + n_1 + n_2 + n_3 + n_4) \; s/m^{1/3}$$

The various values of this formula are given in the following tables

Influence of meandering		a	Bed material	n_0
Moderate	$L/c = 1$ to 1.2	1	Earth	0.020
Considerable	$L/c = 1.2$ to 1.5	1.15	Rock	0.025
			Fine gravel	0.024
Severe	$L/c > 1.5$	1.3	Coarse gravel	0.028

L = length of meander; c = length of the straight line uniting the two ends of the meander

Irregularities of bottom and banks	n_1	Shape variations	n_2	Obstructions	n_3
Negligible	0.0	Progressive	0.0	Negligible	0
Slight	0.005	Infrequent sudden	0.005	Mild	0.010–0.015
Moderate	0.010	Frequent sudden	0.010–0.015	Considerable	0.020–0.030
Important	0.020			Important	0.040–0.060

	Vegetation	n_4
Low:	dense growths of flexible turf grasses or weeds with mean height about 1/2 to 1/3 of the water depth; sparse brushes 1/3 to 1/4 of the water depth	0.005–0.010
Medium:	turf grasses with mean height nearly 1/2 of the water depth; stemmy grasses, weeds or tree seedling with moderate cover with mean height nearly 1/2 to 1/3 of the water depth	0.010–0.025
High:	turf grasses with height nearly that of the water depth; equivalent vegetation.	0.025–0.050
Very high:	turf grasses whose height is more than double the water depth; equivalent vegetation	0.050–0.100

62. Values of the roughness coefficient, K_s, in natural watercourses
According to Parde, quoted by [36]

(1) Small mountain streams, with very irregular bottom and width 10 to 30 m: $K_s = 23$–$26 \; m^{1/3}/s$
(2) Mountain streams, width 30 to 50 m, bottom slope more than 0.002 and bottom of large pebbles (10 to 20 cm in diameter): $K_s = 27$–$29 \; m^{1/3}/s$
(3) Streams of comparable or greater width, with bottom slopes of 0.0008 to 0.002 and pebbles of 8 to 10 cm in diameter: $K_s = 30$–$33 \; m^{1/3}/s$
(4) Ditto, with bottom slope between 0.0006 and 0.0008, pebbles of 4 to 8 cm in diameter: $K_s = 34$–$37 \; m^{1/3}/s$
(5) Ditto, with smaller pebbles: $K_s = 38$–$40 \; m^{1/3}/s$
(6) Ditto, with bottom slope between 0.00025 and 0.0006, gravel or sandy bottom: $K_s = 41$–$42 \; m^{1/3}/s$
(7) Not very turbulent watercourses with bottom slopes between 0.00012 and 0.00025, sandy or muddy bottom: $K_s = 43$–$45 \; m^{1/3}/s$
(8) Large watercourses with bottom slope less than 0.00012 and very smooth bottom: $K_s = 46$–$50 \; m^{1/3}/s$

63. Various channel cross-sections. Geometrical elements
Adapted from [35]

Shape	Cross-section A	Wetted perimeter P	Radius R	Top width L	Mean depth h_m	$A\sqrt{\dfrac{A}{L}}$
	lh	$l+2h$	$\dfrac{lh}{l+2h}$	l	h	$lh^{1.5}$
	$(l+mh)h$	$l+2h\sqrt{1+m^2}$	$\dfrac{(l+mh)h}{l+2h\sqrt{1+m^2}}$	$l+2mh$	$\dfrac{(l+mh)h}{l+2mh}$	$\dfrac{[(l+mh)h]^{1.5}}{\sqrt{l+2mh}}$
	mh^2	$2h\sqrt{1+m^2}$	$\dfrac{mh}{2\sqrt{1+m^2}}$	$2mh$	$\tfrac{1}{2}h$	$\dfrac{\sqrt{2}}{2}mh^{2.5}$
	$\tfrac{2}{3}\times Lh$	$L+\dfrac{8}{3}\dfrac{h^2}{L}$ †	$\dfrac{2L^2h}{3L^2+8h^2}$ †	$\dfrac{3}{2}\dfrac{A}{h}$	$\tfrac{2}{3}h$	$\tfrac{2}{9}\sqrt{6}\,Lh^{1.5}$
	$\left(\dfrac{\pi}{2}-2\right)r^2+(l+2r)h$	$(\pi-2)r+l+2h$	$\dfrac{(\pi/2-2)r^2+(l+2r)h}{(\pi-2)r+l+2h}$	$l+2r$	$\dfrac{(\pi/2-2)r^2}{l+2r}+h$	$\dfrac{[(\pi/2-2)r^2+(l+2r)h]^{1.5}}{\sqrt{l+2r}}$
	$\dfrac{L^2}{4m}-\dfrac{r^2}{m}(1-\dfrac{}{m\tan m})$	$\dfrac{L}{m}\sqrt{1+m^2}-\dfrac{2r}{m}(1-\dfrac{}{m\tan m})$	$\dfrac{A}{P}$	$2\left[m(h-r)+r\sqrt{1+m^2}\right]$	$\dfrac{A}{L}$	$A\sqrt{\dfrac{A}{L}}$
	$\tfrac{1}{8}(\theta-\sin\theta)d^2$	$\tfrac{1}{2}\theta d$	$\tfrac{1}{4}\left(1-\dfrac{\sin\theta}{\theta}\right)d$	$(\sin\tfrac{1}{2}\theta)d$ or $2\sqrt{h(d-h)}$	$\tfrac{1}{8}\left(\dfrac{\theta-\sin\theta}{\sin\tfrac{1}{2}\theta}\right)d$	$\dfrac{\sqrt{2}}{32}\dfrac{(\theta-\sin\theta)^{1.5}}{(\sin\tfrac{1}{2}\theta)^{0.5}}d^{2.5}$

† By making $x=4h/L$, this formula gives satisfactory values of P for $0<x\leqslant 1$.

When $x>1$, the exact expression $P=(L/2)[\sqrt{1+x^2}+1/x\ln(x+\sqrt{1+x^2})]$ must be used.

64. Partly full circular cross-section. Geometrical elements

d = diameter; h = water depth

(a) *Values of area A; of the wetted perimeter P and of the hydraulic radius R.*

$\dfrac{h}{d}$	$\dfrac{A}{d^2}$	$\dfrac{P}{d}$	$\dfrac{R}{d}$	$\dfrac{h}{d}$	$\dfrac{A}{d^2}$	$\dfrac{P}{d}$	$\dfrac{R}{d}$
0,01	0,0013	0,2003	0,0066	0,51	0,4027	1,5908	0,2531
0,02	0,0037	0,2838	0,0132	0,52	0,4127	1,6108	0,2561
0,03	0,0069	0,3482	0,0197	0,53	0,4227	1,6308	0,2591
0,04	0,0105	0,4027	0,0262	0,54	0,4327	1,6509	0,2620
0,05	0,0147	0,4510	0,0326	0,55	0,4426	1,6710	0,2649
0,06	0,0192	0,4949	0,0389	0,56	0,4526	1,6911	0,2676
0,07	0,0242	0,5355	0,0451	0,57	0,4625	1,7113	0,2703
0,08	0,0294	0,5735	0,0513	0,58	0,4723	1,7315	0,2728
0,09	0,0350	0,6094	0,0574	0,59	0,4822	1,7518	0,2753
0,10	0,0409	0,6435	0,0635	0,60	0,4920	1,7722	0,2776
0,11	0,0470	0,6761	0,0695	0,61	0,5018	1,7926	0,2797
0,12	0,0534	0,7075	0,0754	0,62	0,5115	1,8132	0,2818
0,13	0,0600	0,7377	0,0813	0,63	0,5212	1,8338	0,2839
0,14	0,0668	0,7670	0,0871	0,64	0,5308	1,8546	0,2860
0,15	0,0739	0,7954	0,0929	0,65	0,5404	1,8755	0,2881
0,16	0,0811	0,8230	0,0986	0,66	0,5499	1,8965	0,2899
0,17	0,0885	0,8500	0,1042	0,67	0,5594	1,9177	0,2917
0,18	0,0961	0,8763	0,1097	0,68	0,5687	1,9391	0,2935
0,19	0,1039	0,9020	0,1152	0,69	0,5780	1,9606	0,2950
0,20	0,1118	0,9273	0,1206	0,70	0,5872	1,9823	0,2962
0,21	0,1199	0,9521	0,1259	0,71	0,5964	2,0042	0,2973
0,22	0,1281	0,9764	0,1312	0,72	0,6054	2,0264	0,2984
0,23	0,1365	1,0003	0,1364	0,73	0,6143	2,0488	0,2995
0,24	0,1449	1,0239	0,1416	0,74	0,6231	2,0714	0,3006
0,25	0,1535	1,0472	0,1466	0,75	0,6318	2,0944	0,3017
0,26	0,1623	1,0701	0,1516	0,76	0,6404	2,1176	0,3025
0,27	0,1711	1,0928	0,1566	0,77	0,6489	2,1412	0,3032
0,28	0,1800	1,1152	0,1614	0,78	0,6573	2,1652	0,3037
0,29	0,1890	1,1373	0,1662	0,79	0,6655	2,1895	0,3040
0,30	0,1982	1,1593	0,1709	0,80	0,6736	2,2143	0,3042
0,31	0,2074	1,1810	0,1755	0,81	0,6815	2,2395	0,3044
0,32	0,2167	1,2025	0,1801	0,82	0,6893	2,2653	0,3043
0,33	0,2260	1,2239	0,1848	0,83	0,6969	2,2916	0,3041
0,34	0,2355	1,2451	0,1891	0,84	0,7043	2,3186	0,3038
0,35	0,2450	1,2661	0,1935	0,85	0,7115	2,3462	0,3033
0,36	0,2546	1,2870	0,1978	0,86	0,7186	2,3746	0,3026
0,37	0,2642	1,3078	0,2020	0,87	0,7254	2,4038	0,3017
0,38	0,2739	1,3284	0,2061	0,88	0,7320	2,4341	0,3008
0,39	0,2836	1,3490	0,2102	0,89	0,7384	2,4655	0,2996
0,40	0,2934	1,3694	0,2142	0,90	0,7445	2,4981	0,2980
0,41	0,3032	1,3898	0,2181	0,91	0,7504	2,5322	0,2963
0,42	0,3130	1,4101	0,2220	0,92	0,7560	2,5681	0,2944
0,43	0,3229	1,4303	0,2257	0,93	0,7642	2,6061	0,2922
0,44	0,3328	1,4505	0,2294	0,94	0,7662	2,6467	0,2896
0,45	0,3428	1,4706	0,2331	0,95	0,7707	2,6906	0,2864
0,46	0,3527	1,4907	0,2366	0,96	0,7749	2,7389	0,2830
0,47	0,3627	1,5108	0,2400	0,97	0,7785	2,7934	0,2787
0,48	0,3727	1,5308	0,2434	0,98	0,7816	2,8578	0,2735
0,49	0,3827	1,5508	0,2467	0,99	0,7841	2,9412	0,2665
0,50	0,3927	1,5708	0,2500	1,00	0,7854	3,1416	0,2500

64 (Contd.) Partly full circular cross-section. Geometrical elements

d = diameter; h = water depth; K = tabulated values

(b) *Determination of the top width, L*

$L = K d$

$\dfrac{h}{d}$	0	1	2	3	4	5	6	7	8	9
0,0	0,000	0,199	0,280	0,341	0,392	0,436	0,475	0,510	0,543	0,572
1	600	626	650	673	694	714	733	751	768	785
2	800	815	828	842	854	866	877	888	898	908
3	917	925	933	940	947	954	960	966	971	975
4	980	984	987	990	993	995	997	998	999	1,000
0,5	1,000	1,000	0,999	0,998	0,997	0,995	0,993	0,990	0,987	0,984
6	980	975	971	966	960	954	947	940	933	925
7	917	908	898	888	877	866	854	842	828	815
8	800	785	768	751	733	714	694	673	650	626
9	600	572	543	510	575	436	392	341	280	199

(c) *Determination of the mean depth, h_m*

$h_m = Kd$

$\dfrac{h}{d}$	0	1	2	3	4	5	6	7	8	9
0,0	0,000	0,007	0,013	0,020	0,027	0,034	0,040	0,047	0,054	0,061
1	068	075	082	089	096	103	111	118	125	132
2	140	147	155	162	170	177	185	193	200	208
3	216	224	232	240	249	257	265	274	282	291
4	299	308	317	326	335	345	354	363	373	383
0,5	0,393	0,403	0,413	0,423	0,434	0,445	0,456	0,467	0,478	0,490
6	502	514	527	540	553	566	580	595	610	625
7	641	657	674	692	710	730	750	771	793	817
8	842	869	897	928	960	996	1,035	1,078	1,126	1,180
9	1,241	1,311	1,393	1,492	1,613	1,768	1,977	2,282	2,792	3,940

(d) *Determination of the depth, y, of the centre of gravity*

$y = Kd$

$\dfrac{h}{d}$	0	1	2	3	4	5	6	7	8	9
0,0	0,000	0,004	0,008	0,012	0,016	0,020	0,024	0,028	0,032	0,036
1	040	044	049	053	057	061	065	069	073	077
2	082	086	090	094	098	103	107	111	115	119
3	124	128	132	137	141	145	150	154	158	163
4	167	172	176	181	185	189	194	199	203	208
0,5	0,212	0,217	0,221	0,226	0,231	0,235	0,240	0,245	0,250	0,254
6	259	264	269	274	279	284	289	294	299	304
7	309	314	320	325	330	336	341	347	352	358
8	363	369	375	381	387	393	399	405	411	418
9	424	431	438	445	452	459	466	474	482	491

65. Partly full circular cross-section. Values of U and Q

U_h and Q_n — Velocity and discharge corresponding to water depth h
U_d and Q_d — Velocity and discharge corresponding to full cross-section (Table 66)

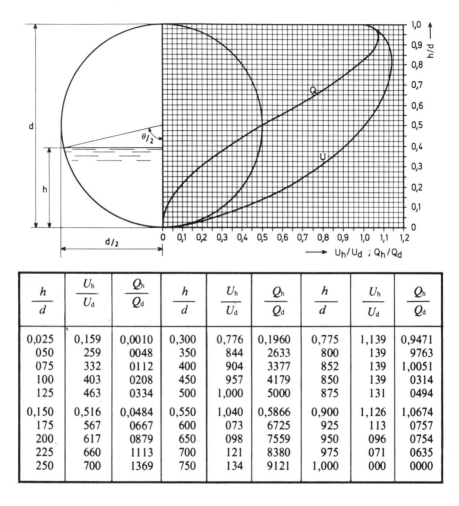

$\dfrac{h}{d}$	$\dfrac{U_h}{U_d}$	$\dfrac{Q_h}{Q_d}$	$\dfrac{h}{d}$	$\dfrac{U_h}{U_d}$	$\dfrac{Q_h}{Q_d}$	$\dfrac{h}{d}$	$\dfrac{U_h}{U_d}$	$\dfrac{Q_h}{Q_d}$
0,025	0,159	0,0010	0,300	0,776	0,1960	0,775	1,139	0,9471
050	259	0048	350	844	2633	800	139	9763
075	332	0112	400	904	3377	852	139	1,0051
100	403	0208	450	957	4179	850	139	0314
125	463	0334	500	1,000	5000	875	131	0494
0,150	0,516	0,0484	0,550	1,040	0,5866	0,900	1,126	1,0674
175	567	0667	600	073	6725	925	113	0757
200	617	0879	650	098	7559	950	096	0754
225	660	1113	700	121	8380	975	071	0635
250	700	1369	750	134	9121	1,000	000	0000

Example: Flow is uniform at a depth, h, of 2.10 m in a sewer of circular cross-section with diameter $d = 3.0$ m and bottom slope $I = 0.005$.

Determine the mean velocity U and the discharge Q ($K_s = 75$ m$^{1/3}$/s).

Solution: $h/d = 2.1/3.0 = 0.7$; from the table we have
      $U_h = 1,121\ U_d$ and $Q_h = 0,838\ Q_d$.
Table 66 gives: $U_d = 61.88\ \sqrt{I} = 61.88\ \sqrt{0.005} = 4.38$ m/s and $Q_d = 437.37\ \sqrt{I} = 437.37$ $\sqrt{0.005} = 30.93$ m^3/s
      Thus $U_h = 4.90$ ms   and $Q_h = 254.92$ m^3/s

66. Completely full circular cross-section

Table 67 must be consulted

$$A = \pi \frac{D^2}{4} = 0.7854 \ D^2 \qquad P = \pi \ D = 3.1416 \ D; \qquad R = 0.25 \ D$$

$D = H$	$D^{2/3}$	$D^{8/3}$	S	P	R	$R^{2/3}$	$SR^{2/3}$	$K_s = 75$	
m	$m^{2/3}$	$m^{8/3}$	m^2	m	m	$m^{2/3}$	$m^{8/3}$	$\dfrac{U}{\sqrt{i}}$	$\dfrac{Q}{\sqrt{i}}$
0,20	0,342	0,0137	0,0314	0,628	0,050	0,136	0,0043	10,20	0,323
0,25	0,397	0,0248	0,0491	0,785	0,063	0,158	0,0078	11,85	0,585
0,30	0,448	0,0403	0,0707	0,942	0,075	0,178	0,0126	13,35	0,945
0,35	0,497	0,0609	0,0962	1,100	0,088	0,198	0,0190	14,85	1,425
0,40	0,543	0,0869	0,1257	1,257	0,100	0,215	0,0270	16,13	2,025
0,50	0,630	0,1575	0,1964	1,571	0,125	0,250	0,0491	18,75	3,683
0,60	0,711	0,2560	0,2827	1,885	0,150	0,282	0,0797	21,15	5,978
0,70	0,788	0,3861	0,3848	2,199	0,175	0,313	0,1204	23,48	9,030
0,80	0,862	0,5517	0,5027	2,513	0,200	0,342	0,1719	25,65	12,893
0,90	0,932	0,7549	0,6362	2,827	0,225	0,370	0,2354	27,75	17,655
1,00	1,000	1,0000	0,7854	3,142	0,250	0,397	0,3118	29,78	23,385
1,10	1,065	1,2887	0,9503	3,456	0,275	0,423	0,4020	31,73	30,150
1,20	1,129	1,6258	1,1310	3,770	0,300	0,448	0,5067	33,60	38,003
1,30	1,191	2,0128	1,3273	4,084	0,325	0,473	0,6278	35,48	47,085
1,40	1,251	2,4520	1,5394	4,398	0,350	0,497	0,7651	37,28	57,383
1,50	1,310	2,9475	1,7672	4,712	0,375	0,520	0,9189	39,00	68,918
1,60	1,368	3,5021	2,0106	5,027	0,400	0,543	1,0918	40,73	81,885
1,70	1,424	4,1154	2,2698	5,341	0,425	0,565	1,2824	42,38	96,180
1,80	1,480	4,7952	2,5447	5,655	0,450	0,587	1,4937	44,03	112,03
1,90	1,534	5,5377	2,8353	5,969	0,475	0,609	1,7267	45,68	129,50
2,00	1,587	6,3480	3,1416	6,283	0,500	0,630	1,9792	47,25	148,44
2,20	1,691	8,1844	3,8013	6,912	0,550	0,671	2,5507	50,33	191,30
2,40	1,792	10,3219	4,5239	7,540	0,600	0,711	3,2165	53,33	241,24
2,60	1,891	12,7832	5,3093	8,168	0,650	0,750	3,9820	56,25	298,65
2,80	1,987	15,5781	6,1575	8,796	0,700	0,788	4,8521	59,10	363,91
3,00	2,080	18,7200	7,0686	9,425	0,750	0,825	5,8316	61,88	437,37
3,20	2,172	22,2413	8,0425	10,053	0,800	0,862	6,9326	64,65	519,95
3,40	2,261	26,1372	9,0792	10,681	0,850	0,897	8,1440	67,28	610,80
3,60	2,349	30,4430	10,179	11,310	0,900	0,932	9,4868	69,90	711,51
3,80	2,435	35,1614	11,341	11,938	0,950	0,966	10,955	72,45	821,625

Examples: $D = 3.0$ m, $i = 0.005$. Determine Q and U corresponding to the cross-section full for $K_s = 60 \ m^{1/3}/s$.

From the table for $D = 3.0$ m, we have $U/\sqrt{i} = 61.88$ m/s and $Q/\sqrt{i} = 437.37 \ m^3/s$ so that for $K_s = 75 \ m^{1/3}/s$ we have $U = 4.38$ m/s and $Q = 30.9 \ m^3/s$. For $K_s = 60 \ m^{1/3}/s$, $U = 4.38 \times 60/70 = 3.5$ m/s and $Q = 30.9 \times 60/75 = 24.7 \ m^3/s$.

67. Completely full circular cross-section

Specially adapted to sewer network design

In a combined system, the sewers will have to be dimensioned for the maximum discharge, Q_M from rainstorms, Q_p and wastewater, Q_s.

$$Q_M = Q_p + Q_s$$

Generally speaking, the term Q_s is very small in relation to Q_p, so that practically $Q_M = Q_p$. In a combined or separate system, it will also be necessary to check whether each sewer, with the waste water discharge Q_s alone, has self-cleaning capacity: it is generally accepted that a velocity of 0.3 m/s is required in separate systems and a velocity of 0.6 m/s in combined systems, it is also accepted that a water depth of $h \geqslant 0.02$ m is required. When this does not occur, the slope of the sewers is increased when this is possible, or automatic flushing devices are provided.

Notes on construction of the graph
(1) $K_s = 75$ m$^{1/3}$/s was accepted, this being valid for sewers in a good state of conservation. For large diameter sewers, with very smooth walls and good transitions, the discharges may be increased by 5%. For sewers in a poor state, this value must be reduced by 10% to 20%, or even more.
(2) The *advisable maximum* was established so that, for each bottom slope, the discharge in critical flow is half the discharge corresponding to the full cross-section.
(3) The *exceptional maximum* is limited by the bottom slope ($I \leqslant 0.04$) and by the velocity ($U \leqslant 5$ m/s).
(4) The *advisable minimum* is limited by the velocity ($U \geqslant 0.6$ m/s) and by the bottom slope ($I \geqslant 0.005$).
(5) The *exceptional minimum* is limited by the velocity ($U \geqslant 0.5$ m/s).

Examples: $Q = 1000$ l/s, $I = 0.006$. From the graph: $D = 0.80$ m; $U = 2.0$ m/s. From the attached table we have: $Q_s = 1.6$ l/s (to ensure a minimum velocity, when it does not rain, of 0.3 m/s); or $Q_s = 11.8$ l/s (to ensure a minimum velocity, when it does not rain, of 0.6 m/s).

67. (Contd.) Minimum discharges of waste water[†]

Q_s: (l/s)

(a) *To ensure a mean velocity of 0.30 m/s and a water depth of h > 0.02 m.*

I m/km \ D m	0,20	0,25	0,30	0,35	0,40	0,45	0,50	0,60	0,70	0,80	0,90	1,00
0,5			14,8	14,6	13,4	12,7	14,3	15,4	16,5	17,6	18,8	19,6
1,0	5,7	5,7	5,9	6,1	6,3	6,5	6,8	7,5	7,9	8,9	9,4	10,2
2,0	2,5	2,5	2,7	2,9	3,1	3,3	3,5	3,8	4,1	4,5	4,9	5,2
4,0	1,2	1,2	1,4	1,4	1,6	1,6	1,8	2,0	2,1	2,4	2,6	2,8
6,0	0,7	0,8	0,9	1,0	1,0	1,1	1,2	1,3	1,5	1,6	1,7	1,9
8,0	0,6	0,6	0,7	0,7	0,8	0,9	0,9	1,0	1,1	1,2	1,3	1,5
10,0	0,5	0,6	0,7	0,7	0,7	0,8	0,8	0,9	0,9	1,0	1,1	1,3
12,0	0,6	0,6	0,7	0,8	0,8	0,9	0,9	1,0	1,0	1,1	1,2	1,3
14,0	0,6	0,7	0,8	0,8	0,9	0,9	1,0	1,1	1,1	1,2	1,3	1,4
16,0	0,7	0,7	0,8	0,9	0,9	1,0	1,1	1,1	1,2	1,2	1,4	1,5
20,0	0,7	0,8	0,9	1,0	1,0	1,1	1,2	1,3	1,3	1,4	1,5	1,7
25,0	0,8	0,9	1,0	1,1	1,2	1,3	1,3	1,4	1,5	1,6	1,7	
30,0	0,9	1,0	1,1	1,2	1,3	1,4	1,5	1,5	1,6	1,7	1,9	
35,0	1,0	1,1	1,2	1,3	1,4	1,5	1,6	1,7	1,7			
40,0	1,0	1,2	1,3	1,4	1,5	1,6	1,7	1,8	1,8			

(b) *To ensure a mean velocity of 0.60 m/s and a water depth of h > 0.02 m.*

I m/km \ D m	0,20	0,25	0,30	0,35	0,40	0,45	0,50	0,60	0,70	0,80	0,90	1,00
0,5										186	184	182
1,0						82,4	72,8	72,4	72,1	77,4	81,7	82,5
2,0			29,6	27,4	27,8	28,9	30,4	31,7	32,9	35,3	37,3	39,1
4,0	12,2	11,4	11,8	12,2	12,9	13,0	13,9	14,6	16,4	17,2	18,9	20,0
6,0	6,8	7,1	7,4	7,7	8,2	8,5	9,0	10,0	10,9	11,8	12,6	13,4
8,0	5,0	5,2	5,5	5,7	6,1	6,5	6,9	7,5	8,2	8,8	9,8	10,3
10,0	3,9	4,0	4,3	4,6	4,9	5,2	5,5	6,0	6,5	7,3	7,9	8,6
12,0	3,2	3,3	3,5	3,9	4,1	4,3	4,6	5,0	5,7	6,1	6,7	7,4
14,0	2,7	2,8	3,0	3,3	3,5	3,8	4,0	4,4	4,9	5,4	5,8	6,4
16,0	2,3	2,4	2,7	2,9	3,1	3,3	3,4	3,9	4,3	4,8	5,3	5,6
20,0	1,8	2,0	2,1	2,3	2,4	2,7	2,9	3,2	3,5	3,9	4,2	4,4
25,0	1,4	1,6	1,7	1,9	2,0	2,2	2,3	2,6	3,0	3,1	3,4	
30,0	1,2	1,3	1,4	1,6	1,7	1,8	2,0	2,2	2,4	2,5	2,8	
35,0	1,0	1,2	1,3	1,4	1,5	1,6	1,8	1,9	2,1			
40,0	1,0	1,2	1,3	1,4	1,5	1,6	1,7	1,8	1,8			

Note: The values above the horizontal lines are determined by the velocity conditions; those under the horizontal lines are determined by the water depth conditions.

[†] Values compiled from *Tabelas Técnicas*, published by A.E.I.S.T. and calculated by P. Celestino da Costa (Civil Engineer).

67 (Contd.) Completely full circular cross-section

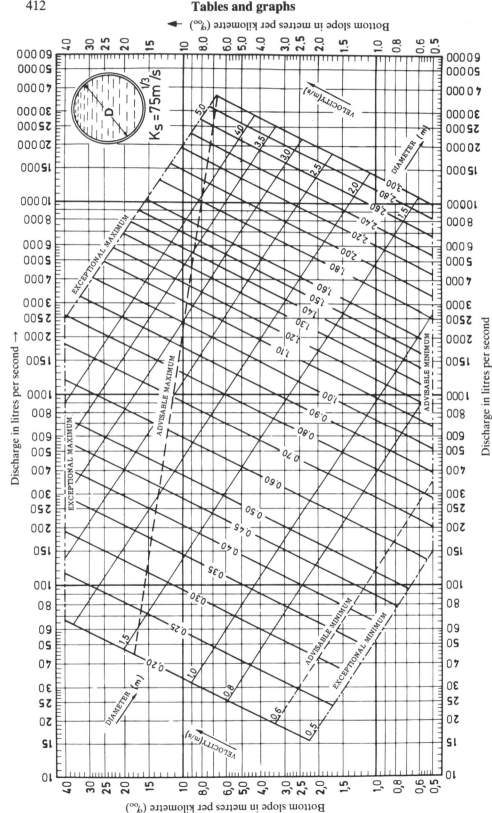

68. Partly full horseshoe cross-section

Geometrical elements

h/d	A/d^2	P/d	R/d	h/d	A/d^2	P/d	R/d
0,01	0,0019	0,2830	0,0066	0,51	0,4466	1,7162	0,2602
0,02	0,0053	0,4006	0,0132	0,52	0,4566	1,7362	0,0630
0,03	0,0097	0,4911	0,0198	0,53	0,4666	1,7562	0,2657
0,04	0,0150	0,5676	0,0264	0,54	0,4766	1,7763	0,2683
0,05	0,0209	0,6351	0,0329	0,55	0,4865	1,7964	0,2707
0,06	0,0275	0,6963	0,0394	0,56	0,4965	1,8165	0,2733
0,07	0,0346	0,7528	0,0459	0,57	0,5064	1,8367	0,2757
0,08	0,0421	0,8054	0,0524	0,58	0,5163	1,8569	0,2781
0,0886	0,0491	0,8482	0,0578	0,59	0,5261	1,8772	0,2804
0,09	0,0502	0,8513	0,0590	0,60	0,5369	1,8976	0,2824
0,10	0,0585	0,8732	0,0670				
0,11	0,670	0,8950	0,0748	0,61	0,5457	1,9180	0,2844
0,12	0,0753	0,9166	0,0823	0,62	0,5555	1,9386	0,2864
0,13	0,0839	0,9382	0,0895	0,63	0,5651	1,9592	0,2884
0,14	0,0925	0,9597	0,0964	0,64	0,5748	1,9800	0,2902
0,15	0,1012	0,9811	0,1031	0,65	0,5843	2,0009	0,2920
0,16	0,1100	1,0024	0,1097	0,66	0,5938	2,0219	0,2937
0,17	0,1188	1,0236	0,1161	0,67	0,6033	2,0431	0,2953
0,18	0,1277	1,0448	0,1222	0,68	0,6126	2,0645	0,2967
0,19	0,1367	1,0658	0,1282	0,69	0,6219	2,0860	0,2981
0,20	0,1457	1,0868	0,1341	0,70	0,6312	2,1077	0,2994
0,21	0,1549	1,1078	0,1398	0,71	0,6403	2,1297	0,3006
0,22	0,1640	1,1286	0,1454	0,72	0,6493	2,1518	0,3018
0,23	0,1733	1,1494	0,1508	0,73	0,6582	2,1742	0,3028
0,24	0,1825	1,1702	0,1560	0,74	0,6671	2,1969	0,3036
0,25	0,1919	1,1909	0,1611	0,75	0,6758	2,2198	0,3044
0,26	0,2013	1,2115	0,1662	0,76	0,6844	2,6844	2,2431
0,27	0,2107	1,2321	0,1710	0,77	0,6929	2,2666	0,3055
0,28	0,2202	1,2526	0,1758	0,78	0,7012	2,2906	0,3060
0,29	0,2297	1,2731	0,1804	0,79	0,7094	2,3149	0,3064
0,30	0,2393	1,2935	0,1850	0,80	0,7175	2,3397	0,3067
0,31	0,2489	1,3139	0,1895	0,85	0,7254	2,3650	0,3067
0,32	0,2586	1,3342	0,1938	0,82	0,7332	2,3907	0,3066
0,33	0,2683	1,3546	0,1981	0,83	0,7408	2,4170	0,3064
0,34	0,2780	1,3748	0,2023	0,84	0,7482	2,4440	0,3061
0,35	0,2878	1,3951	0,2063	0,85	0,7554	2,4716	0,3056
0,36	0,2975	1,4153	0,2103	0,86	0,7625	2,5000	0,3050
0,37	0,3074	1,4355	0,2142	0,87	0,7693	2,5292	0,3042
0,38	0,3172	1,4556	0,2181	0,88	0,7759	2,5595	0,3032
0,39	0,3271	1,4758	0,2217	0,89	0,7823	2,5909	0,3020
0,40	0,3370	1,4959	0,2252	0,90	0,7884	2,6235	0,3005
0,41	0,3469	1,5160	0,2287	0,91	0,7943	2,6573	0,2988
0,42	0,3568	1,5360	0,2322	0,92	0,7999	2,6935	0,2969
0,44	0,3767	1,5761	0,2390	0,94	0,8101	2,7721	0,2922
0,45	0,3867	1,5962	0,2422	0,95	0,8146	2,8160	0,2893
0,46	0,3966	1,6162	0,2454	0,96	0,8188	2,8643	0,2858
0,47	0,4066	1,6362	0,2484	0,97	0,8224	2,9188	0,2816
0,48	0,4166	1,6562	0,2514	0,98	0,8256	2,9832	0,2766
0,49	0,4266	1,6762	0,2544	0,99	0,8280	3,0667	0,2696
0,50	0,4366	1,6962	0,2574	1,00	0,8293	3,2670	0,2538

69. Partly full horseshoe cross-section. Values of U_h and Q_h

U_h and Q_h — Velocity and discharge corresponding to water depth h.
U_d and Q_d — Velocity and discharge corresponding to full cross-section.

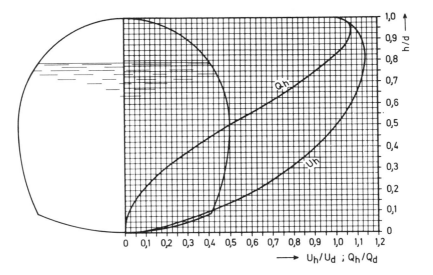

$\dfrac{h}{d}$	$\dfrac{U_h}{U_d}$	$\dfrac{Q_h}{Q_d}$	$\dfrac{h}{d}$	$\dfrac{U_h}{U_d}$	$\dfrac{Q_h}{Q_d}$	$\dfrac{h}{d}$	$\dfrac{U_h}{U_d}$	$\dfrac{Q_h}{Q_d}$
0,02	0,1394	0,0009	0,32	0,8354	0,2605	0,72	1,1222	0,8785
0,04	2213	0039	0,34	8596	2881	0,74	1267	9062
0,06	2889	0096	0,36	8820	3164	0,76	1302	9326
0,08	3492	0177	0,38	9037	3456	0,78	1327	9576
0,10	4116	0292	0,40	9234	3753	0,80	1344	9814
0,12	0,4719	0,0427	0,44	0,9606	0,4364	0,82	1,1342	1,0027
0,14	5243	0583	0,48	9935	4989	0,84	1330	0220
0,16	5717	0758	0,50	1,0092	5311	0,86	1302	0391
0,18	6143	0947	0,54	0377	5964	0,88	1257	0532
0,20	6535	1149	0,58	0629	6617	0,90	1190	0638
0,22	0,6897	0,1362	0,60	1,0736	0,6938	0,92	1,1100	1,0707
0,24	7229	1591	0,64	0933	7576	0,94	0983	0728
0,26	7541	1832	0,66	1020	7889	0,96	0823	0686
0,28	7827	2078	0,68	1095	8195	0,98	0589	0541
0,30	8100	2337	0,70	1162	8496	1,00	0000	0000

70.   Normal oval cross-section ($H = 1.5D$), partly full

U_h and Q_h — Mean velocity and discharge for a water depth h.
U_H and Q_H — Mean velocity and discharge corresponding to full cross-section.

$\dfrac{h}{H}$	$\dfrac{A}{D^2}$	$\dfrac{R}{D}$	$\dfrac{R^{2/3}}{D^{2/3}}$	$\dfrac{AR^{2/3}}{D^{8/3}}$	$\dfrac{U_h}{U_H}$	$\dfrac{Q_h}{Q_H}$
0,0167	0,0037	0,016	0,063	0,0002	0,144	0,0004
0333	0103	032	101	0010	231	0020
0500	0185	046	128	0024	292	0048
0667	0280	060	153	0043	349	0085
0833	0385	076	179	0069	409	0137
0,1000	0,0495	0,083	0,190	0,0094	0,434	0,0187
1333	0753	105	223	0168	509	0334
1667	1048	135	250	0262	571	0521
2000	1348	143	274	0369	626	0734
2333	1688	160	295	0498	674	0990
0,2667	0,2050	0,176	0,314	0,0644	0,717	0,1280
3000	2435	191	332	0808	758	1606
3333	2840	208	351	0997	801	1982
3667	3263	221	366	1194	836	2374
4000	3703	233	379	1403	865	2789
0,4500	0,4385	0,252	0,399	0,1750	0,911	0,3479
5000	5093	269	417	2124	952	4223
6000	6560	299	447	2932	1,021	5829
6500	7308	312	460	3362	050	6684
7000	8058	323	471	3795	075	7545
0,7333	0,8553	0,329	0,477	0,4080	1,089	0,8111
7667	9035	335	482	4355	100	8658
8000	9502	339	486	4618	110	9181
8333	9950	341	488	4856	114	9654
8667	1,0365	341	488	5058	114	1,0056
0,9000	1,0745	0,339	0,486	0,5222	1,110	1,0382
9333	1075	334	481	5327	098	0590
9667	1337	323	471	5340	075	0616
9833	1432	314	462	5282	055	0501
1,0000	1485	290	438	5030	000	0000

Example: Flow is uniform in an oval sewer of dimensions $D \times H = 1.40$ m $\times 2.10$ m with bottom slope $I = 0.003$. Determine the mean velocity of the flow when the discharge Q is 170 l/s ($K_s = 75$ m$^{1/3}$/s).

From Table 72, the velocity corresponding to the full cross-section is $U_H = 41.1 \sqrt{i} = 41.1 \sqrt{0.003} = 2,25$ m/s and $Q_H = 92.52 \sqrt{I} = 5.07$ m^3/s.

Thus $Q_h/Q_H = 0.170/5.07 = 0.0335$ corresponding, according to Table 70, to $h/H = 0.1333$ and $U_h/U_H = 0.509$.

Consequently, the discharge of 170 l/s takes place with a water depth $h = 0.1333\,H = 0.1333 \times 2.10 = 0.28$ m and a velocity $U_h = 0.509\,U_H = 0.509 \times 2.25 = 1.15$ m/s.

71.   Normal oval cross-section ($H = 1.5D$), **partly full**

U_h and Q_h — Mean velocity and discharge for a water depth h.
U_H and Q_H — Mean velocity and discharge corresponding to full cross-section.

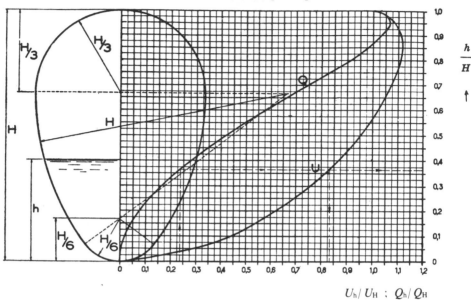

$$U_h/U_H \; : \; Q_h/Q_H$$

Example: Q_h: $Q_H = 0{,}23$. From Graph 71, h: $H = 0{,}37$; U_h: $U_h = 0{,}83$.

72.   Normal oval cross-section ($H = 1.5D$), **completely full**

$D \times H$ (m × m)	$D^{2/3}$ (m²ᐟ³)	$D^{8/3}$ (m⁸ᐟ³)	A (m²)	P (m)	R (m)	$R^{2/3}$ (m²ᐟ³)	$A R^{2/3}$ (m⁸ᐟ³)	$K_s = 75$ m¹ᐟ³/s $\dfrac{U}{\sqrt{i}}$	$\dfrac{Q}{\sqrt{i}}$
0,40 × 0,60	0,543	0,087	0,1838	1,586	0,116	0,238	0,0437	17,850	3,278
0,50 × 0,75	0,630	0,158	0,2871	1,982	0,145	0,276	0,0792	20,700	5,940
0,60 × 0,90	0,711	0,256	0,4135	2,379	0,174	0,312	0,1290	23,400	9,675
0,70 × 1,05	0,788	0,386	0,5628	2,775	0,203	0,345	0,1942	25,875	14,565
0,80 × 1,20	0,862	0,552	0,7350	3,172	0,232	0,378	0,2778	28,350	20,835
0,90 × 1,35	0,932	0,755	0,9303	3,568	0,261	0,408	0,3796	30,600	28,470
1,00 × 1,50	1,000	1,000	1,1485	3,965	0,290	0,438	0,5030	32,850	37,725
1,10 × 1,65	1,065	1,289	1,3897	4,361	0,319	0,467	0,6490	35,025	48,675
1,20 × 1,80	1,129	1,626	1,6538	4,757	0,348	0,495	0,8186	37,125	61,395
1,30 × 1,95	1,191	2,013	1,9410	5,154	0,377	0,521	1,0113	39,075	75,848
1,40 × 2,10	1,251	2,452	2,2511	5,550	0,406	0,548	1,2336	41,100	92,520
1,50 × 2,25	1,310	2,948	2,5841	5,947	0,435	0,574	1,4833	43,050	111,25
1,60 × 2,40	1,368	3,502	2,9402	6,342	0,464	0,599	1,7612	44,925	132,09
1,80 × 2,70	1,480	4,795	3,7211	7,136	0,521	0,648	2,4113	48,600	180,85
2,00 × 3,00	1,587	6,348	4,5940	7,929	0,579	0,695	3,1928	52,125	239,95
2,20 × 3,30	1,691	8,184	5,559	8,722	0,637	0,740	4,1137	55,500	308,53
2,40 × 3,60	1,792	10,322	6,615	9,515	0,695	0,785	5,1928	58,875	389,46
2,60 × 3,90	1,891	12,783	7,764	10,308	0,753	0,828	6,4286	62,100	482,15
2,80 × 4,20	1,987	15,578	9,004	11,101	0,811	0,869	7,8245	65,175	586,84
3,00 × 4,50	2,080	18,720	10,336	11,894	0,869	0,911	9,4161	68,325	706,21

73.   Normal oval cross-section ($H = 1.5\ D$), completely full
Specially adapted to wastewater network design.

Note: The advisable and exceptional maxima and minima with the full cross-section were fixed by comparison with circular sewers. Consult the notes of Graph 67.

Minimum discharges of wastewater, Q (l/s), for ensuring a mean velocity of 0.60 m/s and water depth $h > 0.02$ m.[†]

I $(^o/_{oo})$ \ D (m)	0,40	0,50	0,60	0,70	0,80	0,90	1,00	1,20	1,40	1,60	1,80	2,00
0,5	—	—	—	211,1	210,1	198,5	200,5	200,8	206,1	207,0	217,7	227,2
1,0	—	77,1	78,2	76,7	77,9	76,7	77,4	82,0	88,0	91,6	96,3	101,5
2,0	29,7	31,9	29,2	30,4	31,0	32,6	33,3	36,6	39,2	42,9	46,0	50,2
4,0	11,5	12,2	12,6	13,4	14,1	14,5	15,7	17,8	20,5	22,2	24,7	27,3
6,0	7,1	7,4	8,0	8,6	9,3	10,0	11,0	12,5	14,1	15,9	17,3	18,6
8,0	5,0	5,4	5,9	6,5	7,1	7,6	8,4	9,6	11,0	12,1	13,0	14,2
10,0	3,9	4,3	4,8	5,3	5,8	6,3	6,9	7,8	8,9	9,7	10,7	12,0
12,0	3,2	3,6	4,0	4,4	4,8	5,4	5,9	6,6				
14,0	2,7	3,1	3,5	4,0	4,3	4,8	5,0	5,8				
16,0	2,4	2,7	3,1	3,6	3,8	4,2	4,5	5,1				
20,0	1,9	2,3	2,6	2,9	3,1	3,4	3,7	4,2				
25,0	1,6	1,8	2,1	2,3	2,5	2,7						
30,0	1,3	1,6	1,8	2,0	2,1	2,3						
35,0	1,2	1,3	1,6	1,7								
40,0	1,1	1,2	1,4	1,4								

[†] Values compiled from *Tabelas Técnicas*, published by A.E.I.S.T. and calculated by P. Celestino da Costa (Civil Engineer).

73 (Contd.) Normal oval cross-section ($H = 1.5\ D$), completely full

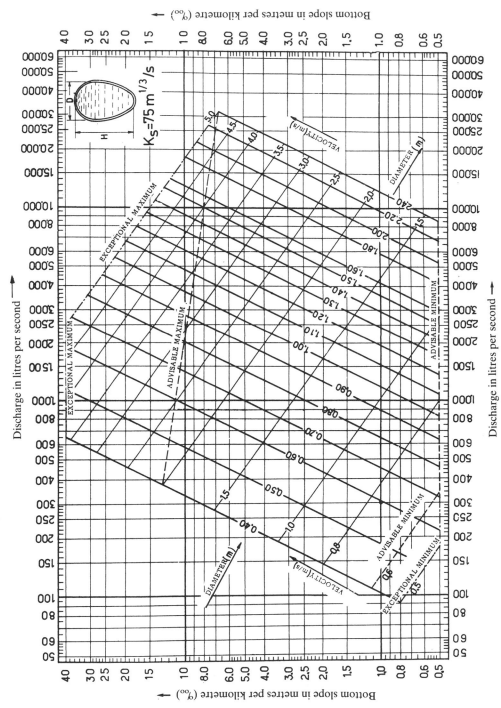

74. Arched gutter cross-section partly full

U_h and Q_h — Mean velocity and discharge for a water depth h.
U_H and Q_H — Mean velocity and discharge corresponding to full cross-section.

$\dfrac{h}{H}$	$\dfrac{A}{D^2}$	$\dfrac{R}{D}$	$\dfrac{R^{2/3}}{D^{2/3}}$	$\dfrac{AR^{2/3}}{D^{8/3}}$	$\dfrac{U_h}{U_H}$	$\dfrac{Q_h}{Q_H}$
0,01	0,0002	0,004	0,0253	0,00001	0,065	0,00003
02	0008	009	043	00003	111	0001
03	0018	013	055	0001	142	0003
04	0032	018	069	0002	178	0007
05	0050	022	079	0004	204	0013
0,06	0,0072	0,027	0,090	0,0006	0,233	0,0020
07	0098	031	099	0010	256	0034
08	0128	036	109	0014	282	0047
09	0162	040	117	0019	302	0064
10	0200	045	127	0025	328	0084
0,15	0,0450	0,067	0,165	0,0074	0,426	0,0249
20	0800	090	201	0161	519	0542
25	1250	112	232	0290	599	0976
30	1750	144	275	0481	711	1619
35	2250	171	308	0693	796	2333
0,40	0,2750	0,194	0,335	0,0921	0,866	0,3100
45	3250	214	358	1164	925	3918
50	3750	232	378	1418	977	4773
55	4250	248	395	1679	021	5651
60	4745	261	408	1936	054	6516
0,65	0,5228	0,272	0,420	0,2196	1,085	0,7391
70	5695	280	428	2437	106	8203
75	6143	287	435	2672	124	8994
80	6560	290	438	2873	132	9670
825	6753	290	438	2958	132	9956
0,850	0,6939	0,290	0,438	0,3039	1,132	1,0229
900	7269	286	434	3155	121	0619
950	7531	276	424	3193	096	0747
1,000	7678	241	387	2971	000	0000

Example: In a cross-section of arched gutter with dimensions $D = H = 3.0$ m, it is wished to know the discharge and velocity for a water depth corresponding to the springing of the arch (K_s: 75 m$^{1/3}$/s; $I = 0.007$).

For the full cross-section, according to Table 76 we have, $U_H = 60.375 \sqrt{I} = 60.375 \sqrt{0.007} = 5.05$ m/s and $Q_H = 417.18 \sqrt{I} = 34.9$ m^3/s.

The springing of the arch corresponds to a ratio $h/H = 0.5$, so that from Table 74 $U_h = 0.977 \, U_H = 0.977 \times 5.05 = 5,4$ m/s and $Q_h = 0.4773 \, Q_H = 0.4773 \times 34.9 = 16.7$ m^3/s.

75. Arched gutter cross-section, partly full

U_h and Q_h — Mean velocity and discharge for a water depth h.
U_H and Q_H — Mean velocity and discharge corresponding to full cross-section.

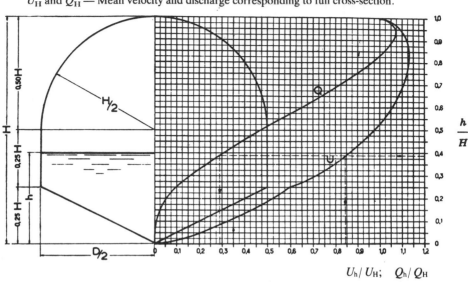

$U_h / U_H; \quad Q_h / Q_H$

Example: $h/H = 0.38$. From Graph 75: $U_h/U_h = 0.845$; $Q_h/Q_H = 0.290$.

76. Arched gutter cross-section, completely full

$D \times H$ (m × m)	$D^{2/3}$ (m$^{2/3}$)	$D^{8/3}$ (m$^{8/3}$)	A (m^2)	P (m)	R (m)	$R^{2/3}$ (m$^{2/3}$)	$AR^{2/3}$ (m$^{8/3}$)	$K_s = 75$ m$^{1/3}$/s $\dfrac{U}{\sqrt{i}}$	$\dfrac{Q}{\sqrt{i}}$
0,40 × 0,40	0,543	0,087	0,1228	1,276	0,096	0,210	0,0258	15,750	1,9350
0,50 × 0,50	0,630	0,158	0,1919	1,595	0,120	0,243	0,0466	18,225	3,4950
0,60 × 0,60	0,711	0,256	0,2764	1,913	0,144	0,275	0,0760	20,625	5,7000
0,70 × 0,70	0,788	0,386	0,3762	2,232	0,169	0,306	0,1151	22,950	8,6325
0,80 × 0,80	0,862	0,552	0,4914	2,551	0,193	0,334	0,1641	25,050	12,3075
0,90 × 0,90	0,932	0,755	0,6219	2,870	0,217	0,361	0,2245	27,075	16,8375
1,00 × 1,00	1,000	1,000	0,7678	3,189	0,241	0,387	0,2971	29,025	22,2825
1,10 × 1,10	1,065	1,29	0,9290	3,508	0,265	0,413	0,3837	30,975	28,7775
1,20 × 1,20	1,129	1,63	1,1056	3,827	0,289	0,437	0,4831	32,775	36,2325
1,30 × 1,30	1,191	2,01	1,2975	4,146	0,313	0,461	0,5981	34,575	44,8575
1,40 × 1,40	1,251	2,45	1,5048	4,465	0,337	0,484	0,7283	36,300	54,6225
1,50 × 1,50	1,310	2,95	1,7274	4,784	0,361	0,507	0,8758	38,025	65,6850
1,60 × 1,60	1,368	3,50	1,9655	5,102	0,385	0,529	1,0397	39,675	77,9775
1,80 × 1,80	1,480	4,79	2,4875	5,740	0,433	0,572	1,4229	42,900	106,7175
2,00 × 2,00	1,587	6,35	3,0710	6,378	0,481	0,614	1,8856	46,050	141,4200
2,20 × 2,20	1,691	8,19	3,7159	7,016	0,530	0,655	2,4339	49,125	182,5425
2,40 × 2,40	1,792	10,32	4,4223	7,654	0,578	0,694	3,0691	52,050	230,1825
2,60 × 2,60	1,891	12,8	5,1900	8,291	0,626	0,732	3,7991	54,900	284,9325
2,80 × 2,80	1,987	15,6	6,0192	8,929	0,674	0,769	4,6288	57,675	347,1600
3,00 × 3,00	2,080	18,7	6,9098	9,567	0,722	0,805	5,5624	60,375	417,1800

77. Arched gutter cross-section, completely full
Specially adapted to sewer network design

Note: The advisable and exceptional maxima and minima with the cross-section full were determined by comparison with circular sewers. Consult the notes of Table 67.

Minimum wastewater discharge

$$Q_s \text{ (l/s)}$$

In order to ensure a mean velocity of 0.6 m/s and water depth $h \geqslant 0.02$ m. Values valid for $h < \dfrac{D}{4}$

i $°/_{oo}$	Q $1/s$	i $°/_{oo}$	Q $1/s$	i $°/_{oo}$	Q $1/s$
0,5	—	8,0	4,3	20,0	1,1
1,0	97,2	10,0	3,1	25,0	0,8
2,0	34,4	12,0	2,3	30,0	0,6
4,0	12,1	14,0	1,9	35,0	0,6
6,0	6,6	16,0	1,5	40,0	0,6

77 (Contd.)　Arched gutter cross-section, completely full

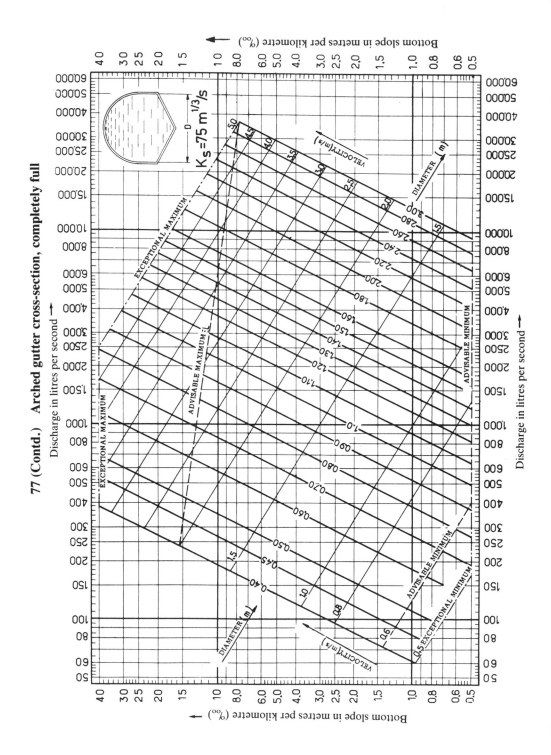

$K_s = 75 \ m^{1/3}/s$

78. Oval cross-section ($H = 1.5\,D$), with narrow sill

A = Cross-section P = Wet perimeter $R = \dfrac{A}{P}$ = Hydraulic radius

U and Q — Mean velocity and discharge of cross-section, full.
U_1 and Q_1 — Mean velocity and discharge of a full circular cross-section, of equal diameter D, for a given slope.

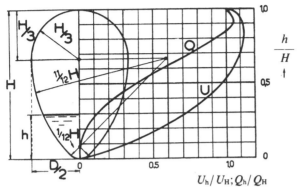

$D = 0,667\,H$
$H = 1,5\,D$

$A = 1,115\,D^2 = 0,496\,H^2$
$P = 3,920\,D = 2,613\,H$
$R = 0,284\,D = 0,189\,H$

$U = 1,091\,U_1$ $\;\|\;$ $U_1 = 0,917\,U$
$Q = 1,548\,Q_1$ $\;\|\;$ $Q_1 = 0,646\,Q$

Cross-section partly full

Cross-section completely full

$\dfrac{h}{H}$	$\dfrac{A}{D^2}$	$\dfrac{R}{D}$	$\dfrac{AR^{2/3}}{D^{8/3}}$	H (m)	A (m²)	R (m)	$AR^{2/3}$ (m^{8/3})	$K_s = 75$ m^{1/3}/s $\dfrac{U}{\sqrt{i}}$	$\dfrac{Q}{\sqrt{i}}$
0,05	0,0128	0,042	0,0015	0,50	0,124	0,095	0,0258	15,60	1,935
0,10	0,0373	0,074	0,0066	0,60	0,178	0,114	0,0418	17,63	3,135
0,15	0,0705	0,102	0,0154	0,70	0,243	0,133	0,0634	19,58	4,755
0,20	0,1120	0,128	0,0284	0,80	0,317	0,152	0,0903	21,38	6,773
0,25	0,1615	0,154	0,0464	0,90	0,401	0,171	0,1235	23,10	9,263
0,30	0,2153	0,177	0,0678	1,00	0,495	0,190	0,1638	24,83	12,285
0,40	0,3388	0,221	0,1240	1,20	0,713	0,228	0,2659	27,98	19,943
0,50	0,4763	0,258	0,1929	1,40	0,970	0,266	0,4016	31,05	30,120
0,60	0,6225	0,290	0,2727	1,60	1,267	0,304	0,5727	33,90	42,953
0,70	0,7723	0,316	0,3583	1,80	1,604	0,342	0,7844	36,68	58,830
0,75	0,8473	0,325	0,4008	2,00	1,980	0,380	1,0395	39,38	77,963
0,80	0,9168	0,332	0,4391	2,20	2,396	0,418	1,3394	41,93	100,46
0,85	0,9828	0,335	0,4737	2,40	2,851	0,456	1,6878	44,45	126,59
0,90	1,0410	0,332	0,4986	2,60	3,346	0,494	2,0913	46,48	156,85
0,95	1,0883	0,324	0,5137	2,80	3,881	0,532	2,5498	49,28	191,24
1,00	1,1150	0,285	0,4828	3,00	4,455	0,570	3,0606	51,53	229,55

79.   Circular cross-section with channel

A = Cross-section P = Wet perimeter $R = \dfrac{A}{P}$ = Hydraulic radius

U and Q — Mean velocity and discharge of cross-section, full.
U_1 and Q_1 — Mean velocity and discharge of a full circular cross-section, of equal diameter D, for a given slope.

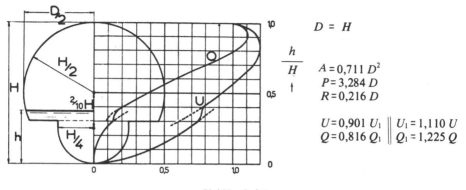

$D = H$

$$A = 0{,}711\,D^2$$
$$P = 3{,}284\,D$$
$$R = 0{,}216\,D$$

$$U = 0{,}901\,U_1 \;\|\; U_1 = 1{,}110\,U$$
$$Q = 0{,}816\,Q_1 \;\|\; Q_1 = 1{,}225\,Q$$

$U_h / U_H ; Q_h / Q_H$

Cross-section partly full				*Cross-section completely full*					
$\dfrac{h}{H}$	$\dfrac{A}{D^2}$	$\dfrac{R}{D}$	$\dfrac{AR^{2/3}}{D^{8/3}}$	H (m)	A (m²)	R (m)	$AR^{2/3}$ (m⁸ᐟ³)	$K_s = 75\ \text{m}^{1/3}/\text{s}$	
								$\dfrac{U}{\sqrt{i}}$	$\dfrac{Q}{\sqrt{i}}$
0,10	0,0280	0,061	0,0043	2,00	2,844	0,432	1,6239	42,83	121,79
0,20	0,0733	0,107	0,0165	2,20	3,441	0,475	2,0957	45,68	157,18
0,30	0,1233	0,139	0,0330	2,40	4,095	0,518	2,6413	48,38	198,10
0,325	0,1463	0,108	0,0332	2,60	4,806	0,562	3,2729	51,08	245,47
0,40	0,2185	0,145	0,0603	2,80	5,574	0,605	3,9854	53,63	298,91
0,50	0,3178	0,186	0,1036	3,00	6,399	0,648	4,7928	56,18	359,46
0,60	0,4170	0,218	0,1510	3,20	7,281	0,691	5,6937	58,65	427,03
0,70	0,5123	0,241	0,1983	3,40	8,219	0,734	6,6903	61,05	501,77
0,80	0,5985	0,254	0,2400	3,60	9,215	0,778	7,7959	63,45	584,69
0,90	0,6695	0,254	0,2685	3,80	10,267	0,821	9,0042	65,78	675,32
1,00	0,7105	0,217	0,2565	4,00	11,376	0,864	10,3180	67,03	773,85

80. Sundry closed cross-sections

A = Cross-section $\qquad P$ = Wet perimeter $\qquad R = \dfrac{A}{P}$ = Hydraulic radius

U and Q — Mean velocity and discharge of cross-section, full.
U_1 and Q_1 — Mean velocity and discharge of a full circular cross-section, of equal diameter D, for a given slope.

1. Inverted normal oval

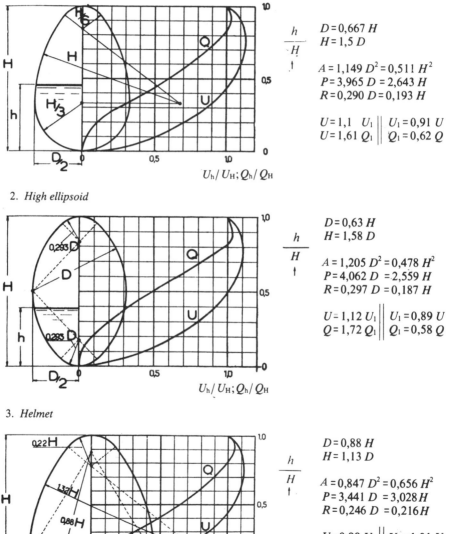

$\dfrac{h}{H}$↑

$D = 0{,}667\,H$
$H = 1{,}5\,D$

$A = 1{,}149\,D^2 = 0{,}511\,H^2$
$P = 3{,}965\,D = 2{,}643\,H$
$R = 0{,}290\,D = 0{,}193\,H$

$U = 1{,}1\;U_1 \;\|\; U_1 = 0{,}91\;U$
$U = 1{,}61\;Q_1 \;\|\; Q_1 = 0{,}62\;Q$

$U_h/U_H; Q_h/Q_H$

2. *High ellipsoid*

$\dfrac{h}{H}$↑

$D = 0{,}63\,H$
$H = 1{,}58\,D$

$A = 1{,}205\,D^2 = 0{,}478\,H^2$
$P = 4{,}062\,D = 2{,}559\,H$
$R = 0{,}297\,D = 0{,}187\,H$

$U = 1{,}12\;U_1 \;\|\; U_1 = 0{,}89\;U$
$Q = 1{,}72\;Q_1 \;\|\; Q_1 = 0{,}58\;Q$

$U_h/U_H; Q_h/Q_H$

3. *Helmet*

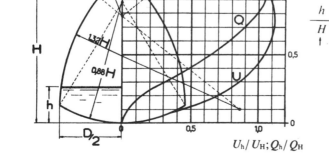

$\dfrac{h}{H}$↑

$D = 0{,}88\,H$
$H = 1{,}13\,D$

$A = 0{,}847\,D^2 = 0{,}656\,H^2$
$P = 3{,}441\,D = 3{,}028\,H$
$R = 0{,}246\,D = 0{,}216\,H$

$U = 0{,}99\;U_1 \;\|\; U_1 = 1{,}01\;U$
$Q = 1{,}06\;Q_1 \;\|\; Q_1 = 0{,}94\;Q$

$U_h/U_H; Q_h/Q_H$

80 (Contd.) Sundry closed cross-sections

A = Cross-section　　P = Wet perimeter　　$R = \dfrac{A}{P}$ = Hydraulic radius

U and Q — Mean velocity and discharge of cross-section, full.
U_1 and Q_1 — Mean velocity and discharge of a full circular cross-section, of equal diameter D, for a given slope.

4. High-circle arch

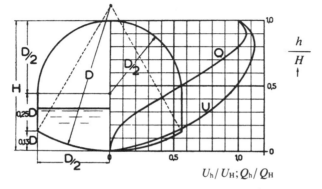

$D = 1{,}13\,H$
$H = 0{,}88\,D$

$A = 0{,}734\,D^2 = 0{,}937\,H^2$
$P = 3{,}118\,D = 3{,}523\,H$
$R = 0{,}236\,D = 0{,}267\,H$

$U = 0{,}96\,U_1$ $\;\big\|\;$ $U_1 = 1{,}04\,U$
$Q = 0{,}90\,Q_1$ $\;\big\|\;$ $Q_1 = 1{,}11\,Q$

$U_h/U_H ; Q_h/Q_H$

5. *Low-circle arch*

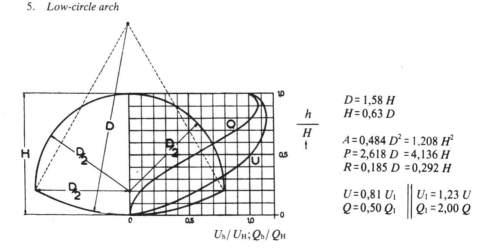

$D = 1{,}58\,H$
$H = 0{,}63\,D$

$A = 0{,}484\,D^2 = 1{,}208\,H^2$
$P = 2{,}618\,D = 4{,}136\,H$
$R = 0{,}185\,D = 0{,}292\,H$

$U = 0{,}81\,U_1$ $\;\big\|\;$ $U_1 = 1{,}23\,U$
$Q = 0{,}50\,Q_1$ $\;\big\|\;$ $Q_1 = 2{,}00\,Q$

$U_h/U_H ; Q_h/Q_H$

Examples:
(1) Determine the mean velocity and the discharge in a helmet cross-section of 1.5 m diameter, with bottom slope of 0.001.
From Graph 67, for a circular conduit of equal diameter and slope, we have:
$Q_1 = 2.2$ m³/s; $U_1 \approx 1.33$ m/s. Thus $U = 0.99$. $U_1 = 1.3$ m/s; $Q = 1.06$. $Q_1 = 2.3$ m³/s.
(2) Determine the dimensions of a low-circle arch sewer, with a bottom slope of 0.001, capable of handling a discharge of 5 m³/s; determine the mean velocity of the flow.
A circular sewer of equal diameter must handle a flow $Q = 2Q = 10$ m³/s. From Graph 67 we have $D_1 = 2.65$ m and $U_1 = 1.8$ m/s. Thus $U = 0.81\,U_1 = 1.5$ m/s. This diameter satisfies the requirements.

81. Pentagonal cross-section

(a) *Cross-section completely full*   (b) *Cross-section partly full*

Q_1 and U_1 are the discharge and velocity in circular cross-sections with equal slope and diameter D.

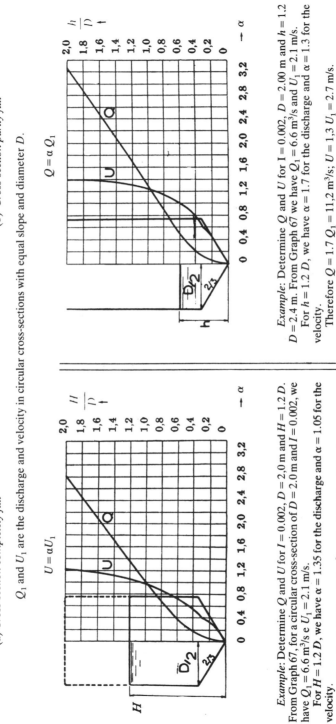

$U = \alpha U_1$

$Q = \alpha Q_1$

Example: Determine Q and U for $I = 0.002$, $D = 2.0$ m and $H = 1.2\,D$. From Graph 67, for a circular cross-section of $D = 2.0$ m and $I = 0.002$, we have $Q_1 = 6.6$ m³/s e $U_1 = 2.1$ m/s.
For $H = 1.2\,D$, we have $\alpha = 1.35$ for the discharge and $\alpha = 1.05$ for the velocity.
Therefore $Q = 1.35\ Q_1 = 8.9$ m³/s; $U = 1.05\ U_1 = 2.2$ m/s.

Example: Determine Q and U for $I = 0.002$, $D = 2.00$ m and $h = 1.2$ $D = 2.4$ m. From Graph 67 we have $Q_1 = 6.6$ m³/s and $U_1 = 2.1$ m/s.
For $h = 1.2\,D$, we have $\alpha = 1.7$ for the discharge and $\alpha = 1.3$ for the velocity.
Therefore $Q = 1.7\ Q_1 = 11.2$ m³/s; $U = 1.3\ U_1 = 2.7$ m/s.

82. Rectangular cross-section

(a) *Cross-section completely full*   (b) *Cross-section partly full*

Q_1 and U_1 are the discharge and velocity in circular cross-sections with equal bottom slope and diameter D.

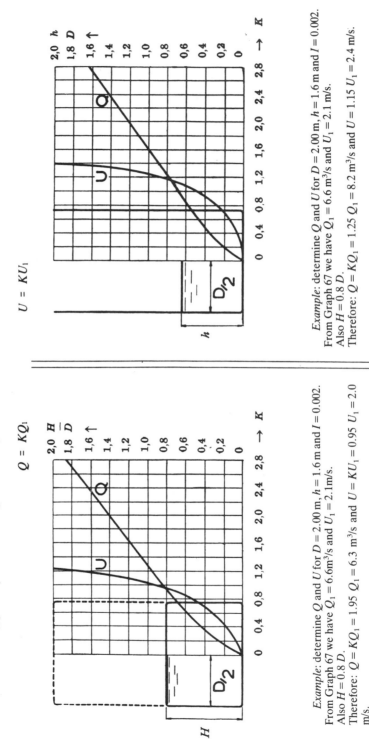

$Q = KQ_1$

$U = KU_1$

Example: determine Q and U for $D = 2.00$ m, $h = 1.6$ m and $I = 0.002$.
From Graph 67 we have $Q_1 = 6.6$ m³/s and $U_1 = 2.1$ m/s.
Also $H = 0.8\,D$.
Therefore: $Q = KQ_1 = 1.25\ Q_1 = 8.2$ m³/s and $U = 1.15\ U_1 = 2.4$ m/s.

Example: determine Q and U for $D = 2.00$ m, $h = 1.6$ m and $I = 0.002$.
From Graph 67 we have $Q_1 = 6.6$ m³/s and $U_1 = 2.1$ m/s.
Also $H = 0.8\,D$.
Therefore: $Q = KQ_1 = 1.95\ Q_1 = 6.3$ m³/s and $U = KU_1 = 0.95\ U_1 = 2.0$ m/s.

83. Trapezoidal channels with 1/1 side slopes
Values of $AR^{2/3}$

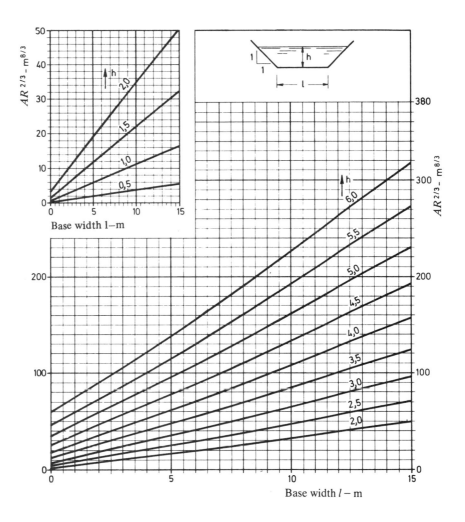

84. Trapezoidal channels with 1/1.5 side slopes

Values of $AR^{2/3}$

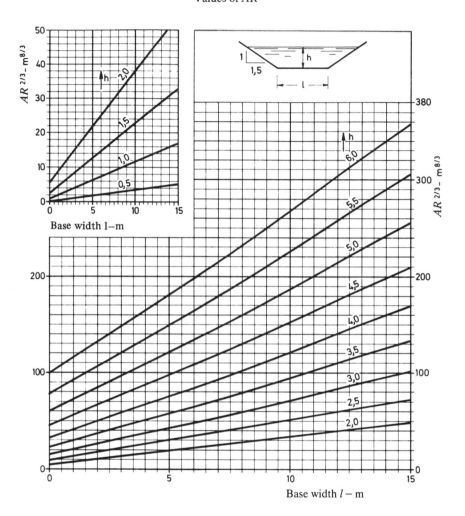

85. Trapezoidal channels with 1/2 side slopes

Values of $AR^{2/3}$

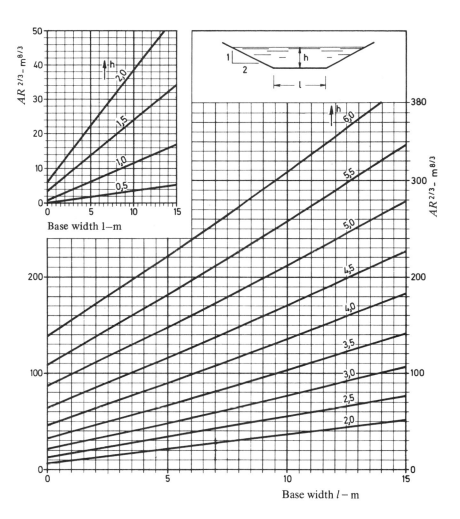

86. Trapezoidal channels with 1/2.5 side slopes

Values of $AR^{2/3}$

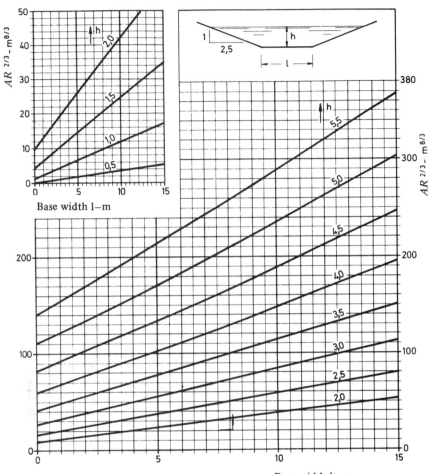

Base width $l-$ m

Base width $l-$ m

87. Trapezoidal channels with 1/3 side slopes
Values of $AR^{2/3}$

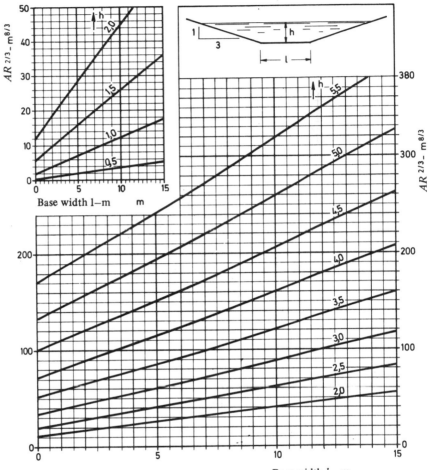

Base width $l-$ m

88. Critical velocities: Practical Data[†]

(a) *Water depths h = 1 m. Straight channels*

1. Non-cohesive materials

Material	Diameter (mm)	Mean velocity (m/s)	Material	Diameter (mm)	Mean velocity (m/s)
Mud	0,005	0,15	Fine gravel	15,0	1,20
Mud	0,05	0,20	Medium gravel	25,0	1,40
Fine sand	0,25	0,30	Coarse gravel	40,0	1,80
Medium sand	1,00	0,55	Coarse gravel	75,0	2,40
Coarse sand	2,50	0,65	Coarse gravel	100,0	2,70
Fine pebbles	5,00	0,80	Coarse gravel	150,0	3,50
Medium pebbles	10,00	1,00	Coarse gravel	200,0	3,90
Coarse pebbles	15,00	1,20			

2. Cohesive materials: U in m/s

Cohesive material of bed \ Type of bed	Hardly compacted with void ratio 2.0 to 1.2	Hardly compacted with void ratio 1.2 to 0.6	Compacted with void ratio 0.6 to 0.3	Highly compacted with void ratio 0.3 to 0.2
Sandy clays (less than 50% sand)	0,45	0,90	1,30	1,80
Soils with clay content	0,40	0,85	1,25	1,70
Clays	0,35	0,80	1,20	1,65
Very fine clays	0,32	0,70	1,05	1,35

(b) *Corrective factor for water depths h ≠ 1m*

Mean depth (m)	0,3	0,5	0,75	1,0	1,5	2,0	2,5	3,0
Corrective factor	0,8	0,9	0,95	1,0	1,1	≈1,1	1,2	≈1,2

(c) *Corrective factor for channels with bends*

Degree of sinuosity	Rectilinear	Not very sinuous	Moderately sinuous	Very sinuous
Corrective factor	1,00	0,95	0,87	0,78

[†] Quoted by [18].

89. Critical velocities

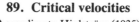

According to Hjulström (1935)

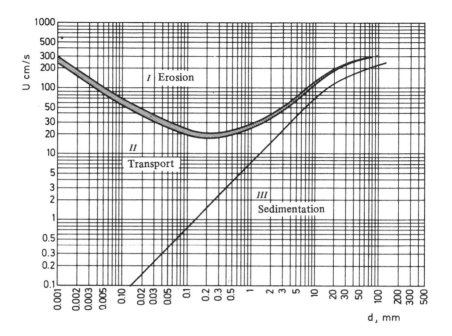

d, mm

Zone I — *Erosion*: The most easily eroded particles will have diameters of between 1.0 and 0.1 mm. For lower values, a certain cohesion is established that impedes erosion; for higher values, the weight of the particle ensures its stability.

Zone II — *Transport*: In this zone the particles put into suspension are transported.

Zone III — *Sedimentation*: For velocities below the line separating Zones II and III the particles in suspension begin to deposit.

Example: A particle of 0.2 mm diameter will start being eroded for $U = 18$ cm/s and will be deposited for velocity $U = 1.5$ cm/s.

Note. See also Neill's formula (equation (5.11)).

90. Distribution of shear stress
According to [18]

τ — Shear stress (N/m²)
γ — Specific weight of the water (N/m³)
h —Water depth (m)
I — Bottom slope of channel (mondim.)

(1) *Channel of indefinite width*
In a channel of indefinite width the average bottom shear stress is $\tau = \gamma h I$

(2) *Trapezoidal channel*
In a straight trapezoidal channel
the distribution of the shear stress
has the aspect shown in the figure.
The maximum bottom shear stress is:

$$\tau_M = K_M \gamma h I$$

The maximum shear stress at the
side slopes is: $\tau'_M = K'_M \gamma h I$
At the bottom, τ_M occurs
in the middle, at the side slopes,
τ'_M occurs at $d = K_d h$ distance
from the bottom.

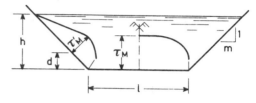

τ_M — Maximum bottom shear stress

τ'_M — Maximum shear stress at side slopes

m $\rightarrow$	2			1,5			0 (rectangular)		
$\dfrac{l}{h}$ ↓	K_M	K'_M	K_d	K_M	K'_M	K_d	K_M	K'_M	K_d
0*	0	0,650	0,3	0	0,565	0,3	0	0	—
1	0,780	0,730	—	0,780	0,695	—	0,372	0,468	1,0
2	0,890	0,760	0,2	0,890	0,735	0,2	0,686	0,686	1,0
3	0,940	0,760	—	0,940	0,743	—	0,870	0,740	1,0
4	0,970	0,770	0,2	0,970	0,750	0,2	0,936	0,744	1,0
6	0,980	0,770	—	0,980	0,755	—	—	—	—
8	0,99	0,770	0,2	0,990	0,760	0,2	—	—	—

* Triangular cross-section

(3) *Triangular channel* (see figure, assuming $l = 0$)

m $\rightarrow$	2	1,5	1	0,667	0,5
K'_M	0,650	0,565	0,480	0,375	0,325
K_d	0,3	0,3	0,5	0,7	0,7

91. Critical shear stresses according to Shields

(a) *Shields curve* $\tau_* = f(\mathbf{R}_{c.})$ Shields (1936).

$$\tau_* = \frac{\tau_0}{(\gamma_s - \gamma)d} \qquad \mathbf{R}_{c.} = \frac{u_* d}{\nu} \qquad u_* = \sqrt{\tau_0/\rho}$$

(b) *Critical shear stress* Shields (1936). Lane (1955)†.

† Quoted by [18].

92. Critical shear stresses for clean water
According to [23]

The values shown here refer to *straight* channels. For slightly sinuous channels with *few bends* (sightly undulating terrain), take 0.90 of the values given; for *moderately sinuous* (reasonably undulating terrain), take 0.75; for *very sinuous channels* with a lot of bends (highly undulating terrain), take 0.60.

1. Coarse non-cohesive materials

At the bottom, take as value of the permissible shear stress, $\tau_{0(crit)}$ (N/m^2) $= 8 \times d_{75}$ (cm) — d_{75} is the diameter for which, 75% of the material by weight, is finer.

At the banks, take $\tau'_{0(crit)} = K\tau_{0(crit)}$ (K is a function of the angle of repose, ψ, of the material and of the angle of the side slopes with horizontal ϕ — see Graph 94.

2. Fine non-cohesive materials: $\tau_{0(crit)}$ in N/m^2

Grain diameter, d_{50}, in mm	0,1	0,2	0,5	1,0	2,0	5,0
Clean water	1,2	1,3	1,5	2,0	2,9	6,0
Water with small amounts of fine sediment	2,4	2,5	2,7	2,9	3,9	8,1
Water with large amounts of fine sediment	3,8	3,8	4,1	4,4	5,4	9,0

3. Cohesive materials: $\tau_{0(crit)}$ in N/m^2

Cohesive material of bed \ Type of bed	Hardly compacted with void ratio 2.0 to 1.2	Hardly compacted with void ratio 1.2 to 0.6	Compacted with void ratio 0.6 to 0.3	Highly compacted with void ratio 0.3 to 0.2
Sandy clays (less than 50% sand)	2,0	7,7	16,0	30,8
Soils with high clay content	1,5	6,9	14,9	27,5
Clays	1,2	6,1	13,7	25,9
Very fine clays	1,0	4,7	10,4	17,3

93. Stability of the banks

(a) *Side slopes*

(Horizontal to vertical)

Type of banks	Slope	Type of banks	Slope
Hard rock, common masonry, concrete	0 to 1/4	Compact alluvia Coarse gravel	1/1 3/2
Cracked rock, dry stone masonry	1/2	Common earth, coarse sand	2/1
Hard clay	3/4	Disturbed earth, normal sand	2,5/1 to 3/1

(b) *Angle of repose of non-choesive material*, ψ. (According to the U.S. Bureau of Reclamation).

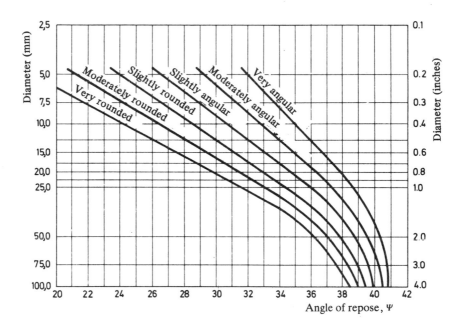

94. Critical shear stress on banks

$$K = \cos \phi \ \sqrt{1 - \frac{\tan^2 \phi}{\tan^2 \Psi}}$$

Ψ — angle of repose

ϕ — angle of side slopes with the horizontal

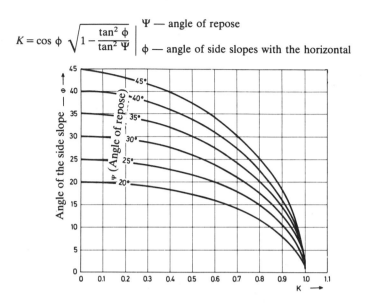

95. Channels lined with grass. Design criteria[†]

(a) *Characteristics of the vegetation cover*

Height of vegetation (cm)		< 5	5–15	15–25	30–60	> 75
Density of coverage	Good	E	D	C	B	A
	Fair	E	D	D	C	B

(b) *Values of UR* (m²/s) *as a function of* K_s (m$^{1/3}$/s)

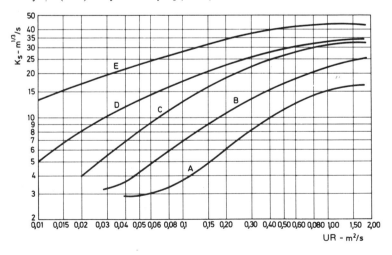

[†] U.S. Soil Conservation Service (1974) *Handbook of Channel Design for Soil and Water Conservation.* Quoted by [35].

95. (Contd.) Channels lined with grass. Design criteria

(c) *Permissible velocities in m/s*

Vegetal coverage		*Slope of channel*	*Permissible velocity m/s*	
G—Gramineous	*L—Leguminous*	*%*	*Soil*	
			Resistant	*Erodible*
1—*Cynodon Dactylon.* Bermuda Grass—G		0—5 5—10 >10	2,5 2,1 1,8	1,8 1,5 1,2
2—*Brachilaria mutica.* Buffalo Grass—G *Poa Pratensis.* Bluegrass—G *Bromus*—G		0—5 5—10 >10	2,1 1,8 1,5	1,5 1,2 1,8
3—*Grass mixture*		0—5 5—10 >10	1,5 1,2 Not to	1,2 1,8 be used
4—*Lespedeza sericea* (Perennial bush clover)—L *Ischaemum* (Mesembryanthemum) *Pueraria lobata* (Kudzu vine)—L *Medreago sativa* (Alfalfa)—L *Digitaria sanguinalis* (Crabgrass)—G		0—5 >5	1,0 Not to	0,75 be used
5—*Annuals* (used on mild slopes until the permanent cover is established) *Lespedeza vulg.* (Common lespedeza) *Sorghum sudomense* (Sudan grass)		0—5 >5	1,0 Not reco	0,75 mmended

The values shown refer to the quality of the cover with some uniformity. Velocities of more than 1.5 m/s must only be used when it is possible to achieve good cover and proper maintenance.

96. Open-channel flows. Steady flow — Index

97. Critical flow basic. Formulae

Shape of cross-section	Critical depth	Critical velocity U_c	Critical discharge Q_c	Critical energy H_c	Critical slope I_c
Any shape (general formulae)	Complies with $$\frac{Q}{\sqrt{g}} = A\sqrt{\frac{A}{L}} = A\sqrt{h_m}$$ thus $$0,319\,Q = A\sqrt{h_m}$$	$$U_c = \sqrt{g\,h_{mc}} = 3,132\sqrt{h_{mc}}$$	Complies with $$\frac{Q}{\sqrt{g}} = A\sqrt{h_m}$$ thus $$0,319\,Q = A\sqrt{h_m}$$	$$H_c = h_{mc} + \frac{h_{mc}}{2}$$	$$I_c = \frac{g\,h_{mc}}{C_c^2\,R_c}$$ $$I_c = \frac{g\,h_{mc}}{K_s^2\,R_c^{4/3}}$$
Rectangular	$$h_c = \sqrt[3]{\frac{Q_c^2}{g\,L_c^2}} = \sqrt[3]{\frac{q_c^2}{g}} = 0,467\,q_c^{2/3}$$	$$U_c = \sqrt{g\,h_c} = 3,132\sqrt{h_c}$$	$$Q_c = L\sqrt{g}\,h_c^{3/2} = 3,132\,L\,h_c^{3/2}$$	$$H_c = \frac{3}{2}\,h_c = 1,5\,h_c$$	For indefinite width $$I_c = \frac{g}{C_c^2} = \frac{g}{K_s^2\,h_c^{1/3}}$$
Isosceles triangle	$$h_c = \sqrt[5]{\frac{2}{m^2 g}\,Q_c^2} = 0,727\sqrt[5]{\frac{Q_c^2}{m}}$$	$$U_c = \sqrt{g\,\frac{h_c}{2}} = 2,215\sqrt{h_c}$$	$$Q_c = \sqrt{\frac{g\,m^2}{2}\,h^{5/2}} = 2,215\,m\,h^{5/2}$$	$$H_c = \frac{5}{4}\,h_c = 1,25\,h_c$$	$$I_c = \frac{g\sqrt{m^2+1}}{m\,C_c^2}$$
Trapezoidal					

The general formulae are applicable. Consult Graph 100.

98. Trapezoidal channels, determination of critical depth
Value of Q^2/gl^5

l/h_c	$m=0,0$	$m=1,0$	$m=1,5$	$m=2,0$	$m=2,5$	$m=3,0$
1,000	1,0000	2,6667	3,9063	5,4000	7,1458	9,1429
1,100	0,7513	1,8550	2,6618	3,6270	4,7491	6,0274
1,200	0,5787	1,3372	1,8834	2,5324	3,2833	4,1353
1,300	0,4552	0,9930	1,3750	1,8262	2,3458	2,9334
1,400	0,3644	0,7560	1,0306	1,3533	1,7234	2,1404
1,500	0,2963	0,5879	0,7901	1,0266	1,2966	1,6000
1,600	0,2441	0,4656	0,6176	0,7945	0,9959	1,2214
1,700	0,2035	0,3747	0,4910	0,6259	0,7788	0,9496
1,800	0,1715	0,3057	0,3962	0,5007	0,6188	0,7504
1,900	0,1458	0,2526	0,3239	0,4060	0,4986	0,6014
2,000	0,1250	0,2109	0,2680	0,3333	0,4068	0,4883
2,100	0,1080	0,1779	0,2240	0,2766	0,3357	0,4010
2,200	0,0939	0,1514	0,1890	0,2319	0,2798	0,3327
2,300	0,0822	0,1298	0,1609	0,1961	0,2354	0,2787
2,400	0.0723	0,1122	0,1380	0,1672	0,1997	0,2354
2,500	0,0640	0,0976	0,1192	0,1436	0,1707	0,2004
2,600	0,0569	0,0854	0,1036	0,1241	0,1469	0,1719
2,700	0,0508	0,0751	0,0906	0,1080	0,1273	0,1483
2,800	0,0456	0,0664	0,0797	0,0945	0,1109	0,1288
2,900	0,0410	0,0590	0,0704	0,0831	0,0972	0,1125

l/h_c	$m=0,0$	$m=1,0$	$m=1,5$	$m=2,0$	$m=2,5$	$m=3,0$
4,000	0,0156	0,0203	0,0232	0,0264	0,0298	0,0335
4,100	0,0145	0,0188	0,0213	0,0242	0,0273	0,0306
4,200	0,0135	0,0174	0,0197	0,0222	0,0250	0,0280
4,300	0,0126	0,0161	0,0182	0,0205	0,0230	0,0257
4,400	0,0117	0,0149	0,0168	0,0189	0,0212	0,0236
4,500	0,0110	0,0139	0,0156	0,0175	0,0196	0,0218
4,600	0,0103	0,0129	0,0145	0,0162	0,0181	0,0201
4,700	0,0096	0,0121	0,0135	0,0151	0,0168	0,0186
4,800	0,0090	0,0113	0,0126	0,0140	0,0156	0,0172
4,900	0,0085	0,0105	0,0117	0,0131	0,0145	0,0160
5,000	0,0080	0,0099	0,0110	0.0122	0,0135	0,0149
5,100	0,0075	0,0093	0,0103	0,0114	0,0126	0,0139
5,200	0,0071	0,0087	0,0096	0,0107	0,0118	0,0129
5,300	0,0067	0,0082	0,0091	0,0100	0,0110	0,0121
5,400	0,0064	0,0077	0,0085	0,0094	0,0103	0,0113
5,500	0,0060	0,0073	0,0080	0,0088	0,0097	0,0106
5,600	0,0057	0,0069	0,0076	0,0083	0,0091	0,0100
5,700	0,0054	0,0065	0,0071	0,0078	0,0086	0,0094
5,800	0,0051	0,0061	0,0067	0,0074	0,0081	0,0088
5,900	0,0049	0,0058	0,0064	0,0070	0,0076	0,0083

98 (Contd.) Trapezoidal channels, determination of critical depth
Value of Q^2/gl^5

l/h_c	$m=0,0$	$m=1,0$	$m=1,5$	$m=2,0$	$m=2,5$	$m=3,0$	l/h_c	$m=0,0$	$m=1,0$	$m=1,5$	$m=2,0$	$m=2,5$	$m=3,0$
3,000	0,0370	0,0527	0,0625	0,0735	0,0856	0,0988	6,000	0,0046	0,0055	0,0060	0,0066	0,0072	0,0078
3,100	0,0336	0,0472	0,0557	0,0653	0,0757	0,0871	6,100	0,0044	0,0052	0,0057	0,0062	0,0068	0,0074
3,200	0,0305	0,0425	0,0499	0,0582	0,0673	0,0772	6,200	0,0042	0,0050	0,0054	0,0059	0,0064	0,0070
3,300	0,0278	0,0383	0,0449	0,0521	0,0601	0,0687	6,300	0,0040	0,0047	0,0051	0,0056	0,0061	0,0066
3,400	0,0254	0,0347	0,0405	0,0468	0,0538	0,0614	6,400	0,0038	0,0045	0,0049	0,0053	0,0058	0,0062
3,500	0,0233	0,0315	0,0366	0,0422	0,0484	0,0550	6,500	0,0036	0,0043	0,0046	0,0050	0,0055	0,0059
3,600	0,0214	0,0287	0,0332	0,0382	0,0436	0,0495	6,600	0,0035	0,0041	0,0044	0,0048	0,0052	0,0056
3,700	0,0197	0,0263	0,0303	0,0347	0,0395	0,0447	6,700	0,0033	0,0039	0,0042	0,0046	0,0049	0,0053
3,800	0,0182	0,0241	0,0276	0,0316	0,0359	0,0405	6,800	0,0032	0,0037	0,0040	0,0043	0,0047	0,0051
3,900	0,0169	0,0221	0,0253	0,0288	0,0326	0,0368	6,900	0,0030	0,0035	0,0038	0,0041	0,0045	0,0048

General expression: $Q^2/gl^5 = (K+m)^3/((K+2m)K^5)$; where $K=l/h_c$.

99. Circular channels. Determination of the critical depth

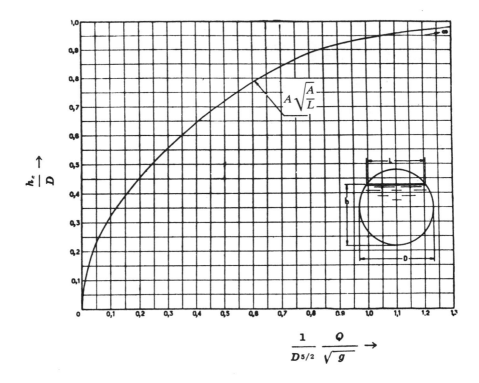

$$\frac{1}{D^{5/2}} \frac{Q}{\sqrt{g}} \rightarrow$$

Example: Determine the critical depth in a circular channel 10 m in diameter, which will carry a discharge of 600 m³/s.

We have: $\dfrac{Q}{\sqrt{g}} = 191.6 \text{ m}^{5/2}; \dfrac{1}{D^{5/2}} = \dfrac{1}{316.23}$

               $\dfrac{Q}{\sqrt{g}} \times \dfrac{1}{D^{5/2}} = 0.61.$

From the graph $\dfrac{h_c}{D} = 0.79.$ Thus $h_c = 7.9$ m.

100. Rectangular and Trapezoidal channels; alternate depths h' and h" of equal specific energy H

B = 0,0

C	k'	k"
0,001	0,032	0,999
0,010	0,106	0,990
0,020	0,154	0,979
0,030	0,193	0,968
0,040	0,228	0,956
0,050	0,257	0,944
0,060	0,290	0,931
0,070	0,321	0,917
0,080	0,351	0,902
0,090	0,381	0,885
0,100	0,412	0,867
0,110	0,445	0,846
0,120	0,480	0,823
0,130	0,521	0,794
0,140	0,571	0,753
0,146	0,619	0,712
0,147	0,632	0,700
0,148	0,654	0,679

B = 0,1

C	k'	k"
0,010	0,104	0,992
0,020	0,151	0,983
0,030	0,188	0,974
0,040	0,221	0,964
0,050	0,252	0,954
0,060	0,281	0,944
0,070	0,308	0,933
0,080	0,336	0,921
0,090	0,363	0,908
0,100	0,390	0,895
0,110	0,417	0,880
0,120	0,445	0,864
0,130	0,475	0,845
0,140	0,507	0,824
0,150	0,544	0,798
0,160	0,590	0,761
0,165	0,622	0,734
0,168	0,683	0,683

B = 0,3

C	k'	k"
0,010	0,102	0,994
0,020	0,146	0,988
0,030	0,182	0,982
0,040	0,211	0,975
0,050	0,239	0,968
0,060	0,264	0,961
0,070	0,288	0,953
0,080	0,311	0,946
0,100	0,356	0,929
0,120	0,399	0,911
0,140	0,442	0,890
0,160	0,487	0,865
0,170	0,511	0,851
0,180	0,537	0,835
0,190	0,565	0,816
0,200	0,598	0,792
0,210	0,643	0,755
0,215	0,701	0,701

B = 0,5

C	k'	k"
0,010	0,010	0,999
0,020	0,142	0,991
0,030	0,750	0,986
0,040	0,203	0,981
0,050	0,229	0,976
0,060	0,252	0,971
0,080	0,293	0,960
0,100	0,331	0,948
0,120	0,368	0,936
0,140	0,403	0,923
0,160	0,437	0,908
0,180	0,472	0,892
0,200	0,508	0,873
0,220	0,547	0,850
0,240	0,591	0,821
0,250	0,618	0,802
0,260	0,651	0,775
0,268	0,716	0,716

B = 1,0

C	k'	k"
0,010	0,096	0,999
0,020	0,134	0,995
0,040	0,187	0,989
0,060	0,227	0,984
0,090	0,276	0,976
0,120	0,318	0,967
0,150	0,356	0,957
0,180	0,391	0,947
0,210	0,424	0,937
0,240	0,456	0,924
0,270	0,488	0,911
0,300	0,520	0,896
0,330	0,553	0,879
0,360	0,589	0,859
0,380	0,612	0,842
0,400	0,646	0,821
0,420	0,686	0,790
0,431	0,740	0,740

B = 2,0

C	k'	k"
0,010	0,089	0,999
0,050	0,181	0,994
0,100	0,244	0,988
0,150	0,291	0,982
0,200	0,329	0,975
0,250	0,363	0,969
0,300	0,394	0,962
0,350	0,422	0,954
0,400	0,449	0,947
0,450	0,474	0,938
0,500	0,500	0,929
0,600	0,550	0,908
0,650	0,575	0,896
0,700	0,601	0,882
0,750	0,630	0,866
0,800	0,662	0,845
0,850	0,703	0,815
0,880	0,762	0,762

B = 4,0

C	k'	k"
0,010	0,079	1,000
0,050	0,151	0,998
0,100	0,197	0,996
0,200	0,257	0,991
0,300	0,298	0,988
0,400	0,333	0,983
0,600	0,388	0,973
0,800	0,434	0,964
1,000	0,476	0,952
1,200	0,514	0,940
1,400	0,551	0,926
1,600	0,588	0,910
1,800	0,627	0,891
1,900	0,647	0,879
2,000	0,669	0,866
2,100	0,694	0,850
2,200	0,726	0,826
2,272	0,779	0,779

B = 6,0

C	k'	k"
0,010	0,072	1,000
0,050	0,134	0,999
0,100	0,171	0,998
0,300	0,252	0,994
0,600	0,321	0,987
0,900	0,371	0,980
1,200	0,412	0,973
1,500	0,447	0,965
1,800	0,479	0,956
2,100	0,511	0,947
2,400	0,539	0,938
2,700	0,568	0,927
3,000	0,596	0,915
3,300	0,624	0,901
3,600	0,656	0,884
3,900	0,690	0,863
4,200	0,737	0,828
4,319	0,784	0,784

General expression: $C = K^2(1-K)(1+BK)^2$, where $C = Q^2/2gH^3l^2$, $B = mH/l$. k' and k'' values presented in the table: $h' = k'H$; $h'' = k''H$

101. Triangular channels:
alternate depths h' and h'' of equal energy H

Q — discharge; m — the side slopes (horizontal to vertical);

K' and K''—Values given by the table.

Thus: $h' = K'\,H$; $h'' = K''\,H$.

$x=\dfrac{Q^2}{2mg^2H^5}$	0	1	2	3	4	5	6	7	8	9
0,00	0,000	0,184	0,225	0,252	0,272	0,290	0,305	0,318	0,331	0,342
	1,000	999	998	997	996	995	994	993	992	991
0,01	0,352	0,362	0,372	0,381	0,389	0,397	0,405	0,412	0,420	0,427
	990	988	987	986	985	984	983	982	981	979
0,02	0,433	0,440	0,446	0,453	0,459	0,465	0,471	0,476	0,482	0,488
	978	977	976	974	973	972	971	969	968	967
0,03	0,493	0,499	0,504	0,509	0,514	0,520	0,525	0,530	0,535	0,539
	965	964	963	961	960	958	957	956	954	953
0,04	0,544	0,549	0,554	0,559	0,563	0,568	0,573	0,578	0,582	0,587
	951	950	948	946	945	943	941	940	938	936
0,05	0,591	0,596	0,601	0,605	0,610	0,614	0,619	0,624	0,629	0,633
	934	932	931	929	927	925	923	921	918	916
0,06	0,638	0,643	0,648	0,653	0,657	0,662	0,667	0,673	0,678	0,683
	914	912	909	907	904	902	899	896	893	890
0,07	0,689	0,694	0,700	0,706	0,712	0,718	0,725	0,733	0,740	0,749
	887	883	880	876	872	868	863	858	852	845
0,8	0,759	0,773	$\dfrac{Q^2}{2gm^2H^5} = 0{,}0819 - K' = K'' = 0{,}800$ (critical flow)							
	837	825								

Examples:

(1) $Q = 10$ m³/s; $m = 2$; $H = 3$ m. Thus,

$$x = \frac{Q^2}{2gm^2\,H^5} = \frac{100}{19071} = 0.00524$$

From the table, by interpolation, $K' = 294$ e $K'' = 0.995$.
Thus: $h' = K'\,H = 0{,}88$ m (supercritical flow); $h'' = K''\,H = 2.99$ m (*subcritical flow*).
(2) Determine the critical depth, for $Q = 10$ m³/s e $m = 2$.

The critical flow is such that $\dfrac{Q^2}{2gm^2\,H^5} = 0.0819$;

hence $\quad H = \sqrt[5]{\dfrac{Q^2}{2gm^2 \times 0.0819}} = \sqrt[5]{15.56} = 1.73$ m

so that $h_c = 0.8\,H = 1.38$ m.

102. Water surface profiles — Bakhmeteff's method
Values of $B(\eta)$

η	$n=$ 2,8	$n=$ 3,0	$n=$ 3,2	$n=$ 3,4	$n=$ 3,6	$n=$ 3,8	$n=$ 4,0	$n=$ 4,2	$n=$ 4,6	$n=$ 5,0	$n=$ 5,4
						$\eta>1$					
1,001	2,399	2,184	2,008	1,856	1,725	1,610	1,508	1,417	1,264	1,138	1,033
005	1,818	1,649	1,506	384	279	188	107	036	0,915	0,817	0,737
010	572	419	291	182	089	007	0,936	1,873	766	681	610
015	428	286	166	065	0,978	0,902	836	778	680	602	537
020	327	191	078	0,982	900	828	766	711	620	546	486
1,03	1,186	1,060	0,955	0,866	0,790	0,725	0,668	0,618	0,535	0,469	0,415
04	086	0,967	868	785	714	653	600	554	477	415	365
05	010	896	802	723	656	598	548	504	432	374	328
06	0,948	838	748	672	608	553	506	464	396	342	298
07	896	790	703	630	569	516	471	431	366	315	273
1,08	0,851	0,749	0,665	0,595	0,535	0,485	0,441	0,403	0,341	0,292	0,252
09	812	713	631	563	506	457	415	379	319	272	234
10	777	681	601	536	480	433	392	357	299	254	218
11	746	652	575	511	457	411	372	338	282	239	204
12	718	626	551	488	436	392	354	321	267	225	192
1,13	0,692	0,602	0,529	0,468	0,417	0,374	0,337	0,305	0,253	0,212	0,181
14	669	581	509	450	400	358	322	291	240	201	170
15	647	561	490	432	384	343	308	278	229	191	161
16	627	542	473	417	369	329	295	266	218	181	153
17	608	525	458	402	356	317	283	255	208	173	145
1,18	0,591	0,509	0,443	0,388	0,343	0,305	0,272	0,244	0,199	0,165	0,138
19	574	494	429	375	331	294	262	235	191	157	131
20	559	480	416	363	320	283	252	226	183	150	125
22	531	454	392	341	299	264	235	209	168	138	114
24	505	431	371	322	281	248	219	195	156	127	104
1,26	0,482	0,410	0,351	0,304	0,265	0,233	0,205	0,182	0,145	0,117	0,095
28	461	391	334	288	250	219	193	170	135	108	088
30	442	373	318	274	237	207	181	160	126	100	081
32	424	357	304	260	225	196	171	150	118	093	075
34	408	342	290	248	214	185	162	142	110	087	069
1,36	0,393	0,329	0,278	0,237	0,204	0,176	0,153	0,134	0,103	0,081	0,064
38	378	316	266	226	194	167	145	127	097	076	060
40	365	304	256	217	185	159	138	120	092	071	056
42	353	293	246	208	177	152	131	114	087	067	052
44	341	282	236	199	169	145	125	108	082	063	049
1,46	0,330	0,273	0,227	0,191	0,162	0,139	0,119	0,103	0,077	0,059	0,046
48	320	263	219	184	156	133	113	098	073	056	043
50	310	255	211	177	149	127	108	093	069	053	040
55	288	235	194	161	135	114	097	083	061	046	035
60	269	218	179	148	123	103	087	074	054	040	030
1,65	0,251	0,203	0,165	0,136	0,113	0,094	0,079	0,067	0,048	0,035	0,026
70	236	189	153	125	103	086	072	060	043	031	023
75	212	177	143	116	095	079	065	054	038	027	020
80	209	166	133	108	088	072	060	049	034	024	017
85	198	156	125	100	082	067	055	045	031	022	015

Tables and graphs

102 (Contd.) Water surface profiles — Bakhmeteff's method
Values of $B(\eta)$

η	$n=$ 2,8	$n=$ 3,0	$n=$ 3,2	$n=$ 3,4	$n=$ 3,6	$n=$ 3,8	$n=$ 4,0	$n=$ 4,2	$n=$ 4,6	$n=$ 5,0	$n=$ 5,4
					$\eta > 1$						
1,85	0,198	0,156	0,125	0,100	0,082	0,067	0,055	0,045	0,031	0,022	0,015
1,90	188	147	117	094	076	062	050	041	028	020	014
1,95	178	139	110	088	070	057	046	038	026	018	012
2,00	169	132	104	082	066	053	043	035	023	016	011
2,10	154	119	092	073	058	046	037	030	019	013	009
2,2	0,141	0,107	0,083	0,065	0,051	0,040	0,032	0,025	0,016	0,011	0,007
2,3	129	098	075	058	045	035	028	022	014	009	006
2,4	119	089	068	052	040	031	024	019	012	008	005
2,5	110	082	062	047	036	028	022	017	010	006	004
2,6	102	076	057	043	033	025	019	015	009	005	003
2,7	0,095	0,070	0,052	0,039	0,029	0,022	0,017	0,013	0,008	0,005	0,003
2,8	089	065	048	036	027	020	015	012	007	004	002
2,9	083	060	044	033	024	018	014	010	006	004	002
3,0	078	056	041	030	022	017	012	009	005	003	002
3,5	059	041	029	021	015	011	008	006	003	002	001
4,0	0,046	0,031	0,022	0,015	0,010	0,007	0,005	0,004	0,002	0,001	0,000
4,5	037	025	017	011	008	005	004	003	001	001	000
5,0	031	020	013	009	006	004	003	002	001	000	000
6,0	022	014	009	006	004	002	002	001	000	000	000
7,0	017	010	006	004	002	002	001	001			
8,0	0,013	0,008	0,005	0,003	0,002	0,001	0,001	000			
9,0	011	006	004	002	001	001	000	000			
10,0	009	005	003	002	001	001	000	000			
20,0	003	003	002	001	000	000	000	000			

η	$n=$ 2,8	$n=$ 3,0	$n=$ 3,2	$n=$ 3,4	$n=$ 3,6	$n=$ 3,8	$n=$ 4,0	$n=$ 4,2	$n=$ 4,6	$n=$ 5,0	$n=$ 5,4
					$\eta \leqslant 1$						
0,00	0,000	0,000	0,000	0,000	0,000	0,000	0,000	0,000	0,000	0,000	0,000
02	020	020	020	020	020	020	020	020	020	020	020
04	040	040	040	040	040	040	040	040	040	040	040
06	060	060	060	060	060	060	060	060	60	60	60
08	080	080	080	080	080	080	080	080	080	080	080
0,10	0,100	0,100	0,100	0,100	0,100	0,100	0,100	0,100	0,100	0,100	0,100
12	120	120	120	120	120	120	120	120	120	120	120
14	140	140	140	140	140	140	140	140	140	140	140
16	160	160	160	160	160	160	160	160	160	160	160
18	180	180	180	180	180	180	180	180	180	180	180
0,20	0,201	0,200	0,200	0,200	0,200	0,200	0,200	0,200	0,200	0,200	0,200
22	221	221	220	220	220	220	220	220	220	220	220
24	241	241	241	240	240	240	240	240	240	240	240
26	262	261	261	261	260	260	260	260	260	260	260
28	282	282	281	281	281	280	280	280	280	280	280

102 (Contd.) Water surface profiles — Bakhmeteff's method
Values of $B(\eta)$

η	$n = 2,8$	$n = 3,0$	$n = 3,2$	$n = 3,4$	$n = 3,6$	$n = 3,8$	$n = 4,0$	$n = 4,2$	$n = 4,6$	$n = 5,0$	$n = 5,4$
					$\eta < 1$						
0,28	0,282	0,281	0,281	0,281	0,281	0,280	0,280	0,280	0,280	0,280	0,280
30	303	302	302	301	301	301	300	300	300	300	300
32	324	323	322	322	321	321	321	321	320	320	320
34	344	343	343	342	342	341	341	341	340	340	340
36	366	364	363	363	362	362	361	361	361	360	360
0,38	0,387	0,385	0,384	0,383	0,383	0,382	0,382	0,381	0,381	0,381	0,380
40	408	407	405	404	403	403	402	402	401	401	400
42	430	428	426	425	424	423	423	422	421	421	421
44	452	450	448	446	445	444	443	443	442	441	441
46	475	472	470	468	466	465	464	463	462	462	461
0,48	0,497	0,494	0,492	0,489	0,488	0,486	0,485	0,484	0,483	0,482	0,481
50	521	517	514	511	509	508	506	505	504	503	502
52	542	540	536	534	531	529	528	527	525	523	522
54	568	563	559	556	554	551	550	548	546	544	543
56	593	587	583	579	576	574	572	570	567	565	564
0,58	0,618	0,612	0,607	0,603	0,599	0,596	0,594	0,592	0,589	0,587	0,585
60	644	637	631	627	623	620	617	614	611	608	606
61	657	650	644	639	635	631	628	626	622	619	617
62	671	663	657	651	647	643	640	637	633	630	628
63	684	676	669	664	659	655	652	649	644	641	638
0,64	0,698	0,690	0,683	0,677	0,672	0,667	0,664	0,661	0,656	0,652	0,649
65	712	703	696	689	684	680	676	673	667	663	660
66	727	717	709	703	697	692	688	685	679	675	672
67	742	731	723	716	710	705	701	697	691	686	683
68	757	746	737	729	723	718	713	709	703	698	694
0,69	0,772	0,761	0,751	0,743	0,737	0,731	0,726	0,722	0,715	0,710	0,706
70	787	776	766	757	750	744	739	735	727	722	717
71	804	791	781	772	764	758	752	748	740	734	729
72	820	807	796	786	779	772	766	761	752	746	741
73	837	823	811	802	793	786	780	774	765	759	753
0,74	0,854	0,840	0,827	0,817	0,808	0,800	0,794	0,788	0,779	0,771	0,766
75	872	857	844	833	823	815	808	802	792	784	778
76	890	874	861	849	839	830	823	817	806	798	791
77	909	892	878	866	855	846	838	831	820	811	804
78	929	911	896	883	872	862	854	847	834	825	817
0,79	0,949	0,930	0,914	0,901	0,889	0,879	0,870	0,862	0,849	0,839	0,831
80	970	950	934	919	907	896	887	878	865	854	845
81	992	971	954	938	925	914	904	895	881	869	860
82	1,015	993	974	958	945	932	922	913	897	885	875
83	039	1,016	996	979	965	952	940	931	914	901	890
0,84	1,064	1,040	1,019	1,001	0,985	0,972	0,960	0,949	0,932	0,918	0,906
85	091	065	043	024	1,007	993	980	969	950	935	923
86	119	092	068	048	031	1,015	1,002	990	970	954	940
87	149	120	095	074	055	039	025	1,012	990	973	959
88	181	151	124	101	081	064	049	035	1,012	994	978

Tables and graphs

102 (Contd.) Water surface profiles — Bakhmeteff's method
Values of $B(\eta)$

η	$n =$ 2,8	$n =$ 3,0	$n =$ 3,2	$n =$ 3,4	$n =$ 3,6	$n =$ 3,8	$n =$ 4,0	$n =$ 4,2	$n =$ 4,6	$n =$ 5,0	$n =$ 5,4
					$\eta < 1$						
0,88	1,181	1,151	1,124	1,101	1,081	1,064	1,049	1,035	1,012	0,994	0,978
0,89	1,216	1,183	1,155	1,131	1,110	1,091	1,075	1,060	1,035	1,015	0,999
90	253	218	189	163	140	120	103	087	060	039	1,021
91	294	257	225	197	173	152	133	116	088	064	045
92	340	300	266	236	210	187	166	148	117	092	072
93	391	348	311	279	251	226	204	184	151	123	101
0,94	1,449	1,403	1,363	1,328	1,297	1,270	1,246	1,225	1,188	1,158	1,134
95	518	467	423	385	352	322	296	272	232	199	172
96	601	545	497	454	417	385	355	329	285	248	217
97	707	644	590	543	501	464	431	402	351	310	275
0,975	773	707	649	598	554	514	479	447	393	348	311
0,980	1,855	1,783	1,720	1,666	1,617	1,575	1,536	1,502	1,443	1,395	1,354
985	959	880	812	752	699	652	610	573	508	454	409
990	2,106	2,017	940	873	814	761	714	671	598	537	487
995	355	250	2,159	2,079	2,008	945	889	838	751	678	617
999	931	788	663	554	457	2,370	2,293	2,223	2,102	2,002	917

Example: In a very wide rectangular channel, bottom slope $I = 0.004$, Manning roughness coefficient $K_s = 40$ m$^{1/3}$/s, there is a discharge per unit width $q = 1.13$ m^2/s. A raising of a sill obliges this discharge to take place in cross-section 0, with a depth $h_0 = 3.05$ m. Determine the distance l from 0, if cross-section 1 is so located that $h_1 = 2.45$ m.

(1) The uniform depth $h_u = 1.22$ m and critical depth $h_c = 0.51$ m have been determined. Since $h_u > h_c$, the channel has a mild slope and type M water surface profile. Since $h_0 > h_u$, the appropriate curve is M_1.

(2) The hydraulic exponent (Table 104) is $n = 3.4$.

(3) The critical slope at 0 is $I_c = 0.0556$, so that $\beta = \dfrac{I}{I_c} = 0,096$ and $I - \beta = 0,904$. This value can be considered constant as far as cross-section 1.

(4) We also have $\eta_0 = \dfrac{h_o}{h_u} = 2,50$ and $\eta_1 = \dfrac{h_1}{h_u} = 2,01$, so that $\eta_1 - \eta_0 = -0,49$.

(5) From Table 102, we have $B(\eta_o) = 0,047$ and $B(\eta_1) = 0,083$, so that $B(\eta_1) - B(\eta_o) = 0,036$. Accordingly:

$$l = \frac{h_u}{I}\left\{ (\eta_1 - \eta_0) - (1 - \beta)\,[B\,(\eta_1) - B\,(\eta_o)] \right\}$$

$$= \frac{1,22}{0,004}(-0,49 - 0,096 \times 0,036) = -162 \text{ m.}$$

That is to say, cross-section 1 is 162 m upstream.

103. Water surface profiles. Approximated Bakhmeteff's solution
Values of φ (η)

η	n = 2,8	n = 3,0	n = 3,2	n = 3,4	n = 3,6	n = 3,8	n = 4,0	n = 4,2
1,001	1,398	1,183	1,007	0,855	0,724	0,609	0,507	0,416
005	0,813	0,644	0,501	379	274	183	102	031
1,010	0,562	0,409	0,281	0,172	0,079	+ 0,003	+ 0,074	+ 0,137
1,015	0,413	0,271	0,151	0,050	+ 0,037	0,113	0,179	0,237
1,02	0,307	0,171	0,058	+ 0,038	0,120	0,192	0,254	0,309
1,03	0,156	0,030	+ 0,075	0,164	0,240	0,305	0,362	0,412
1,04	0,046	+ 0,073	0,172	0,255	0,326	0,387	0,440	0,486
1,05	+ 0,040	0,154	0,248	0,327	0,394	0,452	0,502	0,546
06	112	222	312	388	452	507	554	596
07	174	280	367	440	501	554	599	639
08	229	331	415	485	545	595	639	677
09	278	377	459	527	584	633	675	711
1,10	0,323	0,419	0,499	0,564	0,620	0,667	0,708	0,743
11	364	458	535	599	653	699	738	772
12	402	494	569	632	684	728	766	799
13	438	528	601	662 .	713	756	793	825
14	471	559	631	690	740	782	818	849
1,15	0,503	0,589	0,660	0,718	0,766	0,807	0,842	0,872
16	533	618	687	743	791	831	865	894
17	562	645	712	768	814	853	887	'915
18	589	671	737	792	837	875	908	936
19	616	696	761	815	859	896	928	955
1,20	0,641	0,720	0,784	0,837	0,880	0,917	0,948	0,974
22	689	766	828	879	921	956	985	1,011
24	735	809	869	918	959	992	1,021	045
26	778	850	909	956	995	1,027	055	078
28	819	889	946	992	1,030	061	087	110
1,30	0,858	0,927	0,982	1,026	1,063	1,093	1,119	1,140
32	896	963	1,016	060	095	124	149	170
34	932	998	050	092	126	155	178	198
36	967	1,031	082	123	156	184	207	226
38	1,002	064	114	154	186	213	235	253
1,40	1,035	1,096	1,144	1,183	1,215	1,241	1,262	1,280
42	067	127	174	212	243	268	289	306
44	099	158	204	241	271	295	315	332
46	130	187	233	259	298	321	341	357
48	160	217	261	296	324	347	367	382

103 (Contd.) Water surface profiles. Approximated Bakhmeteff's solution

Values of $\phi\,(\eta)$

η	$n=2,8$	$n=3,0$	$n=3,2$	$n=3,4$	$n=3,6$	$n=3,8$	$n=4,0$	$n=4,2$
1,50	1,190	1,245	1,289	1,323	1,351	1,373	1,392	1,407
55	262	315	356	389	415	436	453	467
60	331	382	421	452	477	497	513	526
65	399	447	485	514	537	556	571	583
70	464	511	547	575	597	614	628	640
1,75	1,538	1,573	1,607	1,634	1,655	1,671	1,685	1,696
80	591	634	667	692	712	728	740	751
85	652	694	725	750	768	783	795	805
90	712	753	783	806	824	838	850	859
95	772	811	840	862	880	893	904	912
2,00	831	868	896	918	934	947	957	965
1	946	981	2,008	2,027	2,042	2,054	2,063	2,070
2	2,059	2,093	117	135	149	160	168	175
3	171	202	225	242	255	265	272	278
4	281	311	332	348	360	369	376	381
2,5	2,390	2,418	2,438	2,453	2,464	2,472	2,478	2,483
6	498	524	543	557	567	575	581	585
7	605	630	648	661	671	678	683	687
8	711	735	752	764	773	780	785	788
9	817	840	856	865	876	882	886	890
3,0	2,922	2,944	2,959	2,970	2,978	2,983	2,988	2,991
5	3,441	3,459	3,471	3,479	3,485	3,489	3,492	3,494
4,0	954	969	978	985	990	993	995	996
5	4,463	4,475	4,483	4,489	4,492	4,495	4,496	4,497
5,0	4,969	4,980	4,987	4,991	4,994	4,996	4,997	4,998
6,0	5,978	5,986	5,991	5,994	5,996	5,998	5,998	5,999
7,0	6,983	6,990	6,994	6,996	6,998	6,998	6,999	6,999
8,0	7,987	7,992	7,995	7,997	7,998	7,999	7,999	
9,0	8,989	8,994	8,996	8,998	8,999	8,999		
10,0	9,991	9,995	9,997	9,998	9,999	9,999		

Example: Use the simplified formula ($\beta = 0$), for the previous example.
(1) In the same way, $h_u = 1.22$ m; $h_c = 0.51$ m; $n = 3.4$; $\eta_0 = 2.50$; $\eta_1 = 2.01$.
(2) From Table 103, $\phi\,(\eta_0) = 2.453$; $\phi\,(\eta_1) = 1.929$; so that $\phi\,(\eta_1) - \phi\,(\eta_0) = -0.524$.
Thus:

$$l = \frac{h_u}{I}[\phi(\eta_0) - \phi(\eta_1)] = -\frac{1.22}{0.004} \times 0.524 = -160 \text{ m}$$

It will be seen that the result practically coincides with the previous one.

104. Water surface profiles — Bakhmeteff's method
Values of the hydraulic exponent n

h — Water depth
l — Base width in trapezoidal cross-sections
D—Diameter in circular cross-sections.

$\dfrac{h}{l}$ or $\dfrac{h}{D}$	Shape of cross-section							
	Rectangular	Trapezoidal with side slopes (hor./vert.)						Circular
		1/2	1/1	1.5/1	2/1	2.5/1	3/1	
0,000 — 0,020	3,4							
0,02 — 0,03	3,4	3,3	3,4	3,4	3,4	3,4	3,4	4,3
0,03 — 0,04	3,3	3,3	3,4	3,4	3,4	3,4	3,4	4,3
0,04 — 0,05	3,2	3,3	3,4	3,4	3,4	3,4	3,4	4,3
0,05 — 0,06	3,2	3,3	3,4	3,4	3,4	3,4	3,5	4,3
0,06 — 0,08	3,2	3,3	3,4	3,5	3,5	3,6	3,6	4,3
0,08 — 0,10	3,1	3,3	3,4	3,5	3,5	3,6	3,7	4,2
0,10 — 0,15	3,0	3,3	3,4	3,5	3,6	3,7	3,8	4,2
0,15 — 0,20	2,9	3,3	3,4	3,6	3,7	3,8	3,9	4,2
0,2 — 0,3	2,8	3,3	3,5	3,7	3,8	3,9	4,0	4,0
0,3 — 0,4	2,8	3,3	3,5	3,8	3,9	4,0	4,1	3,8
0,4 — 0,5	2,8	3,3	3,6	3,9	4,0	4,1	4,2	3,6
0,5 — 0,6	2,7	3,3	3,7	4,0	4,1	4,2	4,3	3,3
0,6 — 0,8	2,6	3,4	3,8	4,1	4,2	4,3	4,4	†
0,8 — 1,0	2,5	3,4	3,9	4,3	4,4	4,4	4,5	
1,0 — 1,5	2,4	3,6	4,2	4,4	4,6	4,6	4,7	
1,5 — 2,0	2,3	3,9	4,4	4,6	4,8	4,8	4,9	
2 — 3	2,2	4,1	4,5	4,7	5,0	5,0	5,0	
3 — 4	2,2	4,4	4,6	4,9	5,2	5,2	5,2	
4 — 5	2,2	4,5	4,7	5,0	5,4	5,4	5,4	
5 — 6	2,2	4,5	4,8	—	—	—	—	
6 — 8	2,2	4,6	4,9	—	—	—	—	
8 — 10	2,2	4,7	5,0	—	—	—	—	

† $\dfrac{h}{D} = 0.6$ to 0.7——$n = 2.9$; $\dfrac{h}{D} = 0.8$——$n = 2,2$.

105. Conjugate depths of the hydraulic jump
According to [37]

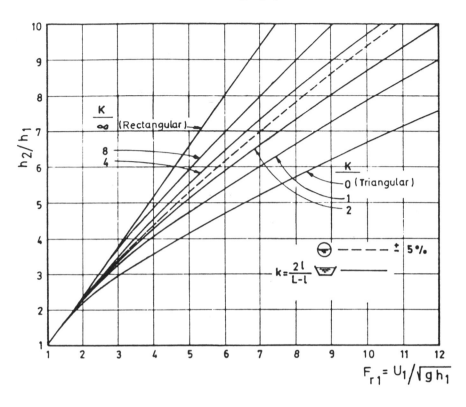

106. Energy loss in the hydraulic jump
According to [37]

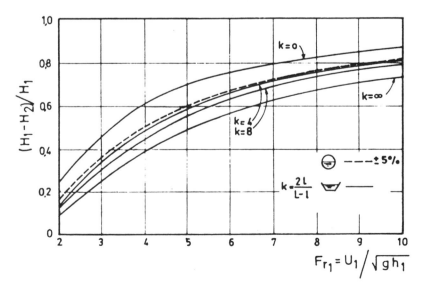

107. Hydraulic jump in a rectangular channel
According to [67]

(a) *Length of the hydraulic jump in sloping channels*

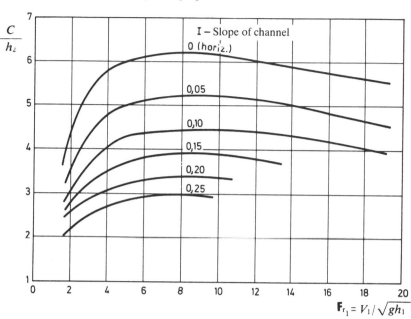

(b) *Conjugate depths*

$$\frac{h_2}{h_1} = \frac{1}{2\cos\theta}\left[\sqrt{\frac{8F_{r_1}^2\cos^3\theta}{1-2\,K\tan\theta}-1}+1\right]$$

θ — Angle of the channel with the horizontal.
Values of K.

$I = \tan\theta$	0,04	0,08	0,12	0,16	0,20	0,24	0,28	0,30
K	3,2	2,7	2,3	2,0	1,80	1,60	1,40	1,35

108. Coefficients of discharge in constrictions

According to *U.S. Geological Survey*. Compiled from [35]
(Values of the coefficients of formula (6.46a))

(a) *Type I*

Vertical embankments; vertical abutments; $e = 1$; $K_e = 1$

$$C = C' \cdot K_F \cdot K_W \cdot K_r \cdot K_\phi$$

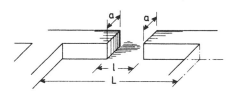

$$\sigma = 100(L - l)/L$$

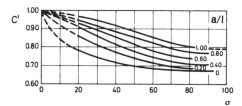

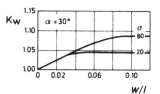

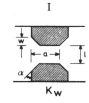

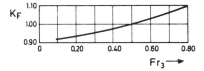

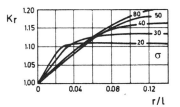

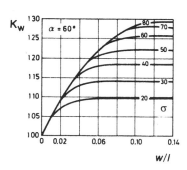

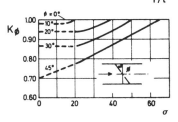

$K_r = 1{,}00$ for scheme I.
$K_w = 1{,}00$ for scheme II.

108 (Contd.) Coefficients of discharge in constrictions
(Values of the coefficients of formula (6.46a))

(b) *Type II*

Sloping embankments; vertical abutments, $F_{r_3} = 0.2$ to 0.7; $e = 1$; $K_e = 1$

$$C = C' \cdot K_F \cdot K_y \cdot K_\phi$$

$$\sigma = 100(L - l)/L$$

Slope of the embankments — 1/1

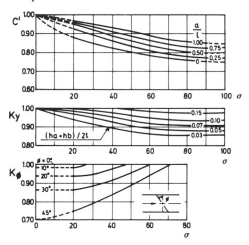

Slopes of the embankments — 2/1 (hor./vert.)

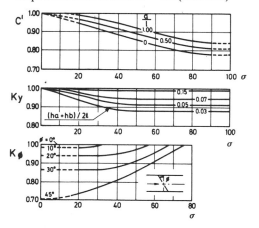

$K_y = 1{,}00$ for $(h_a + h_b)/2l \geqq 0{,}20$.
$K_\phi = 1{,}00$ for $\phi = 0°$.

108 (Contd.)   Coefficients of discharge in constrictions
(Values of the coefficients of formula (6.46a))

(c) *Type III*

Sloping embankment; sloping abutments; $F_{r3} = 0{,}2$ to $0{,}7$; $e = 1$; $K_e = 1$

$$C = C' \cdot K_F \cdot K_\phi \cdot K_x$$

$$\sigma = 100(L - l)/L$$

Slopes of the embankments — 1/1 (hor./vert.)

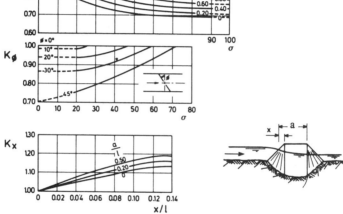

Slopes of the embankments — 2/1 (hor./vert.)

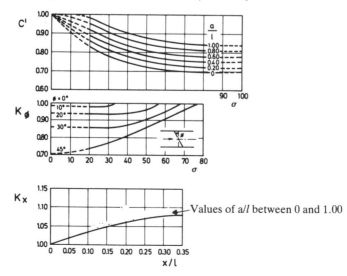

Values of a/l between 0 and 1.00

109. Energy losses at bridge piers: Rehbock–Yarnell formulae
According to [3]

$\varepsilon = 4$†; piers parallel to the current.

(a) *Validity zone*

$$\rightarrow F_{r3}^2 = \frac{U_3^2}{gh_3}$$

(b) *Values of* δ $\left\{\begin{array}{l}\text{- - - - - Rectangular fairing}\\ \text{———— Circular fairing}\end{array}\right.$

$\rightarrow \sigma$

First class of flow — drowned flow (zone of validity); Second class of flow — submerged hydraulic jump; Third class of flow — hydraulic jump.

Example: Take $U_3 = 2$ m/s; $h_3 = 4$ m; distance between axes of the piers, $l_1 = 10$ m.; thickness of the piers, $e = 2$ m; piers with rounded upstream and downstream fairing. Calculate the surcharge Δh.

We have: $l_2 = l_1 - e = 10 - 2 = 8$ m.

$$F_{r3}^2 = \frac{U_3^2}{gh_3} = \frac{4}{4 \times 9,8} = 0,1; \quad \sigma = \frac{l_1 - l_2}{l_1} = \frac{10 - 8}{10} = 0,2.$$

Graph (a) shows that the flow is situated in RehbockClass 1 (drowned flow), so that application of formula (6.47) is valid.

We thus have, Graph (b): $\delta = 2,23$; $\Delta h = (2,3 - 0,2 \times 1,3) \,(0,4 \times 0,2 + 0,2^2 + 9 \times 0,2^4) \times (1 + F_{r3}^2) \dfrac{4}{2 \times 9,8} = 0,060$ m.

† ε is the ratio between the length, a, and the thickness, e, of the piers.

110. Head losses at bends[†]
Formula (6.48)

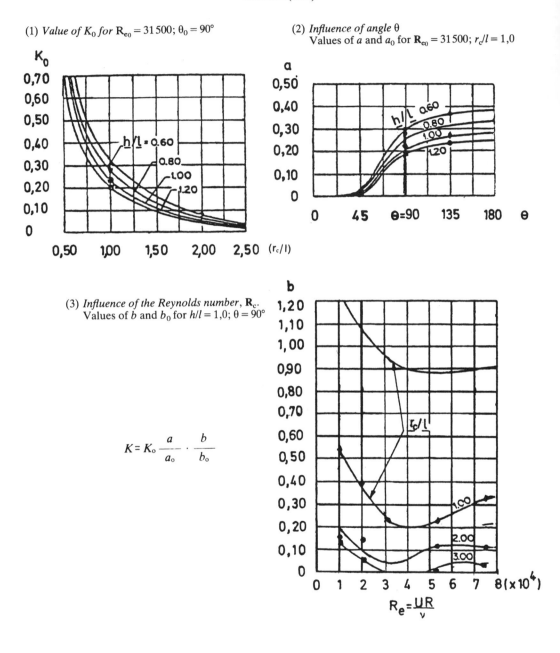

(1) *Value of K_0 for $R_{e_0} = 31\,500$; $\theta_0 = 90°$*

(2) *Influence of angle θ*
   Values of a and a_0 for $R_{e_0} = 31\,500$; $r_c/l = 1,0$

(3) *Influence of the Reynolds number*, R_c.
   Values of b and b_0 for $h/l = 1,0$; $\theta = 90°$

$$K = K_0 \frac{a}{a_0} \cdot \frac{b}{b_0}$$

$$R_e = \frac{UR}{\nu}$$

[†] According to Shukry, A., *Flow around Bends in an Open Flume*, Transactions ASCE, VM115, 1950. Quoted by [43]

111. Overflow spillway crests. W. E. S. profiles

(a) *Geometrical definition* — Complied from [45]

m	n	K	r_1/H_d	r_2/H_d	r_3/H_d	d_1/H_d	d_2/H_d	d_3/H_d
0	1,850	2,000	0,500	0,200	0,040	0,1750	0,2760	0,2818
1/3	1,836	1,936	0,680	0,210	0	0,1390	0,2570	0,2570
2/3	1,810	1,939	0,480	0,220	0	0,1150	0,2140	0,2140
3/3	1,780	1,852	variable-radius curve*					0,200

*Elements for defining the variable–radius curve

0,045 H_d

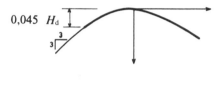

x_1/H_d	y_1/H_d	x_1/H_d	y_1/H_d
0,000	0,000	—0,150	0,0239
—0,040	0,0016	—0,160	0,0275
—0,080	0,0065	—0,170	0,0313
—0,110	0,0125	—0,180	0,0354
—0,130	0,0177	—0,190	0,0399
—0,140	0,0207	—0,200	0,0450

(b) *Coefficients of discharge*, μ — Compiled from [45]

$\dfrac{H}{H_d}$ $\diagdown$ m	0,2	0,4	0,6	0,8	1,0	1,2	1,4
0	0,400	0,433	0,460	0,482	0,500	0,519	0,532
1/3	0,390	0,432	0,466	0,491	0,512	0,529	0,542
2/3	0,392	0,443	0,473	0,497	0,517	0,535	0,551
3/3	0,398	0,440	0,471	0,495	0,516	0,535	0,550

(c) *Variation of μ/μ_0 as a function of Q/Q_0 and H/H_d* — According to [39]

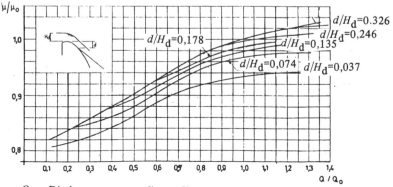

Q_0 — Discharge corresponding to H_d

111. (Contd.) Overflow spillway crests. W. E. S. profiles

(d) *Upper nappe profiles* — Compiled from [25]

$H/H_d = 0,50$		$H/H_d = 1,00$		$H/H_d = 1,33$	
x/H_d	y/H_d	x/H_d	y/H_d	x/H_d	y/H_d
$-1,0$	$-0,490$	$-1,0$	$-0,933$	$-1,0$	$-1,210$
$-0,8$	$-0,484$	$-0,8$	$-0,915$	$-0,8$	$-1,185$
$-0,6$	$-0,475$	$-0,6$	$-0,893$	$-0,6$	$-1,151$
$-0,4$	$-0,460$	$-0,4$	$-0,865$	$-0,4$	$-1,110$
$-0,2$	$-0,425$	$-0,2$	$-0,821$	$-0,2$	$-1,060$
$0,0$	$-0,371$	$0,0$	$-0,755$	$0,0$	$-1,000$
$0,2$	$-0,300$	$0,2$	$-0,681$	$0,2$	$-0,919$
$0,4$	$-0,200$	$0,4$	$-0,586$	$0,4$	$-0,821$
$0,6$	$-0,075$	$0,6$	$-0,465$	$0,6$	$-0,705$
$0,8$	$0,075$	$0,8$	$-0,320$	$0,8$	$-0,569$
$1,0$	$0,258$	$1,0$	$-0,145$	$1,0$	$-0,411$
$1,2$	$0,470$	$1,2$	$0,055$	$1,2$	$-0,220$
$1,4$	$0,705$	$1,4$	$0,294$	$1,4$	$-0,002$
$1,6$	$0,972$	$1,6$	$0,563$	$1,6$	$0,243$
$1,8$	$1,269$	$1,8$	$0,857$	$1,8$	$0,531$

(e) *Cavitation. Minimum of values of* $\dfrac{p}{\gamma H_d}$

Compiled from [40]

m \ $\dfrac{H}{H_d}$	$1,0$	$1,1$	$1,2$	$1,3$	$1,4$
0	0	$-0,16$	$-0,32$	$-0,48$	$-0,64$
$1/3$	$-0,35$	$-0,60$	$-0,86$	$-1,14$	$-1,49$
$2/3$	$-0,19$	$-0,37$	$-0,55$	$-0,74$	$-0,92$
$3/3$	$-0,08$	$-0,23$	$-0,38$	$-0,54$	$-0,69$

(f) *Separation. Maximum values of* H/H_d
Compiled from [40]

m	$\dfrac{H}{H_d}$	Comment
0	$1,40$	—
$1/3$	$1,25$	Slight detachment upstream of crest for $1,10 \leqslant \dfrac{H}{H_d} \leqslant 1,25$
$2/3$	$1,25$	,, ,,
$3/3$	$1,40$	,, ,,

112. Low ogee crests
Compiled from [45]

(a) *Geometrical definition*

m	$\dfrac{h_a}{H_d}$	n	K	r_1/H_d	r_2/H_d	d/H_d	C/H_d
0	0,04	1,851	1,969	0,510	0,212	0,263	0,110
	0,08	1,838	1,953	0,486	0,203	0,242	0,092
	0,12	1,831	1,976	0,458	0,199	0,219	0,075
	0,16	1,830	2,041	0,423	0,196	0,194	0,061
	0,20	1,836	2,146	0,374	0,196	0,165	0,048
1/3	0,04	1,832	1,969	0,552	0,173	0,238	0,085
	0,08	1,818	1,953	0,559	0,190	0,228	0,076
	0,12	1,812	1,976	0,535	0,194	0,212	0,067
	0,16	1,811	2,041	0,476	0,196	0,190	0,056
	0,20	1,817	2,146	0,374	0,196	0,163	0,042
2/3	0,04	1,778	1,881	0,489	0,258	0,213	0,063
	0,08	1,766	1,887	0,499	0,303	0,209	0,058
	0,12	1,762	1,916	0,478	0,357	0,200	0,052
	0,16	1,766	1,970	0,299	∞	0,185	0,045
	0,20	1,773	2,062	0,340	∞	0,162	0,036
3/3	0,04	1,762	1,857	0,460	∞	0,198	0,044
	0,08	1,760	1,869	0,465	∞	0,195	0,042
	0,12	1,747	1,905	0,461	∞	0,190	0,039
	0,16	1,752	1,962	0,446	∞	0,180	0,035
	0,20	1,761	2,060	0,423	∞	0,160	0,028

(b) *Coefficient of discharge*

m	$\dfrac{a}{H_d}$ \ $\dfrac{H}{H_d}$	0,2	0,4	0,6	0,8	1,0	1,2	1,4
0	0,2	0,380	0,401	0,419	0,434	0,446	0,457	0,468
	0,4	0,401	0,423	0,442	0,457	0,470	0,482	0,493
	0,6	0,408	0,430	0,449	0,465	0,478	0,490	0,502
	0,8	0,412	0,435	0,454	0,469	0,483	0,495	0,507
	1,0	0,415	0,438	0,458	0,473	0,487	0,499	0,511
1/3	0,2	0,383	0,404	0,423	0,438	0,450	0,461	0,472
	0,4	0,404	0,426	0,445	0,460	0,473	0,485	0,496
	0,6	0,410	0,432	0,451	0,467	0,480	0,492	0,504
	0,8	0,413	0,436	0,455	0,470	0,484	0,497	0,508
	1,0	0,416	0,439	0,459	0,474	0,485	0,500	0,512
2/3	0,2	0,390	0,412	0,430	0,446	0,458	0,469	0,481
	0,4	0,408	0,431	0,450	0,465	0,478	0,491	0,502
	0,6	0,413	0,436	0,455	0,471	0,484	0,496	0,509
	0,8	0,416	0,439	0,458	0,474	0,488	0,500	0,512
	1,0	0,418	0,441	0,461	0,476	0,491	0,503	0,515
3/3	0,2	0,393	0,415	0,434	0,449	0,462	0,473	0,484
	0,4	0,408	0,431	0,450	0,465	0,478	0,491	0,502
	0,6	0,411	0,433	0,453	0,469	0,482	0,494	0,506
	0,8	0,413	0,436	0,455	0,470	0,484	0,496	0,508
	1,0	0,414	0,437	0,457	0,472	0,486	0,498	0,510

Tables and graphs

113.   Submerged crests

According to [25]

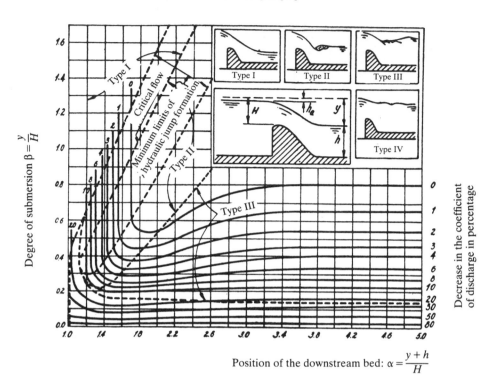

Position of the downstream bed: $\alpha = \dfrac{y + h}{H}$

Examples:

(1) $h = 2$ m; $y = 1$ m; $H = 1.5$ m.

$\alpha = 2$ and $\beta = 0.67$. the flow is Type II and there is no decrease in the coefficient of discharge.

(2) $h = 2.5$ m; $y = 0.5$ m; $H = 1.5$ m. Then $\alpha = 2$ and $\beta = 0.33$. The flow is Type III and there is a 3% decrease in the coefficient of discharge.

114.   Force on the upstream face
Compiled from [25]

h_a/H_s Upstream Face	0	0,02	0,04	0,05	0,06	0,07	0,08	0,09	0,10	0,14
Values of $\Delta p/H_s^2$										
Vertical†	0,087	0,071	0,059	0,055	0,051	0,050	0,048	0,046	0,045	—
1/3	0,113	0,101	0,088	0,083	0,079	0,072	0,067	0,062	0,058	—
2/3	0,130	0,116	0,102	0,097	0,090	0,084	0,079	0,073	0,069	0,051
3/3	0,149	0,138	0,129	0,123	0,118	0,113	0,110	0,105	0,100	0,080
Values of d/H_s										
Vertical†	0,210	0,230	0,245	0,250	0,245	0,240	0,225	0,195	0,150	—
1/3	0,230	0,215	0,185	0,175	0,155	0,135	0,110	0,080	0,040	-
2/3	0,250	0,235	0,220	0,215	0,205	0,190	0,180	0,170	0,155	0,075
3/3	0,290	0,280	0,270	0,265	0,260	0,250	0,240	0,230	0,215	0,120

†Valid also for upstream face with off-set and riser.

115.   Morning-glory spillways. Values of H_s/H_d
Compiled from [45]

a/r	H_d/r								
	0,2	0,3	0,4	0,5	0,6	0,7	0,8	0,9	1,0
2,00 (*)	1,101	1,086	1,072	1,059	1,048	1,038	1,030	1,025	1,022
0,30	1,093	1,077	1,063	1,051	1,039	1,031	1,025	1,021	1,018
0,15	1,072	1,061	1,049	1,038	1,026	1,021	1,017	1,015	1,013

*Negligible approach velocity and aerated nappe.

116.   Morning-glory spillways. Coefficients of discharge
Compiled from [45]

(a) *Values of* μ *for the design heads* (μ_d)

$\dfrac{a}{r}$	$\dfrac{H_d}{r}$								
	0,2	0,3	0,4	0,5	0,6	0,7	0,8	0,9	1,0
2,00 (*)	0,484	0,466	0,444	0,418	0,386	0,346	0,307	0,277	0,253
0,30	0,499	0,481	0,461	0,434	0,404	0,363	0,321	0,290	0,264
0,15	0,495	0,481	0,463	0,441	0,414	0,376	0,333	0,299	0,274

(b) *Values of* μ *for heads other than the design heads*

H/r	0,05	0,10	0,15	0,20	0,25	0,30	0,35	0,40
μ/μ_d	0,88	0,93	0,96	0,98	1,00	1,00	0,99	0,98

*Negligible approach velocity.

117. Morning-glory spillways. Coordinates of the lower nappe profile
Compiled from [45]

	Negligible approach velocity					$a/r = 0{,}15$				
$H_s/r \rightarrow$	0,20	0,25	0,30	0,40	0,50	0,20	0,25	0,30	0,40	0,50
↓ x/H_s	Values of y/H, in the sector OAB, above Ox — see Fig. 6.36.									
0,000	0,0000	0,0000	0,0000	0,0000	0,0000	0,0000	0,0000	0,0000	0,0000	0,0000
010	0133	0130	0129	0122	0121	0120	0120	0115	0110	0105
020	0250	0243	0236	0225	0213	0210	0200	0195	0185	0170
030	0350	0337	0327	0308	0289	0285	0270	0265	0250	0225
040	0435	0417	0403	0377	0351	0345	0335	0325	0300	0265
0,050	0,0506	0,0487	0,0471	0,0436	0,0402	0,0405	0,0385	0,0375	0,0345	0,0300
060	0570	0550	0531	0489	0448	0450	0430	0420	0380	0330
070	0627	0605	0584	0537	0487	0495	0470	0455	0410	0350
080	0677	0655	0630	0578	0521	0525	0500	0485	0435	0365
090	0722	0696	0670	0613	0549	0560	0530	0510	0455	0370
0,100	0,0762	0,0734	0,0705	0,0642	0,0570	0,0590	0,0560	0,0535	0,0465	0,0375
120	0826	0790	0758	0683	0596	0630	0600	0570	0480	0365
140	0872	0829	0792	0705	0599	0660	0620	0585	0475	0345
160	0905	0855	0812	0710	0585	0670	0635	0590	0460	0305
180	0927	0875	0820	0705	0559	0675	0635	0580	0435	0260
0,200	0,0938	0,0877	0,0819	0,0688	0,0521	0,0670	0,0625	0,0560	0,0395	0,0200
250	0926	0850	0773	0596	0380	0615	0560	0470	0265	0015
300	0850	0764	0668	0446	0174	0520	0440	0330	0100	
350	0750	0650	0540	0280		0380	0285	0165		
400	0620	0500	0365	0060		0210	0090			
↓ y/H_s	Values of x/H, in the sector BD, below Ox — see Fig. 6.35.									
0,000	0,554	0,520	0,487	0,413	0,334	0,454	0,422	0,392	0,325	0,253
— 0,020	592	560	526	452	369	499	467	437	369	292
040	627	596	563	487	400	540	509	478	407	328
060	680	630	596	519	428	579	547	516	443	358
080	692	662	628	549	454	615	583	550	476	386
— 0,100	0,722	0,692	0,657	0,577	0,478	0,650	0,616	0,584	0,506	0,412
150	793	762	725	641	531	726	691	660	577	468
200	860	826	790	698	575	795	760	729	639	516
250	919	883	847	750	613	862	827	790	692	557
300	976	941	900	797	648	922	883	843	741	594
— 0,400	1,079	1,041	1,000	0,880	0,706	1,029	0,988	0,947	0,828	0,656
500	172	131	087	951	753	128	1,086	1,040	902	710
600	260	215	167	1,012	793	220	177	129	967	753
800	422	369	312	112	854	380	337	285	1,080	827
— 1,000	564	508	440	189	899	525	481	420	164	878
— 1,200	1,691	1,635	1,553	1,248	0,933	1,659	1,610	1,537	1,228	0,917
400	808	748	653	293	963	780	731	639	276	949
600	918	855	742	330	988	897	843	729	316	973
800	2,024	957	821	358	1,008	2,003	947	809	347	997
— 2,000	126	2,053	891	381	025	104	2,042	879	372	1,013

118. Morning-glory spillways. Coordinates of the upper nappe profile
Compiled from [45]

	Negligible approach velocity					$a/r = 0,15$				
$H_s/r \rightarrow$	0,20	0,25	0,30	0,40	0,50	0,20	0,25	0,30	0,40	0,50
$\downarrow x/H_s$	Values of y/H_s									
−0,40	0,955	0,956	0,959	0,961	0,968	0,957	0,962	0,968	0,978	0,987
−0,20	925	927	929	935	942	917	924	934	949	960
0,00	0,880	0,886	0,892	0,900	0,920	0,870	0,875	0,887	0,909	0,922
20	820	829	838	851	870	800	810	823	850	871
40	740	753	763	787	815	715	727	745	776	807
60	640	658	669	702	748	610	629	648	686	735
80	518	540	556	600		490	511	533	582	
1,00	0,372	0,402	0,420	0,475		0,352	0,377	0,398	0,465	
20	205	240	265	328		187	216	240	337	
40	013	051	081			−0,007	028	055		
60	−0,205	−0,160	−0,122			235	−0,190	−0,155		
80	457	400	357			498	437	388		
2,00	−0,748	−0,678	−0,613			−0,795	−0,710	−0,648		
20	−1,072	981	895			−1,118	−1,023	903		
40	440	−1,315	−1,198			448	350			
60	845	670				800	683			
80	−2,268					−2,148	−2,035			
Point where the upper face meets the 'boil'										
x/H_s			2,410	1,208	0,725			2,222	1,260	0,732
y/H_s			−1,210	0,320	0,696			−0,932	0,295	0,681
Highest point of the 'boil'										
x/H_s				2,545	2,043				2,531	2,009
y/H_s				0,438	0,783				0,458	0,815

119.   Coefficient of contraction at piers
Compiled from [26]

Type H/H_0	1	2	3	4
0,2	0,05	0,09	0,08	0,01
3	04	08	06	00
4	03	06	05	00
5	03	05	04	−0,01
0,6	0,03	0,04	0,03	−0,02
7	03	03	02	02
8	03	03	01	02
9	02	·02	00	02
1,0	0,02	0,02	0,00	−0,02
1	01	01	−0,01	02
2	01	00	0,01	03
3	0	−0,01	0,01	03

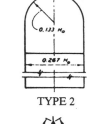

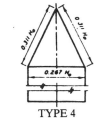

TYPE 1   TYPE 2

TYPE 3   TYPE 4

120.   Overflow spillways with piers.
Upper nappe profiles
Compiled from [26]

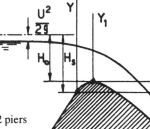

Values of Y_1/H_0 with vertical upstream face and Type 2 piers
(Negligible approach velocity)

$H/H_0 \rightarrow$	0,5		1,00		1,33	
$\downarrow X_1/H_0$	Middle of span	By the pier	Middle of span	By the pier	Middle of span	By the pier
−1,0	−0,482	−0,495	−0,941	−0,950	−1,230	−1,235
0,8	480	492	932	940	215	221
0,6	472	490	913	929	194	209
0,4	457	482	890	930	165	218
0,2	431	440	855	925	122	244
0,0	−0,384	−0,383	−0,805	−0,779	−1,071	−1,103
2	313	265	735	651	015	−0,950
4	220	185	647	545	−0,944	821
6	088	076	539	425	847	689
8	0,075	0,060	389	285	725	549
1,0	0,257	0,240	−0,202	−0,121	−0,564	−0,389
2	462	445	0,015	0,067	356	215
4	705	675	266	286	102	0,011
6	977	925	521	521	0,172	208
8	1,278	1,177	860	779	465	438

121. Height of baffle blocks and end sill (Basin III — USBR)

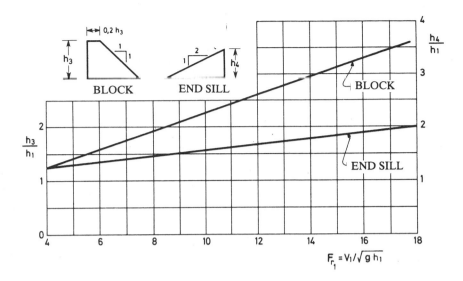

122. Impact type stilling basins

(a) *Basin design* (see Fig. 6.35).

$F_{r_1} = \dfrac{V}{\sqrt{gD}}$	l/D (minimum values)	$F_{r_1} = \dfrac{V}{\sqrt{gD}}$	l/D (minimum values)
0,91	3,1	5,0	7,2
1,5	3,7	5,5	7,6
2,0	4,3	6,0	8,0
2,5	4,9	6,5	8,4
3,0	5,4	7,0	8,8
3,5	5,9	7,5	9,1
4,0	6,3	8,0	9,4
4,5	6,9	9,0	10,0

(b) *Design of downstream riprap protection* (Fig. 6.36)

Diameter of conduit (m)	Diameter of the blocks (m)
0,45	0,10
0,60	0,18
0,75	0,20
0,90	0,23
1,05	0,24
1,20	0,26
1,35	0,30
1,50	0,33
1,80	0,35

123. Design of roller bucket stilling basins
According to [45]

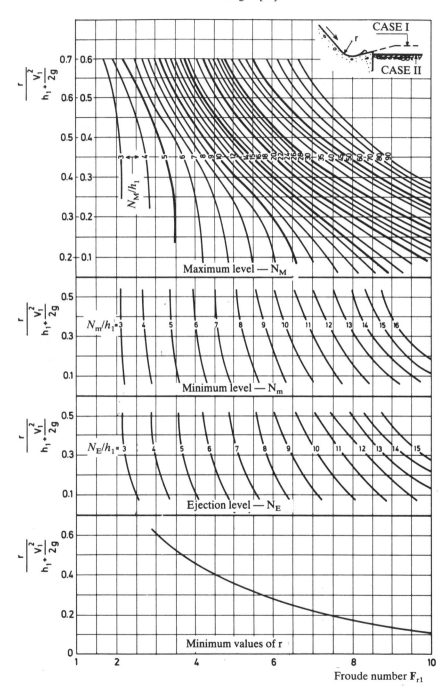

124. Flows in porous media — Index

125. Classification of soils according to grain size

Material		Equivalent diameter d_{10} (mm)
Clay		0.002
Silt:	fine	0.002– 0.005
	medium	0.005– 0.02
	coarse	0.02 – 0.05
Sand:	fine	0.05 – 0.2
	medium	0.2 – 0.5
	coarse	0.5 – 2.0
Very coarse sand		2 – 5
Gravel		5 – 15
Pebbles		15 – 60
Stone		60 –250
Block		> 250

126. Value of the effective porosity, n_e, and relative porosity, n, for some soils and rocks

According to [47]

| Material | | Effective porosity n_e(%) | | | | Relative porosity n(%) | | | | Remarks |
Type	Description	Min	Mean	Max	Mean	Normal Max	Normal Min	Extra-ordinary Max	Extra-ordinary Min	
Soils	Gravel beds	15	25	35	30	40	25	40	20	—
	Sand	10	25	35	35	45	20	—	—	—
	Alluvia	5	15	35	25	40	20	45	15	1
	Silts	2	10	20	40	50	35	—	—	1
	Surface soils	1	10	20	50	60	30	—	—	1
	Loess	0.1	<5	10	45	55	40	—	—	1
	Unconsolidated clays	0.0	2	10	45	60	40	85	30	1
Volcanic rocks	Slag	1	20	50	25	80	10	—	—	1.2
	Vacuolar basalt	1	5	10	12	30	5	—	—	2
	F. compact basalt	0.1	<1	2	2	5	0.1	—	—	3
Sedimentary Rocks (see Compact rocks)	Arenite	0.0	10	20	15	25	3	30	0.5	4
	Detrital limestone	0.5	3	20	10	30	1.5	—	—	—
	Soft sandstone	0.2	1	5	20	50	10	—	—	5
	Argillite	0.0	<2	5	5	15	2	30	0.5	1
Compact rocks	Compact limestone	0.0	<0.5	1	8	15	0.5	20	—	5
	Dolomite	0.0	<0.5	1	5	10	2	—	—	5
	Granite	0.0	<0.2	0.5	0.3	4	0.2	9	0.05	3

1. n_e varies greatly according to circumstances
2. n and n_e diminish with time
3. n and n_e increase with weathering
4. variable according to the degree of cementation and solubility
5. n and n_e increase owing to solution phenomena

127.   Values of relative porosity, n, for some soils†

$$n = \frac{\text{volume of pores}}{\text{Total volume}} = \frac{V_p}{V_t}$$

Soils in their natural state, according to their capacity given by the Standard Penetration Test (SPT) measured in number of blows, or by the (CPT) Penetration Test measured in N/cm²,

Description of the soils	Porosity (n) %
Uniform sand, loose (SPT 4 to 10; CPT 200 to 400)	46
Non-uniform sand, loose (ditto)	40
Uniform sand, dense (SPT 30 to 50; CPT 1200 to 2000)	34
Non-uniform sand, dense (ditto)	30
Soft giacial clay (SPT 4 to 8; CPT 70 to 150)	55
Hard glacial clay (SPT 15 to 30; CPT 300 to 600)	37
Soft clay with low organic matter content	66
Soft clay with high organic matter content	75

†According to [48].

128.   Values of hydraulic conductivity (permeability), K, of some typical soils, for water at 20°C

Type of soil	Hydraulic conductivity K	
	m/s	m/day (approx. value)
Clay	$\leqslant 10^{-8}$	$\leqslant 10^{-3}$
Silt	10^{-7} to 5×10^{-6}	10^{-2} to 0.5
Silty sand	10^{-6} to 2×10^{-5}	0.1 to 2
Fine sand	10^{-5} to 5×10^{-4}	1 to 50
Sand (mixture)	5×10^{-5} to 10^{-4}	5 to 10
Coarse sand	10^{-4} to 10^{-2}	10 to 10^3
Clean gravel	$\geqslant 10^{-2}$	$\geqslant 10^3$

129. Values of the function $F(\delta,\varepsilon) = F(\delta, -\varepsilon)$
for calculating the drawdown in partial wells†

δ \ ε	0,00	0,05	0,10	0,15	0,20	0,25	0,30	0,35	0,40	0,45
0,1	4,298	4,297	4,294	4,287	4,276	4,259	4,232	4,184	4,084	3,605
0,2	3,809	3,806	3,797	3,781	3,756	3,716	3,650	3,525	3,116	
0,3	3,586	3,581	3,566	3,537	3,490	3,425	3,276	2,893		
0,4	3,479	3,471	3,445	3,395	3,312	3,165	2,786			
0,5	3,447	3,433	3,388	3,302	3,145	2,754				
0,6	3,479	3,455	3,374	3,208	2,786					
0,7	3,586	3,538	3,370	2,893						
0,8	3,809	3,688	3,116							
0,9	4,298	3,605								

† Quoted by [47].

130. Values of the well function
Compiled from [51]

$$W(u) = \int_{u}^{\infty} \frac{e^{-u}}{u}\,du$$

u	10^{-15}	10^{-14}	10^{-13}	10^{-12}	10^{-11}	10^{-10}	10^{-9}	10^{-8}
1	33,9616	31,6590	29,3564	27,0538	24,7512	22,4486	20,1460	17,8435
2	33,2684	30,9658	28,6632	26,3607	24,0581	21,7555	19,4529	17,1503
3	32,8629	30,5604	28,2578	25,9552	23,6526	21,3500	19,0474	16,7449
4	32,5753	30,2727	27,9701	25,6675	23,3649	21,0623	18,7598	16,4572
5	32,3521	30,0495	27,7470	25,4444	23,1418	20,8392	18,5366	16,2340
6	32,1698	29,8672	27,5646	25,2620	22,9595	20,6569	18,3543	16,0517
7	32,0156	29,7131	27,4105	25,1079	22,8053	20,5027	18,2001	15,8976
8	31,8821	29,5795	27,2769	24,9744	22,6718	20,3692	18,0666	15,7640
9	31,7643	29,4618	27,1592	24,8566	22,5540	20,2514	17,9488	15,6462

u	10^{-7}	10^{-6}	10^{-5}	10^{-4}	10^{-3}	10^{-2}	10^{-1}	1
1	15,5409	13,2383	10,9357	8,6332	6,3315	4,0379	1,8229	0,2194
2	14,8477	12,5451	10,2426	7,9402	5,6394	3,3547	1,2227	0,0489
3	14,4423	12,1397	9,8371	7,5348	5,2349	2,9591	0,9057	0,0131
4	14,1546	11,8520	9,5495	7,2472	4,9482	2,6813	0,7024	0,0038
5	13,9314	11,6289	9,3263	7,0242	4,7261	2,4679	0,5598	0,0011
6	13,7491	11,4465	9,1440	6,8420	4,5448	2,2953	0,4544	0,0004
7	13,5950	11,2924	8,9899	6,6879	4,3916	2,1508	0,3738	0,0001
8	13,4614	11,1589	8,8563	6,5545	4,2591	2,0269	0,3106	0,0000
9	13,3437	11,0411	8,7386	6,4368	4,1423	1,9187	0,2602	0,0000

131. Values of *a* for determining the headloss resulting from radial flow.
Ernst's equation
Taken from [49]

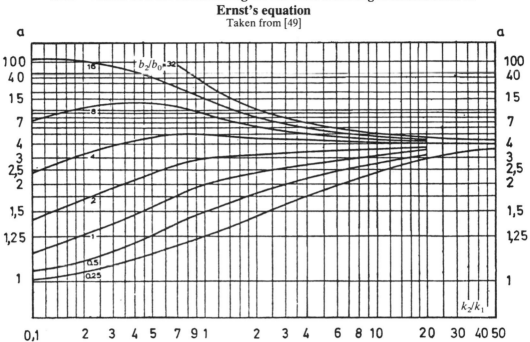

132. Flow measurements — Index

133. Manometric liquids†

Liquid	Specific weight γ (10^4 N/m^3)	Liquid	Specific weight γ (10^4 N/m^3)
Mercury	13.33	Water	0.981
Acteylene tetrabromate	2.92	Benzol	0.857
Bromoform	2.84	Toluene	0.850
Carbon tetrachloride	1.57	Kerosene	≈ 0.795
Carbon sulphide	1.24	Gasoline	≈ 0.726
Ethyl chloracetate	1.14	Ethyl alcohol	0.774

†The values given must be taken for chemically pure substance. In those cases in which great accuracy is required, a direct determination of γ should be carried out, taking account of its variation with the temperature.

134. Values of critical expansion, x_c, for a perfect gas
According to [16]

$$x_c = \frac{p_1 - p_2}{p_1}$$

K \ σ^2	0,00	0,05	0,10	0,15	0,20	0,25	0,30	0,35	0,40	0,45	0,50
1,25	0,445	0,439	0,432	0,425	0,419	0,411	0,403	0,395	0,385	0,375	0,365
1,30	454	448	441	434	427	420	412	403	393	383	372
1,35	463	457	450	443	436	428	420	411	401	391	379
1,40	472	465	458	452	444	437	428	419	409	399	387
1,45	480	473	466	460	452	445	436	427	417	406	395

135.   Values of α in ISA 1932 orifice meter
Compiled from [15] and [24]

$\sigma \rightarrow$	0,05	0,10	0,15	0,20	0,25	0,30	0,35	0,40	0,45	0,50	0,55	0,60	0,65	0,70
	High values of the Reynolds number; sharp-edged orifice; smooth pipes.													
$\alpha \rightarrow$	0,598	0,602	0,608	0,615	0,624	0,634	0,646	0,661	0,677	0,696	0,717	0,742	0,770	0,806
$D \downarrow$ (mm)	Increase of α in % owing to lack of very sharp edge.													
50	2,25	2,0	1,8	1,7	1,6	1,5	1,45	1,40	1,3	1,25	1,2	1,15	1,1	1,05
100	1,75	1,4	1,2	1,0	0,9	0,8	0,70	0,65	0,6	0,55	0,5	0,45	0,4	0,35
200	1,05	0,7	0,5	0,4	0,3	0,2	0,10	0,05	0,0	0,00	0,0	0,00	0,0	0,00
300	0,55	0,3	0,2	0,1	0,0	0,0	0,00	0,00	0,0	0,00	0,0	0,00	0,0	0,00
$D \downarrow$ (mm)	Increase of α in %, for rough pipes.													
50	0,20	0,3	0,45	0,6	0,75	0,85	1,00	1,15	1,30	1,45	1,60	1,75	1,9	2,05
100	0,15	0,2	0,30	0,4	0,50	0,60	0,70	0,80	0,90	1,05	1,15	1,25	1,4	1,50
200	0,10	0,1	0,15	0,2	0,25	0,30	0,35	0,40	0,45	0,50	0,50	0,55	0,60	0,65
300	0,00	0,0	0,00	0,0	0,00	0,00	0,00	0,00	0,00	0,00	0,00	0,00	0,00	0,00

136.   Values of α in ISA 1932 flow nozzles
Compiled from [15] and [24]

$\sigma \rightarrow$	0,05	0,10	0,15	0,20	0,25	0,30	0,35	0,40	0,45	0,50	0,55	0,60	0,65
	High values of the Reynolds number; smooth pipes.												
$\alpha \rightarrow$	0,987	0,989	0,993	0,999	1,006	1,016	1,028	1,041	1,059	1,081	1,108	1,142	1,183
$D \downarrow$ (mm)	Increase of α in %, for rough pipes.												
50	0	0	0	0	0	0	0,1	0,25	0,50	0,75	1,00	1,35	1,70
100	0	0	0	0	0	0	0,0	0,10	0,25	0,45	0,65	0,90	1,15
200	0	0	0	0	0	0	0,0	0,00	0,10	0,20	0,35	0,50	0,65

137.   Values of α in Venturi meters
(see Fig. 8.12)

$\sigma \rightarrow$	0,05	0,10	0,20	0,30	0,35	0,40	0,45	0,50	0,60
$\alpha \rightarrow$	0,986	0,989	1,001	1,020	1,032	1,048	1,067	1,092	1,155

138.   Measurement of discharges at elbows
According to [2]

Values of μ and K (see Section 8.17)

R/D	1,0	1,25	1,50	1,75	2,00	2,25	2,50	2,75	3,00
μ	1,23	1,10	1,07	1,05	1,04	1,03	1,03	1,02	1,02
K	0,570	0,697	0,794	0,880	0,954	1,02	1,08	1,14	1,20

139.   Values of β in ISA 1932 orifice meters and flow nozzles

(a) *Overheated steam* ($K = 1.31$)

(b) *Biatomic perfect gases* ($K = 1.41$)

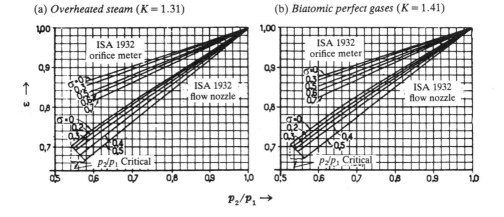

$p_2/p_1 \rightarrow$

(c) *Other gases* (K variable)

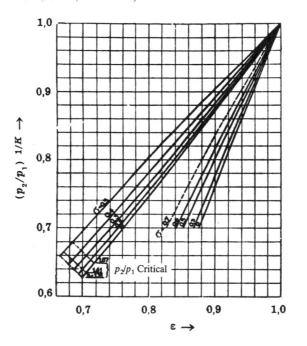

ISA 1932 orifice meters — right-hand band
ISA 1932 flow nozzles — left-hand band

Tables and graphs

140. Coefficients of discharge in circular, square, and rectangular vertical orifices with complete contraction.

h (m)	Circular — Diameter in mm						Square — Side of the square in mm						Rectangular (width 0.30 cm) — Depth of orifice in mm			
	10	15	30	60	120	240	10	15	30	60	120	240	75	150	300	600
0.12	—	0.631	0.618	—	—	—	—	0.637	0.621	—	—	—	0.619	—	—	—
0.15	0.642	627	615	0.600	—	—	0.645	633	619	0.605	0.601	—	618	0.615	—	—
0.30	628	617	608	600	0.596	0.593	633	622	613	605	603	0.600	616	611	0.605	—
0.60	619	610	604	599	598	596	624	615	608	605	605	602	614	609	604	0.609
0.90	615	606	603	599	599	597	620	612	607	605	605	603	612	608	603	607
1.20	612	605	602	599	598	597	617	610	606	605	605	603	610	607	603	606
1.50	611	605	601	598	598	596	616	610	606	604	604	602	609	605	602	605
3.00	605	601	598	597	597	596	610	606	604	603	603	602	604	602	601	602
6.00	600	598	596	596	596	595	605	603	602	602	601	601	604	602	601	602
15.00	596	595	594	594	594	593	601	601	600	600	600	599	607	605	602	606

Note: Head, h, is with reference to the centre of the orifice.

141.   Coefficients of discharge in orifices with partial contraction

(a) *Partial contraction*: not all the contour of the orifice is sharp-edged.

The following ratio then occurs:

$$\frac{\mu'}{\mu} = 1 + K\,\frac{l}{K}$$

in which:

μ' — Effective coefficient of discharge corresponding to partial contraction
μ — Coefficient of discharge that would correspond to complete contraction
K — Constant that takes on the following values:
    circular orifices — 0.128[†]; rectangular orifices — $(0.20 \times 0.10 - 0.157)$[‡]; square orifices
    — 0.152[†]; small rectangular orifices — 0.134[‡]
l — Contour of the orifice in a rounded wall
L — Total countour of the orifice.

(b) *Incomplete or partly suppressed contraction*: the walls of the recipient vessel are very close to the orifice.

Representing the cross-section of the recipient by A' and the cross-section of the orifice by A, the following values of μ'/μ are given:

	$\dfrac{A}{A'}$	0.1	0.2	0.3	0.4	0.5	0.6	0.7	0.8	0.9	1.0
μ'/μ	*Circular* orifices	1.014	1.034	1.059	1.092	1.134	1.189	1.260	1.351	1.471	1.631
μ'/μ	Rectangular orifices	1.019	1.042	1.071	1.107	1.152	1.208	1.278	1.365	1.473	1.608

(c) *Suppressed contraction on one side*: on this side the orifice is supported on a wall of the recipient.
The indications given in (a) are valid. Moreover, for rectangular orifices of width 0.75 m and depth 0.05 to 0.10 m, with heads varying between 0.30 and 0.75 m, the following mean values are given:[§]

— Complete contraction above and suppressed below and at the sides $\mu = 0.607$
— Upper edge rounded and contraction suppressed below and at the sides $\mu = 0.776$
        partly suppressed at another side (sharp-edged), 15 cm from wall of channel $\mu = 0.611$
— As above, with upper edge rounded $\mu = 0.755$
— Other particular cases of suppressed contraction are shown in Table 149.

[†]Bidone, quoted by [19].
[‡]Weisbach, quoted by [19].
[§]Williams, quoted by [17].

142. Coefficients of discharge in rectangular orifices, with different types of contraction.

According to [17]

Type of contraction	Dimensions of the orifice (mm)	Head (mm)		
		0,30	0,90	1,50
	Hor. × Vert.			
Complete contraction	200 × 200	0,598	0,604	0,603
	× 100	616	615	611
	× 50	631	627	620
	× 30	632	628	623
	× 10	652	634	620
Contraction suppressed at bottom[†]	200 × 200	0,620	0,624	0,625
	× 100	649	647	643
	× 50	671	668	666
	× 30	680	677	677
	× 10	710	705	696
Contraction suppressed at both vertical sides	200 × 200	0,632	0,628	0,628
	× 100	637	630	630
	× 50	641	634	635
	× 30	653	643	639
	× 10	682	667	655
Contraction suppressed at bottom and partly on one side[‡]	200 × 200	0,633	0,636	0,637
	× 100	658	656	654
	× 50	676	673	672
	× 30	682	683	681
	× 10	708	705	695
Contraction suppressed at bottom and partly on two sides	200 × 200	0,678	0,664	0,663
	× 100	680	675	672
	× 50	687	680	673
	× 30	693	688	683
	× 10	708	705	698
Contraction suppressed at bottom and on two sides	200 × 200	0,690	0,677	0,672
Contraction totally suppressed	200 × 200		0,950	

† *Contraction suppressed* means that the sides of the channel coincide with the edge of the orifice.

‡ *Contraction partly suppressed* means that the *distance* between the side of the channel and the edge of the orifice is 2 cm.

143. Coefficients of discharge for combined schemes.
According to Mises, quoted by [24]

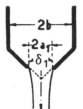

δ_1	$\dfrac{a_1}{b}$					
	≈ 0	0,1	0.3	0,5	0,7	0,9
45°	0,746	0,747	0,748	0,752	0,765	0,829
90°	0,611	0,612	0,622	0,644	0,687	0,781
125°	0,537	0,546	0,569	0,599	0,652	0,761
180°	0,500	0,513	0,544	0,586	0,646	0,760

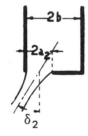

a_2/b	≈ 0	0,1	0,3	0,5	0,7	0,9
δ_2	21°	20° 55′	20° 5′	19°	16° 30′	11° 5′
μ	0,673	0,676	0,686	0,702	0,740	0,842

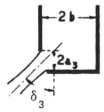

a_3/b	≈ 0	1	2	3	4	5
δ_3	69°	63° 50′	57° 5′	55°	53° 45′	53° 20′
μ	0,673	0,582	0,438	0,320	0,281	0,200

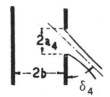

a_4/b	≈ 0	1	2	3	4	5
δ_4	90°	74° 13′	65° 10′	61° 25′	60° 24′	60° 4′
μ	0,611	0,544	0,420	0,319	0,247	0,200

144.   Coefficients of discharge. Vertical flat gates.[†]

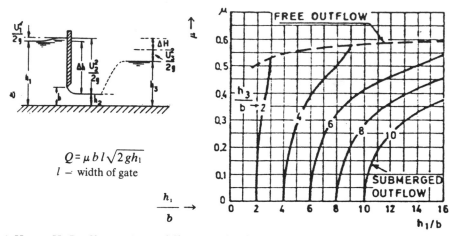

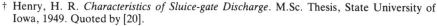

$$Q = \mu \, b \, l \sqrt{2 g h_1}$$
l – width of gate

$$\frac{h_1}{b} \rightarrow$$

† Henry, H. R. *Characteristics of Sluice-gate Discharge*. M.Sc. Thesis, State University of Iowa, 1949. Quoted by [20].

145.   Coefficients of discharge. Tainter gates.[†]

$$Q = \mu \, b \, l \sqrt{2 g h_1}$$

l – width of gate

$$a = h_3 + \frac{U_3^2}{2g} + \Delta H$$

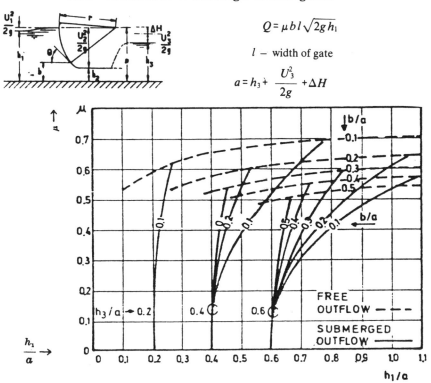

$$\frac{h_1}{a} \rightarrow$$

† Metzler, D. E., *A Model Study of Tainter Gate Operation*, M.Sc. Thesis, State University of Iowa, 1948. Quoted by [20].

146. Coefficients of discharge for sloping flat gates.
According to [11].

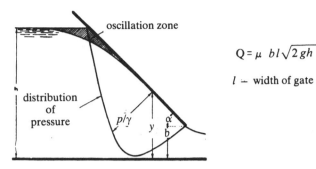

$$Q = \mu \; b l \sqrt{2 gh}$$

l — width of gate

(a) *Coefficient of discharge* μ

$\dfrac{h}{b}$	Slope α								
	15°	20°	30°	40°	50°	60°	70°	80°	90°
2	0,720	0,696	0,659	0,628	0,603	0,583	0,568	0,556	0,549
3	766	745	706	670	639	612	593	577	564
4	796	774	731	693	660	632	608	589	574
5	814	790	747	708	673	644	619	598	580
6	825	802	759	719	683	652	626	604	585
7			0,767	0,726	0,690	0,659	0,633	0,610	0,590
8			774	733	696	664	637	614	593
9			780	737	700	668	642	618	596
10			784	741	703	672	645	621	598

(b) *Distribution of pressures*: values of $\dfrac{p/\gamma}{b}$

$\dfrac{y}{b}$	Values of $\dfrac{h}{b}$							
	2,5	3,0	3,5	4,0	4,5	5,0	5,5	6,0
14,0	0,794	1,122	1,464	1,840	2,226	2,563	2,938	3,316
16,0	0,778	1,126	1,500	1,896	2,302	2,688	3,082	3,744
18,0	0,674	1,036	1,430	1,850	2,274	2,682	3,086	3,508
20,0	0,516	0,914	1,326	1,752	2,178	2,614	3,034	3,466
22,0	0,332	0,768	1,194	1,628	2,060	2,498	2,938	3,372
24,0	0,114	0,606	1,050	1,492	1,928	2,364	2,816	3,254
26,0		0,428	0,890	1,348	1,786	2,216	2,677	3,112
28,0		0,222	0,716	1,119	1,264	2,060	2,516	2,964
30,0			0,522	1,012	1,456	1,894	2,356	2,806
34,0			0,108	0,640	1,098	1,540	2,018	2,478
38,0				0,230	0,714	1,180	1,664	2,128
42,0					0,308	0,806	1,296	1,768
46,0						0,416	0,906	1,384
50,0							0,510	1,004
54,0							0,112	0,614
58,0								0,200

147. Coefficients of discharge in additional pipes.

(a) *Internal pipe* (Borda)   *External pipe*

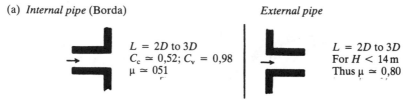

$L = 2D$ to $3D$
$C_c \simeq 0.52; C_v = 0.98$
$\mu \simeq 051$

$L = 2D$ to $3D$
For $H < 14\,m$
Thus $\mu \simeq 0.80$

(b) *Convergent pipes*

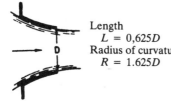

(a)   (b)

Angle θ	0°	5° 45′	11° 15′	22° 30′	45°
μ for Fig. (a)	0,97	0,94	0,92	0,85	
μ for Fig. (b)	0,97	0,95	0.92	0,88	0,75

Length
$L = 0.625D$
Radius of curvature
$R = 1.625D$

h (m)	0,02	0,50	3,5	16,7	100
μ'	0,959	0,967	0,975	0,994	0,994

(c) *Divergent pipes*

The maximum angle of aperture for which the flow remains attached to the pipe (and consequently the coefficients shown are valid) is θ = 8°. The maximum discharge is for $L \simeq 9D$ and $\theta \simeq 5°$.

$\mu \simeq 1.4$ $\mu \simeq 2.0$

(d) *Oblique pipes*

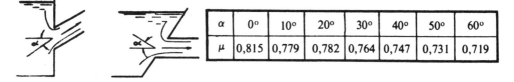

α	0°	10°	20°	30°	40°	50°	60°
μ	0,815	0,779	0,782	0,764	0,747	0,731	0,719

148. Coefficients of discharge for submerged pipes.

Experiments by Stewart, Rogers and Smith, quoted by [17]. Valid for pipes with square cross-section. Probably valid for cross-sections that are not very different.

L = Length of pipe.
P = Wetted perimeter of the transverse cross-section of the pipe.

Intake conditions	L/P							
	0,02	0,06	0,10	0,14	0,20	0,30	0,60	1,00
Sharp edges on whole perimeter	0,61	0,63	0,66	0,69	0,74	0,79	0,80	0,80
Contraction suppressed at bottom	0,63	0,65	0,67	0,69	0,73	0,77	0,80	0,81
Contraction suppressed at bottom and one side	0,68	0,69	0,69	0,71	0,74	0,79	0,81	0,82
Contraction suppressed at bottom and the two sides	0,77	0,76	0,73	0,72	0,73	0,83	0,84	0,85
Contraction suppressed at bottom and the two sides and above	0,95	0,94	0,93	0,92	0,92	0,91	0,90	0,90

149. Coefficients for determining the elevation of jets.

According to Cappa, quoted by [19].

(Formula (8.22))

Type	$h<30$ m				30 m$<h<150$ m			
	d	α	β	γ	d	α	β	γ
Orifice in thin wall	10	1,338	0,012233	0,000256	12,5	0,835	0,028167	0,000153
	15	1,137	0,007000	0,000430	15	1,107	0,015500	0,000180
	20	1,012	0,007900	0,000200	20	1,035	0,010300	0,000094
	25	1,016	0,009200	0,000030	25	0,939	0,012386	0,000009
	30	1,210	0,003550	0,000045	30	1,008	0,010213	0,000047
Conical additional pipe	10	1,112	0,016733	0,000058	12,5	1,136	0,013850	0,000022
	15	1,197	0,002650	0,000265	15	1,191	0,009950	0,000028
	20	1,036	0,009900	0,000010	20	1,219	0,002444	0,000056
	25	1,034	0,004350	0,000095	25	1,064	0,005000	0,000040
	30	1,036	0,003600	0,000090	30	1,088	0,002705	0,000060
Conoidal additional pipe	10	1,097	0,016533	0,000027	12,5	1,333	0,006383	0,000061
	15	1,197	0,002650	0,000265	15	1,175	0,010783	0,000018
	20	1,010	0,009700	0,000020	20	1,113	0,006012	0,000029
	25	1,021	0,005900	0,000010	25	1,089	0,002146	0,000059
	30	1,002	0,005100	0,000050	30	1,105	0,001994	0,000039

150. Coefficients of discharge in culverts.[†]

(a) Circular culverts

L – Length, D – Diameter

$L_{(m)}$ \ $D_{(m)}$	0,15	0,30	0,45	0,60	0,75	0,90	1,05	1,20	1,50	1,80	2,10	2,40
Culverts consisting of circular concrete pipes, with rounded intakes												
3	0,77	0,86	0,89	0,91	0,92	0,92	0,93	0,93	0,94	0,94	0,94	0,94
6	66	79	84	87	89	90	91	91	92	93	93	94
9	59	73	80	83	86	87	89	89	90	91	92	93
12	54	68	76	80	83	85	87	88	89	90	91	92
15	49	65	73	77	81	83	85	86	88	89	90	91
Culverts consisting of circular concrete pipes, with sharp-edged intakes												
3	0,74	0,80	0,81	0,80	0,80	0,79	0,78	0,77	0,76	0,75	0,74	0,73
6	64	74	77	78	78	77	76	75	74	73	72	72
9	58	69	73	75	76	76	76	75	74	74	73	72
12	53	65	70	73	74	74	74	74	74	73	72	71
15	49	62	68	71	72	73	73	73	73	72	72	71

(b) Rectangular culverts

L – Length, R – Hydraulic radius

$L_{(m)}$ \ $R_{(m)}$	0,06	0,09	0,12	0,15	0,18	0,24	0,30	0,36	0,42	0,48	0,54	0,60
Culverts consisting of rectangular concrete pipes with rounded intakes												
3	0,85	0,89	0,92	0,93	0,94	0,95	0,96	0,96	0,96	0,96	0,97	0,97
6	76	83	87	89	91	92	94	94	95	95	96	96
9	70	78	82	85	88	90	92	93	94	94	95	95
12	65	73	79	82	85	88	90	92	93	93	94	94
15	60	70	75	79	82	86	89	90	91	92	93	93
Culverts consisting of rectangular concrete pipes with sharp-edged intakes												
3	0,79	0,82	0,83	0,84	0,84	0,84	0,83	0,83	0,82	0,82	0,82	0,81
6	72	77	79	81	81	82	82	82	82	81	81	81
9	67	73	76	78	79	80	81	81	81	81	81	80
12	62	69	73	76	77	79	80	80	80	80	80	80
15	58	66	71	73	75	77	78	79	79	79	79	79

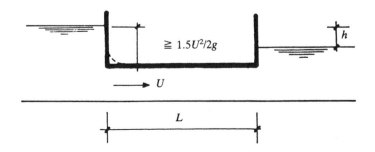

[†]   Yarnell, D. L., *Flow of water through culverts*, University of Iowa, Studies in Engineering, 1926. Quoted by [17].

151. Approximate measurement of discharges by means of the path of the jet.

Compiled from [38].

(a) *Completely full pipe (Purdue method)*
 Value of discharge (Formula (8.26))
 Error — 10 to 15%

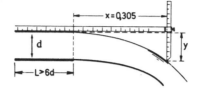

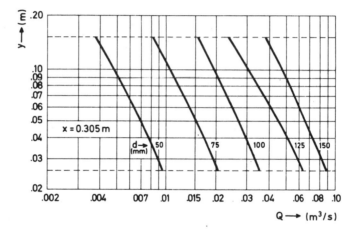

(b) *Partly full pipe (Californian method)*

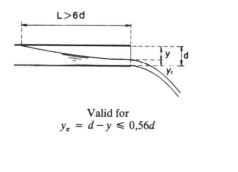

Valid for
$$y_e = d - y \leqslant 0{,}56d$$

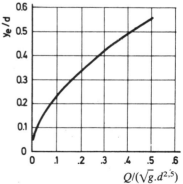

152. Rectangular sharp-crested weir without lateral contraction.

Quoted by [22].

(a) *values of the coefficient of discharge according to Rehbock.*

$$\mu = \frac{2}{3}\left(0.605|+\frac{1}{1050h-3}+0.08\,\frac{h}{a}\right)$$

$h_{(m)}$ \ $a_{(m)}$	0,1	0,2	0,3	0,4	0,6	0,8	1,0	2,0	3,0
0,02	0,451	0,448	0,446	0,445	0,444	0,444	0,443	0,443	0,443
04	443	432	428	427	425	424	423	422	422
06	447	430	426	423	420	419	418	417	416
08	455	433	426	423	422	417	416	414	413
10	464	437	428	424	419	417	416	413	412
0,12	0,472	0,441	0,430	0,425	0,419	0,417	0,415	0,412	0,411
14	483	445	433	427	420	417	416	413	411
16	493	450	436	429	422	418	416	412	410
18	504	456	440	431	423	419	417	412	410
20	513	460	442	433	425	420	416	412	410
0,22		0,466	0,447	0,436	0,426	0,421	0,417	0,412	0,410
24		470	449	439	427	422	419	413	410
26		475	452	439	429	423	420	413	410
28		481	456	443	431	425	421	414	411
30		486	459	446	432	426	422	414	411
0,32			0,462	0,448	0,434	0,427	0,422	0,414	0,411
34			466	451	435	428	423	414	411
36			469	453	437	429	424	415	412
38			473	456	439	430	425	415	412
40			476	458	441	432	426	416	412
0,45				0,464	0,445	0,435	0,429	0,417	0,413
50				471	449	438	431	418	414
55					454	442	434	419	414
60					458	444	436	420	415
65					462	447	439	421	415
0,70					0,466	0,451	0,442	0,423	0,417
75						454	444	424	417
80						457	448	425	418

(b) *Error committed when there is depression*

Head h (m)	0,061	0,091	0,122	0,183	0,245	0,305	0,61	0,91	1,22
1% error for depression of (m)	0,003	0,004	0,006	0,009	0,010	0,013	0,02	0,03	0,04
2% error for depression of (m)	0,006	0,009	0,011	0,015	0,020	0,023	0,04	0,05	0,07

153. Rectangular sharp-crested weir with side contractions.
Kindsvater and Carter's Formula (see Section 8.32). Quoted by [38].

$Q = \mu\sqrt{2g}l_e h_e^{3/2}$
$l_e = l + K_1$ — effective width
$h_e = h + K_h$ — effective head: $K_h \approx 1\,mm$
K_1 and K_h take account of the influence of surface tension and viscosity

$$\mu = \frac{2}{3}\,(\phi + \psi h/a)$$

1. *Values of K_1*

l/L	0	0,2	0,4	0,5	0,6	0,8	1,0
K_1 (mm)	2,4	2,4	2,7	3,0	3,7	4,3	−0,9

2. *Values of ϕ and ψ*

l/L	1,0	0,9	0,8	0,7	0,6	0,5	0,4	0,3	0,2	0,1	0
ϕ	0,602	0,599	0,597	0,595	0,593	0,592	0,591	0,590	0,589	0,588	0,587
ψ	0,0752	0,0640	0,0453	0,0297	0,0178	0,0130	0,0059	0,0018	0,0018	0,0021	0,0023

3. *Limits of application*: $l \geqslant 0.15m$; $h \geqslant 0.03m$; $h/a \leqslant 2$; $a \geqslant 0.10m$; the downstream water level must remain at least 0.05 m below the crest $L - l \geqq 6h$.
4. *Foreseeable error of coefficient* μ: less than 1%.
5. *Foreseeable error in K_b and K_h*: less than 0.3 mm.

154.   V — Notch weir. Triangular weir.

Kindsvater and Carters's Formula, quoted by [38]. (See Section 8.33).

$$Q = \mu' \frac{8}{15} \tan\frac{\alpha}{2} \sqrt{2g}\,(h + K_h)^{5/2}$$

K_h take into account the influence of surface tension and viscosity.

(1) *Values of* μ' *when contraction is complete:* $h/a < 0.4$; $a > 0.45\ m$; $h/L \leqslant 0.20$; $L....0.90\ m$ (error of 1%);

α (degrees)	20	30	40	50	60	70	80	90	100
μ′	0,592	0,587	0,583	0,579	0,578	0,578	0,578	0,579	0,582

(2) *Values of* μ' *when contraction is not complete* (*only for* $\alpha = 90°$) (*error of* 1 *to* 2%)†;

h/a	μ′					
	a/L = 0,1	a/L = 0,2	a/L = 0,4	a/L = 0,6	a/L = 0,8	a/L = 1,0
0,2	0,579	0,579	0,579	0,579	0,579	0,579
0,3	0,579	0,579	0,579	0,579	0,580	0,581
0,4	0,579	0,579	0,579	0,581	0,584	0,590
0,5	0,579	0,579	0,579	0,584	0,591	—
0,6	0,578	0,579	0,580	0,589	0,605	—
0,7	0,578	0,579	0,583	0,594	—	—
0,8	0,577	0,578	0,588	0,605	—	—
0,9	0,577	0,578	0,592	—	—	—
1,0	0,577	0,580	0,598	—	—	—
1,2	0,576	0,582	0,612	—	—	—

(3) *Values of* K_h (*mm*). Water at ordinary temperature.

α (degrees)	20	20	30	40	50	60	70	80
K_h (mm)	3,5	2,3	1,8	1,4	1,2	1,0	0,9	0,8

(4) *Limits of application*
    $0,05\,m \leqslant h \leqslant 0,60\,m$;
    $20° < \alpha < 100°$;
    $a \geqslant 0,1\,m$; $h/a \leqslant 1,2$; $h/L \leqslant 0,4$
    $L > 0,60\,m$.
    The water downstream must be below the vertex of the notch.
    $L - l \geqslant 1,51l$

† Compiled from BS 3680, Part 4A, BSI. 1965.

155. Circular sharp-crested weir.
Compiled from [19].

$$Q = \mu\phi d^{5/2}; \; \mu = 0,555 + \frac{d}{110h} + 0,041\frac{h}{d}$$

Values of ϕ according to $h/d - d$ in dm; Q in l/s.

$\dfrac{h}{d}$	0,05	0,10	0,20	0,30	0,40	0,50	0,60	0,70	0,80	0,90	1,00
ϕ	0,0272	0,1072	0,4173	0,9119	1,5713	2,3734	3,2939	4,3047	5,3718	6,4511	7,4705

Limits of application (see Fig. 8.25):

— the approach velocity must be low: $V^2/2g \approx 0$; $a > r$, with a minimum of 0.10 m; $b > r$; h/d ≥ 0.10; $h > 0.03$ m; the water downstream must be at least $0,05$ m below the crest.

156. Parabolic sharp-crested weir. Sutro weir.
Compiled from [38].

$$Q = \mu \, l \sqrt{2g\,\bar{b}}\,(h - \frac{b}{3})$$

(1) *Values of y/b and x/l, of equation 8.36;*

y/b	x/l	y/b	x/l	y/b	x/l
0,1	0,805	1,0	0,500	10	0,195
0,2	0,732	2,0	0,392	12	0,179
0,3	0,681	3,0	0,333	14	0,166
0,4	0,641	4,0	0,295	16	0,156
0,5	0,608	5,0	0,268	18	0,147
0,6	0,580	6,0	0,247	20	0,140
0,7	0,556	7,0	0,230	25	0,126
0,8	0,536	8,0	0,216	30	0,115
0,9	0,517	9,0	0,205		
1,0	0,500	10,0	0,195		

(2) *μ in a symmetrical Sutro profile. Error < 2%.*

b (m)	l (m)				
	0,15	0,23	0,30	0,38	0,46
0,006	0,614	0,619	0,623	0,6245	0,625
0,015	0,612	0,617	0,621	0,623	0,6235
0,030	0,609	0,614	0,618	0,6195	0,620
0,046	0,607	0,6115	0,616	0,6175	0,618
0,061	0,605	0,610	0,614	0,6155	0,616
0,076	0,604	0,6085	0,6125	0,614	0,6145
0,091	0,603	0,608	0,612	0,6135	0,614

(3) *μ in an unsymmetrical Sutro profile. Error < 2%.*

b (m)	l (m)				
	0,15	0,23	0,30	0,38	0,46
0,006	0,608	0,613	0,617	0,6185	0,619
0,015	0,606	0,611	0,615	0,617	0,6175
0,030	0,603	0,608	0,612	0,6135	0,614
0,046	0,601	0,6055	0,610	0,6115	0,612
0,061	0,599	0,604	0,608	0,6095	0,610
0,076	0,598	0,6025	0,6065	0,608	0,6085
0,091	0,597	0,602	0,606	0,6075	0,608

(4) *Limits of application*
$h \geq 2b$, with a minimum of $0,03$ m
$x \geq 0,005$ m; $b \geq 0,005$ m
$l > 0,15$ m
$l/a > 1$; $L/l > 3$

157. Comparison of errors in full-width rectangular and triangular sharp-crested weirs.

According to [17].

Discharge (l/s)	Error in measurement of the head (cm)	Recommended weir and error in the discharge (%)				
		Full width rectangular			Triangular with $\alpha = \pi/2$	
		Width (m)	Approx. head (cm)	% of error in discharge	Approx. head (cm)	% of error in discharge
1,4	0,03				6	1,2
	0,15	—	—	—		6,1
	0,30					12,2
2,8	0,03				8	0,9
	0,15	—	—	—		4,6
	0,30					9,1
14	0,03	0,30	8	0,5	16	0,5
	0,15			2,7		2,4
	0,30			5,5		4,8
	1,50			27,3		23,8
28	0,03	0,60	8	0,5	21	0,4
	0,15			2,7		1,8
	0,30			5,5		3,6
	1,50			27,3		18,0
70	0,03	0,60	16	0,3	30	0,3
	0,15			1,5		1,2
	0,30			3,0		2,5
	1,50	(1)		14,7		12,4
140	0,03	1,5	13	0,3	40	0,2
	0,15			1,7		0,9
	0,30			3,4		1,9
	1,50	(2)		17,0		9,3
280	0,03	1,5	22	0,2	53	0,1
	0,15			1,1		0,7
	0,30			2,1		1,5
	1,50	(3)		10,6		7,3
700	0,03	1,5	40	0,1		
	0,15			0,6	—	—
	0,30		—	1,1		
	1,50	(4)		5,6		

(1) It is also advisable to use a weir of 1,5 m width. The head will then be about 8 cm. The errors in the discharge are 0,5; 2,7; 5,5; 27,3%, respectively.

(2) It is also advisable to use a weir of 3,0 m width. The head will then be about 8 cm. The errors in the discharge are 0,5; 2,7; 5,5; 27,3%, respectively.

(3) It is also advisable to use a weir of 3,0 m width. The head will then be about 13 cm. The errors in the discharge are 0,5; 1,7; 3,4; 17%, respectively.

(4) It is also advisable to use a weir of 3,0 m width. The head will then be about 25 cm. The errors in the discharge are 0,2; 0.9; 1,8 and 9,1%, respectively.

158. Oblique sharp-crested weir in a rectangular cross-section channel.
Compiled from [36].

Aichel's formula (8.37) $Q = (1 - \beta h/a)Q_n$

(1) *Values of* β

(2) *Limits of application*

$\delta°$	β	$\delta°$	β	$\delta°$	β
15	0,690	35	0,29	60	0,11
20	0,526	40	0,24	65	0,08
25	0,420	45	0,20	70	0,06
30	0,357	50	0,16	75	0,04
		55	0,13	90	0

$\dfrac{h}{a} < 0,62$ and $\delta > 30°$;

or

$\dfrac{h}{a} < 0,46$ and $\delta < 30°$

Q_n is the discharge of a weir of the same width and $\delta = 90°$.

159. Lateral sharp-crested weir.[†]

Dominguez's formula (1945): $Q = \phi \mu l \sqrt{2g} h_0^{3/2}$

(1) *Values of* μ

Mean head (m) $(h_o + h_1)/2$	0,10	0,15	0,20	0,30	0,50	0,70
Thin-plate	0,370	0,360	0,355	0,350	0,350	0,350
Broad crest (rounded edge)	0,315	0,320	0,320	0,325	0,325	0,330
Broad crest (sharp-edge)	0,270	0,270	0,273	0,275	0,276	0,280

(2) *Values of* ϕ

h_1/h_0	ϕ	h_1/h_0	ϕ
0	0,400	0,50	0,659
0,05	0,417	0,60	0,722
0,10	0,443	0,70	0,784
0,20	0,491	0,80	0,856
0,30	0,543	0,90	0,924
0,40	0,598	1,00	1

† Quoted by [36].

160.   Broad-crested weirs.
According to [38]. Values of the coefficient of velocity, C_v
(equation (8.40), Fig. 8.30).

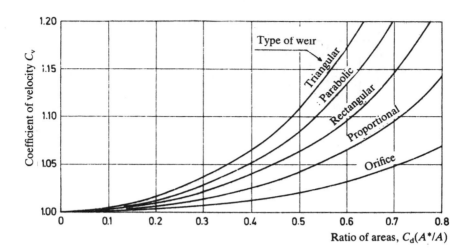

A — wetted area in measurement cross-section corresponding to a water depth of $(h + a)$.
A^* — wetted area in control cross-section, when in it there is a water depth of h.

161. Broad-crested rectangular profile weir without lateral contractions (BS 3680, 1969).

Compiled from [38].

$$Q = 1,7 \, C_d C_v l h^{1,5}$$

1. **Values of** C_v see the previous graph with $A^*/A = h/(h+a)$
2. **Values of** C_d
2.1 *Rectangular-profile weir*

$h/(h+a)$	h/b							
	0,10	0,30	0,50	0,70	0,90	1,10	1,30	1,50
≤ 0,35	0,848	0,848	0,860	0,897	0,940	0,980	1,020	1,050
0,40	0,857	0,857	0,869	0,907	—	—	—	—
0,45	0,868	0,868	0,879	0,916	—	—	—	—
0,50	0,875	0,875	0,889	0,924	—	—	—	—
0,55	0,885	0,885	0,899	0,937	—	—	—	—
0,60	0,898	0,898	0,910	0,950	—	—	—	—

For $h/b \leq 0.08$ $C_d = 0.848$

2.2 *Round-nosed horizontal-crested weir* (Fig. 8.31)
$$C_d = [1 - 2x(b-r)/l][1 - x(b-r)/h]^{3/2}$$
x — function of the roughness of the walls and crest for normal concrete (field works) — $x \simeq 0,005$; for very smooth concrete (laboratory works) — $x \simeq 0,003$.
3. **Errors envolved in** $\mu = C_d C_v$ (%)
3.1 *Rectangular-profile weir*

$h/(h+a)$	h/b							
	0,10	0,30	0,50	0,70	0,90	1,10	1,30	1,50
≤ 0,35	± 2.0	± 2.0	± 2.2	± 2.6	± 3.1	± 3.6	± 4.0	± 4.4
0,40	± 2.1	± 2.1	± 2.3	± 2.7	—	—	—	—
0,45	± 2.2	± 2.2	± 2.4	± 2.8	—	—	—	—
0,50	± 2.3	± 2.3	± 2.5	± 2.9	—	—	—	—
0,55	± 2.4	± 2.4	± 2.6	± 3.0	—	—	—	—
0,60	± 2.6	± 2.6	± 2.7	± 3.2	—	—	—	—

3.2 *Round-nosed horizontal-crested weir* (Fig. 8.31)
$$\varepsilon_\mu = \pm 2(2l - 20C_d)\%$$

4. **Recommended limits of application;**
$h \geq 0,06\,\text{m}; h \geq 5b; a \geq 0,15\,\text{m}; h/a \leq 1,5; 0,05 < h/b < 0,50; l \geq 0,30\,\text{m}; l \geq b/5;$
$l \geq h$ max; $r = 0,2h$ max; $r \geq 0,11h$ max
In order to maintain critical flow, and therefore the validity of the preceding formulae, the values of h'/h must satisfy the following conditions (modular limits).
4.1 *Rectangular-profile weir*
According to BS3680 $h'/h \leq 0.66$
4.2 *Round-nosed horizontal-crested weir* (Fig. 8.31)
Values of h'/h (modular limits)†

h/a	Vertical back face	Sloping 1/5 back face	h/a	Vertical back face	Sloping 1/5 back face
0,2	0,79	0,74	2,0	0,96	0,91
0,4	0,85	0,78	4,0	0,98	0,95
0,6	0,88	0,82	6,0	0,99	0,97
0,8	0,91	0,84	8,0	0,99	0,97
1,0	0,93	0,86	10,0	0,99	0,97

† According to Harrison, A. J. M., *The Streamline Broad-crested Weir*, Proceedings of the Institution of Civil Engineers, 1967. Quoted by [66].

162. Triangular broad-crested weir.
Various experiments synthesized by [38].

(1) *Values of Q (in S.I. units) and C_v*

For $H \leqslant 1.25h_t$ (triangular cross-section) $\qquad Q = 1,27 C_d C_v \tan\frac{\alpha}{2} h^{2,5}$

C_v is given by Graph 160, taking $A^*/A = h^2 \tan\frac{\alpha}{2} \Big/ [l(h + a)]$

For $h > 1,25h_t$ (complete cross-section) $Q = C_d C_v \left[\frac{2}{3}\right]^{0,5} l \left(\frac{h-1}{2}h_t\right)^{1,5}$

C_v is given by Graph 160, taking $A^*/A = \left(h - \frac{1}{2}h_t\right)/(h + a)$

(2) *Values of C_d (mean values)*

h/b	0,1	0,15	0,2	0,25	0,3	0,4	0,6	0,7
C_d	0,90	0,93	0,95	0,96	0,96	0,97	0,99	1,0

(3) *Drowned flow conditions.* Factor f of reduction of C_d, valid if $h/b \leqslant 0.57$ and $h \leqslant 1.25\,h_t$

h'/h	1,00	0,99	0,98	0,96	0,95	0,93	0,89	0,86	0,80
f	0	0,2	0,4	0,6	0,7	0,8	0,9	0,95	1,0

(4) *Limits of application*

$h > 0,05b$ with a minimum of $0,006$ m; $\alpha > 30°$; $a > 0,15$ m; $h/a \geqslant 3,0$
$h < 3,0$ m; $a > 0,15$ m; $h/b < 0,50$; $l > h_{max}$; $l > b/5$; $l > 0,30$ m

(5) *Error of discharge coefficient*: $\varepsilon_\mu = \pm 2(21 - 20C_d)\%$

163. Straight drop structures in a trapezoidal channel with rectangular control cross-section.
According to [38]. (See Section 8.44, Figs. 8.33, 8.34.)

$$Q = 0.385\, C_d C_v \sqrt{2g}\, lh^{1,5}$$

(1) *Values of C_d*

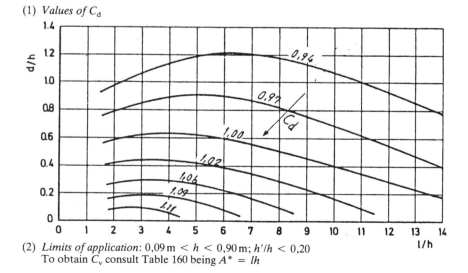

(2) *Limits of application*: $0,09$ m $< h < 0,90$ m; $h'/h < 0,20$
To obtain C_v consult Table 160 being $A^* = lh$

164. Short, broad-crested weirs with trapezoidal profile. Values of μ.

$$Q = \mu l \sqrt{2g} h^{3/2}$$

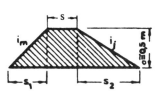

l — width of channel
i_m — upstream face slope (vert. to hor.)
i_j — downstream face slope (ver. to hor)
s — width of crest

(a) *Both faces sloping (low heads)*[†]

i_m	i_j	s (m)	Head h (m)										
			0,06	0,09	0,12	0,15	0,18	0,21	0,24	0,27	0,30	0,36	0,45
2/1	1/1	0,20	0,34	0,35	0,36	0,38	0,39	0,40	0,42	0,43	0,44	0,46	0,48
2/1	1/2	0,20	0,34	0,35	0,35	0,36	0,38	0,39	0,41	0,41	0,42	0,44	0,45
2/1	1/3	0,20	0,34	0,34	0,35	0,36	0,37	0,38	0,39	0,40	0,41	0,42	0,43
2/1	1/4	0,20	0,34	0,34	0,35	0,36	0,37	0,38	0,39	0,40	0,40	0,41	0,42
2/1	1/5	0,20	0,34	0,35	0,36	0,36	0,37	0,38	0,38	0,39	0,40	0,40	0,41
2/1	1/2	0,40	—	0,34	0,35	0,35	0,35	0,35	0,36	0,37	0,37	0,37	0,40
2/1	1/4	0,40	—	0,34	0,35	0,35	0,35	0,36	0,36	0,36	0,37	0,38	0,39
2/1	1/6	0,40	—	—	0,35	0,35	0,35	0,36	0,36	0,36	0,37	0,37	0,38
1/2	1/2	0,20	0,35	0,37	0,38	0,39	0,40	0,41	0,41	0,42	0,43	0,44	0,45
1/1	1/2	0,20	0,34	0,36	0,36	0,38	0,39	0,40	0,41	0,42	0,43	0,44	0,46
3/1	1/2	0,20	0,31	0,33	0,34	0,36	0,37	0,39	0,40	0,41	0,42	0,43	0,44
Vert.	1/2	0,20	0,32	0,32	0,33	0,35	0,36	0,37	0,39	0,40	0,41	0,42	0,44

(b) Both faces sloping (high heads)[‡]

i_m	i_j	s (m)	Head h (m)									
			0,48	0,54	0,60	0,75	0,90	1,05	1,20	1,35	1,50	1,65
1/2	1/2	0,20	0,45	0,44	0,44	0,45	0,45	0,45	0,45	0,46	0,46	0,46
1/2	1/5	0,10	0,45	0,44	0,44	0,43	0,43	0,43	0,43	0,44	0,45	0,45

(c) *Upstream face sloping, downstream face vertical*[‡]

i_m	s (m)	Head h (m)								
		0,30	0,45	0,60	0,75	0,90	1,05	1,20	1,35	1,50
1/2	0,10	0,48	0,48	0,47	0,47	0,47	0,47	0,46	0,46	0,45
1/2	0,20	0,43	0,45	0,46	0,46	0,46	0,47	0,47	0,47	0,47
1/3	0,20	—	—	0,45	0,45	0,45	0,45	0,45	0,45	0,45
1/4	0,20	—	—	0,43	0,43	0,43	0,43	0,43	0,43	0,43
1/5	0,20	—	—	0,42	0,42	0,42	0,42	0,42	0,42	0,42

† Bazin's experiments, quoted by [22].
‡ Experiments by U.S. Department Waterway Board, quoted by [17].

165. Short, broad-crested weirs with triangular profile. Values of μ.

$$Q = \mu l \sqrt{2g} h^{3/2}$$

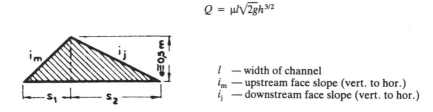

l — width of channel
i_m — upstream face slope (vert. to hor.)
i_j — downstream face slope (vert. to hor.)

The values shown for μ are only valid if the nappe remains adherent, otherwise the behaviour of the weir is close to that with a sharp crest.

(a) *Both faces sloping*[†]

i_m	i_j	Head h (m)										
		0,06	0,09	0,12	0,15	0,18	0,21	0,24	0,27	0,30	0,36	0,45
1/1	1/1	—	0,53	0,52	0,52	0,51	0,51	0,51	0,51	0,51	0,49	0,47
1/1	1/2	0,48	0,47	0,47	0,47	0,47	0,48	0,48	0,48	0,48	0,48	0,48
1/1	1/3	—	0,44	0,44	0,43	0,43	0,43	0,43	0,43	0,43	0,43	0,43
1/2	1/2	0,48	0,48	0,48	0,48	0,48	0,48	0,48	0,48	0,48	0,48	0,48
1/1	1/2	0,49	0,48	0,47	0,47	0,47	0,48	0,48	0,48	0,48	0,48	0,48
2/1	1/2	0,47	0,46	0,46	0,46	0,46	0,47	0,47	0,47	0,47	0,47	0,46
3/1	1/2	0,46	0,45	0,45	0,45	0,46	0,46	0,46	0,46	0,46	0,46	0,46
Vert.	1/2	0,44	0,43	0,43	0,44	0,44	0,45	0,45	0,45	0,45	0,45	0,47

(b) *Upstream face vertical*[†]

Slope of downstream weir face	Height of the weir (m)	Head h (m)										
		0,06	0,09	0,12	0,15	0,18	0,21	0,24	0,27	0,30	0,36	0,45
1/1	0,75	0,48	0,48	0,48	0,48	0,48	0,48	0,48	0,48	0,48	0,488	0,48
1/2	0,75	0,43	0,43	0,44	0,44	0,44	0,44	0,44	0,44	0,44	0,444	0,44
1/2	0,50	0,44	0,43	0,43	0,44	0,44	0,45	0,45	0,45	0,45	0,48	0,45
1/3	0,50	—	0,36	0,39	0,40	0,41	0,42	0,42	0,42	0,42	0,43	0,43
1/5	0,75	—	0,38	0,38	0,38	0,38	0,38	0,39	0,39	0,39	0,39	0,39
1/10	0,75	—	0,35	0,35	0,35	0,36	0,36	0,36	0,36	0,36	0,36	0,37

(c) *Upstream face vertical*[‡]
Mean values, valid for $h > 0,20$ m

i_j	μ	i_j	μ	i_j	μ
1/1	0,48	1/6	0,38	1/12	0,36
1/2	0,44	1/7	0,38	1/14	0,35
1/3	0,42	1/8	0,37	1/16	0,34
1/4	0,40	1/9	0,37	1/18	0,34
1/5	0,39	1/10	0,36	1/20	0,34

† Bazin's experiments, quoted by [17].
‡ Extrapolation by King, from Bazin's experiments [17].

166. Short, broad-crested weirs. Sundry short profiles.
According to [9].

Ratio μ/μ_o between the coefficient of discharge, μ, of the different shapes and the coefficient of discharge, μ_o, of the sharp-crested weir for equal head.

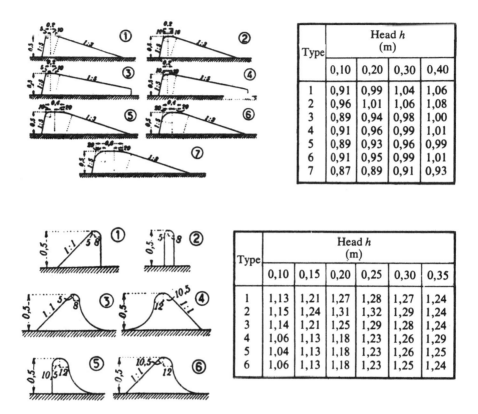

Type	Head h (m)			
	0,10	0,20	0,30	0,40
1	0,91	0,99	1,04	1,06
2	0,96	1,01	1,06	1,08
3	0,89	0,94	0,98	1,00
4	0,91	0,96	0,99	1,01
5	0,89	0,93	0,96	0,99
6	0,91	0,95	0,99	1,01
7	0,87	0,89	0,91	0,93

Type	Head h (m)					
	0,10	0,15	0,20	0,25	0,30	0,35
1	1,13	1,21	1,27	1,28	1,27	1,24
2	1,15	1,24	1,31	1,32	1,29	1,24
3	1,14	1,21	1,25	1,29	1,28	1,24
4	1,06	1,13	1,18	1,23	1,26	1,29
5	1,04	1,13	1,18	1,23	1,26	1,25
6	1,06	1,13	1,18	1,23	1,25	1,24

Example: One a Type 2 weir a flow exists, with an upstream depth over the crest of 0.30 m. The width of the crest is = 1.5 m.

Calculate the discharge Q.

μ/μ_o = 1,06. From Table 160(a), for h = 0.30 m and p = 0.5 m, we have μ_o = 0.439; thus μ = 0.465. The discharge is $Q = \mu l \sqrt{2g} h^{3/2}$ = 0.465 × 1.5 × 4.429 × 0.30$^{3/2}$ = 0.508 m^3/s.

167.   Short, broad-crested weirs with triangular profile (Crump weir).
According to [38].

$$Q = 2{,}95C_dC_vlh_e^{3/2} \text{ (SI units)}$$

l being the width of the channel

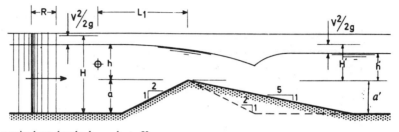

h_e, equivalent depth: $h_e = h + K$
$K = 0{,}0003$ m for a 5/1 slope of the downstream face
$K = 0{,}00025$ for a 2/1 slope of the downstream face.
The head h must be measured at distances $L_1 = 6a$ and $L = 4a$ from the weirs with downstream faces at 5/1 and 2/1, respectively.

(1) *Values of C_v* — Graph 160 with $A^*/A = \left(\dfrac{h}{h+a}\right)$

(2) *Values of C_d in non-submerged conditions (modular limits)*[†]

H/a'	Values of C_d	
	1:2/1:2 profile	1:2/1:5 profile
0 to 0,9	0,723	0,674
1,00	0,722	0,674
1,25	0,719	0,674
1,50	0,710	0,674
1,75	0,698	0,674
2,00	0,678	0,674
2,50	—	0,673
3,00	—	0,670

(3) *Modular limits. Maximum values of H'/H*
   (a) For 1:2/1:2 profile weir

H/a'	0,2	0,4	0,6	0,8	1,0	1,5	2,0
H'/H	0.2	0.3	0.4	0.5	0.6	0.7	0.89

   (b) For 1:2/1:5 profile weir
      $H'/H < 0.75$ for $0.5 < H/a < 1.6$.

(4) *Error committed in $\mu C_d C_v$*
   $\varepsilon_\mu = \pm(10\,C_v - 9)\%$

(5) *Limits of application*
   $h \geqslant 0.03$ m for a metal crest;
   $h \geqslant 0.06$ for a concrete or similar crest;
   $h/a \leqslant 3.0$; $a \geqslant 0.06$ m; $l \geqslant 0.30$ m; $l/H \leqslant 2.0$.

† Compiled from [69].

168. Parshall flumes. Geometrical definition.
According to [38].

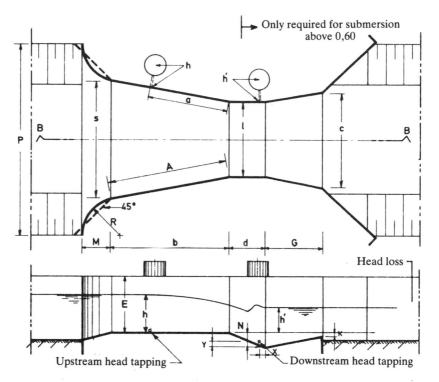

Only required for submersion above 0,60

Head loss

Upstream head tapping

Downstream head tapping

Dimensions (mm)

"	l mm	A	a	b	c	S	E	d	G	K	M	N	P	R	X	Y
1"	25,4	363	242	356	93	167	229	76	203	19	—	29	--	—	8	13
2"	50,8	414	276	406	135	214	254	114	254	22	—	43	—	—	16	25
3"	76,2	467	311	457	178	259	457	152	305	25	—	57	—	—	25	38
6"	152,4	621	414	610	394	397	610	305	610	76	305	114	902	406	51	76
9"	228,6	879	587	864	381	575	762	305	457	76	305	114	1080	406	51	76
1'	304,8	1372	914	1343	610	845	914	610	914	76	381	229	1492	508	51	76
1'6"	457,2	1448	965	1419	762	1026	914	610	914	76	381	229	1676	508	51	76
2'	609,6	1524	1016	1495	914	1206	914	610	914	76	381	229	1854	508	51	76
3'	914,4	1676	1118	1645	1219	1572	914	610	914	76	381	229	2222	508	51	76
4'	1219,2	1829	1219	1794	1524	1937	914	610	914	76	457	229	2711	610	51	76
5'	1524,0	1981	1321	1943	1829	2302	914	610	914	76	457	229	3080	610	51	76
6'	1828,8	2134	1422	2092	2134	2667	914	610	914	76	457	229	3442	610	51	76
7'	2133,6	2286	1524	2242	2438	3032	914	610	914	76	457	229	3810	610	51	76
8'	2438,4	2438	1626	2391	2743	3397	914	610	914	76	457	229	4172	610	51	76
10'	3048	—	1829	4267	3658	4756	1219	914	1829	152	—	343	—	—	305	229
12'	3658	—	2032	4877	4470	5607	1524	914	2438	152	—	343	—	—	305	229
15'	4572	—	2337	7620	5588	7620	1829	1219	3048	229	—	457	—	—	305	229
20'	6096	—	2845	7620	7315	9144	2134	1829	3658	305	—	686	—	—	305	229
25'	7620	—	3353	7620	8941	10668	2134	1829	3962	305	—	686	—	—	305	229
30'	9144	—	3861	7925	10566	12313	2134	1829	4267	305	—	686	—	—	305	229
40'	12192	—	4877	8230	13818	15481	2134	1829	4877	305	—	686	—	—	305	229
50'	15240	—	5893	8230	17272	18529	2134	1829	6096	305	—	686	—	—	305	229

169. Parshall flumes. Discharge characteristics
According to [38]

Discharge — $Q = Kh^u$
K and u — given in table below
h — upstream head

b		Limit of Q (m³/s)		Constants in the formula $Q = Kh^u$		Limits of h (m)		Limits of h'/h
"	(mm)	Min	Max	K	u	Min	Max	
1"	25,4	$0,09 \times 10^{-3}$	$5,4 \times 10^{-3}$	0,0604	1,55	0,015	0,21	0,50
2"	50,8	0,18	13,2	0,1207	1,55	0,015	0,24	0,50
3"	76,2	0,77	32,1	0,1771	1,55	0,03	0,33	0,50
6"	152,4	1,50	111	0,3812	1,58	0,03	0,45	0,60
9"	228,6	2,50	251×10^{-3}	0,5354	1,53	0,03	0,61	0,60
1'	304,8	3,324	0,457	0,6909	1,522	0,03	0,76	0,70
1'6"	457,2	4,80	0,695	1,056	1,538	0,03	0,76	0,70
2'	609,6	12,1	0,937	1,428	1,550	0,046	0,76	0,70
3'	914,4	17,6	1,427	2,184	1,566	0,046	0,76	0,70
4'	1219,2	35,8	1,923	2,953	1,578	0,06	0,76	0,70
5'	1524,0	44,1	2,424	3,732	1,587	0,06	0,76	0,70
6'	1828,8	74,1	2,929	4,519	1,595	0,076	0,76	0,70
7'	2133,6	85,8	3,438	5,312	1,601	0,076	0,76	0,70
8'	2438,4	$97,2 \times 10^{-3}$	3,949	6,112	1,607	0,076	0,76	0,70
10'	3048	0,16	8,28	7,463	1,60	0,09	1,07	0,80
12'	3658	0,19	14,68	8,859	1,60	0,09	1,37	0,80
15'	4572	0,23	25,04	10,96	1,60	0,09	1,67	0,80
20'	6096	0,31	37,97	14,45	1,60	0,09	1,83	0,80
25'	7620	0,38	47,14	17,94	1,60	0,09	1,83	0,80
30'	9144	0,46	56,33	21,44	1,60	0,09	1,83	0,80
40'	12192	0,60	74,70	28,43	1,60	0,09	1,83	0,80
50'	15240	0,75	93,04	35,41	1,60	0,09	1,83	0,80

170.   Parshall flumes. Submerged flow

According to [46]

Q — Table 178: $Q' = Q - \triangle Q$: $\triangle Q$ — given below

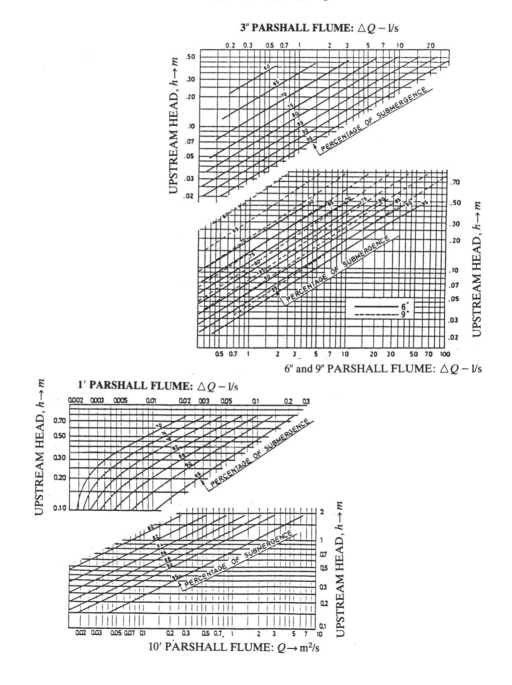

3″ PARSHALL FLUME: $\triangle Q - l/s$

6″ and 9″ PARSHALL FLUME: $\triangle Q - l/s$

1′ PARSHALL FLUME: $\triangle Q - l/s$

10′ PARSHALL FLUME: $Q \rightarrow m^2/s$

171. Parshall flues. Head losses in metres

According to [38]

η — degree of submersion = h''/h

Parshall	1'		2'		3'	
$\eta(\%)$ / Q m³/s	60	95	60	95	60	95
0.02	0.04	0.006	0.03	0.004	0.022	0.003
0.05	0.08	0.012	0.05	0.008	0.04	0.006
0.10	0.15	0.02	0.09	0.012	0.065	0.01
0.20	0.24	0.03	0.15	0.02	0.12	0.015
0.30	0.28	0.04	0.18	0.025	0.15	0.02

Parshall	4'		6'		8'	
$\eta(\%)$ / Q m³/s	60	95	60	95	60	95
0.30	0.15	0.016	0.10	0.012	0.075	0.009
0.60	0.23	0.025	0.18	0.018	0.14	0.015
1.00	0.30	0.037	0.23	0.025	0.18	0.02
1.50	—	0.05	0.30	0.035	0.23	0.026
3.00	—	0.08	—	0.058	—	0.044

Parshall	10'		12'		15'	
$\eta(\%)$ / Q m³/s	80	95	80	95	80	95
1.00	0.08	0.03	0.075	0.028	0.06	0.023
2.00	0.15	0.05	0.13	0.045	0.10	0.038
4.00	0.22	0.08	0.19	0.07	0.16	0.06
6.00	0.28	0.10	0.26	0.09	0.22	0.08
8.00	—	0.13	0.30	0.11	0.26	0.09

Parshall	20'		30'		40'	
$\eta(\%)$ / Q m³/s	80	95	80	95	80	95
4.00	0.14	0.046	0.09	0.033	0.068	0.025
8.00	0.20	0.08	0.15	0.054	0.10	0.04
10.00	0.23	0.09	0.17	0.06	0.13	0.045
20.00	0.40	0.15	0.27	0.10	0.20	0.08
40.00	0.60	0.23	0.45	0.16	0.31	0.12

172. Water hammer and protection of pipelines — Index

173.   Values of the elastic modulus, E

Pipe material	$E(10^{10}\ \mathrm{N/m^2})$
Steel	20.0–22.0
Aluminium	6.8–7.0
Concrete	1.4–3.0
Prestressed concrete	4.8
Copper	11.0–13.4
Cast iron	8.0–17.0
Asbestos cement	2.4–3.0
Plastics	
Rigid PVC	2.0–3.0
Polyethylene	0.1–0.2
Nylon	1.0–2.0
Polyester	1.8–2.5
Plexiglass	0.5
Perspex	0.6
Glass	4.6–7.3

174.   Velocity of propagation of an elastic wave. Influence of the shape of cross-section in steel pipes.[†]

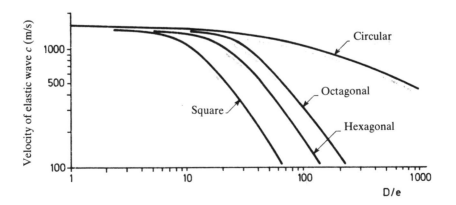

† According to Thorley A. R. D. *et al*. [51].

175.   Closing of valves — linear motion.[†]

(Null head losses in the pipe)

θ — Total valve closing time Δh_M — maximum suppresion

$t = L/c$ $\Delta h_j = cU_o/g$

(a) *Globe valve*

$\dfrac{\sigma}{\sigma_o} = \dfrac{Z}{Z_o}$ (linear motion)

$A^* = \dfrac{cU_o}{gh_o}$

(b) *Gate valve*

$$\dfrac{\sigma}{\sigma_o} = 1 - \dfrac{2}{\pi}\left[\arccos\left(\dfrac{Z}{Z_o}\right) - \dfrac{Z}{Z_o}\sqrt{1-\left(\dfrac{Z}{Z_o}\right)^2}\right]$$

$Z/Z_o = (1 - t/T)$

(c) *Rectangular sluice valve*

$$\dfrac{\sigma}{\sigma_o} = 1 - \dfrac{1}{\pi}\left[\arccos\left(2\dfrac{Z}{Z_o} - 1\right) - 2\left(\dfrac{Z}{Z_o} - 1\right)\sqrt{1-\left(2\dfrac{Z}{Z_o} - 1\right)^2}\right]$$

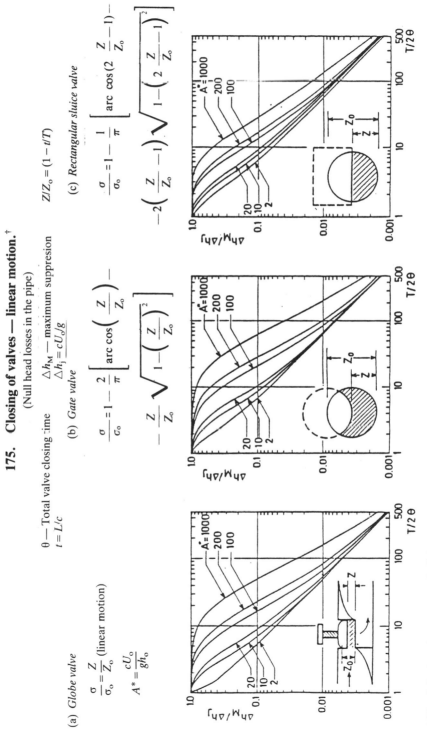

† Wood, D. J. and Jones, S. E. [52] (Kindly authorized by ASCE)

175 (Contd.)   Closing of valves — linear motion

(Null head losses in the pipe)

$$A^* = \frac{cU_o}{gh_o}$$

$$Z/Z_o = (1 - t/T)$$

(d) *Needle valve*

$$\frac{\sigma}{\sigma_o} = 2\frac{Z}{Z_o}\left(\frac{Z}{Z_o}\right)^2$$

(e) *Butterfly valve*

$$\frac{\sigma}{\sigma_o} = 1 - \cos\left(\frac{\pi}{2} - \alpha\right)$$

(f) *Ball valve*

$$\frac{\sigma}{\sigma_o} = \frac{1+\cos\alpha}{2} + \frac{\cos\alpha}{\pi}\left\{\arcsin(-x)+\right.$$

$$0.5 \, \mathrm{sen}\left[2\arcsin(-x)\right]\right\} \frac{1}{\pi}\left\{\arcsin(x)+\right.$$

$$\left.0.5 \, \mathrm{sen}\left[2\arcsin x\right]\right\} \, ; \, x = \frac{\sin\alpha\sqrt{\left(\frac{B}{b}\right)^2}-1}{1+\cos\alpha}$$

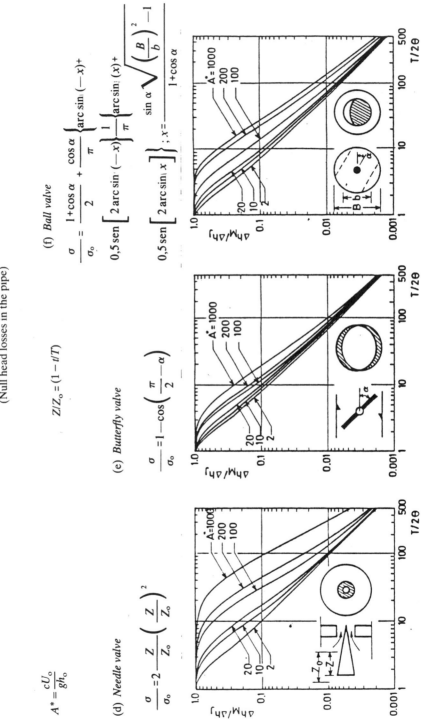

176. Closing of valves — accelerated motion. Gate valve.

(Null head losses in the pipe).[†]

$$A^* = \frac{cU_o}{gh_o} \qquad\qquad z/z_o = 1 - \left(\frac{t}{T}\right)^2$$

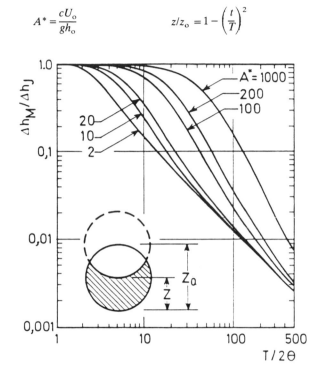

† Same reference and authorization as for previous graph.

177.   Minimum and maximum pressure in a pipeline after sudden stoppage of the pumps

(Taking account of head losses in the pipeline).[†]

(a) *Maximum pressure reduction, $\triangle h_m$, below the static level*

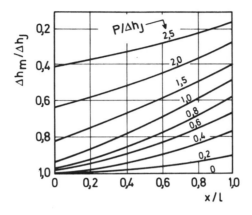

(b) *Maximum overpressure, $\triangle h_M$, above the static level*

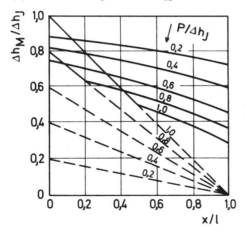

P — Head loss in the pipeline;
– – – Initial piezometric line in steady flow
—— Limit of maximum pressures

$$\triangle h_j = \frac{cU_o}{g}$$

† Stephenson, D. [58]. (Kindly authorized by the author and publisher.)

178. Transmission of an elastic wave downstream of a shaft without throttled orifice at the base

According to [54] (Kindly authorized by the author and publisher.)

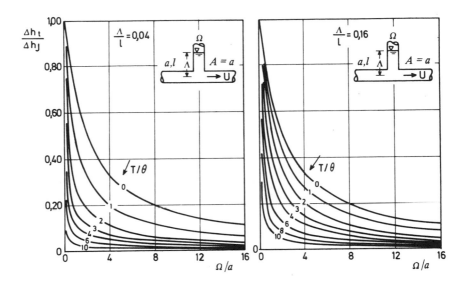

$$\triangle h_J = \frac{cu_o}{g} \qquad \theta = l/c$$

T — Closing time of the valve or pump.

a,l — Area and length of the pipeline directly connected to the valve or pump (pipeline not protected).

c — Velocity of the elastic waves in the unprotected pipeline.

u_o — Initial velocity of flow in the unprotected pipeline.

$\triangle h_t$ — Part of wave transmitted to pipeline protected by the surge shaft.

179. Transmission of an elastic wave downstream of a shaft with throttled orifice at the base.

According to [54]. (Kindly authorized by the author and publisher.)

(a) $\Omega = a$ and $A = a$

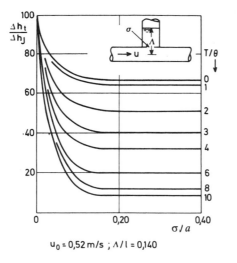

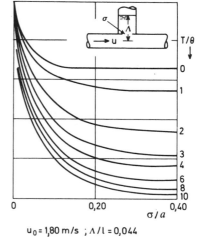

(b) $\Omega = 16a$ and $A = a$

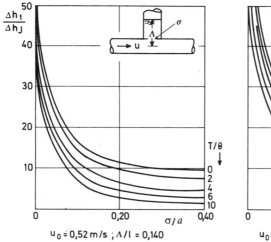

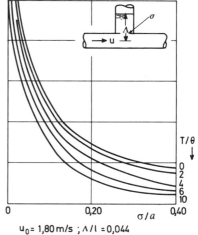

180. Influence of shaft length when $a = A = \Omega$

According to [54]. (Kindly authorized by the author and publisher.)

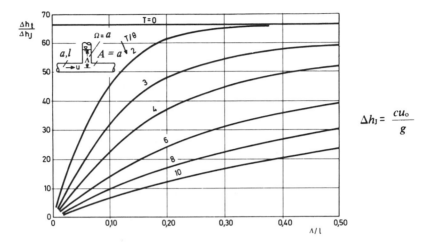

$$\Delta h_J = \frac{cu_o}{g}$$

181. Surge shaft with constant cross-section without throttled orifice at the base[†]

A surge shaft located upstream of the pipeline is considered (which is the most frequent situation in pumping systems). In the case of a shaft located downstream of the pipe (which is the most frequent situation in power systems) the drawdowns will be rises and vice versa.

(a) *Instantaneous total stoppage.* Values of the first drawdown, Z_m, first rise and second drawdown.

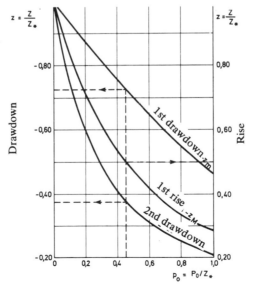

† Compiled from Calamé and Gaden. Quoted by [55].

181 (Contd.)   Surge shaft with constant cross-section without throttled orifice at the base

(b) *Instantaneous total start-up*. Values of the first drawdown and first rise.

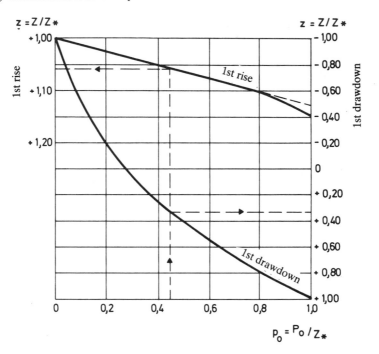

$$p_o = P_o / Z_*$$

(c) *Partial instantaneous start/up from n% to 100%*. Values of maximum rise: $Z_M = z_M \cdot Z_X$

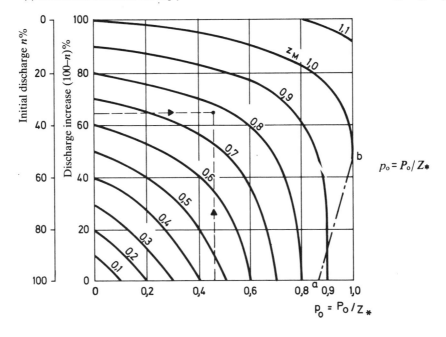

$$p_o = P_o / Z_*$$

181 (Contd.)   Surge shaft with constant cross-section without throttled orifice at the base

(d) *Linear start-up in the time T.* Values of maximum rise

$$Z_M = z_M Z_*$$

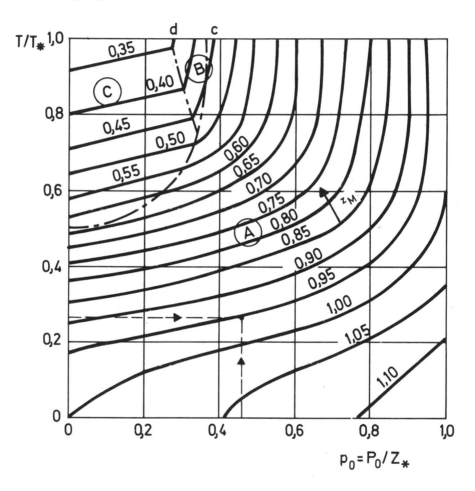

Zone A — The maximum occurs after the end of start-up.
Zone B — There are two maxima, one before and the other after the end of the start-up, the latter being greater.
Zone C — The maximum occurs during the start-up; after the end of the start-up another peak can occur but with a lesser value.

182.   Surge shaft with constant cross-section with throttled orifice at the base[†]

A surge shaft located upstream of the pipeline is considered (which is the most frequent situation in pumping systems). In the case of a shaft located downstream of the pipeline to be protected the drawdowns will be rises and vice versa.

(a) *Instantaneous total stoppage*

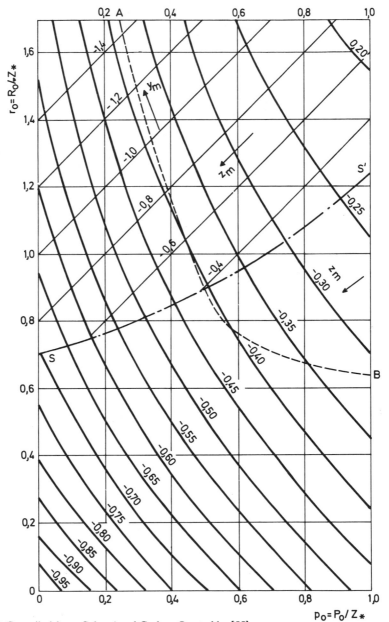

† Compiled from Calamé and Gaden. Quoted by [55].

182 (Contd.)   **Surge shaft with constant cross-section with throttled orifice at the base**

(b) *Instantaneous total start-up*

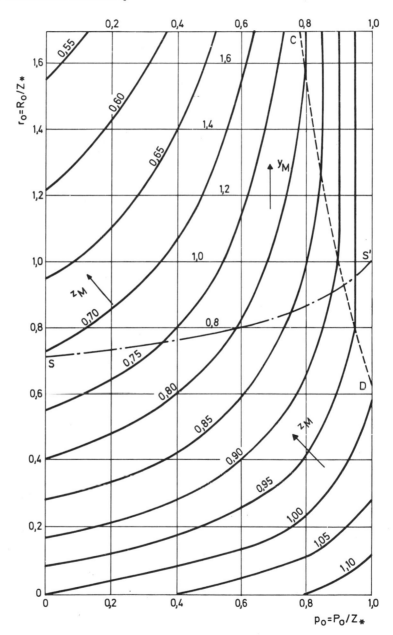

183. Influence of inertia. Maximum and minimum pressures near the pump after the power failure

According to Kinno and Kennedy [59] quoted by Stephenson [58]. (Kindly authorized by the author and publisher.)

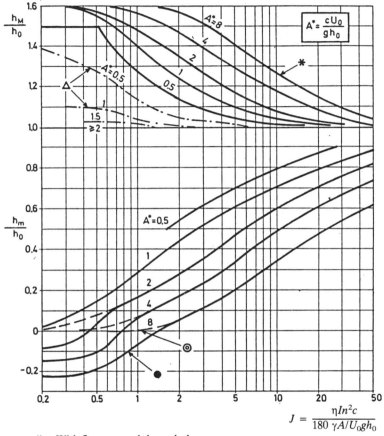

$$J = \frac{\eta l n^2 c}{180 \; \gamma A/U_0 g h_0}$$

∗ - With flow reversal through the pump
△ - With flow reversal but reverse rotation of pump prevented
◎ - At time of flow reversal
● - At time $2\,l/t$

184. Air vessels

According to Dubin and Guéneau [60]

$$A^* = \frac{cU_o}{g/h_o^*}\;; \qquad B^* = \frac{U_o^2}{g/h_o^*}$$

(a) *Minimum pressures in the pipeline*

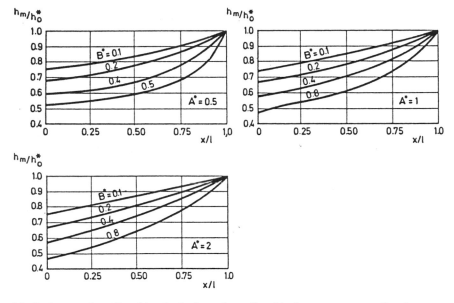

(b) *Optimum value of head loss in the flap valve orifice. Maximum pressure at* $x/L = 1$

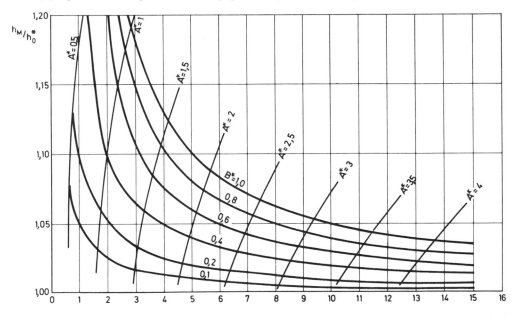

Head loss on the throttled orifice $\alpha w^2/h_o^*$

185. Cavitation velocities in sharp-edged orifices

According to [56]

U_i — incipient cavitation velocity, $U_i = CU_i^*$
U_c — critical cavitation velocity, $U_c = CU_c^*$
U_{ch} — choking cavitation velocity, $U_{ch} = CU_{ch}^*$

Head loss $\triangle h = KU^2/2g$

(a) U_i^*, U_c^* and U_{ch}^* values

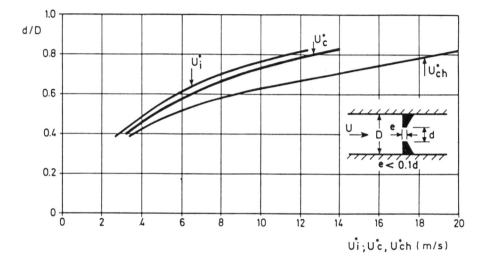

(b) *K and C values*

d^2/D^2	K	C values					
		$d = 0.05$ m	$d = 0.1$ m	$d = 0.2$ m	$d = 0.4$ m	$d = 0.6$ m	$d = 1.0$ m
0.86	0.2	1.15	0.91	0.80	0.70	0.61	0.55
0.77	0.5	1.14	0.92	0.80	0.71	0.63	0.57
0.69	1.0	1.13	0.93	0.82	0.73	0.65	0.60
0.60	2.0	1.12	0.93	0.83	0.75	0.69	0.62
0.47	5.0	1.11	0.95	0.85	0.79	0.73	0.66
0.38	10.0	1.10	0.95	0.86	0.81	0.76	0.70
0.29	20.0	1.09	0.96	0.88	0.83	0.80	0.75
0.20	50.0	1.08	0.97	0.90	0.83	0.83	0.80
0.16	100.0	1.07	—	0.92	0.90	—	0.83
0.12	200.0	1.06	—	0.93	0.90	—	0.84
0.10	500.0	1.05	—	0.95	0.91	—	0.85
0.08	1000.0	1.03	—	0.95	0.92	—	0.87

186. Feed tanks
According to [58]. (Kindly authorized by the author and publisher.)

$$\alpha = \frac{\forall gh}{AlU_o^2}$$

(a) One feed tank with a check valve in the pipe

Maximum overpressure: $\triangle h_M$ *Minimum volume required*: $\forall$

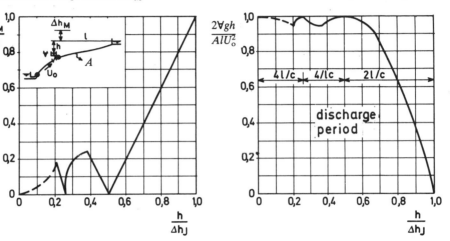

(b) One feed tank without a check valve in the pipe

Maximum overpressure: $\triangle h_M$ *Minimum volume required*: $\forall$

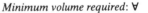
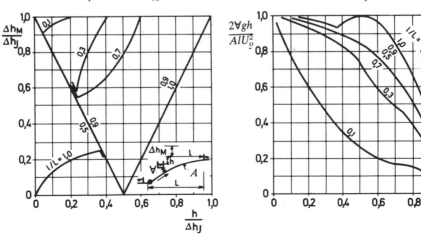

186 (Contd.) Feed tanks

(c) Two feed tanks with check valve upstream of each tank

Maximum overpressure: $\triangle h_M$

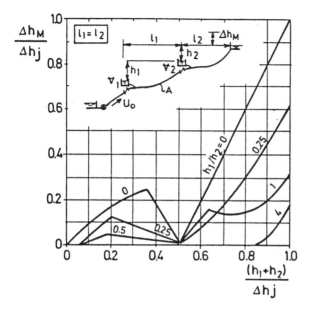

Minimum volume requires: $\forall$

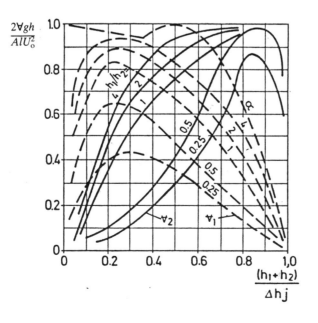

Bibliographical references to the tables and graphs

[1] Accvcdo, M. L., *Semejanza Mecanica y Experimentacion con Modelos de Buques*, Canal de Experiencias Hidrodinamicas, Madrid, 1943.

[2] Addison, H., *Hydraulic Measurements*, Chapman & Hall, London, 1949.

[3] Alves Costa, M., *Formulário de Hidráulica Geral*, Lisbon 1961.

[4] ASCE, *Hydraulic Models*, New York, 1961.

[5] Bakhmeteff, B. A., *Hydraulics of Open Channel Flow*, McGraw-Hill, New York, 1932.

[6] Bas, L., *Agenda del Quimico*, Aguilar, Editor, Madrid.

[7] Bradley, J. N. and Thompson, L. R., *Friction Factors for Large Conduits Flowing Full*, U.S. Department of the Interior-Bureau of Reclamation, Engineering Monograph, no. 7, March, 1951.

[8] Crausse, E., *Hydraulique des Canaux Découverts*, Editions Eyrolles, Paris, 1951.

[9] Forcheimer, P., *Tratado de Hidráulica*, Editorial Labor, Buenos Aires, 1939.

[10] Fosdick, E. R., *Diversion Losses in Pipe Bends*. Monthly Bulletin of the State College of Washington. Vol. 20, No. 7, December, 1937.

[11] Gentilini, B., *Effuso dalle Luci Soggiacenti ale Paratoie Piane Inclinate e a Settore*, Memorie e Studi dell'Istituto di Idraulica e Constrzioni Idrauliche, Milano, 1941.

[12] US Hydraulic Institute, *Standards of the Hydraulic Institute*, New York, 1951.

[13] *National Research Council International Critical Tables*, Vol. III, IV, V, McGraw-Hill, New York, 1928–29.

[14] Vanomi, A., *Sedimentation Engineering*, ASCE, New York, 1977.

[15] ISA, *Rules for Measuring the Flow of Fluids by Means of Nozzles and Orifice Plates*, Bulletin 9.

[16] ISO, *Proposition de Rédaction d'une Norme International de Mesure*

de Débit des Fluides au Moyen de Diaphragmes, Tuỳeres ou Tubes de Venturi, 1954.

[17] King, H. W., *Handbook of Hydraulics,* McGraw-Hill, New York, 1939.

[18] Lane, E. W., *Design of Stable Channels.* Trans. A.S.C.E., Vol. 120, 1955.

[19] Puppini, U., *Idraulica,* Nicola Zanichelli, Editore, Bologna, 1947.

[20] Rouse, H., *Enginering Hydraulics,* Chapman & Hall, London, 1950.

[21] Rouse, H. & Hassen, M. M., *Cavitation free Inlets and Contractions,* Mechanical Engineering, Vol. 71, 3, 1949.

[22] Scimemi, E., *Compendio di Idráulica,* Liberia Universitaria de G. Randi, Padova, 1952.

[23] Roark, R. J., *Fórmulas de Resistencia de Materiales, Esfuerzos y Deformaciones,* Aguilar, S. S. de Ediciones, Madrid, 1952.

[24] Schlag, A., *La Normalisation des Mesures de Débits de Fluides,* Standards, no. 7, 1934.

[25] U.S. Department of the Interior-Bureau of Reclamation, *Studies of Crests for Overaffl Dams,* Bulletin 3, Denver, Colorado, 1948.

[26] U.S. Army Corps of Engineers, *Hydraulic Design Criteria.*

[27] Farinha, J. S. B., *Tabelas Técnicas para Engenharia Civil.* Técnica, 5th Ed., Lisbon, 1962.

[28] Colebrook, C. F., White, C. M., *The Reduction of Carrying Capacity of Pipes with Age.* Journal of the Institution of Civil Engineers, Nr. 1, 1937–38.

[29] Lamont, P., *Formulae for Pipeline Calculations.* International Water Supply Association, Third Congress, London, 1955.

[30] Ckimitski, *Coefficient of Contraction of Tainter Gates.* Hydrotechnical Construction (in Russian), Nr. 11, 1964.

[31] Levin, L., *Formulaire des Conduites Forcées, Oléoducs et Conduites d'Aeration,* Dunod, Paris, 1968.

[32] Lautrich, R., *Tables et Abaques pour le Calcul Hydraulique des Canalisations sous Pression, Egouts et Caniveaux,* Eyrolles, Paris, 1971.

[33] Idel'cik, I. E., *Memento des Pertes de Charge,* Eyrolles, paris, 1969.

[34] Quintela, A., *Perdas de Carga Continuas no Escoamento de Liquidos Incompressiveis,* Revista de Fomento, Lisbon, 1973.

[35] Chow, V.-T., *Hydraulics of Open Channels.* McGraw-Hill, New York, 1959.

[36] Carlier, M., *Hydraulique Générale et Appliquée,* Eyrolles, Paris, 1972.

[37] Davis, C. V., *Handbook of Applied Hydraulics,* McGraw-Hill, New York, 1969.

[38] IRLI, *Discharge Measurement Structures,* International Institute for Land Reclamation and Improvement, Wageningen, 1976.

[39] Abecassis, F., *Soleiras Descarregadoras,* LNEC, Memo no. 165, Lisbon, 1961.

[40] Lemos, F. O., *Critérios para o Dimensionamento Hidráulico de*

Barragens Descarregadoras, Relatório, LNEC, Lisbon, September, 1978.

[41] Fuller, W. E., *Loss of Head in Bends.* Journal of New Eng. Water Works Assoc., December, 1913.

[42] Berezinsky, A., *Head Losses Through Trashracks.* Hydrotechnical Construction, Nr. 5, Moscow, 1958.

[43] Levin, L., *Étude Hydraulique des Grilles des Prises d'Eau.* 7th Congress of IAHR, 1957.

[44] Novikov, A., *Head Losses in Trashracks.* Hydrotechnical Construction, No. 10, Moscow, 1957.

[45] USDI-Bureau of reclamation, *Design of Small Dams,* U.S. Government Printing Office, Washington, 1973.

[46] Scheneebeli, G., *Hydraulique Souterraine,* Collection du Centre de Recherches et d'Essais de Chatou, Eyrolles, Paris, 1966.

[47] Custodio, E., *Hidrologia Subterranea,* Ediciones Omega, Barcelona, 1976.

[48] Terzaghi, K., Peck, R. B., *Soil Mechanics in Engineering Practice.* John Wiley and Sons, New York, 1967.

[49] IRLI, *Drainage Principles and Applications,* International Institute for Land Reclamation and Improvement, Publication 16, Wageningen. The Netherlands, 1973.

[50] Horn, J. W., *Principes Fondamentaux du Drainage des Terres,* Annual Bulletin of the International Commission onIrrigation and Drainage, 1964.

[51] Thorley, A. R. D., *et al., Control and Suppression of Pressure Surges in Pipelines and Tunnels,* CIRIA, 1979.

[52] Wood, D. S. and Jones, S. E., *Water hammer Charts for Various Types of Valves,* ASCE, 99, HY1, 1973.

[53] Martin, C. S., *Entrapped Air in Pipelines,* Second International Conference on Pressure Surges, London, 1976.

[54] Bernhart, H. M., *The Dependence of Pressure Wave Transmission through Surge Tanks of the Valve Closure Time.* Proc. 2nd. International Conference on Pressure Surges, London, BHRA, 1976.

[55] Stucky, M., *Chambres d'Equilibre,* École Polytechnique de l'Université de Lausanne, 1951.

[56] Miller, D. S., *Internal Flows Systems,* BHRA. 1978.

[57] Almeida, A. B., *Manual de Protecção contra o Golpe de Ariete em Condutas Elevatórias,* LNEC, Lisbon, 1981.

[58] Stephenson, D., *Pipeline Design for Water Engineers,* Elsevier, Amsterdam, 1976.

[59] Kinno, H., Kennedy, J. F., *Water Hammer Charts for Centrifugal Pump Systems.* Proc. ASCE, Vol. 91, HY3, 1965.

[60] Dubin, C., Guéneau, A., *Détermination des Dimensions Caracteristiques d'un Réservoir d'Air sur une Installation Elévatoire.* La Houille Blanche, Nr. 6, 1955.

[61] Levin, L., *Étude Hydraulique de Huit Revêtements Interieures de Conduites Forcées.* La Houille Blanche, No. 4, 1972.

[62] Henderson, F. M., *Open Channel Flow*. Collier MacMillan Publishers, London, 1966.

[63] Saraiva, J. G., Ramos, C. M., *Hydrodynamic Loads and Vibrations of Trashrack Elements*. Proc. of XX Congress of IAHR, Vol. 3, Moscow, 1983.

[64] Bradley, J. N., Peterka, A. J., *The Hydraulic Design of Stilling Basins: Stilling Basins with Sloping Apron (Basin V)*. Proc. ASCE. Vol. 83, HY5, October, 1957, pp. 1–32.

[65] Terzaghi, K., Peck, R. B., *Soil Mechanics in Engineering Practice*. John Wiley & Sons, New York, 1957.

[66] Ackers, P., *et al., Weirs and Flumes for Flow Measurement*. John Wiley & Sons, New York, 1978.

[67] Lefebvre, J., *Mesure des Débits et des Vitesses de Fluides*. Masson, S. A., Paris, 1986.

[68] ASCE, *Sedimentation Manual*. Proc. ASCE, Vol. 12. HY2, March, 1966.

[69] White, W. R., Burgess, J. S., *Triangular Profile Weir with 1:2 Upstream and Downstream Slopes*. Hidr. Res. Sta., Wallingford, England, Oct. 1967. Rep. Nr. INT 64.

[70] Crump, E. S., *A New Method of Gauging Steam Flow with Little Afflux by Means of a Submerged Weir of Triangular Profit*. Proc. Inst. Civil Engrs., Part 1, Vol. 1, March, 1952.

List of symbols

1. The following table lists the most common symbols used in this book.
2. $\mathbf{G}$ and G representing the vector $\mathbf{G}$ and its modulus G respectively are common notations used in the book.

Symbol	Concept	Dimensions
Roman Symbols		
A	Area (usually wetted area)	L^2
A	Geometric parameter	—
A^*	Pipeline parameter	1
B	Torque	$ML^2 T^{-2}$
B	Geometric parameter	—
B^*	Air vessel parameter	1
C	Chezy coefficient	$L^{1/2} T^{-1}$
C	Constant: Generic coefficient	Various
C	Length of the hydraulic jump	L
C^*	Throttling parameter	1
$\mathbf{C}_a$	Cauchy number	1
C_c	Contraction coefficient	1
C_d	Discharge coefficient (broad crested weirs)	1
C_v	Velocity coefficient	1
D	Hydraulic diameter	L
E	Modulus of elasticity	$ML^{-1} T^{-2}$
E	Specific energy	L
E_b	Modulus of elasticity of the material of the bars	$ML^{-1} T^{-2}$
$\mathbf{E}_u$	Euler number	1
F	Force	MLT^{-2}
F_r	Froude number	1
$F(\ )$	Function	—

Symbol	Concept	Dimensions
G	Weight flowrate	MLT^{-3}
G	Centroid	—
H	Specific energy in open channel; total head (pumps); total head over a weir crest	L
H_a	Total suction head	L
H_c	Total discharge head	L
H_d	Design head (spillways)	L
H_o	Net positive suction head	L
ΔH	Energy loss; head loss	L
I	Moment of inertia	ML^2
I	Buoyancy force	MLT^{-2}
I	Slope of channel bottom	1
I	Langelier index	L^4
I	Bottom slope	1
J	Slope of the free surface	1
J	Inertia parameter	1
K	Head loss coefficient	1
K	Hydraulic conductivity	L^2
K	Radius of gyration	L
K_h	Horizontal hydraulic conductivity	LT^{-1}
K_s	Strickler's roughness coefficient (Manning formula)	$L^{1/3} T^{-1}$
K_v	Vertical hydraulic conductivity	LT^{-1}
L	Length of a crest spillway; pipe length; width of liquid surface in open channel	L
M	Fixity factor (bars)	1
$\mathbf{M}$	Mach number	1
M	Flux of momentum	$ML\,T^{-1}$
N	Elevation of surface level	L
P	Wetted perimeter	L
P_a	Power absorbed by a pump	$ML^2\,T^{-3}$
P_u	Effective power of a pump	$ML^2\,T^{-3}$
PD^2	Moment of inertia	$ML^2\,T^{-3}$
Q	Discharge	$L^3\,T^{-1}$
R	Hydraulic radius	L
R	Throttled head loss coefficient	1
$\mathbf{R}_e$	Reynolds number	1
S	Storage coefficient	1
$\mathbf{S}_t$	Strouhal number	1
T	Averaging time; time interval	T
T	Temperature	$°C$
T	Transmissivity	$L^2\,T^{-1}$
U	Mean velocity in a cross-section	LT^{-1}
U_i	Incipient cavitation velocity	LT^{-1}
V	Velocity at a point; instantaneous velocity	LT^{-1}

Symbol	Concept	Dimensions
$\overline{V}$	Mean velocity at a point	LT^{-1}
V'	Velocity fluctuation	LT^{-1}
V_r	Radial component of velocity	LT^{-1}
V_t	Tangential component of velocity	LT^{-1}
V_z	Axial component of velocity	LT^{-1}
$\overline{V}$	Volume	L^3
W	Energy	$MLT^2\,T^{-2}$
$\mathbf{W}_e$	Weber number	1
$W(u)$	Well function	—
Y	Static head; piezometric head (surge shafts)	L
Z	Elevation; water level	L
Z^*	Maximum amplitude of oscillation	L
a	Acceleration	LT^{-2}
a	Constant or exponent; geometric parameter	Various
a	Cross section area	L^2
a	Weir height	L
b	Thickness of an aquifer	L
b	Constant or exponent; geometric parameter	Various
c	Velocity of elastic wave (celerity)	LT^{-1}
c	Constant or exponent	1
d	Diameter, especially of pipe; distance	L
d_{65}	Particle size for which 65% by weight of sediment is finer	L
d_e	Equivalent diameter	L
e	Void ratio	1
e	Thickness	L
e	Base of natural logarithm, ln	1
f	Friction factor	1
f_b	Natural frequency	T^{-1}
f_t	Frequency of vortex shedding	1
$f(\ )$	Function	—
g	Acceleration of gravity	LT^{-2}
h	Pressure head; total discharge head; depth of liquid below a free surface head related to weir crest	L
H_a	Atmospheric pressure head; suction head	L
h_c	Critical depth; discharge pressure head	L
h_m	Hydraulic mean depth	L
h_r	Seepage depth	L
h_v	Vapour pressure head	L
i	Energy gradient	1
$\mathbf{i}$	Unit vector in x-direction	L
$\mathbf{j}$	Unit vector in y-direction	L
$\mathbf{k}$	Unit vector in z-direction	L

Symbol	Concept	Dimensions
l	Length; length of well screen; mixing length in turbulent flow	L
m	Bank slope	1
m	Coefficient	1
m	Mass	M
n	Unit vector normal to streamline	L
n	Porosity; adiabatic constant	1
n	Rotational speed (rpm)	T^{-1}
n_e	Effective porosity	1
n_s	Specific speed	$L^{3/4}\,T^{-3/2}$
n_r	Field capacity (specific retention)	1
p	Pressure; gauge pressure	$ML^3\,T^{-2}$
p_a	Absolute pressure	$ML^3\,T^{-2}$
p_o	Atmospheric pressure	$ML^3\,T^{-2}$
Δp	Pressure loss	$ML^3\,T^{-2}$
Δpd	Pressure drop	$ML^3\,T^{-2}$
q	Discharge per unit width of open channel	$L^2\,T^{-1}$
r	Geometric radius	L
$\mathbf{r}$	Unit vector in radio coordinate (cylindrical or spherical coordinates)	L
$\mathbf{s}$	Unit vector a long streamline	L
s	Drawdown of water table	L
t	Time	T
u	Velocity	T
u_x	Velocity in x-direction	LT^{-1}
u_y	Velocity in y-direction	LT^{-1}
u_z	Velocity in z-direction	LT^{-1}
u_*	Friction velocity ($\sqrt{\tau_w/\rho}$)	LT^{-1}
v_g	Volume of water	L^3
v_p	Volume of grains	L^3
v_ϕ	Volume of pores	L^3
v_t	Total volume	L^3
w	Complex velocity	LT^{-1}
x	Cartesian or cylindrical coordinate, usually parallel to main flow direction	L
y	Cartesian coordinate, usually perpendicular to main flow	L
y	Depth of the centre of gravity	L
z	Cartesian or cylindrical coordinate, usually pointing upward	L
z	Complex variable	L
z	Elevation above specified datum	L
z^*	Maximum amplitude of oscillation	L

Symbol	Concept	Dimensions
Greek symbols		
α	Kinetic energy correction factor (Coriolis coefficient)	1
α	Vertical compressibility	$M^{-1}LT^2$
α	Angle	—
β	Momentum correction factor (Boussinesq coefficient)	1
β	Volumetric compressibility	$M^{-1}LT^2$
Γ	Circulation	L^2T^{-1}
γ	Specific weight	$ML^{-2}T^{-2}$
δ	Relative density	1
δ( )	Small change ( )	Various
Δ	An increment of	—
Δ( )	Change of ( )	Various
ε	Bulk modulus (modulus of elasticity)	$ML^{-1}T^{-2}$
ε	Surface roughness	L
η	Efficiency	1
θ	Angle, angular coordinate in cylindrical coordinate system	—
θ	Time parameter of a conduit ($\theta = l/c$)	T
μ	Dynamic (absolute) viscosity	$ML^{-1}T^{-1}$
μ	Coefficient of discharge	1
ν	Kinematic viscosity	LT^{-2}
π	Non-dimensional parameter	1
π	3.14159	—
ρ	Density	ML^{-3}
σ	Surface tension	MT^{-2}
σ	Coefficient of saturation	1
σ	Area	—
τ	Shear stress	$ML^{-1}T^{-2}$
φ	Piezometric potential	L
Φ	Velocity potential function	L^2T^{-1}
Ψ	Stream function	L^2T^{-1}
ω	Angular velocity	T^{-1}
ω	Kinematic capillarity	L^3T^{-2}
Λ	Distance (in surge shafts)	L
Δ	Laplacian operator	—

Subject index

WITHDRAWAL